Household Service Robotics

Household Service Robotics

Yangsheng Xu
Huihuan Qian
Xinyu Wu

ELSEVIER

Amsterdam • Boston • Heidelberg • London
New York • Oxford • Paris • San Diego
San Francisco • Singapore • Sydney • Tokyo
Academic Press is an imprint of Elsevier

Academic Press is an imprint of Elsevier
32 Jamestown Road, London NW1 7BY, UK
525 B Street, Suite 1800, San Diego, CA 92101-4495, USA
225 Wyman Street, Waltham, MA 02451, USA
The Boulevard, Langford Lane, Kidlington, Oxford OX5 1GB, UK

Chapter 2.2 was originally published in Journal of Intelligent and Robotic Systems, 2009, Vol. 55 Issue 4–5 and is reprinted with permission of Springer Science + Business Media

Chapter 2.3 was originally published in Autonomous Robots, 2006, Vol. 21 Issue 1 and is reprinted with permission of Springer Science + Business Media

Chapter 2.5 was originally published in International Journal of Information Acquisition, 2013, Vol. 9 Issue 2 and is reprinted with permission of World Scientific, Singapore

Chapter 3.3 was originally published in Robotics and Autonomous Systems, 2013, Vol. 60 Issue 4 and is reprinted with permission of Elsevier

Chapter 3.4 was originally published in Robotics and Autonomous Systems, 2013, Vol. 60 Issue 6 and is reprinted with permission of Elsevier

Chapter 3.5 was originally published in IEEE Transaction on Robotics, 2011, Vol. 27 Issue 1 and is reprinted with permission of Chi-Pang Lam, Chen-Tun Chou, Kuo-Hung Chiang and Li-Chen Fu

Chapter 4.2 was originally published in IEEE Robotics and Automation Magazine, 2011, Vol. 18 Issue 2 and is reprinted with permission of Ciocarlie M., Pantofaru C., Hsiao K., Bradski G., Brook P. and Dreyfuss E.

Chapter 4.4 was originally published in Journal of Intelligent and Robotic Systems, 2012, Vol. 68 Issue 2 and is reprinted with permission of Springer Science + Business Media

Chapter 5.2 was originally published in International Journal of Robotics Research, 2013, Vol. 31 Issue 2 and is reprinted with permission of SAGE Publications

Chapter 5.3 was originally published in International Journal of Robotics Research, 2013, Vol. 22 Issue 10–11 and is reprinted with permission of SAGE Publications

Chapter 6.2 was originally published in Advanced Robotics, 2011, Vol. 28 Issue 18 and is reprinted with permission of Taylor & Francis Ltd

Chapter 6.3 was originally published in Proceedings of the IEEE, 2012, Vol. 100 Issue 8 and is reprinted with permission of Juan Fasola and Maja J Matarić

Chapter 6.4 was originally published in Robotics and Computer-Integrated Manufacturing, 2013, Vol. 19 Issue 6 and is reprinted with permission of Elsevier

ISBN: 978-0-12-800881-2

British Library Cataloguing-in-Publication Data
A catalogue record for this book is available from the British Library

Library of Congress Cataloging-in-Publication Data
A catalog record for this book is available from the Library of Congress

For information on all Academic Press publications
visit our website at http://store.elsevier.com/

Typeset by TNQ Books and Journals
www.tnq.co.in

Printed and bound in the USA

Contents

Part IV Object Recognition

Part V Grasping and Manipulation

Part VI Human–Robot Interaction

Preface

Technological breakthroughs in computer hardware, sensors, actuators, controllers, and algorithms have facilitated the increased use of highly integrated robotic technologies in a range of disciplines, enabling robots to step out of their structured manufacturing industries and into our homes and daily lives. This is an exciting trend, and one that seems to hold the promise of science-fiction characters such as R2-D2 and C-3PO of *Star Wars* fame finding work in unstructured household environments.

Before that promise can be fulfilled, however, a number of research questions need to be answered. How can we design household robots to fit specific functional requirements? How can robots navigate unknown and dynamic environments without colliding into moving obstacles? Can robots ever understand what they see and feel? Can they agilely grasp objects to satisfy the increasingly demand of human needs? Can robots be friendly and efficient in their interactions with human beings? The questions posed by both users and researchers are seemingly never-ending, constituting the driving force behind technological advancements in the highly integrated robotic technology arena, and leading to the motivation for this book.

This book is the result of research efforts by scholars and scientists in many distinguished universities and institutions. Its aim is to provide engineers, postgraduate students, and researchers in relevant disciplines with a collection of the latest research developments. Researchers can draw upon the applications presented herein to devise additional household robotic concepts, thereby bringing us a step closer to incorporating robots into our daily lives.

We are grateful to all of the authors who contributed chapters, and thus important insights, to this book. We would also like to acknowledge the support of grants from the Publishing Funding of the National Science and Technology of China, the Chinese Academy of Engineering (Academician Consultation Project 2013-XZ-17-3), the Hong Kong Innovation and Technology Fund (Projects GHP/009/11GD and GHP/007/13SZ), and the research funding from the Chinese University of Hong Kong-Smart China Center for Research on Robotics and Smart-cities, Guangdong-Hong Kong Technology Cooperation Funding (Project 2011A091200001), and the Guangdong Innovative Research Team Program (No. 201001D0104648280).

We would also like to express special thanks to our many colleagues for their contributions to this book in the form of ideas, discussions, proofreading, and other support. They include Dan Xu, Zhimei Lin, Yuandong Sun, Long Han, Borislav Dzodzo of the Chinese University of Hong Kong, Dr Guoyuan Liang of the Chinese Academy of Sciences, Prof Dezhen Song and Yan Lu of Texas A&M University, and others. Finally, particular gratitude goes to our families for their patience and moral support during the course of this work.

Yangsheng Xu
Huihuan Qian
Xinyu Wu
April 2014

PART 1

Introduction

CHAPTER 1.1

Introduction

Huihuan Qian[1,2], Xinyu Wu[2], Yangsheng Xu[1,2]
[1]The Chinese University of Hong Kong, Hong Kong, China; [2]Shenzhen Institutes of Advanced Technology, Chinese Academy of Sciences, Shenzhen, China

Chapter Outline

With the significant development of robotic technologies, the applications of robots have been extended broadly from their traditional industry role to military, medical field, and even daily services. Robotics has gradually become a very large industry with tremendous influence on the sustainable development of the world.

In February 25, 2004, the International Robot Fair 2004 issued the World Robot Declaration [1] in Fukuoka, Japan. The declaration made a confident statement of the future development of robotic technology and of the numerous contributions that robots will make to all humankind. It elaborated on the expectations for the next generation robots as well as the efforts toward the creation of new markets through the next generation robots. It can be expected that the next generation robots will have a

Household Service Robotics. http://dx.doi.org/10.1016/B978-0-12-800881-2.00001-3

partnership with human beings, to assist human beings both physically and psychologically and to contribute to the realization of a safe and peaceful society.

As Bill Gates envisioned [2], robots will be ubiquitous in every home, similar to the development trend of the computer business that began over 30 years ago. He has mentioned some researchers as the "world's best minds" trying to solve the "toughest problems of robotics, such as visual recognition, navigation and machine learning." He also pointed out that the robotics industry was facing the challenges of nonexistence of standard operating software and standard processors, limited hardware, and the incompatibility of one programming code in another robot. Household service robots, although really appealing to everyone, still have a long way to go before entering every home and making contributions, such as has happened with personal computers.

In as early as 1995, Kawaruma [3] addressed the design philosophy for service robots, emphasizing limited autonomy, which was the balance between the low autonomy in industrial robots and full autonomy in field robots. The major reasons for this were the participation of users, whom service robots are working for, and the consideration of affordable cost and system complexity. He reasonably suggested the balanced design philosophy:

1. Environmental modification: By means of low-cost environmental modifications, such as wit beacons, infrared markers, etc., the navigation problem becomes easier.
2. User—robot communication: Balance should be achieved to both take advantage of the user's intelligence and prevent tedious and exhausting tele-operations.
3. Robot intelligence: The robot should have limited autonomy in the paradigm that it has a rich set of robust, reactive behaviors, while the user can selectively activate behaviors to control. The safety mechanisms should be always operated independently and in parallel.

In 1994, Kawaruma claimed [4] that the key R&D issues in service robots for the disabled and elderly include tele-operation, motion programming, planning, dexterity, sensing, and safety.

Karlsson [5], in 2004, from the perspective of the robotic industry, presented three core technologies for service robots, including an object recognition system, a vision-based navigation system, and a flexible and rich software platform assisting rapid design and prototyping of robotic applications.

Most of the key issues remain similar, in household service robotics as in service robots, for the disabled and elderly as presented by Kawaruma in 1994 because disabled and elderly people are one of the major groups for whom household service robots are designed to serve. Karlsson's point of view also provides a contributive subset in the key technologies for household service robots from the developer's point of view.

This book provides a large number of case studies on household service robotics on a systematic level from five functionality perspectives, as shown:

1. Overall system design: presents some highlighted cases of household service robots with hardware and software structures, so that the readers can study and analyze the overall robotic system.
2. Mapping, localization, and navigation: elaborates on the problem of environment sensing, path planning, and mobility execution of household service robots. The environment may or may not need to be modified based on the different sensors resulting in different system costs.
3. Object recognition: addresses how the robots can perceive and understand the object which it should handle. This reflects the intelligence of the robotic systems.
4. Grasping and manipulation: showcases the examples on how to manipulate the objects or carry out tasks using the robot arm or other actuation structures. This reflects the dexterity of the robots. A complex task can also be planned and divided into several smaller subtasks for better efficiency.
5. Human–robot interaction: focuses on how to serve the human subject so that he/she may enjoy the convenience of and feel more natural in using the robot.

1.1.1 Work Environments for Household Service Robots

This book will illustrate the robot work scenarios in a household environment composed of living/office rooms, corridors/narrow paths, kitchens, and so forth. These environments are highly unstructured, causing more involved challenges than the traditional structured environments. Some of the figures in the next sections show these typical scenarios.

1.1.1.1 Living Rooms/Office Rooms

As a coworker, partner, or butler in a house, service robots have to adapt to the main working environment in living rooms or office rooms. The figures below illustrate some examples (Figure 1).

1.1.1.2 Corridors/Narrow Paths

The connection between different rooms, corridors, or narrow paths are also transitive environments with different features for robots to work as shown in Figure 2.

1.1.1.3 Kitchen

As one of the recent emerging application venues, service robots have more complex work in kitchens for meal preparation, dish grasping, drawer opening, etc. (Figure 3).

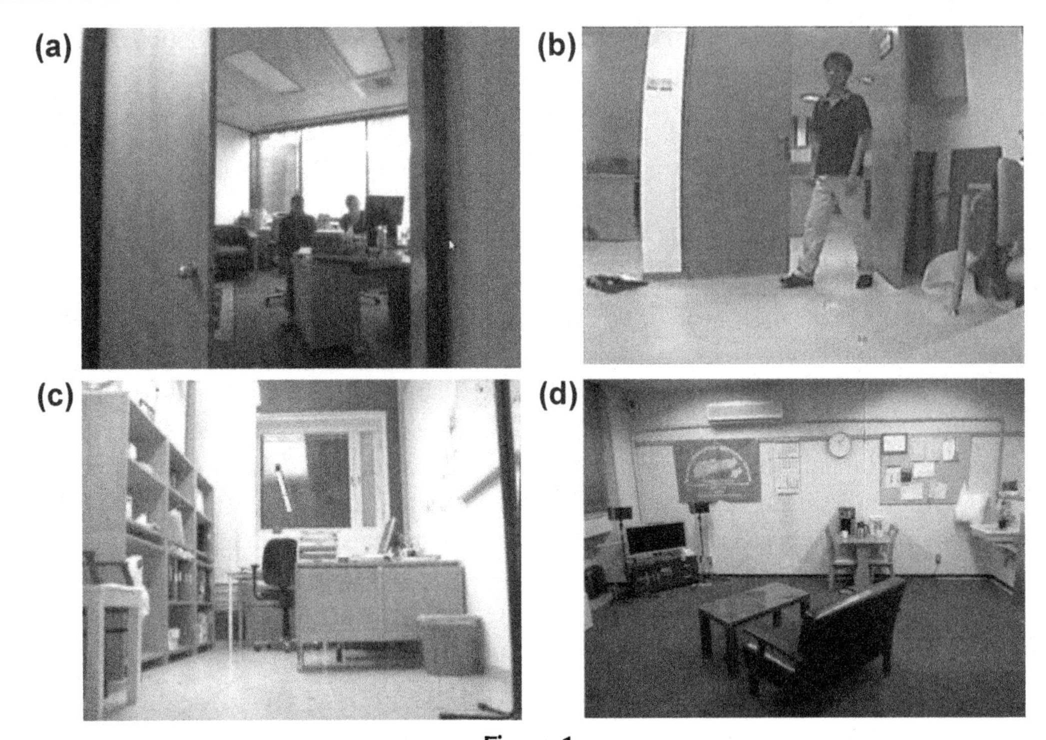

Figure 1
Living rooms and office rooms. (a) Office room 1 [6], (b) Office room 2 [7], (c) Office room 3 [8], and (d) Living room [9].

1.1.2 Functionalities of Household Service Robots

Within all these scenarios, what are the required and feasible functionalities that household robots should have? This book will present the R&D achievements that are highlighted below. Due to the great passion in household service robotic research, the included robots are far from a complete set, but they provide good case studies as research references.

1.1.2.1 Human Detection and Recognition

As household service robots are aimed to serve human beings, they should be able to detect and recognize humans to ensure that they are serving the right person (Figure 4).

1.1.2.2 Communication with Humans

Communication between humans and robots is a bidirectional channel, and it should be conducted in a natural way, such as with speech, facial expressions, gestures, and so forth (Figure 5).

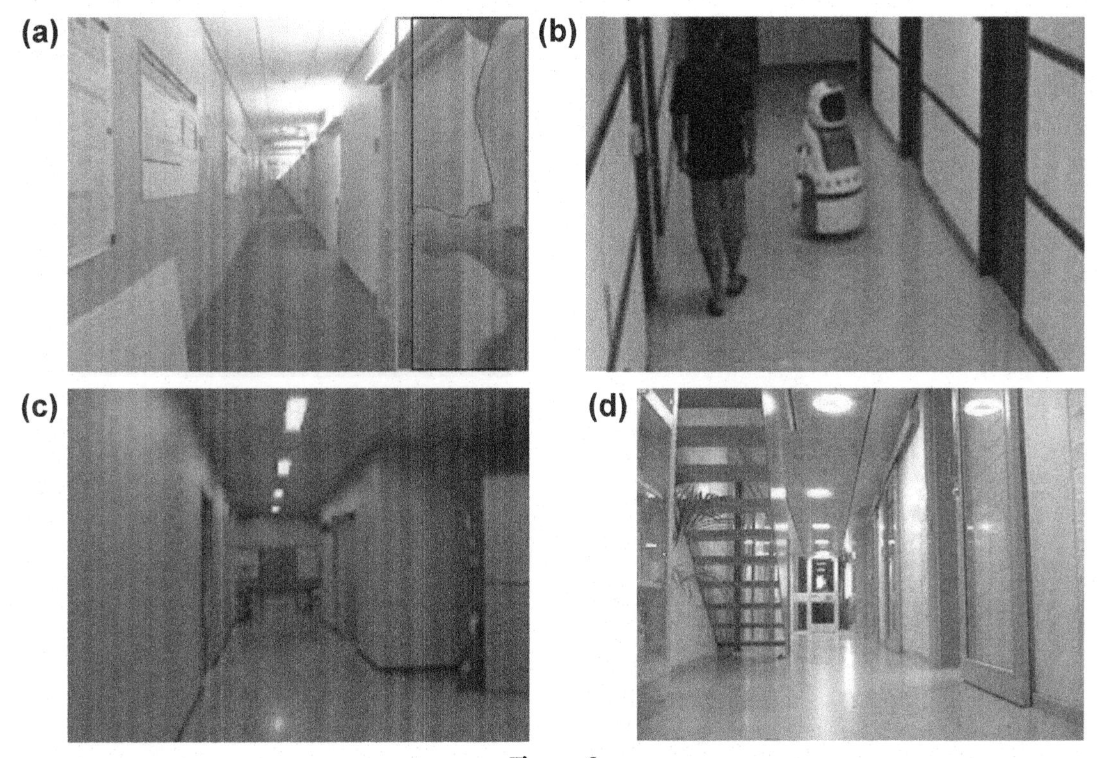

Figure 2
Corridors and narrow paths. (a) Corridor 1 [6], (b) Corridor 2 [10], (c) Corridor 3 [11], and (d) Corridor 4 [8].

1.1.2.3 Abnormal Event Detection

For the security of the household environment, it is necessary for some robots to be able to classify abnormal situations from normal ones and activate an alarm as appropriate (Figure 6).

1.1.2.4 Floor Cleaning

Floor cleaning, as a routine and tedious work, may be one of the first functional tasks for commercialized household service robots. iRobot has sold its 500 series Roomba around the world (Figure 7).

1.1.2.5 Object Pick and Place

Object pick and place is one of the basic manipulation tasks for household service robots (Figure 8).

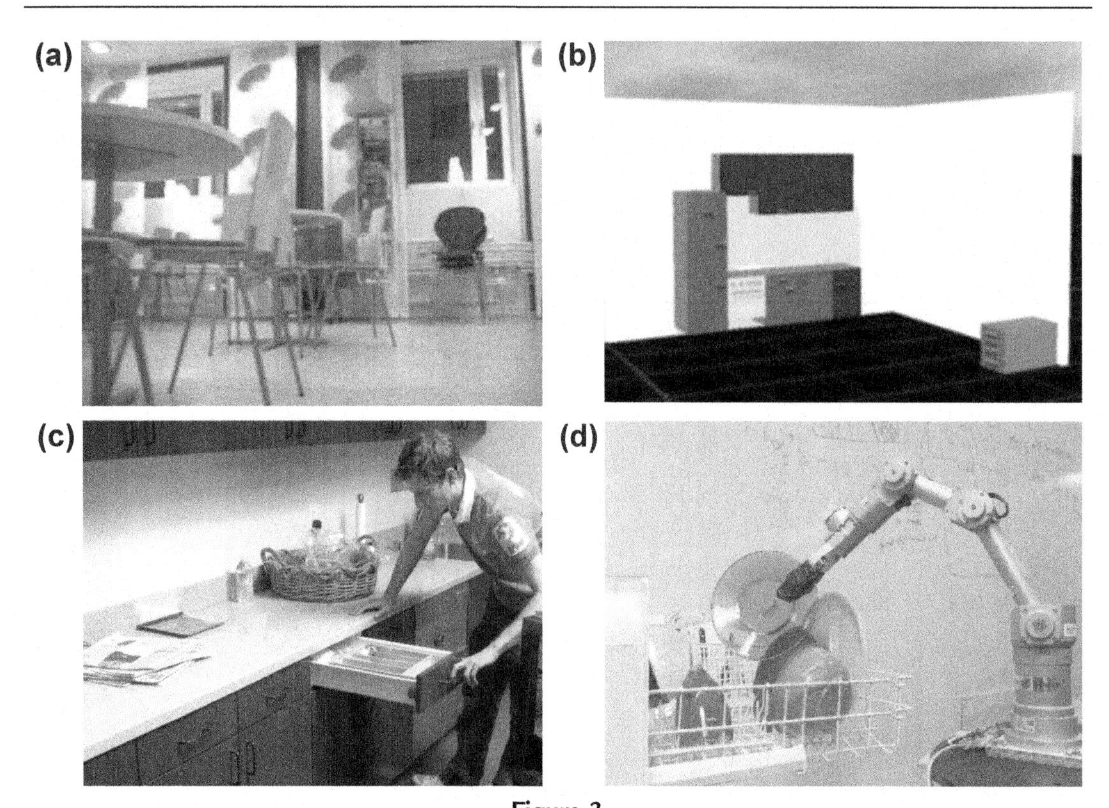

Figure 3
Kitchen environments. (a) Kitchen 1 [8], (b) Kitchen 2 [12], (c) Kitchen 3 [13], and (d) Kitchen 4 [14].

1.1.2.6 Door Opening

For navigation through the domestic environment, robots need to be able to open locked doors for access into another room (Figure 9).

1.1.2.7 Self-Charging

Energy, as a key problem for all robots, should be self-sustainable. Hence, intelligence household service robots need to support themselves by self-charging (Figure 10).

1.1.2.8 Meal Preparation

As one regular daily task in a kitchen, meal preparation has been tested and achieved by some robots (Figure 11).

Figure 4
Human detection [6].

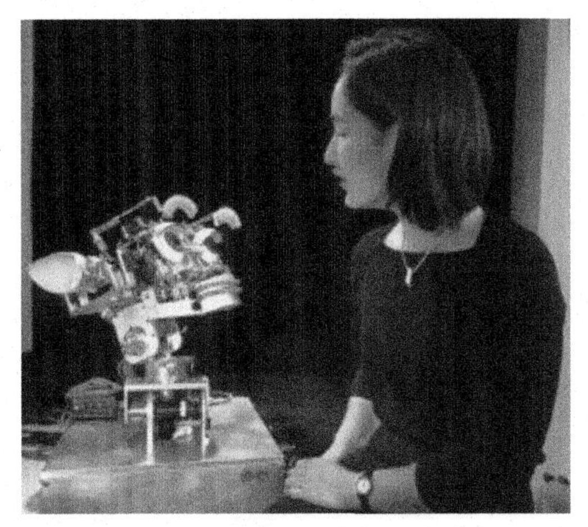

Figure 5
Robot—human communication [15].

1.1.2.9 Laundry Work

With simple object grasping and pick-and-place operations, robots can perform more sophisticated laundry tasks, such as folding cloths and categorizing clothes (Figure 12).

1.1.2.10 Object Carrying with Humans

Household service tasks inevitably include coworking with humans, in order to take advantage of human intelligence, while reducing the workload. Object carrying with human assistance is one such task, tested at RoboCup@Home (Figure 13).

Figure 6
Abnormal event detection [7].

Figure 7
iRobot, 500 series Roomba [16].

1.1.2.11 Feeding People

As needed by many aged or disabled persons, feeding is one of the significant work tasks service robots can provide. Both safety and efficiency are required (Figure 14).

1.1.2.12 Coaching for Elderly

Exercise at home for elderly people can significantly improve their health or assist in the rehabilitation process; with the shortage of household nurses, household robots can serve in this role (Figure 15).

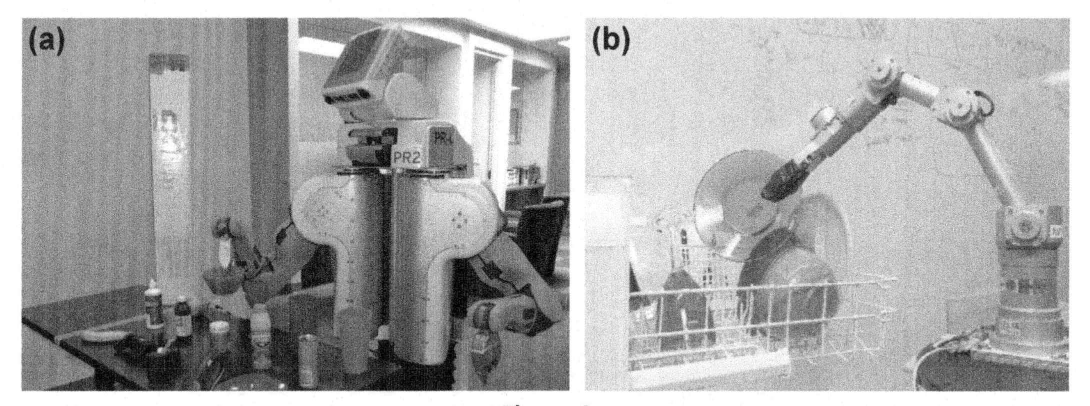

Figure 8
Robotic pick and place. (a) Table object pick and place [6] and (b) kitchen object pick and place [14].

Figure 9
Door opening [17].

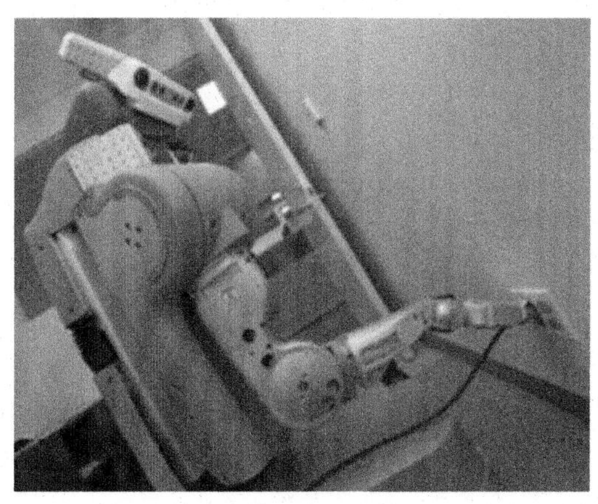

Figure 10
Self-charging [17].

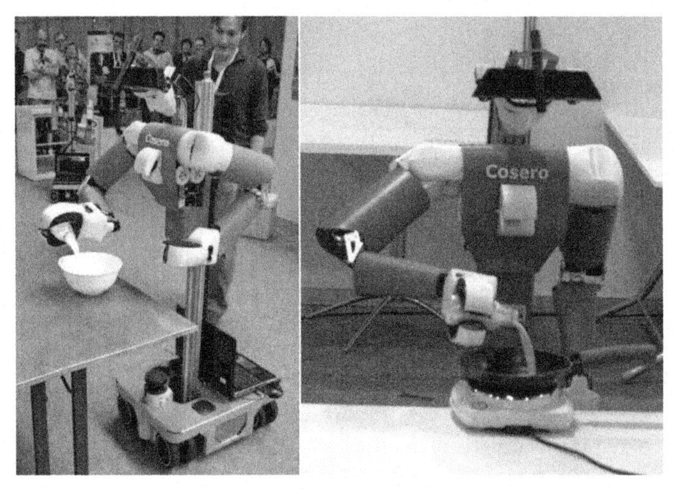

Figure 11
Meal preparation [18].

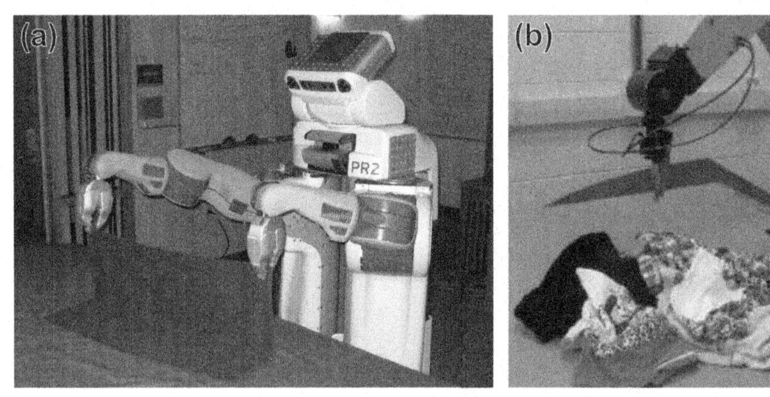

Figure 12
Laundry work. (a) Cloth folding [19] and (b) Cloth categorizing [20].

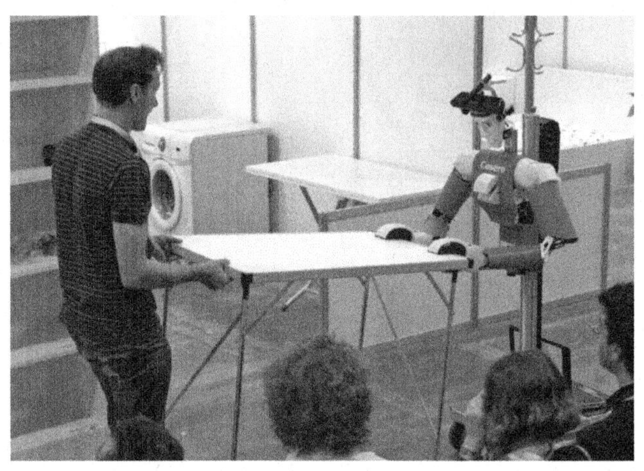

Figure 13
Robot, with a human, carrying a table [18].

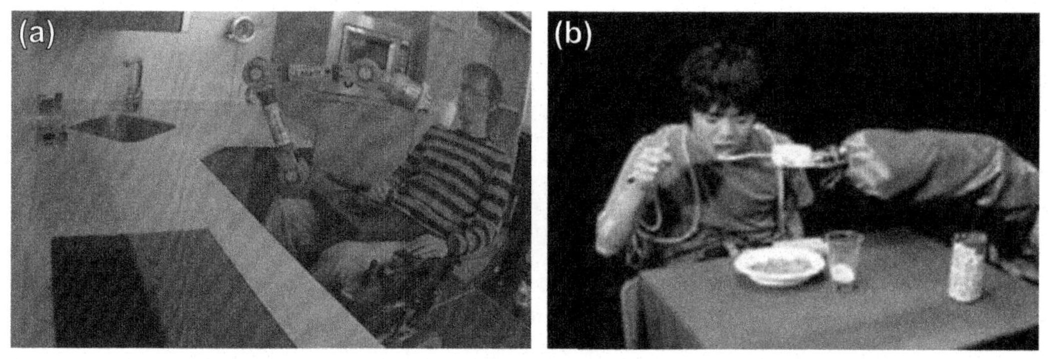

Figure 14
Robot feeding people. (a) Feeding a drink [21] and (b) Feeding a spoonful of food [22].

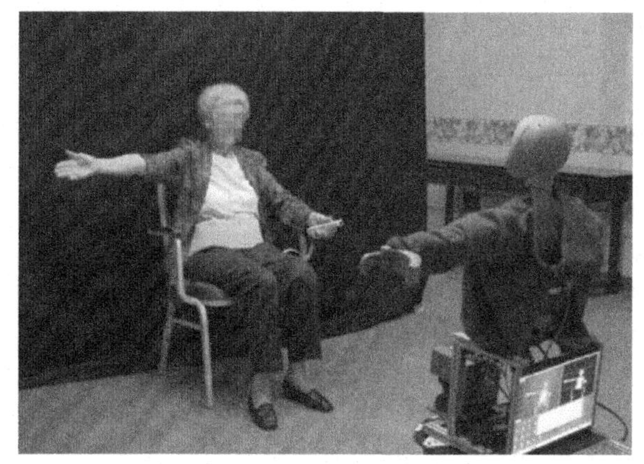

Figure 15
Coaching robot for elderly [23].

Figure 16
Plot for robotic play with humans [24].

Figure 17
Game "rock–paper–scissors" played between a robot and human [24].

1.1.2.13 Play with a Human

For the purpose of entertainment, one possibility is to have robots playing with humans in various activities, e.g., hide and seek, draw, play blocks, play cards, etc. Although not completely realized yet, it is a worthy area to try out. Some games such as rock–paper–scissors have been implemented [24] (Figures 16 and 17).

This book aims to provide readers with a large variety of the latest technological advancements in household service robotics so they can implement, compare, improve, and contribute to the field of household service robotics and finally put the dream of "a robot in every home" into reality in the near future.

References

[1] http://www.prnewswire.co.uk/news-releases/world-robot-declaration-from-international-robot-fair-2004-organizing-office-154289895.html.

[2] B. Gates, A robot in every home, Sci. Am. (January 2007). http://www.scientificamerican.com/magazine/sa/2007/01-01/.

[3] K. Kawaruma, R.T. Pack, M. Iskrous, Design philosophy for service robots, IEEE Int. Conf. Syst. Man Cybern. 4 (1995) 3736–3741.

[4] K. Kawaruma, M. Iskarous, Trends in service robots for the disabled and the elderly, IEEE/RSJ/GI Int. Conf. Intell. Robots Syst. 3 (1994) 1647–1654.

[5] N. Karlsson, M.E. Munich, L. Goncalves, J. Ostrowski, E.D. Bernardo, P. Pirjanian, Core technologies for service robotics, IEEE/RSJ Int. Conf. Intell. Rob. Syst. 3 (2004) 2979–2984.

[6] M. Ciocarlie, C. Pantofaru, K. Hsiao, G. Bradski, P. Brook, E. Dreyfuss, A side of data with my robot, IEEE Rob. Autom. Mag. 18 (2) (2011) 44–57.

[7] X. Wu, H. Gong, P. Chen, Z. Zhong, Y. Xu, Surveillance robot utilizing video and audio information, J. Intell. Rob. Syst. 55 (4–5) (2009) 403–421.

[8] A. Pronobis, O. Martínez Mozos, B. Caputo, P. Jensfelt, Multi-modal semantic place classification, Int. J. Rob. Res. 29 (2–3) (2010) 298–320.

[9] T. Nakamura, K. Sugiura, T. Nagai, N. Iwahashi, T. Toda, H. Okada, T. Omori, Learning novel objects for extended mobile manipulation, J. Intell. Rob. Syst. 66 (1–2) (2012) 187–204.

[10] C.-P. Lam, C.-T. Chou, K.-H. Chiang, Li-C. Fu, Human-centered robot navigation—towards a harmoniously human–robot coexisting environment, IEEE Trans. Rob. 27 (1) (2011) 99–112.

[11] J. Ido, Y. Shimizu, Y. Matsumoto, T. Ogasawara, Indoor navigation for a humanoid robot using a view sequence, Int. J. Rob. Res. 28 (2) (2009) 315–325.
[12] R.B. Rusu, Z.C. Marton, N. Blodow, M.E. Dolha, M. Beetz, Functional object mapping of kitchen environments, 2008 IEEE/RSJ Int. Conf. Intell. Rob. Syst. (2008) 3525–3532.
[13] J. Sturm, K. Konolige, C. Stachniss, W. Burgard, Vision-based detection for learning articulation models of cabinet doors and drawers in household environments, 2010 IEEE Int. Conf. Rob. Autom. (2010) 362–368.
[14] A. Saxena, J. Driemeyer, A.Y. Ng, Robotic grasping of novel objects using vision, Int. J. Rob. Res. 27 (2) (2008) 157–173.
[15] C. Breazeal, Regulation and entrainment in human–robot interaction, Int. J. Rob. Res. 21 (10–11) (2002) 883–902.
[16] http://www.irobot.com/us/.
[17] W. Meeussen, M. Wise, S. Glaser, et al., Autonomous door opening and plugging in with a personal robot, 2010 IEEE Int. Conf. Rob. Autom. (2010) 729–736.
[18] J. Stückler, I. Badami, D. Droeschel, K. Gräve, D. Holz, M. McElhone, M. Nieuwenhuisen, M. Schreiber, M. Schwarz, S. Behnke, NimbRo@Home: winning team of the RoboCup@Home competition 2012, in: RoboCup, Robot Soccer World Cup, vol. XVI, Springer, LNCS, 2012.
[19] S. Miller, J.van den Berg, M. Fritz, T. Darrell, K. Goldberg, A geometric approach to robotic laundry folding, Int. J. Rob. Res. 31 (2) (2011) 249–267.
[20] B. Willimon, S. Birchfield, I. Walker, Classification of clothing using interactive perception, 2011 IEEE Int. Conf. Rob. Autom. (2011) 1862–1868.
[21] A. Jardón, M.F. Stoelen, F. Bonsignorio, C. Balaguer, Task-oriented kinematic design of a symmetric assistive climbing robot, IEEE Trans. Rob. 27 (6) (2011) 1132–1137.
[22] N. Bu, M. Okamoto, T. Tsuji, A hybrid motion classification approach for EMG-based human–robot interfaces using Bayesian and neural networks, IEEE Trans. Rob. 25 (3) (2009).
[23] J. Fasola, M.J. Matarić, Using socially assistive human–robot interaction to motivate physical exercise for older adults, Proc. IEEE 100 (8) (2012) 2512–2526.
[24] K. Abe, A. Iwasaki, T. Nakamura, T. Nagai, A. Yokoyama, T. Shimotomai, H. Okada, T. Omori, Playmate robots that can act according to a child's mental state, 2012 IEEE/RSJ Int. Conf. Intell. Rob. Syst. (2012) 4460–4467.

PART 2

Service Robotic System Design

CHAPTER 2.1

The State of the Art in Service Robotic System Design

Huihuan Qian[1,2], Xinyu Wu[2], Yangsheng Xu[1,2]
[1]The Chinese University of Hong Kong, Hong Kong, China; [2]Shenzhen Institutes of Advanced Technology, Chinese Academy of Sciences, Shenzhen, China

Chapter Outline

With the emerging need for household applications, service robotics has become a hot spot attracting researchers globally. Universities, institutes, and industrial companies, either independently or collaboratively, have been focusing on research and development (R&D) for household service robots, resulting in interesting and well-performing household service robotic systems. In this chapter, we first give a categorization of the various applications, followed by an elaboration of some highlighted cases with their overall system designs.

Kawaruma provided some taxonomy as early as 1994 on service robotics for disabled and elderly people [1], by classifying the service robots into five categories including workstation-based systems, stand-alone manipulator systems, wheelchair-based systems, mobile robot systems, and collaborative robotic aid systems. Inheriting and further developing this taxonomy, we can draw a more up-to-date picture for the state of the arts.

2.1.1 Stationary Service Robotic Systems

The stationary robotic systems are not able to move on their base. Although these robots were represented quite early in robotic design, they initialized the spark for robotic assistance in household service. One subgroup is the workstation-based system, primarily composed of a robot arm, mounted onto a bench or a table, and a workstation. This type

Household Service Robotics. http://dx.doi.org/10.1016/B978-0-12-800881-2.00002-5

of robotic system was relatively cheap in the 1990s. However, the user was limited to being within the working cell. Examples of such systems are the following:

HANDY1 was developed by the University of Keele [2,3], UK. It can feed severely disabled people and help with their drinking, shaving, and teeth cleaning.

DeVAR (Desktop Vocational Assistant Robot) [4–6], developed by Stanford University and the Palo Alto VA Medical Center, adopted the PUMA-260 robot and a modified Otto-bock Greifer prosthetic hand. It is aimed at office applications in handling papers and floppy disks, picking up and using a telephone, and retrieving medication. It has a voice recognition system to command the arm. The voiced "stop," a shout, pressing a "panic switch," or its encountering a resistance of 2.3 kg or more will stop the robot.

RAID (Robot for Assisting the Integration of the Disabled) [2,7] employs a modified RTX robot (SCARA configuration) for office applications. It is operated via a joystick on a wheelchair.

Meal Assistance Robot Systems [8] has a specially designed arm (simple, light, and small), with a spoon and spatula. The user gives command through an optical pointer on his/her head and a sensor panel.

The other subgroup is stand-alone manipulator systems, which typically use large arms not attached to any objects, e.g., a table. These systems have more flexibility and a larger work space.

Tou [9], developed by the Universitat Politècnica de Catalunya in Spain, has an arm built from foam-rubber modules, a personal computer, and a communication interface (including a speech recognition system, a joystick, and a special keyboard). The arm is directly controlled by the user, without need for environmental sensing. Tasks include picking up objects and rearranging a bed.

ISAC (Intelligent Soft Arm Control), developed by Vanderbilt University in the United States, uses pneumatic actuators ("rubbertuators," like human muscles) for compliance. The human–robot communication is through a voice recognition system. Its accomplished tasks include feeding soup and similar foods, assisting with drinking, and using a napkin.

2.1.2 Attached Mobile Service Robotic Systems

In this type of service robotic system, an arm is assembled to a mobile base, generally a wheelchair. It can achieve higher mobility for larger work spaces, but the mobility of the base is controlled by the driver. Hence the autonomy is in a relatively local region only.

MANUS [2] has an electrically powered arm. The user directly controls the arm in an unstructured and previously unknown environment, through interactive procedures.

Figure 1
ASIBOT attached to a wheelchair [10].

The UNVENTAID arm system [2] adopts an arm using novel pneumatic and electric actuators. It has the benefits of compliance and low weight, but it requires a small compressor.

The ASIBOT robot [10], developed by the Universidad Carlos III de Madrid (UC3M), is a 5-DOF (degrees of freedom) symmetric robotic arm, which can be attached to a wheelchair through a low-cost docking station, so as to serve in a kitchen test bed at UC3M (Figure 1).

2.1.3 Mobile Household Service Robotic Systems

This type of household service robotic system has the highest mobility and autonomy. It has become the mainstream in recent research.

WALKY [11] has a Scorbot ER VII robot mounted on a Labmate mobile base. The user points to an AutoCAE drawing for the destination command, and WALKY navigates and avoids obstacles based on ultrasonic sensors.

There are collaborative systems between a number of mobile robots, or between a mobile robot and a stationary robot. ISAC-HERO [12] is a multirobotic system further developed by Kawaruma at Vanderbilt University, using ISAC (the stationary robot arm) and HERO-2000, which is a mobile robot with an arm to fetch an object and hand it to ISAC (Figure 2).

More recently, the blossoming worldwide research interest and the reduced cost of key components have resulted in a large number of mobile robot systems. We can classify them into: (1) mobile manipulator robots, (2) mobile base robots, and (3) limited mobility robots. A mobile-manipulator robot system consists of a mobile platform for mobility and at least one manipulator for manipulation (or potential capability for manipulation).

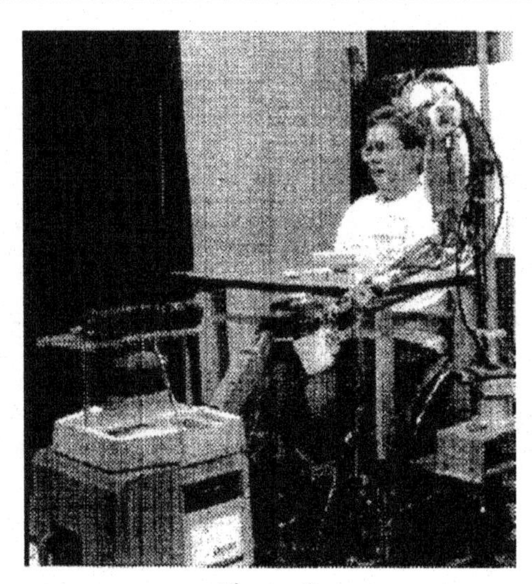

Figure 2
ISAC-HERO exchanges a soda can [1].

It attracts the major research interest owing to its capability for generalized manipulation tasks. The majority of these systems have a wheeled mobile base, owing to its easiness to control, whereas some others have two-leg humanoid configurations. Figure 3 illustrates a subset of mobile manipulator robots, including:

a. STAIR 2 platform, equipped with a 7-DOF Barrett arm and a three-fingered hand, by Stanford University [13];
b. Bandit, a biomimetic anthropomorphic robot platform by the University of Southern California [14];
c. Nexi by the Massachusetts Institute of Technology Media Lab [15];
d. HERB developed by Carnegie Mellon University [16];
e. TUM-Rosie by Technische Universität München [17];
f. Hubo+ humanoid robot developed by the Korea Advanced Institute of Science and Technology [18].

Some other references can be made to examples such as Snackbot, developed by Carnegie Mellon University [19]; BIRON II [20], by Bielefeld University; a daily assistive robot by the University of Tokyo [21]; Julia, by National Taiwan University [22]; T-Rot, by Korea University [23]; and so on. But the household robots listed here are far from complete, because of the zealous research efforts still going on around the world.

Mobile base robots consist of only a mobile platform, without any robot arm for manipulation. Their main function is interaction with humans, such as security alarm, companion, etc. Some can use their mobility to push objects for path clearing or vacuum cleaning, etc.

Figure 3
Examples of mobile manipulator service robots.

Figure 4 showcases such systems, including:

a. Robotic platform by the University of Tokyo [24];
b. Household surveillance robot by Shenzhen Institute of Advanced Technology [25];
c. The mobile home robot companion (pre-final version) developed by the Companion-Able project [26].

Limited-mobility robots have limited mobility or are not mobile in the base at all. Their difference from stationary service robotic systems is that the size is much smaller, with the primary purpose of communication, edutainment, or therapy, rather than manipulation. Figure 5 shows some examples. Many of them are already commercialized, because the function requirements are not sophisticated and thus the technology has already matured:

a. iSobot (commercialized toy robot) by Tomy [27];
b. NAO robot (commercialized education robot) by Aldebaran-Robotics [28];
c. Paro (commercialized therapeutic robot) by PARO Robots U.S., Inc. [29].

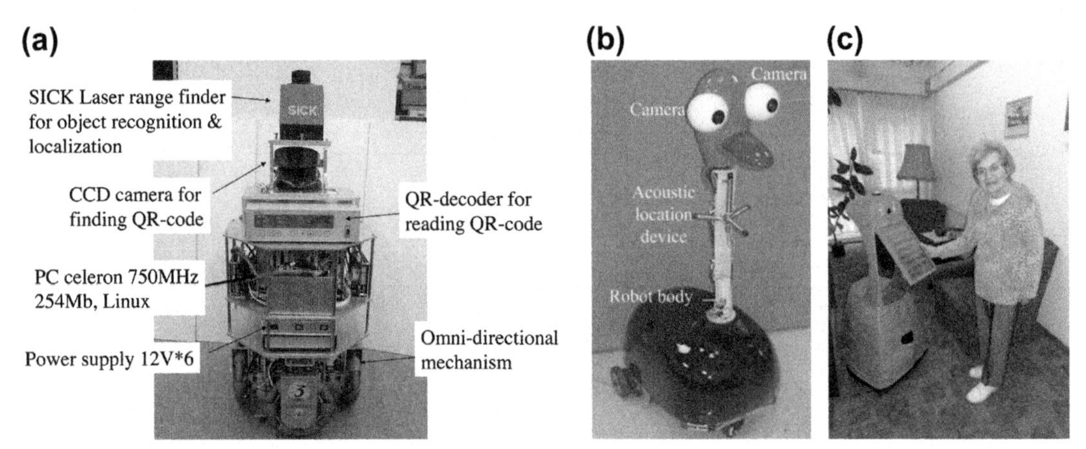

Figure 4
Examples of mobile base service robots.

Figure 5
Limited-mobility service robot.

2.1.3.1 RoboCup@Home

In the survey of mobile household service robots, it is necessary to mention the worldwide competition in this area, i.e., RoboCup@Home. As a part of the RoboCup initiative, RoboCup@Home, established in 2006, focuses on domestic applications for robots. It is the largest annual international competition for domestic robotics. It provides a platform for competition at a technologically challenging but achievable level. Many tests are conducted on the robots' functionalities for benchmarking the technologies developed by participating teams. Almost all the household service robots in RoboCup@Home can be categorized into the group of mobile manipulator robots, owing to the task requirements of mobility and manipulation.

Each year, RoboCup@Home is guided by three committees:

- The Executive Committee: consisting of members of the board of trustees and representatives of each activity area
- The Technical Committee: in charge of the rules for each league
- The Organizing Committee: responsible for organizing the competition

Because of the intrinsic competitive nature of RoboCup@Home, the league uses a set of benchmark tests for evaluating the robots' abilities and performance in realistic nonstandardized home environment settings. The detailed tasks vary from year to year, with focus being placed in the following domains but not limited to:

- Human–robot interaction and cooperation
- Navigation and mapping in dynamic environments
- Computer vision and object recognition under natural light conditions
- Object manipulation
- Adaptive behaviors
- Behavior integration
- Ambient intelligence
- Standardization and system integration

Table 1 summarizes the RoboCup@Home competition host cities, participation team numbers, and results. The events have been held in Europe, USA, and Asia, and the number of participating teams increased from about 10 to over 20 in recent years, showing

Table 1: RoboCup@Home competition

Year	Place	Team No.	First Place	Second Place	Third Place	Innovation Award	Technical Challenge
2006	Bremen, Germany	12	AllemaniACs	CMAssist	RoboCare		
2007	Atlanta, USA	11	AllemaniACs	UT Austin Villa	Pumas	UChileHome-breakers	
2008	Suzhou, China	14	eR@sers	AllemaniACs	b-it-bots	UChileHome-breakers	
2009	Graz, Austria	18	b-it-bots	eR@sers	NimbRo	NimbRo	
2010	Singapore	24	eR@sers	NimbRo	b-it-bots	homer@Uni-Koblenz	
2011	Istanbul, Turkey	19	NimbRo	eR@sers	ToBI		
2012	Mexico City, Mexico	21	NimbRo	WrightEagle	b-it-bots		homer@Uni-Koblenz
2013	Eindhoven, The Netherlands	21	NimbRo	WrightEagle	Tech United Eindhoven	Golem	

Figure 6
Some examples of participating robots in the RoboCup@Home.

the passion that universities and institutes have for household service robotic research. Figure 6 shows some examples of participating robots [30,31], including:

a. AllemaniACs ([32] RWTH Aachen University, Germany);
b. CMAssist ([33] Carnegie Mellon University, USA);
c. RoboCare ([34] Institute for Cognitive Science and Technology, Italy);
d. UT Austin Villa ([35] University of Texas at Austin, USA);
e. eR@sers ([36], including two robots, i.e., tam@home and DiGORO; developed by Tamagawa University, National Institute of Information and Communications Technology, and The University of Electro-Communications, Japan);

f. b-it-bots ([37,38], including two robots, i.e., Johnny and Jenny; developed by the University of Applied Sciences Bonn-Rhein-Sieg, Germany);
g. NimbRo ([39], including two robots, i.e., Cosero and Dynamaid; developed by the University of Bonn, Germany);
h. ToBI ([40] Bielefeld University, Germany);
i. WrightEagle ([41] University of Science and Technology of China, China).

2.1.3.2 Personal Robot-2 and ASIMO

Aside from the continuous research efforts at universities and institutes, industrial companies are also passionate in their R&D tasks for household service robots. Two notable robots are Personal Robot-2 (PR2) from Willow Garage and ASIMO from Honda.

The first prototype of Personal Robot, i.e., PR1 [42], developed by Stanford University, had its major considerations in human and object safety in the human–robot coexisting environment. Because software cannot guarantee there will be no malfunctions, the mechanical design requires a trade-off between performance and safety.

PR1 (as shown in Figure 7) is composed of a 2-DOF differentially driven base, a rotational torso, and two 7-DOF manipulators mounted on the upper torso. The payload of each manipulator is 5 kg. A gravity compensation system based on springs makes possible the reduction of the actuation power, so as to guarantee safety while keeping the payload.

Encouraged by PR1, Willow Garage, located in Silicon Valley near Stanford, further developed the prototype into PR2. The mobile base is upgraded to an omnidirectional base with four steered and driven wheel sets. The arm payload is reduced to 1.8 kg each. More sensors are equipped, including cameras (a Microsoft Kinect, a 5-megapixel global shutter

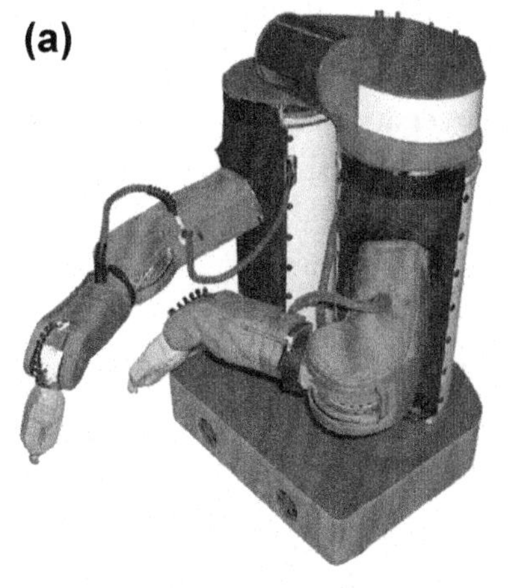

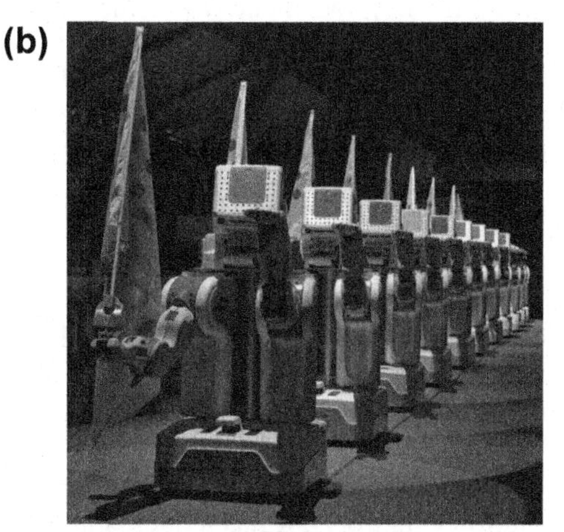

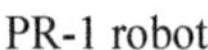

PR-1 robot   11 PR-2 robots

Figure 7
Two types of personal robot. (a) PR-1 robot and (b) 11 PR-2 robots.

color camera, a wide-angle global shutter color stereo camera, a narrow-angle global shutter monochrome stereo camera, etc.), laser scanners (Hokuyo UTM-30LX laser scanner and tilting Hokuyo UTM-30LX laser scanner), and IMU (Microstrain 3DM-GX2 IMU). The software platform of PR2 is the robot operating system (ROS) [43], which is open-sourced.

In May 2010, Willow Garage announced that it would give 11 PR2's for free to 11 institutions, including:

- Stanford University, USA
- University of California, Berkeley, USA
- Massachusetts Institute of Technology, USA
- Georgia Institute of Technology, USA
- University of Southern California, USA
- University of Pennsylvania, USA
- Albert-Ludwigs-Universität Freiburg, Germany
- Bosch, USA
- Technische Universität München, Germany
- Katholieke Universiteit Leuven, Belgium
- University of Tokyo, Japan

Some encouraging achievements have been made, and some of the case studies are included in later parts of this book.

Another significant contribution in household service robots is the humanoid robot ASIMO [44] from Honda. This humanoid robot dates back to the two-leg robot E0 in 1986. Thereafter, prototypes went through E1, E2, E3, E4, E5, E6, P1, P2, and P3 and finally achieved ASIMO in 2000. Figure 8 shows all the robot prototypes.

Table 2 shows the key specifications of the latest ASIMO version since 2011.

ASIMO-based research is conducted at universities, institutes, and companies, including Carnegie Mellon University (USA) [45], Columbia University (USA) [46], Bielefeld University (Germany) [47], Honda Research Institute (USA) [48], and so on.

2.1.4 Summary of Case Studies

The remaining portion of this section elaborates on four case studies in household service robotic system design.

- Household surveillance robot [25]
  Wu et al. developed a household surveillance robot with acquisition of video and audio information (Figure 9). It has a duck-shaped outlook, with two cameras to track upper body motion for abnormal behavior detection and an acoustic microphone array to

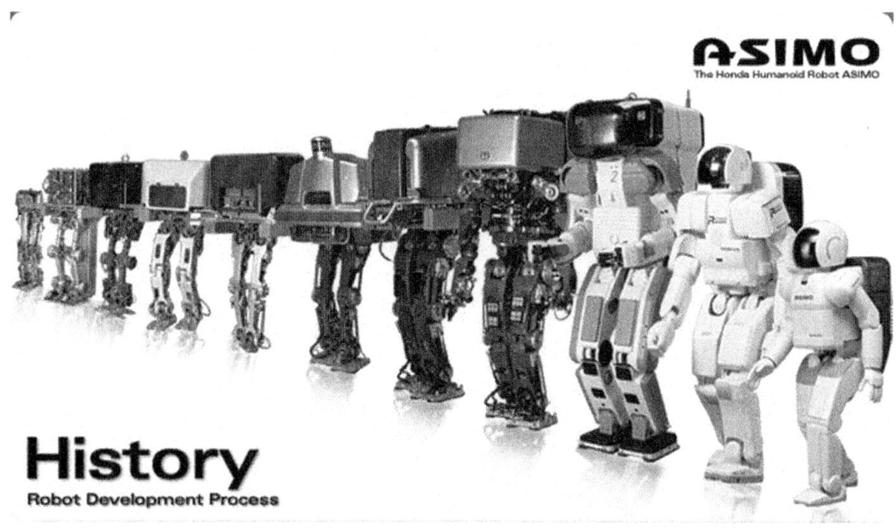

Figure 8
Prototype series leading to ASIMO [44].

Table 2: Key specifications of ASIMO (since 2011) [48]

Size	
Height	130 cm
Weight	48 kg (decreased by 6 kg from previous model)
Running Speed	
Max	9 km/h
Operating Degrees of Freedom	
Head	3 DOF
Arm	7 DOF × 2
Hands	13 DOF × 2
Hip	2 DOF
Legs	6 DOF × 2
Total	57 DOF (increase of 23 DOF from previous model)

detect abnormal audio information such as crying, groaning, and gun-shooting. The passive acoustic device also directs the robot to the location of the abnormal events, so that the robot can move to examine with its visual sensor.

- Way-finding robot for the visually impaired [49]
  Kulyukin et al. developed a robot-assisted way-finding system for the visually impaired in the indoor environment, such as homes, office buildings, supermarkets, etc. (Figure 10). It is composed of a Pioneer 2DX mobile base and sensors, including an RFID reader and laser range finder.

Figure 9
Household security robot [25].

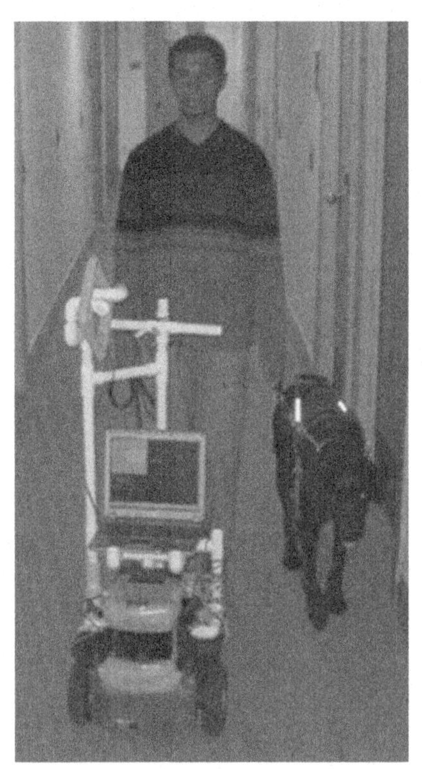

Figure 10
Way-finding robot [49].

- Service robot for elders [50]
  Mei et al. designed and developed a robot for taking care of elders, which has the functions of autonomous mobility in the household environment, manipulation for small commodities, recognition of speech, safety-oriented movement detection, and some basic nursing abilities (Figure 11).
- Household service butler [51]
  Han et al. designed and developed a dual-armed household service robotic butler (Figure 12). It consists of a mobile base, two arms, a wide-angle head camera, and a

Figure 11
Service robot for elders [50].

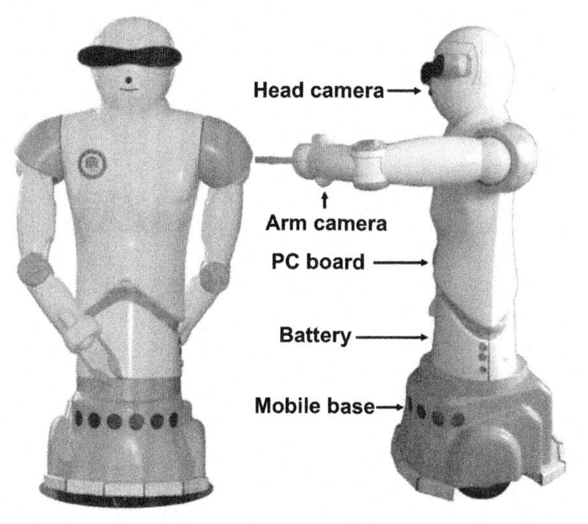

Figure 12
Household service robotic butler [51].

narrow-angle arm camera. With a command from a cell phone, the robot is capable of fetching a Coke® from the fridge.

References

[1] K. Kawaruma, M. Iskarous, Trends in service robots for the disabled and the elderly, IEEE/RSJ/GI Int. Conf. Intell. Rob. Syst. 3 (1994) 1647–1654.
[2] R.D. Jackson, Robotics and its role in helping disabled people, Eng. Sci. Edu. J. 2 (6) (December 1993) 267–272.
[3] M. Whittaker, HANDY 1 robotic aid to eating: a study in social impact, RESNA Int. (June 1992) 589–594.
[4] J. Hammel, K. Hall, D. Lees, L. Leifer, M. Van der Loos, I. Perkash, R. Crigler, Clinical evaluation of a desktop assistant, J. Rehabil. Res. Dev. 26 (3) (1989) 1–16.
[5] Rehabilitation Robotics Newsletter, no. 1, The Rehabilitation Robotics Research Program, Applied Science and Engineering Laboratories, 1994.
[6] M. Kassler, Robotics for health care: a review of the literature, Robotica 11 (1993) 495–516.
[7] TIDE: Technology Initiative for Disabled and Elderly People, Pilot Action Synopses, Commission of the European Communities, Brussels, March 1993.
[8] S. Ishii, F. Hiramatsu, S. Tanaka, Y. Amari, I. Masuda, A meal assistance robot system for handicapped people's welfare, in: Conference on Robots and Mechatronics, Japan Society of Mechanical Engineers, 1991.
[9] A. Casals, R. Villa, D. Casals, A soft assistant arm for tetraplegics, in: Proceedings of the First TIDE Congress, April 1993, pp. 103–107.
[10] A. Jardón, M.F. Stoelen, F. Bonsignorio, C. Balaguer, Task-oriented kinematic design of a symmetric assistive climbing robot, IEEE Trans. Rob. 27 (6) (2011) 1132–1137.
[11] Rehabilitation Robotics Newsletter, no. 4, The Rehabilitation Robotics Research Program, Applied Science and Engineering Laboratories, 1993.
[12] K. Kawamura, M. Cambron, K. Fujiwara, J. Barile, A cooperative robotic aid system, in: Proceedings of the Virtual Reality Systems, Teleoperation and beyond Speech Recognition Conference, 1993.
[13] A. Saxena, J. Driemeyer, A.Y. Ng, Robotic grasping of novel objects using vision, Int. J. Rob. Res. 27 (2) (2008) 157–173.
[14] J. Fasola, M.J. Matarić, Using socially assistive human–robot interaction to motivate physical exercise for older adults, Proc. IEEE 100 (8) (2012) 2512–2526.
[15] C. Breazeal, Role of expressive behaviour for robots that learn from people, Philos. Trans. R. Soc. Lond. B Biol. Sci. 364 (2009) 3527–3538.
[16] S.S. Srinivasa, D. Ferguson, C.J. Helfrich, D. Berenson, A. Collet, R. Diankov, G. Gallagher, G. Hollinger, J. Kuffner, M. Vande Weghe, HERB: a home exploring robotic butler, Auton. Rob. 28 (2010) 5–20.
[17] http://ias.cs.tum.edu/robots/tum-rosie.
[18] M. Zucker, Y. Jun, B. Killen, T.-G. Kim, P. Oh, Continuous trajectory optimization for autonomous humanoid door opening, in: 2013 IEEE International Conference on Technologies for Practical Robot Applications, 2013, pp. 1–5.
[19] M.K. Lee, J. Forlizzi, S. Kiesler, P. Rybski, J. Antanitis, S. Savetsila, Personalization in HRI: a longitudinal field experiment, in: 7th ACM/IEEE International Conference on Human–Robot Interaction, 2012, pp. 319–326.
[20] F. Yuan, L. Twardon, M. Hanheide, Dynamic path planning adopting human navigation strategies for a domestic mobile robot, in: The 2010 IEEE/RSJ International Conference on Intelligent Robots and Systems, 2010, pp. 3275–3281.
[21] Y. Kimitoshi, U. Ryohei, N. Shunichi, M. Yuto, M. Toshiaki, H. Naotaka, O. Kei, I. Masayuki, System integration of a daily assistive robot and its application to tidying and cleaning rooms, in: The 2010 IEEE/RSJ International Conference on Intelligent Robots and Systems, 2010, pp. 1365–1371.

[22] C.-P. Lam, C.-T. Chou, K.-H. Chiang, L.-C. Fu, Human-centered robot navigation—towards a harmoniously human—robot coexisting environment, IEEE Trans. Rob. 27 (1) (2011) 99—112.
[23] H.-D. Yang, A.-Y. Park, S.-W. Lee, Gesture spotting and recognition for human—robot interaction, IEEE Trans. Rob. 23 (2) (2007) 256—270.
[24] Y. Fukazawa, C. Trevai, J. Ota, T. Arai, Acquisition of intermediate goals for an agent executing multiple tasks, IEEE Trans. Rob. 22 (5) (2006) 1034—1040.
[25] X. Wu, H. Gong, P. Chen, Z. Zhong, Y. Xu, Surveillance robot utilizing video and audio information, J. Intell. Rob. Syst. 55 (4—5) (2009) 403—421.
[26] H.-M. Gross, Ch. Schroeter, S. Mueller, M. Volkhardt, E. Einhorn, A. Bley, Ch. Martin, T. Langner, M. Merten, Progress in developing a socially assistive mobile home robot companion for the elderly with mild cognitive impairment, in: 2011 IEEE/RSJ International Conference on Intelligent Robots and Systems, 2011, pp. 2430—2437.
[27] http://www.isobotrobot.com/.
[28] http://www.aldebaran-robotics.com/en.
[29] http://www.parorobots.com.
[30] T. Wisspeintner, T. van der Zant, L. Iocchi, S. Schiffer, Robocup@home 2008: Analysis of Results, Technical Report (2008). http://www.ai.rug.nl/crl/Publications/Publications
[31] http://wiki.robocup.org/wiki/@Home_League.
[32] http://robocup.rwth-aachen.de/athome.
[33] P. Rybski, K. Yoon, J. Stolarz, M. Veloso, CMAssist: A RoboCup@Home Team, Technical Report CMU-RI-TR-06—47, Robotics Institute, Carnegie Mellon University, October 2006.
[34] http://robocare.istc.cnr.it/robocup.htm.
[35] http://www.time.com/time/photogallery/0,29307,1642800_1406458,00.html.
[36] H. Okada, T. Omori, N. Watanabe, T. Shimotomai, N. Iwahashi, K. Sugiura, T. Nagai, T. Nakamura, Team eR@sers 2012 in the @Home league team description paper, Technical Report, 2012.
[37] http://www.b-it-bots.de/%40Home.html.
[38] T. Breuer, G.R.G. Macedo, R. Hartanto, N. Hochgeschwender, D. Holz, F. Hegger, J. Zha, C. Müller, J. Paulus, M. Reckhaus, J.A. Álvarez Ruiz, P.G. Plöger, G.K. Kraetzschmar, Johnny: an autonomous service robot for domestic environments, J. Intell. Rob. Syst. 66 (1—2) (2012) 245—272.
[39] http://www.ais.uni-bonn.de/nimbro/@Home/.
[40] http://www.cit-ec.de/ToBI.
[41] http://www.wrighteagle.org/en/robocup/atHome/media.php.
[42] K.A. Wyrobek, E.H. Berger, H.F.M. Van der Loos, J.K. Salisbury, Towards a personal robotics development platform: rationale and design of an intrinsically safe personal robot, in: IEEE International Conference on Robotics and Automation, 2008, pp. 2165—2170.
[43] S. Cousins, ROS on the PR2, IEEE Rob. Autom. Mag. 17 (3) (2010) 23—25.
[44] S. Shigemi, Y. Kawaguchi, T. Yoshiike, K. Kawabe, N. Ogawa, Development of new ASIMO, Honda R&D Technical Review 18 (1) (2006) 38—44.
[45] J. Chestnutt, P. Michel, J. Kuffner, T. Kanade, Locomotion among dynamic obstacles for the Honda ASIMO, in: Proceedings of the 2007 IEEE/RSJ International Conference on Intelligent Robots and Systems, 2007, pp. 2572—2573.
[46] S.Y. Okita, V. Ng-Thow-Hing, Learning together: ASIMO developing an interactive learning partnership with children, in: The 18th IEEE International Symposium on Robot and Human Interactive Communication, 2009, pp. 1125—1130.
[47] M. Rolf, J.J. Steil, M. Gienger, Efficient exploration and learning of whole body kinematics, in: IEEE 8th International Conference on Development and Learning, 2009, pp. 1—7.
[48] http://world.honda.com/ASIMO/technology/2011/specification/index.html.
[49] V. Kulyukin, C. Gharpure, J. Nicholson, G. Osborne, Robot-assisted wayfinding for the visually impaired in structured indoor environments, Autonom. Rob. 21 (1) (2006) 29—41.

[50] T. Mei, M. Luo, X. Ye, J. Cheng, L. Wang, B. Kong, R. Wang, Design and implementation of a service robot for elders, in: 12th International Conference on Intelligent Autonomous Systems, 2012.
[51] L. Han, X. Wu, Y. Ou, Y.-L. Chen, C. Chen, Y. Xu, Household service robot with cellphone interface, Int. J. Inf. Acquis. 9 (2) (2013).

CHAPTER 2.2

Surveillance Robot Utilizing Video and Audio Information[1]

Xinyu Wu[1,2], Haitao Gong[2], Pei Chen[2], Zhi Zhong[1], Yangsheng Xu[1,2]

[1]Department of Mechanical and Automation Engineering, The Chinese University of Hong Kong, Hong Kong SAR, China; [2]Shenzhen Institutes of Advanced Technology, Chinese Academy of Sciences, Shenzhen, China

Chapter Outline

For an aging population, surveillance in household environments has become more and more important. In this chapter, we present a household robot that can detect abnormal events by utilizing video and audio information. In our approach, moving targets can be detected by the robot using a passive acoustic location device. The robot then tracks the targets by employing a particle filter algorithm. To adapt to different lighting conditions, the target model is updated regularly based on an update mechanism. To ensure robust tracking, the robot detects abnormal human behavior by tracking the upper body of a

[1] With kind permission from Springer Science + Business Media: Journal of Intelligent and Robotic Systems, Surveillance Robot Utilizing Video and Audio Information, vol. 55, 2009, pp. 403–421, Xinyu Wu.

Household Service Robotics. http://dx.doi.org/10.1016/B978-0-12-800881-2.00003-7

person. For audio surveillance, Mel frequency cepstral coefficients (MFCC) is used to extract features from audio information. Those features are input to a support vector machine classifier for analysis. Experimental results show that the robot can detect abnormal behavior such as "falling down" and "running." Also, an 88.17% accuracy rate is achieved in the detection of abnormal audio information like "crying," "groaning," and "gun shooting." To lower the false alarms by the abnormal sound detection system, the passive acoustic location device directs the robot to the scene where the abnormal events are occuring and the robot can employ its camera to further confirm the occurrence of the events. At last, the robot will send the image captured by the robot to the mobile phone of its master.

2.2.1 Introduction

There are many global issues that could be eased by the correct usage of video surveillance, most of which occur in public places such as elevators, banks, airports, and public squares. However, surveillance could also be used in the home. With the increasing numbers of aged people living alone, household surveillance systems could be used to help the elderly live more safely. The fundamental problem with surveillance systems is the intelligent interpretation of human events in real-time. In this chapter, we present a household surveillance robot combining video and audio surveillance that can detect abnormal events.

The testing prototype of the surveillance robot is composed of a pan/tilt camera platform with two cameras and a robot platform (Figure 1). One camera is employed to track the target and detect abnormal behavior. The other is planned to detect the face and to determine facial recognition, which is not discussed in this chapter. Our robot can first detect a moving target by sound localization and then track it across a large field of vision using a pan/tilt camera platform. It can detect abnormal behavior in a cluttered environment, such as a person suddenly running or falling down to the floor. By teaching the robot the differences between normal and abnormal sound information, the computational action models built inside the trained support vector machines can automatically identify whether newly received audio information is normal. If abnormal audio information is detected, then the robot can employ its camera to further check the events directed by a passive acoustic location device.

A number of video surveillance systems for detecting and tracking multiple people have been developed, such as *W4* in [1], TI's system in [2], and the system in [3]. Occlusion is a significant obstacle for such systems and good tracking often depends on correct segmentation. Furthermore, none of these systems is designed to detect abnormal behavior as their main function. Radhakrishnan et al. [4] presented a systematic framework for the detection of abnormal sounds that may occur in elevators.

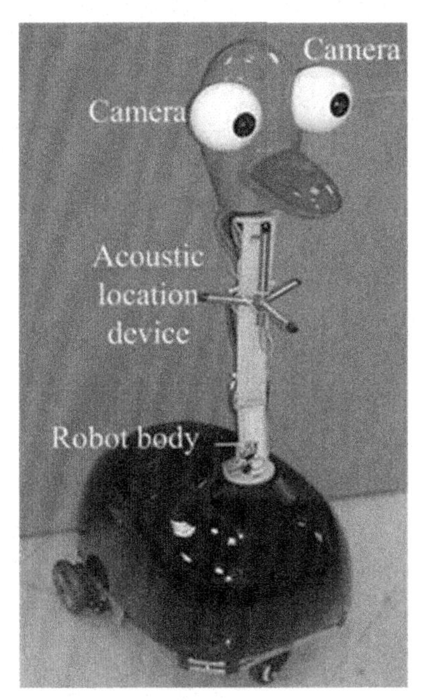

Figure 1
The testing prototype of surveillance robot.

Luo [5] built a security robot that can detect a dangerous situation and provide a timely alert, focusing on fire detection, power detection, and intruder detection. Nikos [6] presented a decision-theoretic strategy for surveillance as a first step toward automating the planning of the movements of an autonomous surveillance robot.

The overview of the system is shown in Figure 2. In the initialization stage, two methods are employed to detect moving objects. One is to pan the camera step by step and employ the frame differencing method to detect moving targets during the static stage. The other method uses a passive acoustic location device to direct the camera at the moving object, keeping the camera static and employing the frame differencing method to detect foreground pixels. The foreground pixels are then clustered into labels and the center of each label is calculated as the target feature, which is used to measure the similarity in the particle filter tracking. In the tracking process, the robot camera tracks the moving target using a particle filter tracking algorithm, and updates the tracking target model at the appropriate time. To detect abnormal behavior, upper body (which is more rigid) tracking is implemented that uses the vertical position and speed of the target. At the same time, with the help of a learning algorithm, the robot can detect abnormal audio information, such as crying or groaning, even in other rooms.

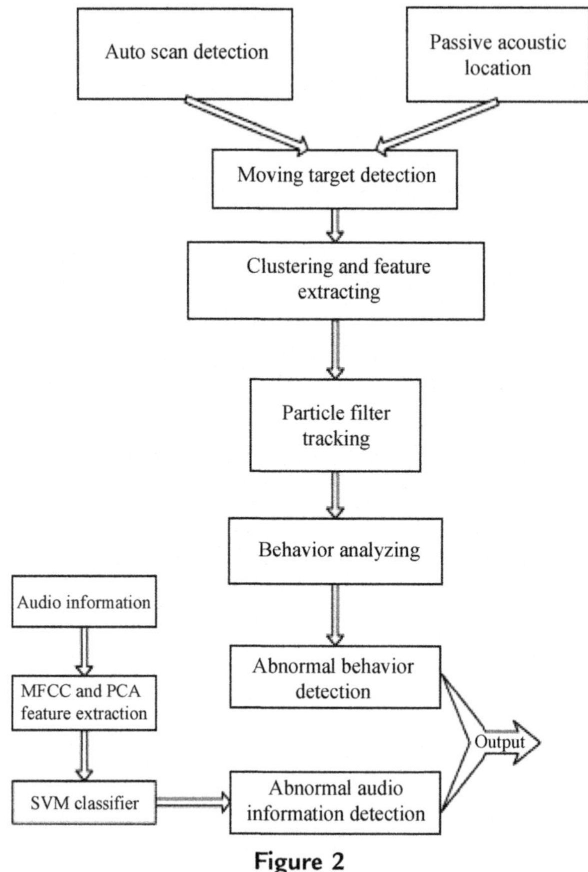

Figure 2
Block diagram of the system.

The rest of this chapter is organized as follows. Section 2.2.2 introduces the passive acoustic location device, frame differencing method and feature selection and segmentation. Section 2.2.3 presents how to employ the particle filter algorithm to track a moving target and detect abnormal behaviors. In Section 2.2.4, we present how to employ a support vector machine to detect abnormal audio information. Section 2.2.5 shows the experimental results utilizing video and audio information before we conclude in Section 2.2.6.

2.2.2 System Initialization

In many surveillance systems, the background subtraction method is used to find the background model of an image so that moving objects in the foreground can be detected by simply subtracting the background from the frames. However, our surveillance robot cannot use this method because the camera is mobile and we must therefore use a slightly different approach. When a person speaks or makes noise, we can locate the position of the person with a passive acoustic location device and rotate the camera to the correct direction. The frame differencing method is then employed to detect movement. If the passive acoustic location device does not detect any sound, then the surveillance robot turns the camera 30° and employs the differencing method to detect moving targets. If the robot does not detect any moving targets, then the process is repeated until the robot finds a moving target or the passive acoustic device gives it a location signal.

2.2.2.1 Passive Acoustic Location

An object producing an acoustic wave is located and identified by the passive acoustic location device. Figure 3 shows the device which comprises four microphones installed in an array. The device uses the time-delay estimation method, which is based on the time differences in sound reaching the various microphones in the sensor array. The acoustic source position is then calculated from the time-delays and the geometric position of the microphones. To obtain this spatial information, three independent time-delays are needed, and therefore the four microphones are set at different positions on the plane. Once the direction result has been obtained, the pan/tilt platform will move so that the moving object is included in the camera view field.

The precision of the passive acoustic location device depends on the distances between the microphones and the precision of the time-delays. We test the device and the passive acoustic location error is about 10° in the x−y plane. See Figure 4, the camera angle is about 90° and much larger than the passive acoustic location error. When the passive acoustic location device provides a direction, the robot turns the camera and keeps the direction in the center of the camera view. Thus the location error could be ignored.

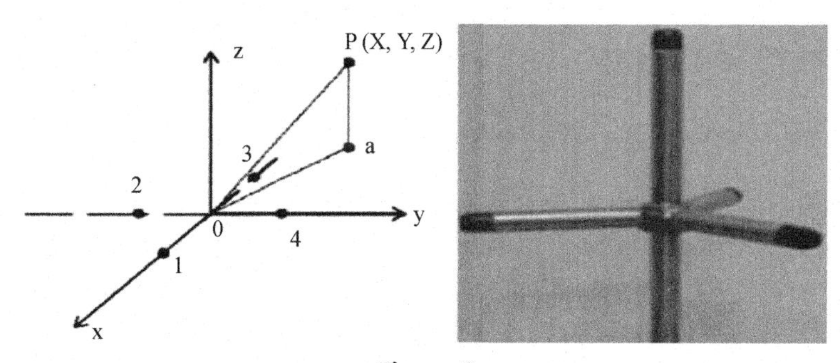

Figure 3
Passive acoustic location device.

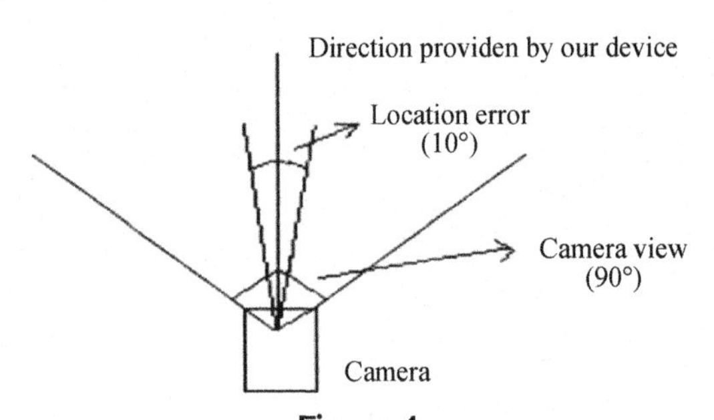

Figure 4
The method to solve location error.

2.2.2.2 Target Detection Using the Frame Differencing Method

We employ the frame differencing method to detect a target, as this only requires the camera to be kept static for a while. Frame differencing is the simplest method for moving object detection, because the background model is simply equal to the previous frame. After performing a binarization process with a predefined threshold with the differencing method, we can find the target contour and the target blob is obtained through the contour filling process. However, sometimes the blob contains too many background pixels when the target is moving very fast, or the blob may lose part of the target information when the target is moving slowly. It is impossible to obtain pure foreground pixels when using the frame differences as the background model, but by using the following method, we can remove the background pixels and retrieve more foreground pixels, on the condition that the color of the foreground is not similar to the pixels of the nearby background.
By separately segmenting the foreground and background in a rectangular area, we can label and cluster the image in the rectangle area again to obtain a more accurate foreground blob.

Figure 5 shows the foreground detection process in a frame differencing method, where (a) and (b) show the target detection result using the frame differencing method and the blob filling result, respectively. In Figure 5(c), we can see the left leg of the target is lost. After a labeling and clustering process, we can retrieve the left leg (see Figure 5(d)).

2.2.2.3 Feature Selection and Segmentation

Feature selection is very important in tracking applications. Good features will result in excellent performance, whereas poor features will restrict the ability to distinguish the target from the background in the feature space. In general, the most desirable property of a visual feature is its uniqueness in the environment. Feature selection is closely related to object representation, in which the object edge or shape feature is used as the feature for contour-based representation and color is used as a feature for histogram-based appearance

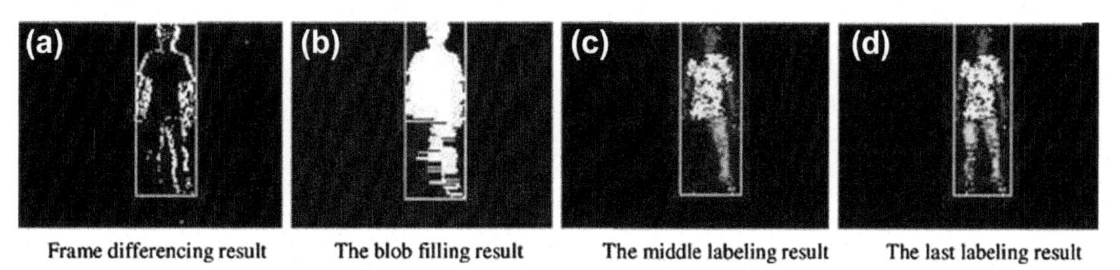

Figure 5
Foreground detection process (a–d).

representations. Some tracking algorithms use a combination of these features. In this chapter, we use color-spatial information for the feature selection. The apparent color of an object is influenced primarily by the spectral power distribution of the illuminant and the surface reflectance properties of the object. The choice of color space also influences the tracking process. Usually, digital images are represented in the RGB (red, green, blue) color space, but the RGB space is not a perceptually uniform color space because the differences between the colors do not correspond to the color differences perceived by humans. Additionally, the RGB dimensions are highly correlated. HSV (Hue, Saturation, Value) is an approximately uniform color space, that is more similar to human perception and we therefore select this color space for this research. Using the color information is not sufficient, but if we consider combining the color-spatial distribution information, then the selected features will become more discriminative.

The main task is to segment the image using this feature. We choose the SMOG method to model the appearance of an object and define the Mahalanobis distance and similarity measure [7]. We then employ the K-means algorithm followed by a standard EM algorithm to cluster the pixels. The difference of our approach is that we do not cluster and track the whole region in a rectangle, but only the moving target in the rectangle, as described in the previous section. Figure 6 shows the clustering and tracking results for the

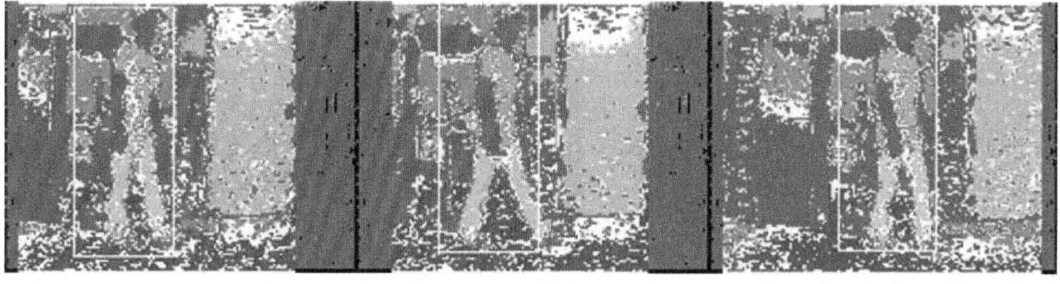

Figure 6
Clustering and tracking results on the whole region in rectangle.

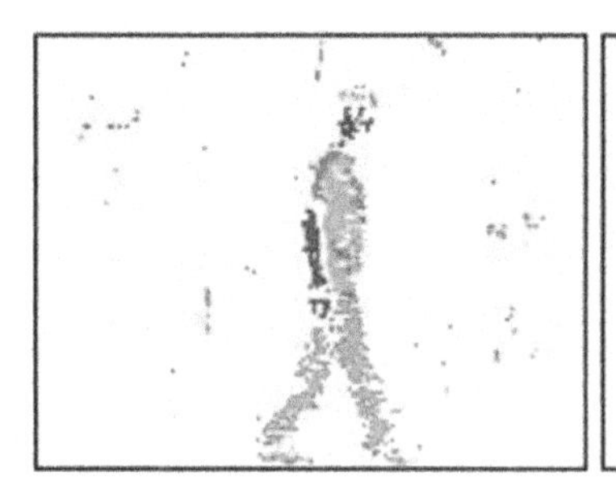
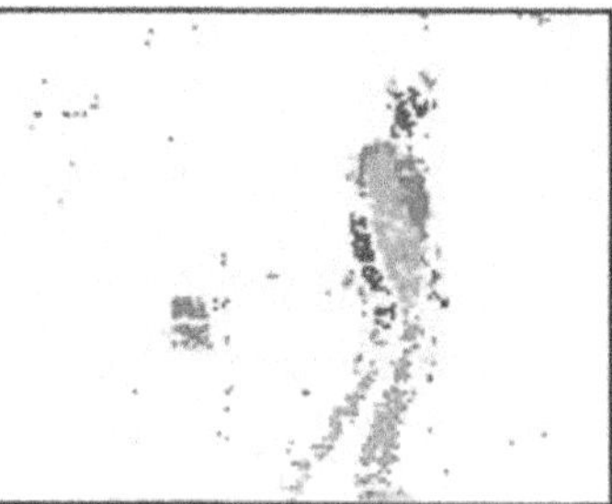
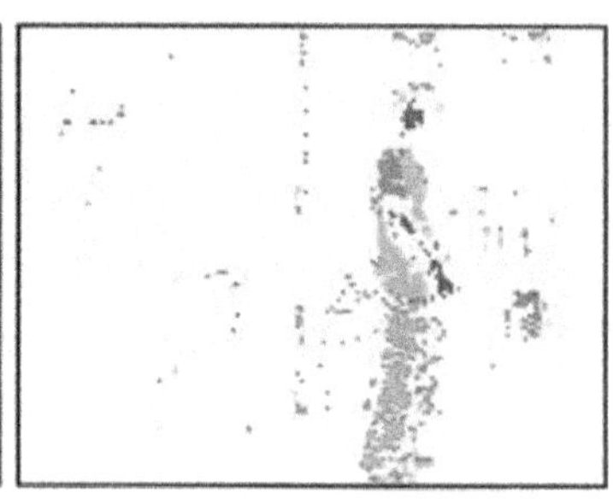

Figure 7
Clustering and tracking results on moving target.

whole region in the rectangle, and Figure 7 shows the clustering and tracking results of the moving target.

If we employ the standard method to cluster and track the whole region in the rectangle, it may track properly but requires more particles and computation time. It is because the whole region in the rectangle contains many background pixels. When we choose a new particle at any place in the image, the similarity coefficient is likely high. Thus more particles are required to find good candidate particles from the complicated background. We test the standard method to track a target in a real case and the frame rate can reach 10 frames per second at a resolution of 160×120.

On the other hand, if we cluster and track the moving target only, few particles and less time are needed. The frame rate can reach 15 frames when we track a target employing our method. To save computation time, therefore, the robot clusters and tracks the moving target only.

2.2.3 Video Surveillance

2.2.3.1 Tracking Using Particle Filter

We propose a tracking strategy that always keeps the target in the scene. Here we do not want and do not always need to keep the target in the exact center of the scene, because this needs the camera to move frequently and thus it is hard to obtain an accurate speed of the target. See Figure 8, when the target center is in the place between the left edge and right edge, the camera and the mobile platform both remain static. When the target moves and the target center reaches the left edge or right edge, the robot moves to keep the target in the center of the scene according to the predicted results. When the camera moves, a particle filtering algorithm is employed to perform the tracking because it can overcome the difficulty of background changes.

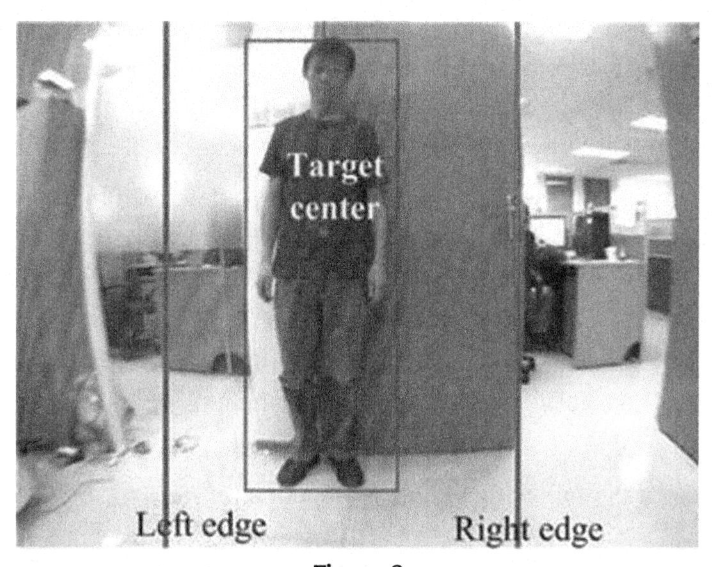

Figure 8
The left edge and right edge for the tracking strategy.

Sequential Monte Carlo techniques, which are also known as particle filtering and condensation algorithms [8–10], have been widely applied in visual tracking in recent years. The general concept is that if the integrals required for a Bayesian recursive filter cannot be solved analytically, then the posterior probabilities can be represented by a set of randomly chosen weighted samples. The posterior state distribution $p(x_k|Z_k)$ needs to be calculated at each time step. In the Bayesian sequential estimation, the filter distribution can be computed by the following two-step recursion.

Prediction step:

$$p(x_k|Z_{k-1}) = \int p(x_k|x_{k-1})p(x_{k-1}|Z_{k-1})dx_{k-1} \tag{1}$$

Filtering step:

$$p(x_k|Z_k) \propto p(z_k|x_k)p(x_k|Z_{k-1}) \tag{2}$$

Based on a weighted set of samples $\{x_{k-1}^{(i)}, \omega_k^{(i)}\}_{i=1}^{N}$ approximately distributed according to $p(x_{k-1}|Z_{k-1})$, we draw particles from a suitable proposal distribution, i.e., $x_k^{(i)}: q_p(x_k|x_{k-1}^{(i)}, z_k),\ \ i = 1, \ldots, N$. The weights of new particles become:

$$\mathrm{w}_k^{(i)} \propto \omega_{k-1}^{(i)} \frac{p\left(z_k\middle|x_k^{(i)}\right)p\left(x_k^{(i)}\middle|x_{k-1}^{(i)}\right)}{q_p\left(x_k\middle|x_{k-1}^{(i)}, z_k\right)} \tag{3}$$

The observation likelihood function $p(z_k|x_k)$ is important because it determines the weights of the particles and thereby significantly influences the tracking performance. Our observation likelihood function is defined by the SMOG method, which combines spatial layout and color information, as explained below.

Suppose that the template is segmented into k clusters, as described in the previous section. For each cluster, calculate its histograms, vertically and horizontally; and make a new histogram of $\tilde{q}^i$, by concatenating the vertical and horizontal histograms and then being normalized.

As for as a particle is concerned, you need to first classify the pixels into the k clusters above the template. Then, calculate its normalized histograms of $\{q_t^i(x_t)\}$, as done with the template.

As in [11], we employ the following likelihood function for $p(z_k|x_k)$:

$$p(z_k|x_k) \propto \prod_i^k \exp - \lambda D^2 \left[\tilde{q}^i, q_t^i(x_t)\right]$$

where λ is fixed as 20 [11], and D is the Bhattacharyya similarity coefficient between two normalized histograms $\tilde{q}^i$ and $q_t^i(x_t)$:

$$D\left[\tilde{q}^i, q_t^i(x_t)\right] = \left[1 - \sum_k \sqrt{\tilde{q}^i(k) q_t^i(k; x_t)}\right]^{\frac{1}{2}}$$

The steps in the particle sample and updating process are as follows.

Step 1 Initialization: draw a set of particles uniformly.
Step 2 (1) Sample the position of the particle from the proposal distribution. (2) Find the feature of the moving object. (3) Update the weight of the particles. (4) Normalize the weight of the particles.
Step 3 Output the mean position of the particles that can be used to approximate the posterior distribution.
Step 4 Resample the particles with probability to obtain independent and identically distributed random particles.
Step 5 Go to the sampling step.
Computationally, the crux of the PF algorithm lies in the calculation of the likelihood.

2.2.3.2 Target Model Update

The target model obtained in the initialization process cannot be used for the whole tracking process due to changes in lighting, background environment, and target gestures.

We therefore need to update the tracking target model in time. But if we update the target model at an improper time, such as when the camera is moving and the image is not clear, then the tracking will fail. Figure 9 shows a new target model updating process. In our camera control strategy, the camera remains static when the target is in the center of the camera view. When the camera is static, the frame differencing method is employed to obtain the target blob, and the similarity between the current blob and the initial blob is calculated. If the similarity property is larger than a given threshold, then we update the target model, otherwise, we move on to the frame differencing step. How to choose the threshold is an interesting problem.

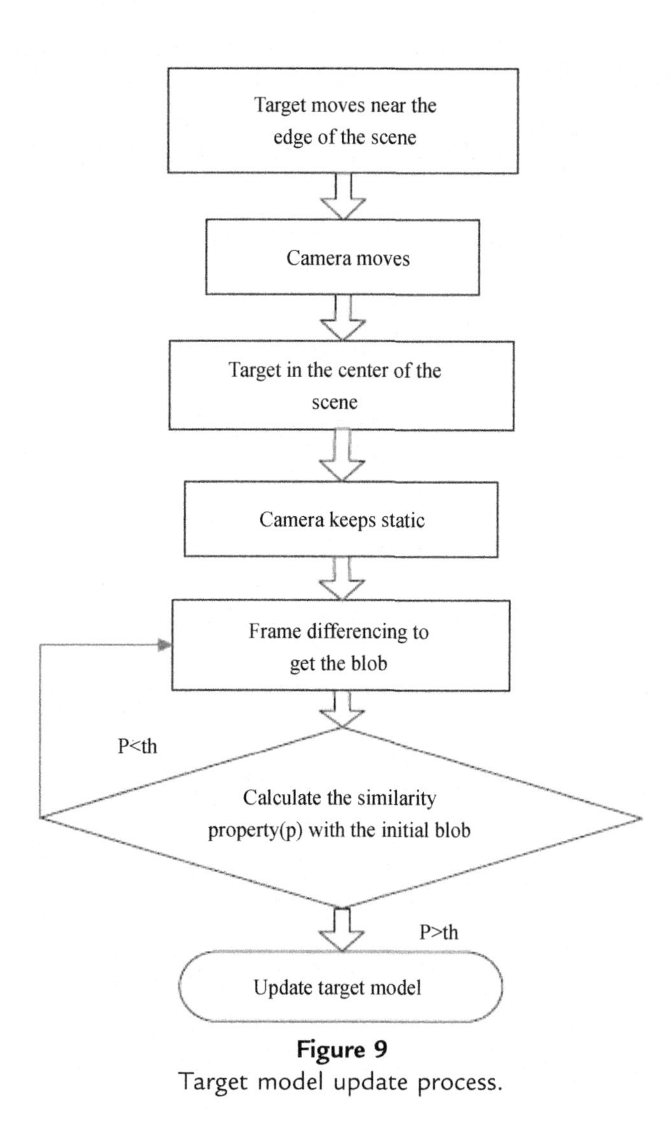

Figure 9
Target model update process.

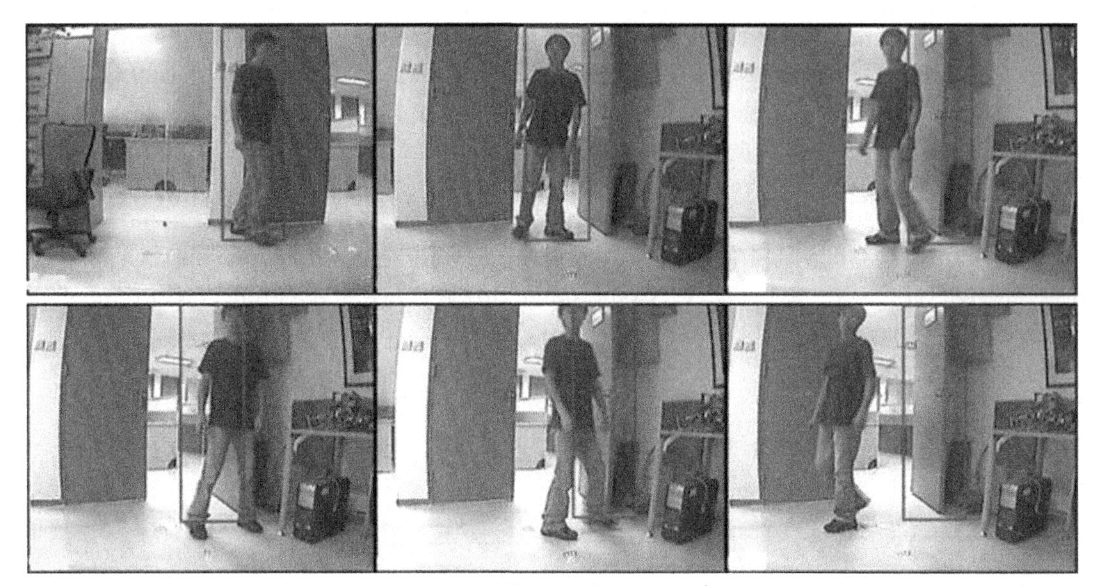

Figure 10
Tracking results without update.

If the threshold is very large, the similarity coefficient may be easily lower than the threshold. This will cause continual updating and consume much computation time. If the threshold is very low, the wrong detection will happen. To balance the computation time and tracking results, we choose the threshold as 0.6 according to many experiments in the real environment.

Figure 10 shows failed tracking results without updating in the lighting changes environment. Figure 11 shows robust tracking results employing our update process in the lighting changes environment.

2.2.3.3 Experimental Results on Abnormal Behavior Detection

Nowadays, abnormal behavior detection is a popular research topic, and many studies have presented methods to detect abnormal behavior. Our surveillance robot mainly focuses on the household environment, so it is important to detect abnormal behaviors such as people falling down and people running.

It is not easy to track the whole body of a person because of the large range of possible body gestures, which can lead to false tracking. To solve this problem, we propose a method that only tracks the upper body of a person (Figure 12), which does not vary much with gesture changes. We take the upper half rectangle as the upper body of a target. It may contain some part of legs or it may lose some part of the upper body. We can obtain a pure upper body by using the clustering method mentioned above. Based on this robust

Figure 11
Tracking results employing our update process.

tracking system, we can obtain the speed of the target, the height and width of the target. Through the speed of the upper body and the thresholds selected by different experiments, the running movement can be successfully detected. Also, based on the height and width of the target, we can detect falling down movement through shape analysis.

Figure 13(a–d) shows the robot moving to the proper direction to detect the target on receiving a direction signal from the passive acoustic location device. Figure 13(a) and 13(b) are not clear because the camera is moving very quickly. Whereas in Figure 13(e) and 13(f), the robot tracks the target and keeps it in the center of the camera view.

Figure 12
Clustering results on the upper body.

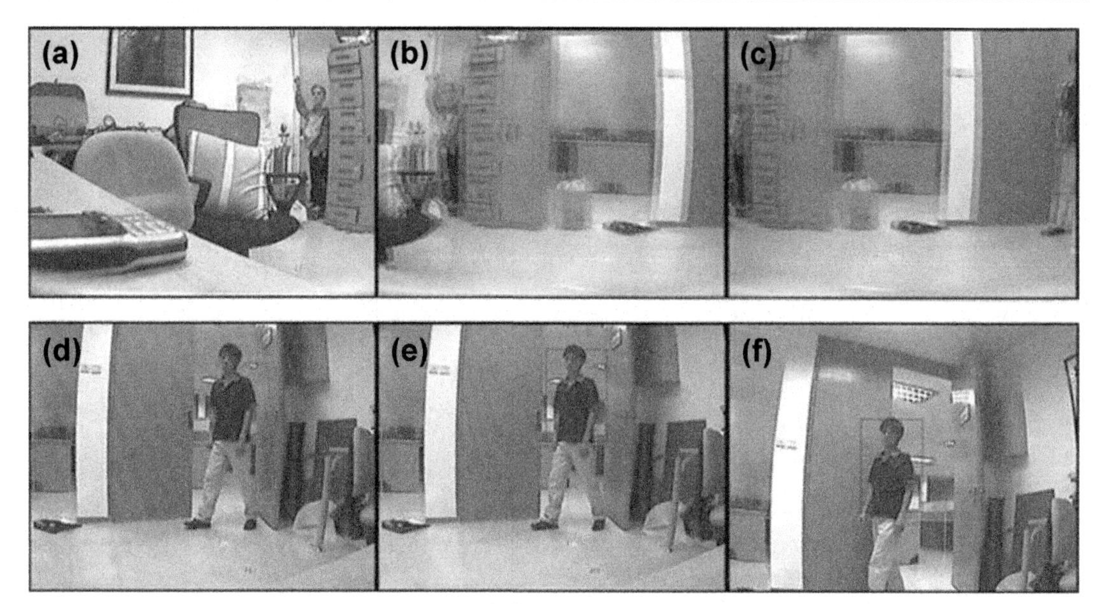

Figure 13
Initialization and tracking results (a–f).

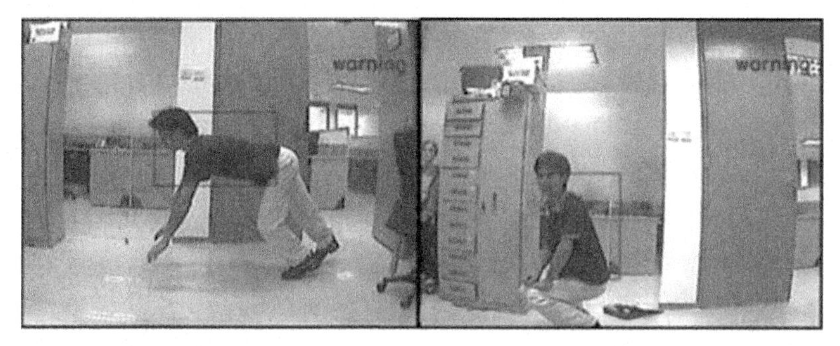

People falling down   People bending down

People running   People running

Figure 14
Abnormal behavior detection results.

Figure 14 shows the detection of the abnormal behavior such as people falling down, bending down, and running in the household environment based on the tracking results.

2.2.4 Abnormal Audio Information Detection

Compared with video surveillance, audio surveillance does not need to directly "watch" the scene. The effectiveness of audio surveillance is not influenced by the occlusions which may cause the failure of the video surveillance system. Especially, in a house or storehouse, some areas may be occluded by moving objects or static objects. Also, the robot and people may not be in the same room if there are several rooms in a house.

We propose a supervised learning based approach to audio surveillance in the household environment [12]. Figure 15 shows the training framework of our approach. First, we collected a sound-effect dataset (See Table 1) which includes many sound effects collected from a household environment. Secondly, we manually labeled these sound effects as abnormal samples (e.g., screaming, gun shooting, glass breaking sound and banging sound) or normal samples (e.g., speech, footstep, shower sound and phone ringing). Thirdly, MFCC features were extracted from a 1.0 s waveform of each sound effect sample. Finally, we trained a classifier using support vector machine. For detecting, when a new 1.0 s waveform was received, the MFCC feature was extracted from the waveform; then the classifier was employed to determine whether this sound sample is normal or abnormal.

2.2.4.1 MFCC Feature Extraction

To discriminate normal sounds from abnormal sounds, a meaningful acoustic feature must be extracted from the waveform of the sound.

Many audio feature extraction methods have been proposed for different audio classification applications. For speech and music discrimination tasks, the spectral centroid, zero-crossing rate, percentage of "low-energy" frames, and spectral "flux"

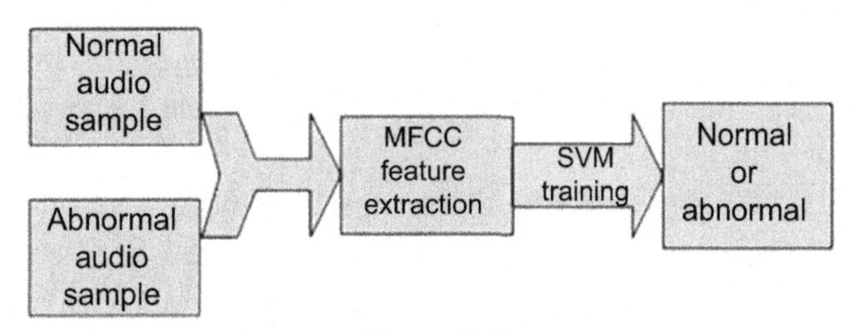

Figure 15
Training framework.

Table 1: The sound effect dataset

Normal Sound Effects	Abnormal Sound Effects
Normal speech	Gun shot
Boiling water	Glass breaking
Dish-washer	Screaming
Door closing	Banging
Door opening	Explosions
Door creaking	Crying
Door locking	Kicking a door
Fan	Groaning
Hair dryer	
Phone ringing	
Pouring liquid	
Shower	
...	...

methods [13] have been used. Spectral centroid represents the "balancing point" of the spectral power distribution. Zero-crossing rate measures the dominant frequency of a signal. The percentage of "Low-Energy" frames describes the skewness of the energy distribution. Spectral "flux" measures the rate of change of the sound. For automatic genre classification, timbral features, rhythmic features, and pitch features [14], which describe the timbral, rhythmic and pitch characteristics of the music, respectively, have been proposed.

In our approach, the MFCC feature is employed to represent audio signals. The idea of the MFCC feature is motivated by perceptual or computational considerations. As the feature captures some of the crucial properties used in human hearing, it is ideal for general audio discrimination. The MFCC feature has been successfully applied to speech recognition [15], music modeling [16], and audio information retrieval [17], and more recently, has been used in audio surveillance [18].

The steps to extract MFCC features from the waveform are as follows:

Step 1 Normalize the waveform to the range $[-1.0, 1.0]$ and window the waveform with a hamming window;
Step 2 Divide the waveform into N frames, i.e., $\frac{1000}{N}$ ms for each frame;
Step 3 Take the Fast Fourier Transform (FFT) of each frame for getting the frequency information of each frame;
Step 4 Convert the FFT data into filter bank outputs. Since the lower frequencies are perceptually more important than the higher frequencies, the 13 filters allocated below 1000 Hz are linearly spaced (133.33 Hz between center frequencies) and the 27 filters

allocated above 1000 Hz are spaced logarithmically (separated by a factor of 1.0711703 in frequency). Figure 16 shows the frequency response of the triangular filters;
Step 5 Since the perceived loudness of a signal has been found to be approximately logarithmic, we take the log of the filter bank outputs;
Step 6 Take the cosine transform to reduce dimensionality. Since the filter bank outputs calculated for each frame are highly correlated, we take the cosine transform which approximates the principal components analysis to decorrelate the outputs and reduce dimensionality. 13 (or so) cepstral features are obtained for each frame by this transform method. If we divide the waveform into 10 frames, the total dimensionality of the MFCC feature for the 1.0 s waveform is 130.

2.2.4.2 Support Vector Machine

After extracting the MFCC features from the waveform, we employed a classifier trained by a support vector machine (SVM) to determine whether this sound is normal or abnormal [19–21].

Our goal is to separate sounds into two classes, normal and abnormal, according to a group of features. There are many types of neural networks that can be used for such a binary classification problem, such as SVMs, Radial Basis Function Networks, Nearest Neighbor Algorithm, Fisher Linear Discriminant and so on. SVM is chosen as audio classifiers because it has a stronger theory-interpretation and better generalization than previously mentioned neural networks. Compared to other neural network classifiers, SVM

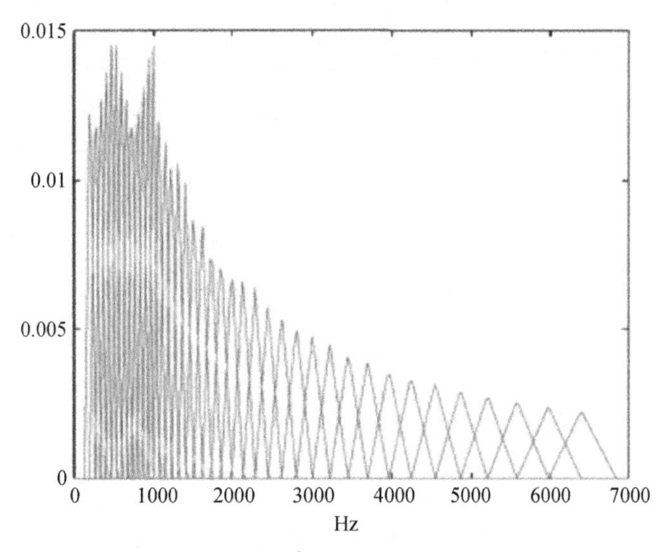

Figure 16
Frequency response of the triangular filters.

has three distinct characteristics. First, it estimates a classification using a set of linear functions that are defined in a high-dimensional feature space. Second, SVM carries out the classify estimation by risk minimization, where risk is measured using Vapnik's ε-insensitive loss function. Third, it implements the Structural Risk Minimization principle, which minimizes the risk function that consists of the empirical error and a regularized term.

2.2.4.3 Experimental Results on Abnormal Sound Detection

To evaluate our approach, we collected 169 sound effects samples from the internet (http://www.grsites.com) including 128 normal samples and 41 abnormal samples (Most of them were collected in the household environment).

For each sample, the first 1.0 s waveform was used for training and testing. The rigorous jack-knifing cross-validation procedure, which reduces the risk of overstating the results, was used to estimate the classification performance. For a dataset with M samples, we chose $M - 1$ samples to train the classifier, then tested the performance using the left sample. This procedure was then repeated for M times. The final estimation result was obtained by averaging the M accuracy rates. To train a classifier by using SVM, we apply a polynomial kernel where the kernel parameter $d = 2$, and the adjusting parameter C in the loss function is set to 1.

Table 2 shows the accuracy rates using the MFCC feature trained with different frame sizes. Table 3 shows the accuracy rates using MFCC and PCA features trained with different frame sizes. PCA is employed to reduce the data dimension. For example, PCA reduces the data dimension from 637 to 300 when the frame size is 20 ms and from 247 to 114 when the frame size is 50 ms. Comparing the results in Tables 2 and 3, we found that it is unnecessary to use PCA. We obtained the best accuracy rate of 88.17% with a 100 ms frame size. The data dimension of this frame size is 117 which does not need reducing.

Table 2: Accuracy rates (%) using the MFCC feature trained with 20, 50, 100, and 500 ms frame sizes

Frame Size	Accuracy
20 ms	86.98
50 ms	86.39
100 ms	88.17
500 ms	79.88

Table 3: Accuracy rates (%) using the MFCC and PCA features trained with 20, 50, 100, and 500 ms frame sizes

Frame Size	Accuracy
20 ms	84.62
50 ms	78.70
100 ms	82.84
500 ms	76.33

2.2.5 Experimental Results Utilizing Video and Audio Information

We test our abnormal sound detection system in a real working environment. The sound consists of normal speeches, a door opening, pressing keyboard, etc. which are all normal sounds. The system gives eight false alarms in 1 h. On the other hand, a sound box is used to play abnormal sounds like a gun shooting, crying, etc. The accuracy rate is 83% among 100 abnormal sound samples. The accuracy is lower compared with previous experiments for noise in a real environment.

To lower the false alarms by the abnormal sound detection system, the passive acoustic location device directs the robot to the scene where abnormal events occur and the robot can employ its camera to further confirm the occurrence of the events. In Figure 17, we can see the process of abnormal behavior detection utilizing video and audio information when two persons are fighting in the other room. First, the abnormal sound detection system detects abnormal sound (fighting and shouting). At the same time, the passive acoustic location device obtains the direction. Then the robot turns to the abnormal direction and captures images to check if abnormal behavior is occuring. Here we can employ an optical flow method to detect fighting, which we presented in a previous paper [22].

There are some cases in which our system cannot detect abnormal behaviors. For example, a person is already falling down before the robot turns to the abnormal direction. To solve this problem, the robot will send the image captured by the robot to the mobile phone of the master (Figure 18).

2.2.6 Conclusions

In this chapter, we described a household surveillance robot that can detect abnormal events combining video and audio surveillance. Our robot first detects a moving target by sound localization, and then tracks it across a large field of vision using a pan/tilt camera platform. It can detect abnormal behavior in a cluttered environment, such as a person

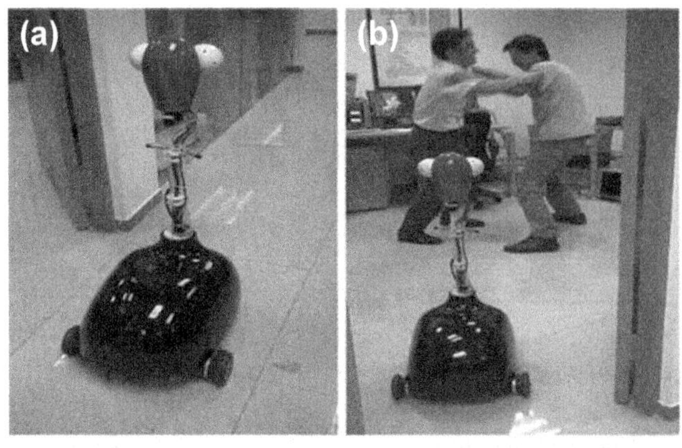

The images are captured by us using a digital video

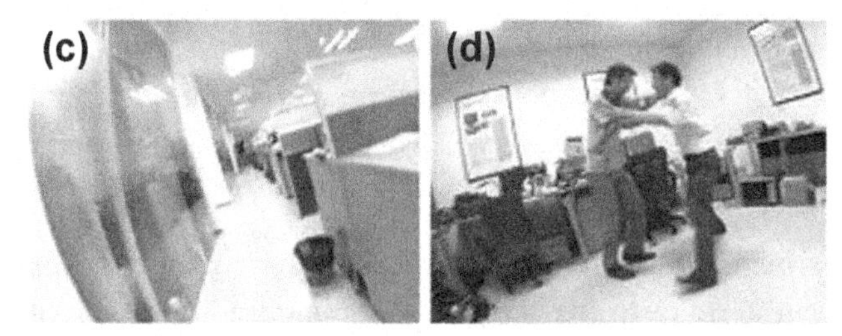

The images are captured by the robot

Figure 17

The process of abnormal behavior detection utilizing video and audio information: (a) The initial state of the robot; (b) The robot turning to the abnormal direction; (c) The initial image captured by the robot; (d) The image captured by the robot after turning to the abnormal direction.

suddenly running or falling down on the floor, and can also detect abnormal audio information and employ its camera to further check the event.

There are three main contributions in our research: (1) an innovative strategy for the detection of abnormal events by utilizing video and audio information; (2) an initialization process that employs a passive acoustic location device to help the robot detect moving targets; and (3) an update mechanism to update the target model regularly.

However, there remains a lot of work to be completed. In the future, the abnormal audio information needs to be realized within the hardware, such as arm systems. To promote the abnormal events detection accuracy rate, the fusion of video and audio information needs more investigation.

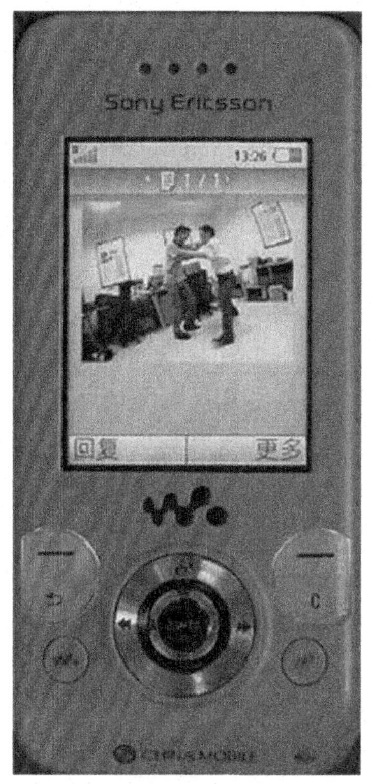

Figure 18
The robot sends the image to the mobile phone of the master.

Acknowledgments

The authors would like to thank Mr. Y.J. Liang, Mr. Deng Lei, Mr. Qin Jianzhao, and Mr. Fang Zhou for their valuable contribution to this project. The authors would also wish to acknowledgment Mr. Shi Xi with his help on the SVM classification process.

The work described in this chapter is partially supported by a grant from the Ministry of Science and Technology, The Peoples Republic of China (International Science and Technology Cooperation Projects 2006DFB73360), and by the National Basic Research Program of China (No.2007cb311005).

References

[1] I. Haritaoglu, D. Harwood, L.S. Davis, W4: real-time surveillance of people and their activities, IEEE Trans. Pattern Anal. Mach. Intell. 22 (8) (2000).
[2] T. Olson, F. Brill, Moving object detection and event recognition algorithms for smart cameras, in: Proc. DARPA Image Understanding Workshop, May 1997, pp. 159–175. New Orleans.
[3] T. Zhao, R. Nevatia, Tracking multiple humans in complex situations, IEEE Trans. Pattern Anal. Mach. Intell. 26 (9) (2004).

[4] R. Radhakrishnan, A. Divakaran, Systematic acquisition of audio classes for elevator surveillance, in: Proc. of SPIE, 24–26 May 2005, pp. 64–71. Austin.
[5] R.C. Luo, K.L. Su, A multiagent multisensor based real-time sensory control system for intelligent security robot, in: Proceedings of International Conference on Robotics and Automation, 14–19 September 2003. Taiwan.
[6] N. Massios, F. Voorbraak, Hierarchical decision-theoretic planning for autonomous robotic surveillance, in: Massios (Ed.), Advanced Mobile Robots, 1999 Third European Workshop, 6–8 September 1999, pp. 219–226. Zurich.
[7] H. Wang, D. Suter, K. Schindler, C. Shen, Adaptive object tracking based on an effective appearance filter, IEEE Trans. Pattern Anal. Mach. Intell. 29 (9) (2007).
[8] Z. Khan, T. Balch, F. Dellaert, MCMC-based particle filtering for tracking a variable number of interacting targets, IEEE Trans. Pattern Anal. Mach. Intell. 27 (11) (2005).
[9] J. Carpenter, P. Clifford, P. Fernhead, An Improved Particle Filter for Non-Linear Problems, Technical Report, Department of Statistics, University of Oxford, 1997.
[10] M.S. Arulampalam, S. Maskell, N. Gordon, T. Clapp, A tutorial on particle filters for online nonlinear/non-Gaussian Bayesian tracking, IEEE Trans. Signal Process. 50 (2) (2002).
[11] P. Perez, C. Hue, J. Vermaak, M. Gangnet, Color-based probabilistic tracking, in: European Conference on Computer Vision, 27 May–2 June 2002, pp. 661–675. Copenhagen.
[12] X.Y. Wu, J.Z. Qin, J. Cheng, Y.S. Xu, Detecting audio abnormal information, in: The 13th International Conference on Advanced Robotics, 21–24 August 2007, pp. 550–554. Jeju.
[13] E. Scheirer, M. Slaney, Construction and evaluation of a robust multi feature speech/music discriminator, in: Proc. of International Conference on Acoustics, Speech, and Signal Processing, 21–24 April 1997, pp. 1331–1334. Munich.
[14] G. Tzanetakis, P. Cook, Musical genre classification of audio signals, IEEE Trans. Speech Audio Process. 10 (5) (2002) 293–302.
[15] J.N. Holmes, W.J. Holmes, Speech Synthesis and Recognition, second ed., Taylor & Francis CRC, London, 2001.
[16] B.T. Logan, Mel frequency cepstral coefficients for music modeling, in: Proceedings of the First International Symposium on Music Information Retrieval, Bloomington, 15–17 October 2001.
[17] J. Foote, An overview of audio information retrieval, Multimedia Syst. 7 (1) (1999) 2–10.
[18] R. Radhakrishnan, A. Divakaran, P. Smaragdis, Audio analysis for surveillance applications, in: IEEE Workshop on Application of Signal Processing to Audio and Acoustics, 16–19 October 2005, pp. 158–161. New Paltz.
[19] N. Cristianini, J. Shawe-Taylor, A Introduction to Support Vector Machines and Other Kernel based Learning Methods, Cambridge University Press, Cambridge, 2000.
[20] S. Bernhard, C.J.C. Burges, A. Smola, Advanced in Kernel Methods Support Vector Learning, MIT, Cambridge, 1998.
[21] Y. Ou, X.Y. Wu, H.H. Qian, Y.S. Xu, A real time race classification system, in: Information Acquisition, 2005 IEEE International Conference. Hong Kong, 27 June–3 July 2005.
[22] Y.S. Ou, H.H. Qian, X.Y. Wu, Y.S. Xu, Real-time surveillance based on human behavior analysis, Int. J. Inf. Acquis. 2 (4) (December 2005) 353–365.

CHAPTER 2.3

Robot-Assisted Wayfinding for the Visually Impaired in Structured Indoor Environments[1]

Vladimir Kulyukin[1], Chaitanya Gharpure[1], John Nicholson[1], Grayson Osborne[2]
[1]Computer Science Assistive Technology Laboratory (CSATL), Department of Computer Science, Utah State University, Logan, UT, USA; [2]Department of Psychology, Utah State University, Logan, UT, USA

Chapter Outline

We present a robot-assisted wayfinding system for the visually impaired in structured indoor environments. The system consists of a mobile robotic guide and small passive RFID sensors embedded in the environment. The system is intended for use in indoor environments, such as office buildings, supermarkets and airports. We describe how the

[1] With kind permission from Springer Science + Business Media: Autonomous Robots, Robot assisted wayfinding for the visually impaired in structured indoor environments, Vol. 21, 2006, pp. 29–41, Vladimir Kulyukin.

Household Service Robotics. http://dx.doi.org/10.1016/B978-0-12-800881-2.00004-9

system was deployed in two indoor environments and evaluated by visually impaired participants in a series of pilot experiments. We analyze the system's successes and failures and outline our plans for future research and development.

2.3.1 Introduction

Since the adoption of the Americans with Disabilities Act of 1990 that provided legal incentives for improvement in universal access, most of the research and development (R&D) has focused on removing *structural* barriers to universal access, e.g., retrofitting vehicles for wheelchair access, building ramps and bus lifts, improving wheelchair controls, and providing access to various devices through specialized interfaces, e.g., sip and puff, haptic, and Braille.

For the 11.4 million visually impaired people in the United States [1], this R&D has done little to remove the main *functional* barrier: the inability to navigate dynamic and complex environments. This inability denies the visually impaired equal access to many private and public buildings, limits their use of public transportation, and makes the visually impaired a group with one of the highest unemployment rates (74%) [1]. Thus, there is a significant need for systems that improve the wayfinding abilities of the visually impaired, especially in unfamiliar environments, where conventional aids, such as white canes and guide dogs, are of limited use. In the remainder of this chapter, the term *unfamiliar* is used only with respect to visually impaired individuals.

2.3.1.1 Assisted Navigation

Over the past three decades, considerable R&D effort has been dedicated to navigation devices for the visually impaired. Benjamin et al. [2] built the C-5 Laser Cane. The cane uses optical triangulation with three laser diodes and three photo-diodes as receivers. Bissit and Heyes [3] developed the Nottingham Obstacle Detector (NOD), a handheld sonar device that gives the user auditory feedback with eight discrete levels. Shoval et al. [4] developed the NavBelt, an obstacle avoidance wearable device equipped with ultrasonic sensors and a wearable computer. The NavBelt produces a 120-degree wide view ahead of the user. The view is translated into stereophonic audio directions. Borenstein and Ulrich [5] built Guide-Cane, a mobile obstacle avoidance device for the visually impaired. Guide-Cane consists of a long handle and a ring of ultrasonic sensors mounted on a steerable two-wheel axle.

More recently, a radio-frequency identification (RFID) navigation system for indoor environments was developed at the Atlanta VA Rehabilitation Research and Development Center [6,7]. In this system, the blind users' canes are equipped with RFID receivers, while RFID transmitters are placed at hallway intersections. As the users pass through transmitters, they hear over their headsets commands like *turn left*, *turn right*, and *go straight*. The Haptica Corporation has developed Guido©, a robotic walking frame for people with impaired vision and reduced mobility (www.haptica.com). Guido© uses the onboard sonars to scan the immediate environment for obstacles and communicates detected obstacles to the user via speech synthesis.

While the existing approaches to assisted navigation have shown promise, they have had limited success for the following reasons. First, many existing systems increase the user's navigation-related physical load, because they require that the user wear additional and, oftentimes substantial, body gear [4], which contributes to physical fatigue. The solutions that attempt to minimize body gear, e.g., the C-5 Laser Cane [2] and the Guide-Cane [5]; require that the user effectively abandon his/her conventional navigation aid, e.g., a white cane or a guide dog, which is not acceptable to many visually impaired individuals. Second, the user's navigation-related cognitive load remains high, because the user makes all final wayfinding decisions. While device-assisted navigation enables visually impaired individuals to avoid immediate obstacles and gives them simple directional hints, it provides little improvement in wayfinding over white canes and guide dogs. Limited communication capabilities also contribute to the high cognitive load. Finally, few assisted navigation technologies are deployed and evaluated in their target environments over extended time periods. This lack of deployment and evaluation makes it difficult for assistive technology (AT) practitioners to compare different solutions and choose the one that best fits the needs of a specific individual.

2.3.1.2 Robotic Guides

The idea of robotic guides is by no means novel. Horswill [8] used the situated activity theory to build Polly, a mobile robot guide for the MIT AI Lab. Polly used lightweight vision routines that depended on textures specific to the lab.

Thrun et al. [9] built MINERVA, an autonomous tour guide robot that was deployed in the National Museum of American History in Washington, D.C. The objective of the MINERVA project was to build a robot capable of educating and entertaining people in public places. MINERVA is based on Markov localization and uses ceiling mosaics as its main environmental cues. Burgard et al. [10] developed RHINO, a close sibling of MINERVA, which was deployed as an interactive tour guide in the Deutsches Museum in Bonn, Germany. The probabilistic techniques for acting under uncertainty that were used in RHINO and MINERVA were later used in Pearl, a robotic guide for the elderly with cognitive and motor disabilities, developed by Montemerlo et al. [11].

Unfortunately, these robotic guides do not address the needs of the visually impaired. The robots depend on the users' ability to maintain visual contact with them, which cannot be assumed for the visually impaired. Polly has very limited interaction capabilities: the only way users can interact with the system is by tapping their feet. To request a museum tour from RHINO [10], the user must identify and press a button of a specific color on the robot's panel. Pearl also assumes that the elderly people interacting with it do not have visual impairments.

The approach on which Polly is based requires that a robot be evolved by its designer to fit its environment not only in terms of software, but also in terms of hardware. This makes it difficult to produce replicable solutions that work out of the box in a variety of environments. Autonomous solutions like RHINO, MINERVA, and Pearl also require substantial investments in customized engineering and training to become and, more importantly, to remain operational. While the software and hardware concerns may be alleviated as more onboard computer power becomes available with time, collisions remain a concern [10].

Mori and Kotani [12] developed HARUNOBU-6, a robotic travel aid to guide the visually impaired on the Yamanashi University campus in Japan. HARUNOBU-6 is a motorized wheelchair equipped with a vision system, sonars, a differential GPS, and a portable GIS. Whereas the wheelchair is superior to the guide dog in its knowledge of the environment, as the experiments run by the HARUNOBU6 research team demonstrate, the wheelchair is inferior to the guide dog in mobility and obstacle avoidance. The major source of problems was vision-based navigation, because the recognition of patterns and landmarks was greatly influenced by the time of day, weather, and season. In addition, the wheelchair is a highly customized piece of equipment, which negatively affects its portability across a broad spectrum of environments.

2.3.1.3 Robot-Assisted Wayfinding

Any R&D endeavor starts with the basic question: is it worthy of time and effort? We believe that with respect to robot-assisted wayfinding for the visually impaired this question can be answered in the affirmative. We offer several reasons to justify our belief. First, robot-assisted wayfinding offers feasible solutions to two hard problems perennial to wearable assisted navigation devices for the visually impaired: hardware miniaturization and a portable power supply. The amount of body gear carried by the user is significantly minimized, because most of it can be mounted on the robot and powered from onboard batteries. Therefore, the navigation-related physical load is reduced. Second, since such key wayfinding capabilities as localization and navigation are delegated to the robotic guide, the user is no longer responsible for making all navigation decisions and, as a consequence, can enjoy a smaller cognitive load. Third, the robot can interact with other people in the environment, e.g., ask them to yield or receive instructions. Fourth, robotic guides can carry useful payloads, e.g., suitcases and grocery bags. Finally, the user can use robotic guides in conjunction with his/her conventional navigation aids, e.g., white canes and guide dogs.

In the remainder of this chapter, we will argue that robot-assisted wayfinding is a viable universal access strategy in structured indoor environments where the visually impaired face wayfinding barriers. We begin, in Section 2.3.2, with an ontology of environments that helps one analyze their suitability for robot-assisted wayfinding. In Section 2.3.3, we describe our robotic guide for the visually impaired. We specify the scope limitations of our project and present the hardware and software solutions implemented in the robotic guide. Section 2.3.4 discusses robot-assisted wayfinding and the instrumentation of the environments. In Section 2.3.5, we describe the pilot experiments conducted with and without visually impaired participants in two structured indoor environments. We analyze our successes and failures and outline several directions for future R&D. Section 2.3.6 contains our conclusions.

2.3.2 An Ontology of Environments

Our ability to operate in a given environment depends on our familiarity with that environment and the environment's complexities [13,14]. When we began our work on a robotic guide, we soon found ourselves at a loss as to what criteria to use in selecting target environments. This lack of an analytical framework caused us to seek an *operational* ontology of environments. After conducting informal interviews with visually impaired individuals on environmental accessibility and analyzing system deployment options available to us at the time, we decided to classify environments in terms of their *familiarity* to the user, their *complexity*, and their *containment.*

In terms of user familiarity, the ontology distinguishes three types of environments: *continuous*, *recurrent*, and *transient*. Continuous environments are environments with which the user is closely familiar, because he/she continuously interacts with them. For example, the office space in the building where the user works is a continuous environment. Recurrent environments are environments with which the user has contact on a recurrent but infrequent basis, e.g., a conference center where the user goes once a year or an airport where the user lands once or twice a year. Recurrent environments may undergo significant changes from visit to visit and the user may forget most of the environment's topology between visits. Transient environments are environments with which the user has had no previous acquaintance, e.g., a supermarket or an airport the user visits for the first time.

Two types of environmental complexities are distinguished in the ontology: *structural* and *agent-based*. Structural complexity refers to the physical layout and organization of a given environment, e.g., the number of halls, offices, and elevators, the number of turns on a route from A to B, and the length of a route from A to B. Agent-based complexity refers to the complexity caused by other agents acting in the environment, and is defined in terms of the number of static and dynamic obstacles, e.g., boxes, pieces of furniture, and closed doors, and the number of people en route.

Our ontology describes environmental complexity in terms of two discrete values: *simple* and *complex*. Hence, in terms of environmental complexity, the ontology distinguishes four types of environments: (1) simple structural, simple agent-based; (2) simple structural, complex agent-based; (3) complex structural, simple agent-based; and (4) complex structural, complex agent-based. It should be noted that, in terms of its agent-based complexity, the same environment can be simple and complex at different times. For example, the agent-based complexity of a supermarket at 6:00 am on Monday is likely to be much less complex than at 11:00 am on Saturday. Similarly, the agent-based complexity of a student center at a university campus changes significantly, depending on whether or not the school is in session.

In terms of containment, the ontology distinguishes two types of environment: *indoor* and *outdoor*. Thus, our ontology distinguishes a total of eight environments: the four above types classified according to environmental complexity, each of which can be either indoor or outdoor.

Given this ontology, we proceed to the next basic question: are all environments suitable for robot-assisted wayfinding? We do not think so. There is little need for such guides in *continuous* environments, i.e., environments with which the user is very familiar. As experience shows [15], conventional navigation aids, such as white canes and guide dogs, are quite adequate in these environments, because either the user or the user's guide dog has an accurate topological map of the environment.

We do not think that robotic guides are suitable for outdoor environments either. The reason is twofold. First, outdoor environments are not currently within reach of robots unless the robots are teleoperated, at least part of the time [16]. To put it differently, the state of the art in outdoor robot navigation technology does not yet allow one to reliably navigate outdoor environments. Second, the expense of deploying and maintaining such systems may be prohibitive not only to individuals, but to many organizations as well. Naturally, as more progress is made in vision-based outdoor navigation, this outlook is likely to change.

We believe that recurrent or transient indoor environments, e.g., supermarkets, airports, and conference centers, are both feasible and socially valid for robot-assisted navigation [17]. Guide dogs, white canes, and other navigation devices are of limited use in such environments, because they cannot help their users localize and find paths to useful destinations. Furthermore, as we argue below, such environments can be instrumented with small sensors that make robot-assisted wayfinding feasible.

2.3.3 RG-I: A Robotic Guide

In May 2003, the Computer Science Assistive Technology Laboratory (CSATL) at the Department of Computer Science of Utah State University (USU) and the USU Center for Persons with Disabilities (CPD) launched a collaborative project whose objective is to build an indoor robotic guide for the visually impaired. We have so far built, deployed and tested one prototype in two indoor environments. Our guide's name is RG-I, where "RG" stands for "robotic guide." We refer to the approach behind RG-I as *non-intrusive instrumentation of environments*. Our basic objective is to alleviate localization and navigation problems of completely autonomous approaches by instrumenting environments with inexpensive and reliable sensors that can be placed in and out of environments without disrupting any indigenous activities [18]. Additional requirements are: (1) that the instrumentation be reasonably fast and require only commercial off-the-shelf (COTS) hardware components; (2) that sensors be inexpensive, reliable, easy to maintain (no external power supply), and provide accurate localization; (3) that all computation be run onboard the robot; and (4) that human–robot interaction be both reliable and intuitive from the perspective of the visually impaired users.

2.3.3.1 Scope Limitations

Several important issues are beyond the scope of our project. First, robotic guides prototyped by RG-I are not meant for individual ownership. Rather, we expect institutions, e.g., airports, supermarkets, conference centers, and hospitals, to operate such guides on their premises in the future. One should think of RG-I as a step toward developing robotic navigational redcap services for the visually impaired in airport-like environments.

Second, it is important to emphasize that robotic wayfinding assistants, prototyped by RG-I, are not intended as re-placements for guide dogs. Rather, these service robots are designed to complement and enhance the macro-navigational performance of guide dogs in the environments that are not familiar to the guide dogs and/or their handlers.

Third, we do not address the issue of navigating large open spaces, e.g., large foyers in hotels. While some references in the localization literature suggest that ultrasonic sensors could be used to address this issue [19], the proposed solutions are sketchy, have been deployed in small, carefully controlled lab environments, and do not yet satisfy the COTS hardware requirement. In addition, the ultrasonic sensors used in these evaluations must have external power sources, which make both maintenance and deployment significantly harder. Thus, we currently assume that all environments in which RG-I operates are structured indoor environments, i.e., have walls, hallways, aisles, rows of chairs, T- and X-intersections, and solid and static objects, e.g., vending machines and water fountains, that the robot's onboard sensors can detect.

2.3.3.2 Hardware

RG-I is built on top of the Pioneer 2DX commercial robotic platform (See Figure 1) from the ActivMedia Corporation. The platform has three wheels, two drive wheels in the front

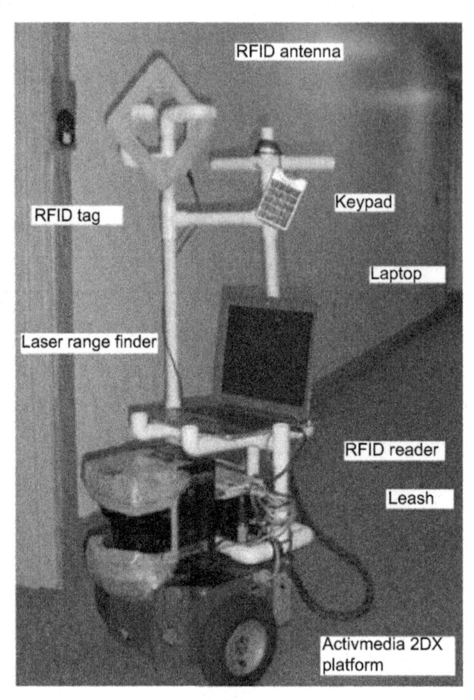

Figure 1
RG-I: A robotic guide.

and a steering wheel in the back, and is equipped with three rechargeable Power Sonic PS-1270 onboard batteries. What turns the platform into a robotic guide is a Wayfinding Toolkit (WT) mounted on top of the platform and powered from the onboard batteries. As shown in Figure 1, the WT resides in a polyvinyl chloride (PVC) pipe structure attached to the top of the platform. The WT includes a Dell Ultralight X300 laptop connected to the platform's microcontroller, a SICK LMS laser range finder from SICK, Inc., and a TI Series 2000 radio-frequency identification (RFID) reader from Texas Instruments, Inc.

The laptop interfaces to the RFID reader through a USB-to-serial cable. The reader is connected to a square 200 mm by 200 mm RFID RI-ANT-GO2E antenna that detects RFID sensors (tags) placed in the environment. Figure 2 shows several TI RFID Slim Disk tags. These are the only types of tags currently used by the system. These tags can be attached to any objects in the environment or worn on clothing. They do not require any external power source or direct line of sight to be detected by the RFID reader. They are activated by the spherical electromagnetic field generated by the RFID antenna with a radius of approximately 1.5 m.

Several research efforts in mobile robotics have also used RFID technology in robot navigation. Kantor and Singh used RFID tags for robot localization and mapping [20].

RFID tag at a door

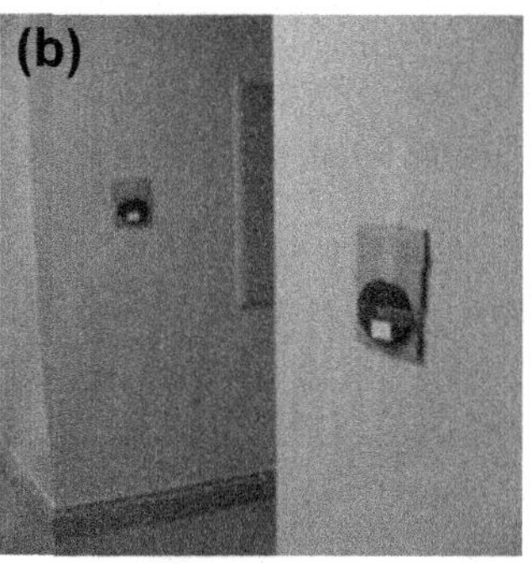

RFID tag at a turn

Figure 2
Deployed RFID tags.

Once the positions of the RFID tags are known, their system uses time-of-arrival information to estimate the distance from detected tags. Tsukiyama [21] developed a navigation system for mobile robots using RFID tags. The system assumes perfect signal reception and measurement. Hähnel et al. [22] developed a robotic mapping and localization system to analyze whether RFID can be used to improve the localization of mobile robots in office environments. They proposed a probabilistic measurement model for RFID readers that accurately localize RFID tags in a simple office environment.

2.3.3.3 Software

RG-I has a modular hybrid architecture that consists of three main components: a path planner, a behavior manager, and a user interface (UI). The UI has two input modes: haptic and speech. The haptic mode uses inputs from a handheld keypad; the speech mode accepts inputs from a wireless wearable microphone coupled to Microsoft's SAPI 5.1 speech recognition engine. The UI's output mode uses non-verbal audio beacons and speech synthesis.

The UI and the planner interact with the behavior manager through a socket communication. The planner provides the robot with path plans from start tags to destination tags on demand. The behavior manager executes the plans and detects plan execution failures.

This architecture is inspired by and partially realizes Kupiers' Spatial Semantic Hierarchy (SSH) [23]. The SSH is a framework for representing spatial knowledge. It divides spatial knowledge of autonomous agents, e.g., humans, animals, and robots, into four levels: the control level, causal level, topological level, and metric level. The control level consists of low-level mobility laws, e.g., trajectory following and aligning with a surface. The causal level represents the world in terms of views and actions. A view is a collection of data items that an agent gathers from its sensors. Actions move agents from view to view.

For example, a robot can go from one end of a hallway (start view) to the other end of the hallway (end view). The topological level represents the world's connectivity, i.e., how different locations are connected. The metric level adds distances between locations.

The path planner realizes the causal and topological levels of the SSH. It contains the declarative knowledge of the environment and uses that knowledge to generate paths from point to point. The behavior manager realizes the control and causal levels of the SSH. Thus, the causal level is distributed between the path planner and the behavior manager.

The control level is implemented with the following low-level behaviors all of which run on the WT laptop: *follow-hallway*, *turn-left*, *turn-right*, *avoid-obstacles*, *go-thru-doorway*, *pass-doorway*, and *make-u-turn*. These behaviors are written in the behavior programming language of the ActivMedia Robotics Interface for Applications (ARIA) system from ActivMedia Robotics, Inc. Further details on how these behaviors are implemented can be found in [24] and [25].

The behavior manager keeps track of the global state of the robot. The global state is shared by all the modules. It holds the latest sensor values, which include the laser range finder readings, the latest detected RFID tag, current velocity, current behavior state, and battery voltage. Other state parameters include: the destination, the command queue, the plan to reach the destination, and internal timers.

2.3.4 Wayfinding

Visually impaired individuals follow RG-I by holding onto a dog leash. The leash is attached to the battery bay handle on the back of the platform. The upper end of the leash is hung on a PVC pole next to the RFID antenna's pole. Figure 3 shows a visually impaired guide dog user following RG-I. RG-I always moves closer to the right wall to follow the flow of traffic in structured indoor environments.

2.3.4.1 Instrumentation of Environments

In instrumenting indoor environments with RFID tags, the following four guidelines are followed:

1. Every tag in the environment is programmed with a unique ID and placed on a non-metallic padding to isolate it from metallic substances in the walls;
2. Every door is designated with a tag;
3. Every object in the environment that can serve as a destination, e.g., a soda machine or a water fountain, is also tagged;
4. Every turn is designated with a tag; the tags are placed about a meter away from each turn.

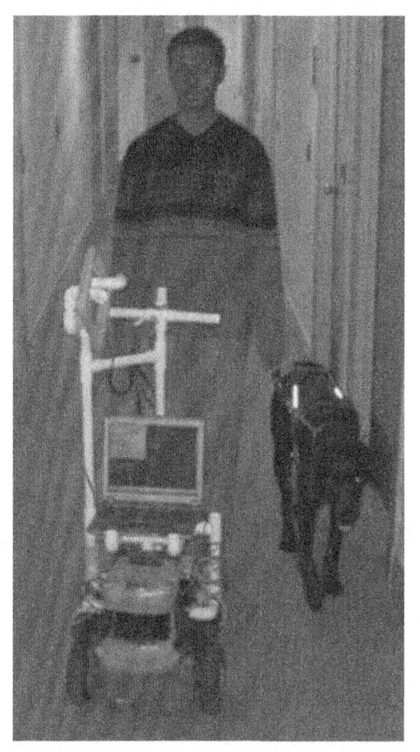

Figure 3
RG-I leading a guide dog user.

After the tags are deployed in the environment, the knowledge base of the environment is manually engineered. The knowledge base represents an aerial view of the environment in which RG-I operates. The knowledge base for an environment consists of a tag connectivity graph, tag to destination mappings, and low-level behavior scripts associated with specific tags. Figure 4 shows a subgraph of the connectivity graph used in RG-I. The path is a *tag-behavior-tag* sequence. In the graph, *f*, *u*, *l* and *r* denote *follow-hallway*, *make-u-turn*, *turn-left*, and *turn-right*, respectively.

The planner uses the standard breadth first search (BFS) to find a path from the start tag to the destination tag. For example, if the start tag is 4 and the destination tag is 17, *(4 l 5 f 6 r 7 f 8 f 15 f 16 r 17)* is a path plan. The plan is executed as follows. The robot detects tag 4, executes a left turn until it detects tag 5, follows the hallway until it detects tag 6, executes a right turn until it detects tag 7, follows the hallway until it detects tag 8, follows the hallway until it detects tag 15, follows the hallway until it detects tag 16, and executes a right turn until it detects tag 17. Given the tag connectivity graph, there are only two ways the robot can localize the user in the environment: (1) the user is approaching X, where X is the location tagged by the next tag on the robot's current path;

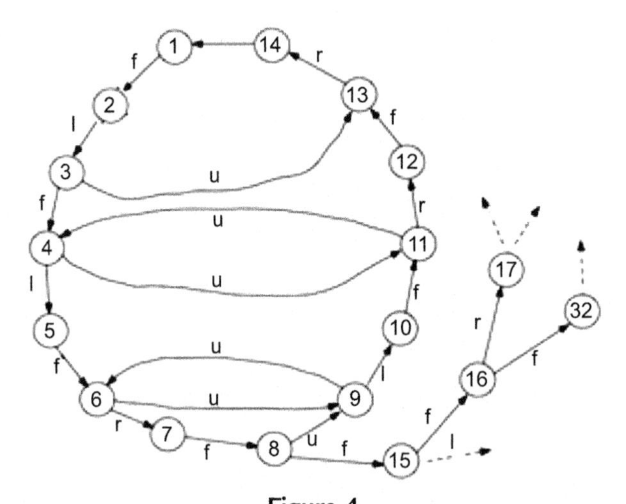

Figure 4
A subgraph of a connectivity graph used in RG-I.

and (2) the user is at X, where X is the location tagged by the tag that is currently visible by the robot's antenna.

As another example, consider Figure 5. The figure shows a map of the USU CS Department with a route that the robot is to follow. Figure 6 shows a path returned by the planner. Figure 7 shows how this path is projected on the map of the environment as it is followed by the robot. Figure 7 also shows how the robot switches from one navigational behavior to another as its RFID antenna detects the tags on the path.

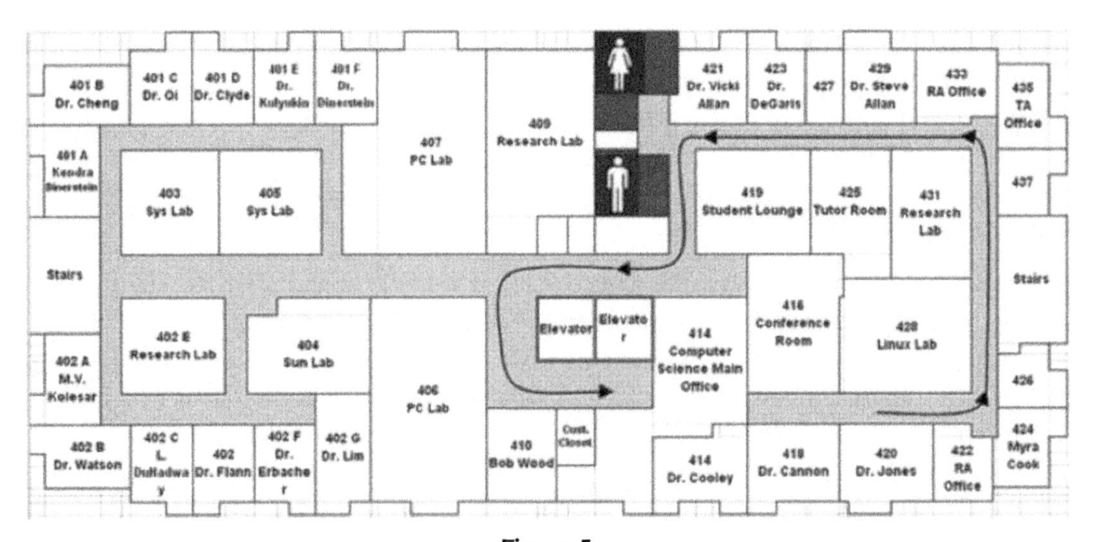

Figure 5
USU CS Department.

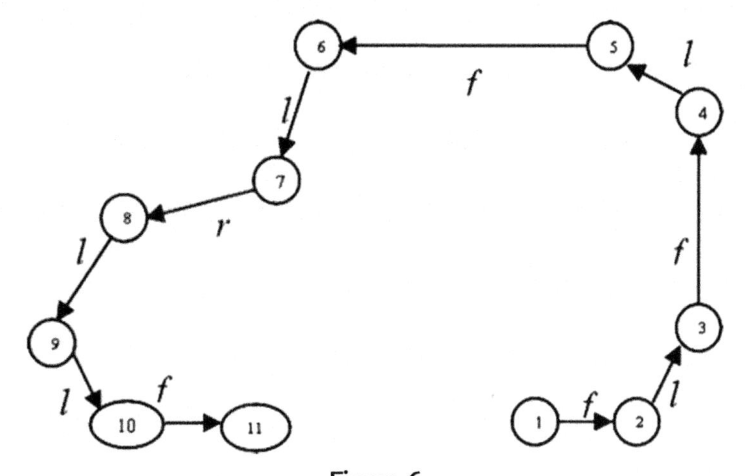

Figure 6
A path of RFID tags and behaviors.

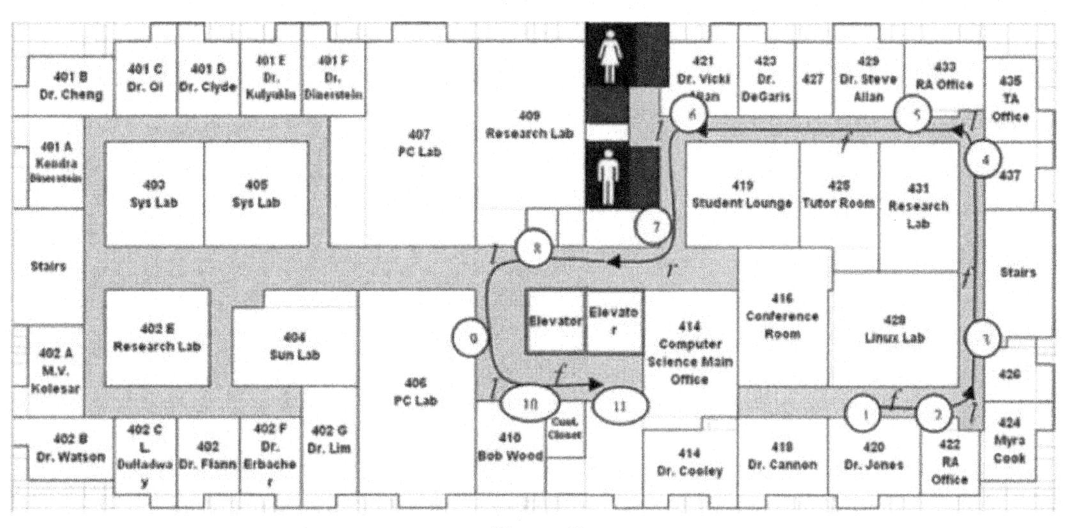

Figure 7
Run-time behavior switches.

2.3.4.2 Obstacle Avoidance

Obstacle avoidance is critical to robots navigating dynamic, structurally complex environments. Over the past two decades, several obstacle avoidance techniques have been developed and tried on mobile robots in a variety of environments. The most prominent of those techniques are the potential fields approach (PF) [26], the dynamic window approach (μdWA) [10], the vector field histogram (VFH) [27], and the curvature velocity method (CVM) [28].

While navigation in RG-I utilizes PF techniques, the overall approach to navigation implemented in RG-I differs from the above approaches in three respects. First, our approach does not focus on *interpreting* obstacles detected by sensors and generating motion commands as a result of that interpretation. Instead, we focus on *empty spaces* that define the navigational landmarks of many indoor environments: hallways, turns, X- and T-intersections, etc. The robot itself cannot interpret these landmarks but can navigate them by following paths induced by empty spaces. Second, navigation in RG-I is not egocentric: it is distributed between the robot and the environment insomuch as the environment is instrumented with sensors that assist the robot in its navigation. Third, navigation in RG-I is *orientation-free*, i.e., the robot's sensor suite does not include any orientation sensor, such as a digital compass or an inertia cube; nor does the robot infer its orientation from external signals through triangulation or trilateration.

The robot's PF is a 10×30 egocentric grid. Each cell in the grid is 200×200 mm. The grid covers an area of 12 square meters (2 m in front and 3 m on each side). The grid is updated continuously with each scan of the laser range finder. A 180° laser scan is taken in front of the robot. The scan consists of a total of 90 laser range finder readings, taken at every 2°. A laser scan is taken every 50 ms, which is the length of time for an average action cycle in the ARIA task manager.

The exact navigation algorithm for determining the direction of travel executed by RG-I is as follows:

1. Do the front laser scan.
2. Classify each cell in the grid as free, occupied, or unknown and assign directions and magnitudes to the vectors in occupied cells.
3. Determine the maximum empty space.
4. Assign directions to the vectors in free cells.

The robot's desired direction of travel is always in the middle of the *maximum empty space*, a sector of empty space in front of the robot. Further details on the local navigation algorithms used in RG-I can be found in [29] and [25].

2.3.4.3 Dealing with Losses

We distinguish two types of losses: *recoverable* and *irrecoverable*. A recoverable loss occurs when the robot veers from a given path but reaches the destination nonetheless. In graph-theoretic terms, a veering event means that the original path is replaced with a different path. An irrecoverable loss occurs when the robot fails to reach the destination regardless of how much time the robot is given.

As shown in Figure 8, there are two situations in which RG-I gets lost: (1) failure to determine the correct direction of travel and (2) RFID malfunction. The first situation

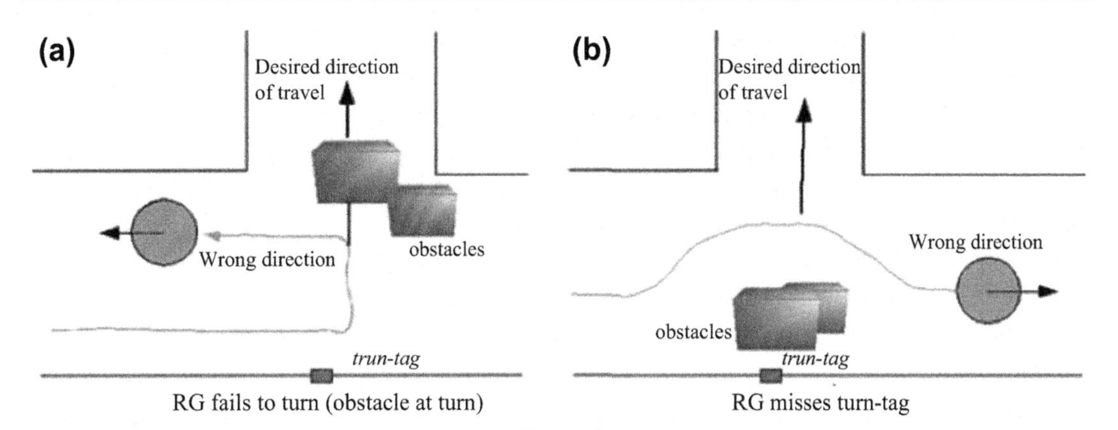

Figure 8
Two situations leading to a loss. (a) RG fails to turn (obstacle at turn) and (b) RG misses turn-tag.

occurs when the robot, due to its current orientation, finds the maximum empty space that causes it to veer from the correct path. In Figure 8(a), RG-I detects the turn tag and, as prescribed by the plan, first executes the left turn behavior and then moves in the desired direction. However, since the hallway is blocked, RG-I veers away from the correct path. In Figure 8(b), the turn tag is blocked by obstacles, which triggers the obstacle avoidance behavior. While avoiding the obstacle, the robot fails to detect the turn tag, because the tag falls outside the range of the RFID antenna, and the robot does not make the left turn. The second situation that causes a loss, the RFID reader's malfunction, arises when the reader misreads the tag's ID or fails to activate the tag due to some interference in the environment. In our target environments, the second situation was rare.

A loss occurs when too many invalid tags are detected. In general, RG-I always reaches *B* from *A* if the following assumptions are true: (1) the robot's batteries have sufficient power (above 8 V); (2) there is an actual path from *A* to *B* in the current state of the world; and (3) the *critical* tags on the path from *A* to *B* are not incapacitated. By critical tags we mean the start tag, the destination tag, and the turn tags. If either the second or third assumptions do not hold and the robot is lost, the loss is irrecoverable. To be more exact, the robot will keep trying to recover from the loss until its power drops down to 8 V, which will cause a complete shutdown.

2.3.5 Pilot Experiments

Our pilot experiments focused on robot-assisted navigation and human-guide interaction. The robot-assisted navigation experiments evaluated the ability of visually impaired individuals to use RG-I to navigate unfamiliar environments as well as the ability of the robot to navigate on its own. The human–robot interaction experiments investigated how visually impaired individuals can best interact with robotic guides.

2.3.5.1 Robot-Assisted Navigation

We deployed our system for a total of approximately 70 h in two indoor environments: the Assistive Technology Laboratory (ATL) of the USU Center for Persons with Disabilities and the USU CS Department. The ATL occupies part of a floor in a building on the USU North Campus. The area occupied by the ATL is approximately 4270 square meters and contains six laboratories, two bathrooms, two staircases, and an elevator. The CS Department occupies an entire floor in a multi-floor building. The floor's area is 6590 square meters. The floor contains 23 offices, seven laboratories, a conference room, a student lounge, a tutor room, two elevators, several bathrooms, and two staircases.

Forty RFID tags were deployed at the ATL and 100 tags were deployed at the CS Department. It took one person 20 min to deploy the tags and about 10 min to remove them at the ATL. The same measurements at the CS Department were 30 and 20 min, respectively. The tags, which were placed on small pieces of cardboard to insulate them from the walls, were attached to the walls with regular masking tape. The creation of the connectivity graphs took 1 h at the ATL and about two and a half hours at the CS Department. One member of our research team first walked around the areas with a laptop and recorded tag-destination associations and then associated behavior scripts with tags.

RG-I was first repeatedly tested in the ATL, the smaller of the two environments, and then deployed for pilot experiments at the USU CS Department. We ran three sets of navigation experiments. The first and third sets did not involve visually impaired participants. The second set did. In the first set of experiments, we had RG-I navigate three types of hallways in the CS Department: narrow (1.5 m), medium (2.0 m) and wide (4.0 m), and we evaluated its navigation in terms of two variables: path deviations and abrupt speed changes. We also observed how well the robot's RFID reader detected the tags.

To estimate path deviations, in each experiment we first computed the ideal distance that the robot has to maintain from the right wall in a certain width type of hallway (narrow, medium, and wide). As shown in Figure 12, the ideal distance was computed by running the robot in a hallway of the type being tested with all doors closed and no obstacles en route. RFID tags were placed along the right wall of every route every 2 m to help with interpolation and graphing. During each run, the distance read by the laser range finder between the robot and the right wall was recorded every 50 ms. The ideal distance was computed as the average of the distances taken during the run. Once the ideal distances were known, we ran the robot three times in each type of hallway. The hallways in which the robot ran were different from the hallways in which the ideal distances were computed. Obstacles, e.g., humans walking by and open doors, were allowed during the test runs. The average of all the readings for each set of three runs, gives the average distance the robot maintains from the right wall in a particular type of hallway.

Figures 9–11 give the distance graphs of the three runs in each hallway type. The vertical bars in each graph represent the robot's width. As can be seen from Figure 9, there is almost no deviation from the ideal distance in narrow hallways; nor is there any oscillation. Figures 10 and 11 show some insignificant deviations from the ideal distance. The deviations were caused by people walking by and by open doors. However, there is no oscillation, i.e., sharp movements in different directions. In both environments, we observed several tag detection failures, particularly near or on metallic door frames. However, after we insulated the tags with thicker pieces of cardboard, the tag detection failures stopped.

Figures 13–15 give the velocity graphs for each hallway type (x-axis is time in seconds, y-axis is velocity in mm/sec). The graphs show that the narrow hallways cause short abrupt changes in velocity. In narrow hallways even a slight disorientation, e.g., 3°, in the robot causes changes in velocity because less empty space is detected in the grid. In medium and wide hallways, the velocity is generally smooth. However, several speed changes occur when the robot passes or navigates through doorways or avoids obstacles.

The mean and standard deviation numbers for the hallway experiments were as follows: in wide hallways, $\mu = 708.94$, $\sigma = 133.32$; in medium hallways, $\mu = 689.19$, $\sigma = 142.32$; in narrow hallways, $\mu = 670.43$, $\sigma = 166.31$. It should be noted that the means are influenced by the fact that the robot always started at 0 velocity. Thus, since the mean, as a statistical measure, is influenced by outliers, these means may be slightly skewed.

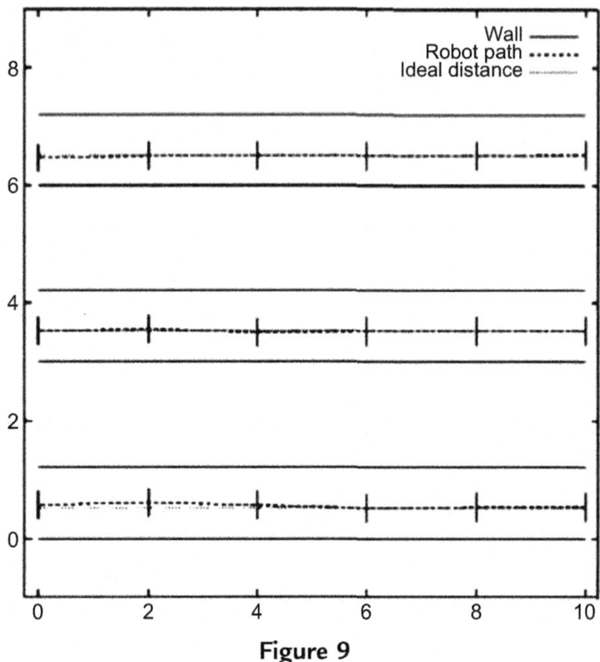

Figure 9
Path deviations in narrow hallways.

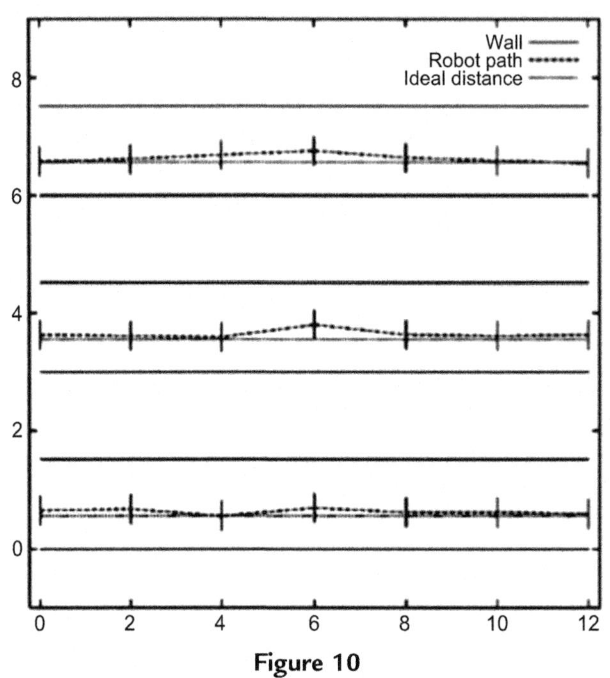

Figure 10
Path deviations in medium hallways.

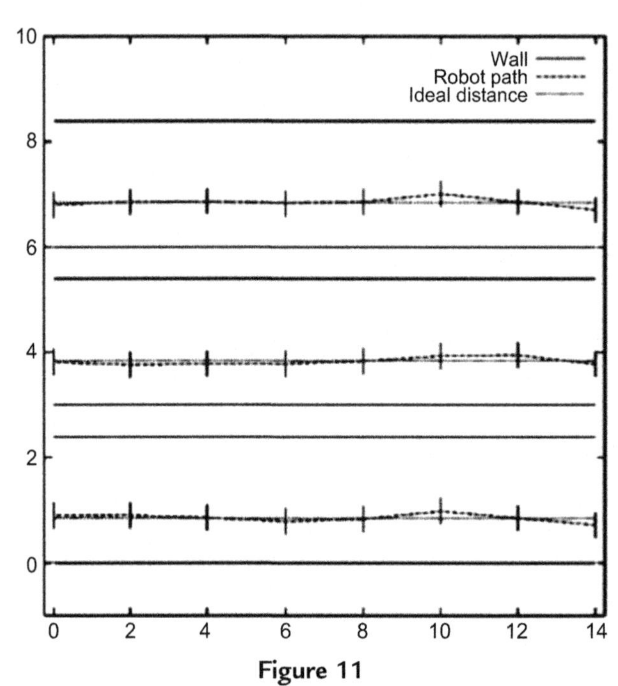

Figure 11
Path deviations in wide hallways.

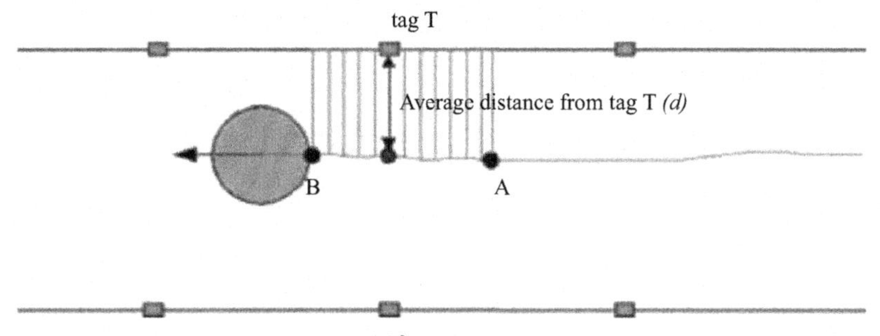

Figure 12
Computing an ideal distance in hallways.

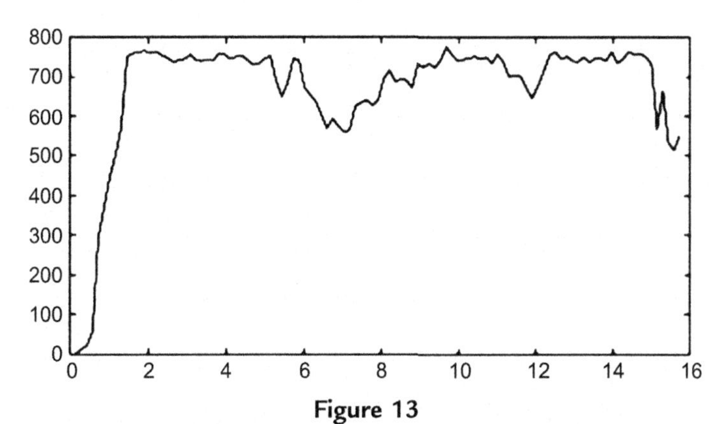

Figure 13
Velocity changes in narrow hallways.

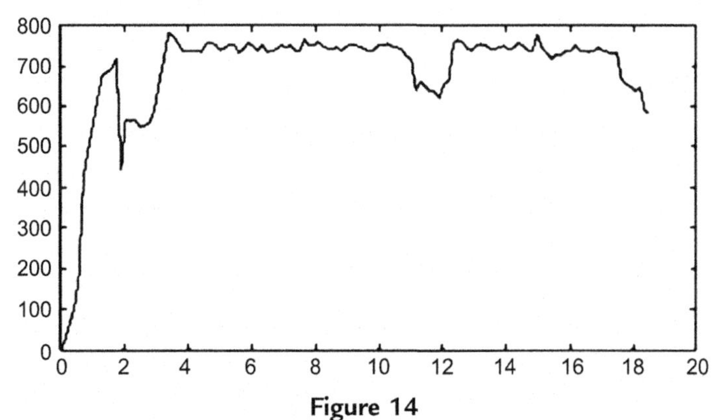

Figure 14
Velocity changes in medium hallways.

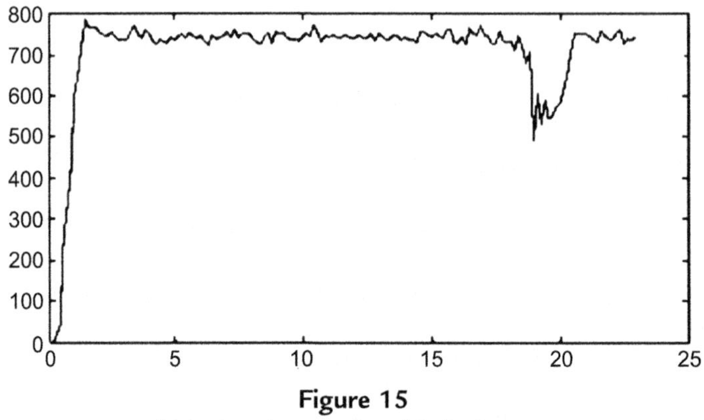

Figure 15
Velocity changes in wide hallways.

The second set of pilot experiments involved five visually impaired participants, one participant at a time, over a period of two months. Three participants were completely blind and two participants could perceive only light. The participants had no speech impediments, hearing problems, or cognitive disabilities. Two participants were dog users and the other three used white canes. The participants were asked to use RG-I to navigate to three distinct locations (an office, a lounge, and a bathroom) at the USU CS Department. All participants were new to the environment and had to navigate approximately 40 m to get to all destinations. Thus, in the experiments with visually impaired participants, the robot navigated approximately 200 m. All participants had to use a wireless wearable microphone to interact with the robot: at the beginning of a run, each participant would speak the destination he or she wanted to reach. All participants reached their destinations. In their exit interviews, all participants said they liked the fact that they did not have to give up their white canes and/or guide dogs to use RG-I. Most complaints were about the human–robot interaction aspects of the system. For example, all of them had problems with the speech recognition system and often had to repeat destinations several times before the robot understood them [29,30].

Another problem with speech recognition occurs when the person guided by RG-I stops and engages in conversation with someone. Since speech recognition runs continuously, some phrases said by the person during a conversation may be erroneously recognized as route directives, thereby causing the robot to start moving. For example, once RG-I erroneously recognized a directive and started pulling its user away from his interlocutor until the user's stop command pacified it. In another situation, RG-I managed to run a few meters away from its user, because the user hung the leash on the PVC pole when he stopped to talk to a friend in a hallway. Thus, after saying "Stop," the user had to grope his way along a wall to the robot that was standing a few meters away.

In the third navigation experiment, RG-I was made to patrol the entire area of the USU CS Department on its own. This experiment focused on recoverable and irrecoverable losses on two types of routes: (1) simple structural, simple agent-based and (2) complex structural, simple agent-based. The first type was operationally defined as routes having 0, 1, or 2 turns, of less than 40 m in length with no or few people or obstacles. The second type was operationalized as routes with more than 2 turns, of more than 40 m in length, with no or few people or obstacles. A total of 18 routes (9 routes of the first type and 9 routes of the second type) were set up. The robot continuously navigated these routes until its battery voltage reached 8 V. The robot navigated a total of 6 km for 4 h and had five recoverable and zero irrecoverable losses. All losses occurred on complex structural and simple agent-based routes. During several separate trial runs with visually impaired participants on different types of routes, the robotic guide suffered several recoverable losses. While we did not collect statistics on the participants' reactions to recoverable losses, we observed that the participants did not complain. Two participants said that they would not have known that the robot had to take a different route if the robot had not announced it to them.

2.3.5.2 Human–Robot Interaction

After we tested speech-based interaction in navigation experiments and received negative feedback from the participants, we decided to evaluate the feasibility of using speech more systematically. Therefore, our first human-guide experiment tested the feasibility of using speech as a means of input for humans to communicate with the robot. Each participant was asked to speak approximately sixty phrases while wearing a headset that consisted of a microphone and one headphone. The phrase list was a list of standard phrases that a person might say to a robotic guide in an unfamiliar environment, e.g., "go to the bathroom," "where am I?" etc. Each phrase was encoded as a context-free command and control grammar rule in SAPI's XML-based grammar formalism. Each participant was positioned in front of a computer running SAPIThe test program was written to use SAPI's text-to-speech engine to read the phrases to the participant one by one, wait for the participant to repeat a phrase, and record a recognition result (speech recognized vs. speech not recognized) in a database.

The speech feasibility experiment was repeated in two environments: noise-free and noisy. The noise-free environment did not have any ambient sounds other than the usual sounds of a typical office. To simulate a noisy environment, a long audio file of a busy bus station was played on another computer in the office very close to where each participant was sitting. All five participants were native English speakers and did not train SAPI's speech recognition engine on sample texts.

We found that the average percentage of phrases recognized by the system in the noise-free environment was 38%, while the average percentage of recognized phrases in the noisy environment was 40.2%. Although the level of ambient noise in the environment did not seem to affect the system's speech recognition, in both environments fewer than 50% of phrases were correctly recognized. Even worse, some non-phrases were incorrectly recognized as phrases. For example, when one participant made two throat clearing sounds, the system recognized the sound sequence as the phrase "men's room."

The statistics were far better for the participants understanding phrases spoken by the computer. The average percentage of speech understood in the noise-free environment was 83.3%, while the average percentage of phrases understood in the noisy environment was 93.5%. Clearly, in the second trial (the noisy environment), the participants were more used to SAPI's speech recognition and synthesis patterns. These results suggest that speech appears to be a better output medium than input [30].

Audio perception experiments were conducted with all five participants to test whether they preferred speech to audio icons, e.g., a sound of water bubbles, to signify different objects and events in the environment and how well participants remembered their audio icon selections. A simple GUI-based tool was built that allows visually impaired users to create their own audio associations. The tool was used to associate events and objects, e.g., water cooler to the right, approaching left turn, etc., with three audio messages: one speech message and two audio icons. A small number of objects were chosen to eliminate steep learning curves. All in all, there were seven different objects, e.g., elevator, vending machine, bathroom, office, water cooler, left turn, and right turn.

Each object was associated with two different events: *at* and *approaching*. For example, one can be at the elevator or approaching the elevator. The audio icons available for each event were played to each participant at selection time. The following statistics were gathered:

1. Percentage of accurately recognized icons;
2. Percentage of objects/events associated with speech;
3. Percentage of objects/events associated with audio icons;
4. Percentage of objects/events associated with both.

The averages for these experiments were:

1. Percentage of accurately recognized icons—93.3%;
2. Percentage of objects/events associated with speech—55.8%;
3. Percentage of objects/events associated with icons—32.6%;
4. Percentage of objects/events associated with both—11.4%.

The analysis of the audio perception experiments showed that two participants were choosing audio preferences essentially at random, while the other three tended to follow a pattern: they chose speech messages for *at* events and audio icons for *approaching* events or vice versa. The experiments also showed that the participants tended to go either with speech or with audio icons, but rarely with both. The experiments did not give a clear answer as to whether visually impaired individuals prefer to be notified of objects/events via speech or audio icons. It is important to keep in mind, however, that our objective was to collect preliminary descriptive statistics on the perception of audio cues in robot-assisted navigation. No attempt was made to make statistically significant inferences. Further work is needed on a larger and more representative sample to answer this question on a statistically significant level.

2.3.6 Conclusions

From our experiences with RG-I, we can make the following preliminary observations.

First, orientation-free RFID-based navigation guarantees reach ability at the expense of optimality [31]. If the path to the destination is not blocked and all critical tags are in place, the robot reaches the destination. The obvious tradeoff is the optimality of the path, because the actual path taken by the robot may be suboptimal in terms of time and distance due to a recoverable loss.

Second, the instrumentation of the environments with RFID sensors is reasonably fast and requires only commercial off-the-shelf (COTS) hardware and software components. RFID tags are inexpensive (15 USD a piece), reliable, and easy to maintain, because they do not require external power supplies. RFID tag reading failures are rare, and can be recovered from as long as a large number of tags are placed in the environment. The placement of RFID tags in the environment does not seem to disrupt any indigenous activities. People who work in the environment do not seem to mind the tags due to their small size.

Third, an alternative technique for recognizing such landmarks as left turns, right turns, and T- and X-intersections will make the navigation behavior more robust even when critical tags are not detected. We are investigating several landmark recognition techniques that work on laser range finder signatures. Another problem that we plan to address in the future is the detection of irrecoverable losses. As of now, RG-I cannot detect when the loss is irrecoverable.

Fourth, the robot is able to maintain a moderate walking speed during most of the route, except at turns and during obstacle avoidance. The robot's motion is relatively smooth, without sideways jerks or abrupt speed changes. However, obstacles that block a critical tag may cause the robot to miss the tag due to obstacle avoidance and fail to trigger an appropriate behavior. In addition, at intersections, RG-I can select a wrong open space due to obstacles blocking the correct path. Adding other landmark recognition techniques is likely to improve the robot's navigation at intersections.

Fifth, at this point in time, speech, when recognized by Mircrosoft's SAPI 5.1 with no user training, does not appear to be a viable input mode. As we indicated elsewhere [30,32], it is unlikely that speech recognition problems can be solved on the software level until there is a substantial improvement in the state-of-the-art speech recognition software. Our pilot experiments suggest that speech appears to be a viable output mode. We believe that for the near future wearable hardware solutions may offer reliable input modes. We are currently exploring human–robot interaction through a wearable keypad. The obvious advantage is that keypad-based interaction eliminates the input ambiguity problems of speech recognition. Additional experiments with human participants are needed to determine the feasibility of various wearable hardware devices for human–robot interaction.

The previous conclusion is not to be construed as an argument that speech-based HRI is not an important venue of research. It is. However, it is important not to confuse interaction itself with a specific mode of interaction. Speech is just one mode of interaction. Typed text, eye gaze, sipping and puffing, gesture and touch are also valid interaction modes. As assistive technology researchers, we are interested, first and foremost, in effective and safe communication between a disabled person and an assistive device. Consequently, to the extent that speech introduces ambiguity, it may not be appropriate as an interaction mode in some assistive robots. Finally, the SSH construction is done manually. The obvious question is can it be completely or partially automated? We are currently investigating a tool that would allow one to generate a tag connectivity graph and associate tags with behavior scripts through drag and drop GUIs. Another possibility that we are contemplating is equipping RG-I with an orientation sensor and manually driving the robot with a joystick on the previously chosen routes in an environment with deployed RFID tags. As the robot is driven through the environment, it senses the RFID tags and turns and associates the detected tags with behavior scripts. In effect, the SSH is first constructed as the robot is driven through the environment and is subsequently edited by a knowledge engineer.

Acknowledgments

The first author would like to acknowledge that this research has been supported, in part, through the NSF CAREER Grant (IIS-0346880), two Community University Research Initiative (CURI) grants from the State of Utah, and a New Faculty Research grant from Utah State University. The authors would like to thank the reviewers of this chapter for their helpful and instructive comments.

References

[1] M. LaPlante, D. Carlson, Disability in the United States: Prevalence and Causes, National Institute of Disability and Rehabilitation Research, U.S. Department of Education, Washington, DC, 2000.

[2] J.M. Benjamin, N.A. Ali, A.F. Schepis, A laser cane for the blind, in: Proceedings of San Diego Medical Symposium, 1973.

[3] D. Bissit, A. Heyes, An application of biofeedback in the rehabilitation of the blind, Appl. Ergonomics 11 (1) (1980) 31–33.
[4] S. Shoval, J. Borenstein, Y. Koren, In mobile robot obstacle avoidance in a computerized travel for the blind, in: Proceedings of the IEEE International Conference on Robotics and Automation, San Diego, CA, 1994.
[5] J. Borenstein, I. Ulrich, The guidecane—a computer, in: Proceedings of the IEEE International Conference on Robotics and Automation, San Diego, CA, 1994.
[6] D. Ross, Implementing assistive technology on wearable computers, IEEE Intell. Syst. 16 (3) (2001) 47–53.
[7] D. Ross, B.B. Blasch, Development of a wearable computer orientation system, IEEE Pers. Ubiquitous Comput. 6 (2002) 49–63.
[8] I. Horswill, Polly:A vision-based artificial agent, in: Proceedings of the Conference of the American Association for Artificial Intelligence (AAAI-1993), Washington, DC, 1993.
[9] S. Thrun, M. Bennewitz, W. Burgard, A.B. Cremers, F. Dellaert, D. Fox, et al., Minerva: a second generation mobile tour-guide robot, in: Proceedings of the IEEE International Conference on Robotics and Automation (ICRA-1999), Antwerp, Belgium, 1999.
[10] W. Burgard, A. Cremers, D. Fox, D. Hähnel, G. Lakemeyer, D. Schulz, W. Steiner, S. Thrun, Experiences with an interactive museum tour-guide robot, Artif. Intell. 114 (1999) 3–55.
[11] M. Montemerlo, J. Pineau, N. Roy, S. Thrun, V. Verma, Experiences with a mobile robotic guide for the elderly, in: Proceedings of the Conference of the American Association for Articial Intelligence (AAAI-2002), Edmonton, AB, 2002.
[12] H. Mori, S. Kotani, Robotic travel aid for the blind: HARUNOBU-6, in: Second European Conference on Disability, Virtual Reality, and Assistive Technology, Sövde, Sweden, 1998.
[13] P. Agre, The Dynamic Structure of Everyday Life (Ph.D. Thesis), MIT Artificial Intelligence Laboratory, 1988.
[14] N. Tinbergen, Animal in its World: Laboratory Experiments and General Papers, Harvard University Press, 1976.
[15] C. Pfaffenberger, J.P. Scott, J. Fuller, B.E. Ginsburg, S.W. Bielfelt, Guide Dogs for the Blind: Their Selection, Development, and Training, Elsevier Scientific Publishing, Amsterdam, Holland, 1976.
[16] T. Fong, C. Thorpe, Vehicle teleoperation interfaces, Auton. Robot. 11 (2) (2001) 9–18.
[17] V. Kulyukin, C. Gharpure, J. Nicholson, RoboCart: toward robot-assisted navigation of grocery stores by the visually impaired, in: Proceedings of the IEEE/RSJ International Conference on Intelligent Systems and Robots (IROS-2005), Edmonton, Alberta, Canada, 2005.
[18] V. Kulyukin, M. Blair, Distributed tracking and guidance in indoor environments, in: Proceedings of the Rehabilitation Engineering and Assistive Technology Society of North America (RESNA-2003), Atlanta, GA, avail. on CD-ROM, 2003.
[19] M. Addlesee, R. Curwen, S. Hodges, J. Newman, P. Steggles, A. Ward, Implementing a sentient computing system, IEEE Comput. 34 (8) (August 2001) 2–8.
[20] G. Kantor, S. Singh, Preliminary results in range-only localization and mapping, in: Proceedings of the IEEE Conference on Robotics and Automation (ICRA-2002), Washington, DC, 2002.
[21] T. Tsukiyama, Navigation system for the mobile robots using RFID tags, in: Proceedings of the IEEE Conference on Advanced Robotics, Coimbra, Portugal, 2003.
[22] D. Hähnel, W. Burgard, D. Fox, K. Fishkin, M. Philipose, Mapping and Localization with RFID Technology, Intel Research Institute, Seattle, WA, 2003.
[23] B. Kupiers, The spatial semantic hierarchy, Artif. Intell. 119 (2000) 191–233.
[24] V. Kulyukin, P. Sute, C. Gharpure, S. Pavithran, Perception of audio cues in robot-assisted navigation, in: Proceedings of the Rehabilitation Engineering and Assistive Technology Society of North America (RESNA-2004), Orlando, FL, avail. on CDROM, 2004.
[25] C. Gharpure, Orientation Free RFID-Based Navigation in a Robotic Guide for the Visually Impaired (Masters Thesis). Department of Computer Science, Utah State University, 2004.
[26] O. Khatib, Real-time obstacle avoidance for manipulators and mobile robots, in: Proceedings of the IEEE International Conference on Robotics and Automation (ICRA-1985), St. Louis, MI, 1985.

[27] J. Borenstein, Y. Koren, Real-time obstacle avoidance for fast mobile robots, IEEE T. Syst. Man Cy. 19 (1989) 1179–1187.
[28] R. Simmons, The curvature-velocity method for local obstacle avoidance, in: Proceedings of the IEEE International Conference on Robotics and Automation (ICRA-1996), Minneapolis, MN, 1996.
[29] V. Kulyukin, C. Gharpure, N. De Graw, Human-robot interaction in a robotic guide for the visually impaired, in: Proceedings of the AAAI Spring Symposium on Interaction between Humans and Autonomous Systems over Extended Operation, Stanford, CA, 2004.
[30] P. Sute, Perception of Audio Cues in Robot-assisted Navigation for the Visually Impaired, Masters Report, Department of Computer Science, Utah State University, 2004.
[31] V. Kulyukin, C. Gharpure, J. Nicholson, S. Pavithran, RFID in robot-assisted indoor navigation for the visually impaired, in: Proceedings of the IEEE/RSJ International Conference on Intelligent Systems and Robots (IROS-2004), Sendai, Japan, 2004.
[32] V. Kulyukin, Human-robot interaction through gesture-free spoken Dialogue, Auton. Robot. 16 (3) (2004) 239–257.

CHAPTER 2.4

Design and Implementation of a Service Robot for Elders

Tao Mei[1], **Minzhou Luo**[1], **Xiaodong Ye**[1], **Jun Cheng**[2], **Lan Wang**[2], **Bin Kong**[3], **Rujin Wang**[3]

[1]Institutes of Advanced Manufacturing Technology, Hefei Institutes of Physical Science, CAS Changzhou, China; [2]Shenzhen Institutes of Advanced Technology, Shenzhen, China; [3]Institute of Intelligent Machines, Hefei Institutes of Physical Science, CAS Hefei, China

Chapter Outline

A prototype of a service robot for elders is described in this chapter. The robot has the functions of autonomous mobility in the home environment, manipulation for small commodities, speech recognition and synthesis in specific backgrounds, and movement detection for elders' safety and basic nursing ability. There are three ways for the elders to interact with the robot: touching and watching the LCD screen, speech conversation with the robot, and motion detection. The experiments in daily care, emotional company, and identifying the falling alarm scenarios demonstrate that robots are useful for elders living alone.

Household Service Robotics. http://dx.doi.org/10.1016/B978-0-12-800881-2.00005-0

2.4.1 Introduction

Service robots have been regarded as final solutions to help elders at home or at rest homes since the lack of manpower is becoming a worldwide problem [1,2]. However, many technical and economic challenges exist in using service robots for elders. Human–robot interaction is one of the most important challenges since the service robots work in unconstructed environments. Although the keyboard, mouse, joystick, and touch screen could be utilized to control the robot, it is not an easy task for nonprofessional users, especially for elders, to program the robots to meet users various requests and needs. Therefore, natural behavior understanding is required for these service robots. Furthermore, the elders need to interact psychologically with the outside in a natural language, posture, emotion, and so forth.

To overcome the key technical problems in service robots for elders, a joint research by the authors was undertaken. A prototype of a service robot for elders has been developed under the support of the Chinese Academy of Sciences.

2.4.2 Robot System

2.4.2.1 System Function

Multimode input technology is used to integrate senses of hearing and vision, location, speed, and so forth. The utilization of multimode input technology makes the robots provide better humanity service and increases the enjoyment in the robot operating processes. Five main functions, such as autonomous mobility in a home environment, manipulation for small commodities, speech recognition and synthesis in specific backgrounds, movement detection for elders' safety, and basic nursing ability, are designed for the service robot for elders. Navigation, localization, obstacle detection, and ground movement abilities are required for autonomous mobility in home environments, while the ability of self-adaption to versatile shapes is required for gripping small commodities. Simple conversation ability between the user and the robot is provided by speech recognition and synthesis in specific backgrounds. Movement detection ability is necessary to monitor the elder's movement states and to prevent or provide an alarm to prevent falling over oneself. Basic nursing ability includes reminding the elder to take pills, helping the elder to arrange meal schedules, and accompanying the elder to do sitting-up exercises and games.

2.4.2.2 System Architecture

To implement the five functions, the robot system is divided into seven subsystems including mechanical system, control system, perception system, speech system, multichannel interface system, decision-making system, and motion detection system. The system architecture of the service robot is shown in Figure 1.

There are five computers in the robot. The PC0 computer is the central controller of the robot and controls the other four computers, the haptic interfaces, and the display. Most tasks are distributed into the other four subcomputers based on manually determining the calculation load. A PC1 computer is used for image acquisition, data processing, and stereovision perception. A PC2 computer performs three tasks, including data acquisition and processing for nonvision sensors, environment modeling, and decision making. A PC3 computer is a motion controller for the 24-DOF robot. A PC4 computer is used for speech recognition and motion detection.

The perception system has two charge-coupled device (CCD) cameras, a scanning laser range finder, a magnetic orientation sensor, and four rotation angle sensors in the driving motors of the robot base. Two cameras are the robot's eyes to provide stereovision. The laser scanner is fixed on the front part of the robot base. The scanning range is 20–5600 mm with a 240° scanning angle. The magnetic orientation sensor is assembled over the head to avoid magnetic disturbance from the body. The wall and obstacles in the moving direction of the robot can be detected and located by matching the scanning map and the environment model of the room.

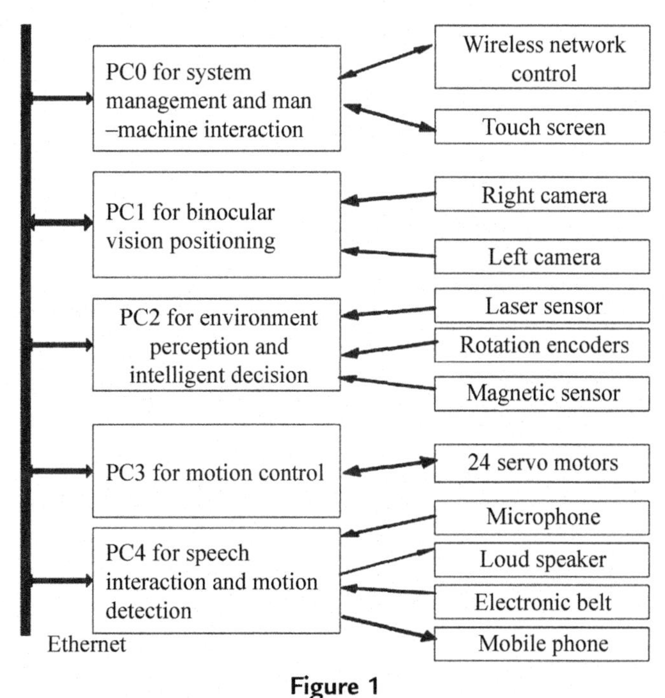

Figure 1
System architecture of the robot.

A new data fusion method and intelligent decision system based on a support vector machine are used to process the multichannel information such as the robot's motivation, body control, and stereovision information. The decision-making system includes a task plan module for subtask decomposition and distribution based on knowledge, path plan module for obstacle avoidance by generating checkpoint sequences, voice interaction module for dialogue feedback based on knowledge background, and an entertainment module for multimedia game and video/audio player.

The speech system, multichannel interface system, and motion detection system are further relevant to human–robot interaction described in the next section.

2.4.2.3 Hardware Structure

The robot has 24 degree of freedom (24-DOF) including 2-DOF for the head, 6 × 2 DOF for the two arms, 3 × 2 DOF for the two hands, and 4-DOF for the base. The mechanism and shape of the robot are shown in Figure 2. The robot's shape was designed by a group of undergraduates who were the winners of a design competition for robot appearance organized by the Institute of Advanced Manufacturing Technology from the industry design and art colleges. Appearance is an important part for the service robots since the elders will face and interact with the machine 24 h per day and also day by day.

Two underactuated hands were developed to provide the self-adaptive ability to grasp commodities of versatile shapes [3,4]. Each hand has three fingers, and each finger has three joints driven by only one motor. The structure of the underactuated fingers and hand

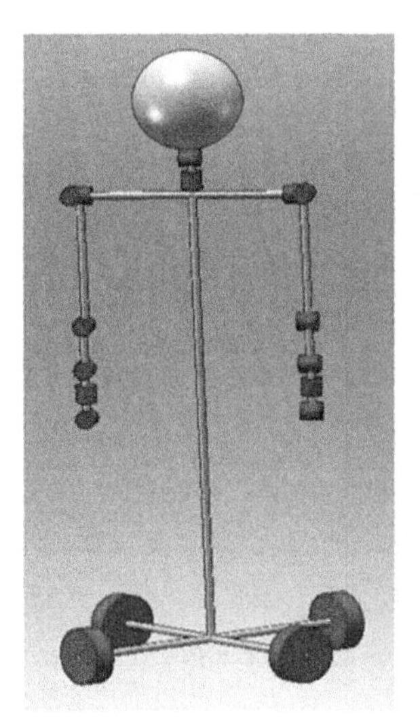

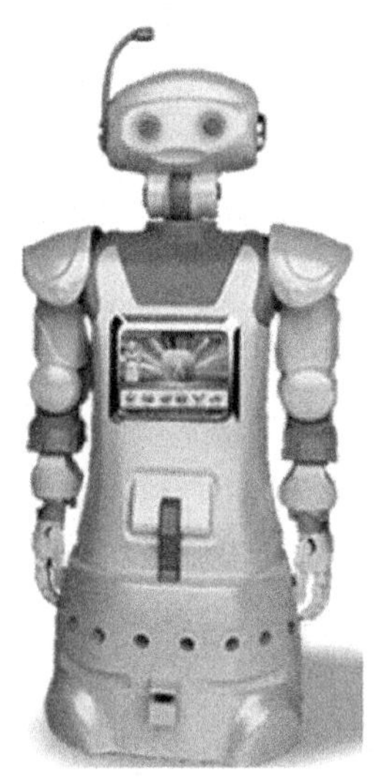

Figure 2
Mechanism and shape of the robot.

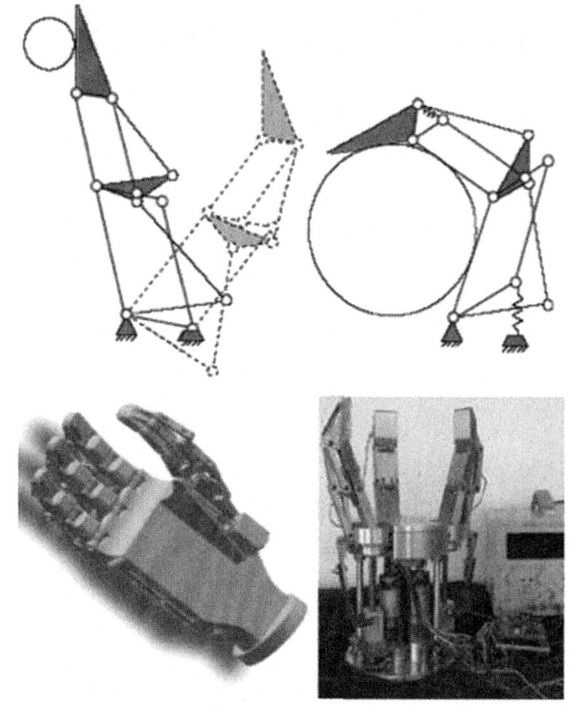

Figure 3
Underactuated finger and hand for versatile grasping.

for versatile grasping is shown in Figure 3. The flexible contact surface with force perception is used to mimic the soft finger contact mode of a human's hands, grabbing commodities with friction. These fingers have the ability of acquiring force information and stable grasping.

2.4.3 Human–Robot Interaction

The speech system, multichannel interface system, and motion detection system are three ways for human–robot interaction in the service robot for elders.

2.4.3.1 Touch and Watch Interaction

First, the elders can control the robot by pushing the function buttons displayed on the touch screen fixed on the chest of the robot. The touch and watch interaction is the simplest way to command the robot. To fit elders' status, a friendly display interface shows the robot's working environment and its own state information as much as possible and has been designed by using graphic symbols and large fonts to make the operation easy for the elders.

The multichannel interface system receives users' information from the perception system, speech system, and motion detection system, and accepts the users' commands from the keyboard, mouse, and touch screen. The multichannel information is fused and distributed to the relative subsystems.

2.4.3.2 Speech Interaction

The elders can talk to the robot in Chinese. The robot can understand the elder's voice commands and answer his/her simple questions. The speech system performs speech recognition and synthesis tasks. A set of software for voice detection, noise filtering, and meaning understanding were developed to accept the user's commands in natural languages [5].

The speech recognition process is shown in Figure 4. To improve the recognition rate, a set of words in the applied scenarios are imported into the database for training. Three demonstration scenarios are designed, such as daily care, emotional company, and a falling alarm. Experimental results show that the recognition rate is 90.15%.

2.4.3.3 Motion Interaction

The motion detection system was developed to monitor the elder's motion state by using an accelerometer and a magnetic orientation sensor fixed in a belt worn on the elder's waist. The sensor signals are sent to the PC4 computer through wireless communication. If the user falls accidently, the computer will send an alarm message to his relatives' mobile phones or will directly call an ambulance.

Figure 5 shows the detection process for elders falling down. In the step of "Press button command," the robot sends a voice to awake the elder. If the elder is all right, he/she will

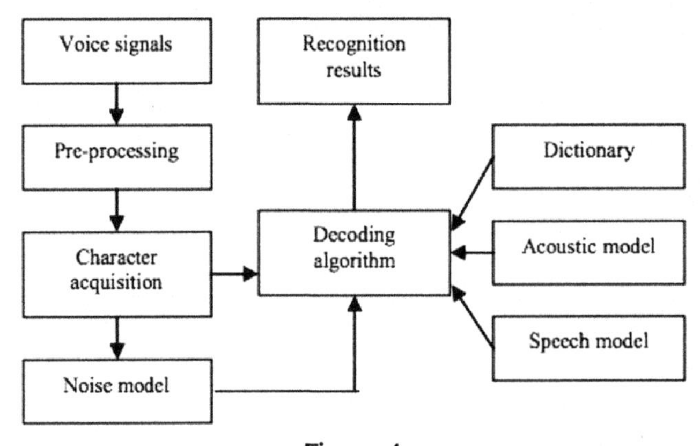

Figure 4
Speech recognition process.

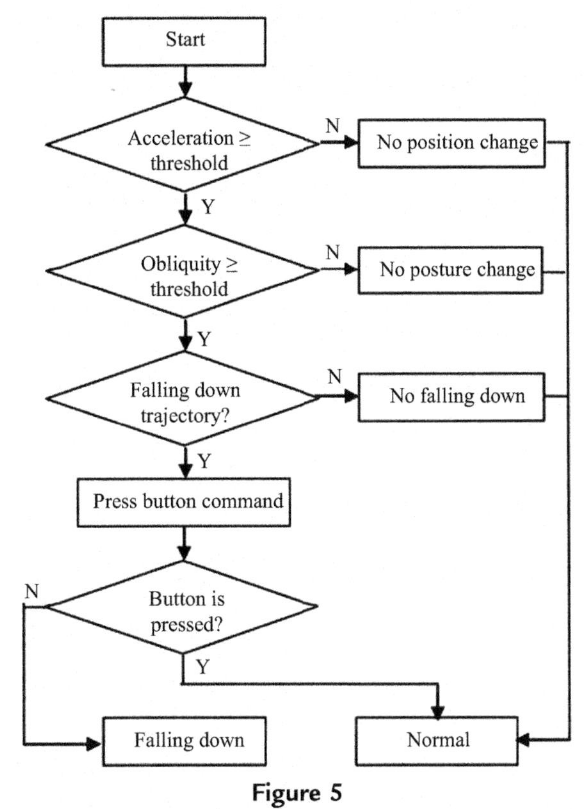

Figure 5
The detection process of falling down.

be able to press a button on the electronic belt. If the button signal is not received in a certain time duration, the robot will judge that the elder has fallen down. The correct rate of estimating the elders' falling states is 94.83%.

2.4.4 Experiments

Three scenarios including daily care, emotional company, and falling alarm were used to demonstrate the functions of the robot. The demonstration was done in the laboratory and at China Hi-Tech Fair in Shenzhen.

2.4.4.1 Daily Care Scenario

In the daily care scenario, three tasks have been designed to provide simple daily service for the users. First, the robot is required to remind the elder to take pills according to the prescribed timetable. Then, the robot can fetch a cup of water for the elder from a drinking trough at the request of the elder (Figure 6). The robot takes nine steps to fetch the water in the following sequence: finding the empty cup, grasping the cup, moving to the position near the drinking trough, finding the water tap, receiving the water, going

Figure 6
The robot is fetching a cup of water.

back to the position near the elder, finding a table, putting the cup on the table, and telling the elder to drink water. The third task is to help the elder to make a dietetic plan by selecting the food in the electronic menu. The food nutrition of the dietetic plan could be calculated automatically.

Based on the equidistance interpolation algorithm, a software program of motion control for the manipulator is designed and used in the motion control of the manipulator mounted on "Service Robot for the Elderly." The position of the target given in the vision system is transformed to the coordinate information in the manipulator coordinate system through a coordinate transformation. According to the coordinate information, the inverse kinematics solution is calculated for the system of the manipulator and the solution is sent in order to drive the motor of the manipulator to make sure it is working as required. Therefore, the manipulator of "Service Robot for the Elderly" can do daily care tasks.

When the manipulator is executing real tasks, we use the NDI Optotrak 3D-optical measuring system to measure the actual position. A part of the statistical results of the manipulator errors are presented in Table 1. From the table, we can find that the manipulator's accuracy of positioning is in the range of plus and minus 3 mm, and the relative error is in the range of 1%. It achieves a high pointing accuracy, so the algorithm has great application value in practice (Table 2).

2.4.4.2 Emotional Company Scenario

In the emotional company scenario, the robot plays a role of an accompanier. The robot can chat and play chess with the elder, dance, and play music for the elder. Although the tasks in this scenario are simple, the functions are important for elders living alone (Figure 7).

Table 1: Manipulator experimental precision analysis

ID	Space Targets Coordinate (x,y,z) (mm) / Manipulator End Coordinate (x,y,z) (mm)			Absolute Error (mm)	Relative Error (%)	Δ_{max} (mm)
1	312.11	181.94	209.56	3.35	0.80	2.70
	313.70	180.75	212.26			
2	242.03	404.05	219.02	2.80	0.54	−2.58
	240.96	401.47	218.78			
3	318.77	−94.16	170.31	3.39	0.91	−2.20
	320.21	−96.30	168.11			
4	319.80	−124.66	170.75	2.81	0.73	−1.95
	321.41	−125.89	168.80			
5	314.95	269.15	210.06	2.68	0.58	−1.70
	313.39	270.52	208.36			

Table 2: The accuracy of falling down judgment and the misstatement rate of daily performance

Tester / Experiment	Xiang Yang Gao	Yuan Xue	Chan Xie	Po Li	Zhuo Wen	Ensemble Average
Accuracy of falling down judgment	91.6%	90%	90%	95%	93.33%	91.99%
The misstatement rate for daily performance	1.67%	1.67%	0.83%	0%	0.83%	1%

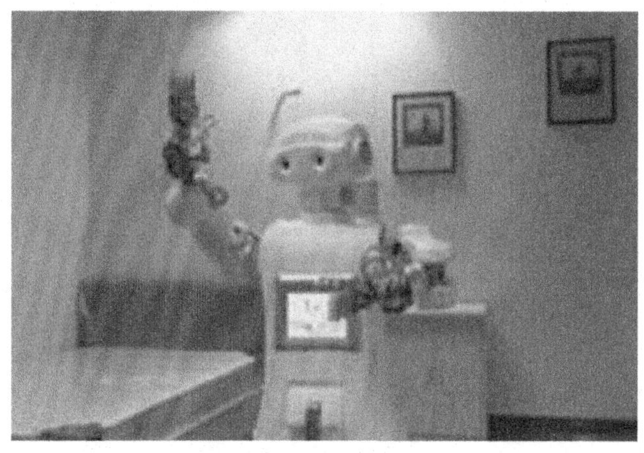

Figure 7
The robot is dancing to entertain the lonely elder.

2.4.4.3 Falling Alarm Scenario

In the falling alarm scenario, the robot will send short messages to the relatives' mobile phones when the user wears the sensor belt and falls down quickly (Figure 8). The detection process has been described in Figure 5.

The falling alarm system is designed to detect the safety for the daily life of the elders. When the elder falls down, the robot can figure it out and use an appropriate alarm, such as send a short message to relatives and/or contact the rescue service center.

During our experiment, we tested the function of the falling alarm system as follows: the accuracy of the falling down judgment and the misstatement rate of daily performance. A part of the statistical results of the accuracy of falling down judgment and the misstatement rate of daily performance are given in Table 1. From the table, we can find that the accuracy of the falling down judgment for a normal activity is 90%, while for a strenuous exercise it is 80%.

2.4.5 Conclusion

We have successfully developed a prototype of a service robot for elders. The advanced technologies, such as multimode input technology, new type of humanoid hands using underactuated fingers technology, and an intelligent decision system, and so forth, are used in order to realize the robot's five main functions: autonomous mobility in the home environment, manipulation of small commodities, speech recognition and synthesis in a specific background, movement detection for elders' safety, and basic nursing abilities.

Figure 8
The robot is monitoring the falling user.

Nevertheless, this service robot has not become widespread because of the high cost of the robot development process and low reliability of the robot operation. In the future, we will conduct research on breakthrough key technologies of low cost and high reliability to realize the service robots entry into every household.

Acknowledgments

The work is jointly accomplished by six research groups led by the authors. These groups are the Research Center for Robot Technology and the Research Center for Digital Design and Manufacturing in the Institute of Advanced Manufacturing Technology, the Laboratory on Machine Vision and the Laboratory on Artificial Intelligence in the Institute of Intelligent Machines, and the Laboratory on Intelligent System and the Laboratory on Semiconductor Equipment in the Shenzhen Institute of Advanced Technology. More than 40 research staff, students, and engineers contributed to the project. The authors gratefully acknowledge their hard work and innovative contributions.

The work is financially supported by a grant from the Chinese Academy of Sciences for the "Service Robot for Elders" project.

The authors gratefully acknowledge Chinese Academy of Sciences for the grant of "Service Robot for Elders" project.

References

[1] M. Pollack, S. Engberg, J.T. Matthews, S. Thrun, L. Brown, D. Colbry, C. Orosz, B. Peintner, S. Ramakrishnan, J. Dunbar-Jacob, C. McCarthy, M. Montemerlo, J. Pineau, N. Roy, Pearl: a mobile robotic assistant for the elderly, in: Workshop on Automation as Caregiver: The Role of Intelligent Technology in Elderly Care (AAAI), August 2002. http://www.ri.cmu.edu/publication_view.html?pub_id=4829.

[2] B. Graf, M. Hans, R. Schraft, Care-O-bot II development of a next generation robotic home assistant, Auton. Robots 16 (2004) 193–205.

[3] M. Luo, T. Mei, Grasp characteristics of an underactuated robot hand, in: Proceedings of the 2004 IEEE International Conference on Robotics and Automation, April 26–May 1, 2004, New Orleans, LA, pp. 2236–2241.

[4] M. Luo, T. Mei, etc., Autonomous grasping of a space robot multisensory gripper, in: Proceeding of 2006 IEEE/RSJ International Conference on Intelligent Robots and Systems, October 9–15, 2006, Beijing, China.

[5] J. Chen, L. Wang, Automatic lexical stress detection for Chinese learners of English, ISCSLP November 29, 2010–December 03, 2010, pp. V1–V338.

CHAPTER 2.5

A Household Service Robot with a Cellphone Interface[1]

Long Han[1,2], Xinyu Wu[1,2], Yongsheng Ou[1], Yen-Lun Chen[1], Chunjie Chen[1], Yangsheng Xu[1,2]

[1]Guangdong Provincial Key Laboratory of Robotics and Intelligent System, Shenzhen Institutes of Advanced Technology, Chinese Academy of Sciences, Shenzhen, China; [2]Department of Mechanical and Automation Engineering, The Chinese University of Hong Kong, Hong Kong, China

Chapter Outline

In this chapter, an efficient and low-cost cellphone-commandable mobile manipulation system is described. Aiming at home use and elderly caring, this system can be easily commanded through a common cellphone network to efficiently grasp objects in a household environment, utilizing several low-cost off-the-shelf devices. Unlike the visual servo technology using a high quality vision system with the associated high cost, the household-service robot would not be able to afford such a high quality vision servo system, and thus it

[1] First published in IJIA Vol. 9. No. 2., 2013. World Scientific, Singapore.

Household Service Robotics. http://dx.doi.org/10.1016/B978-0-12-800881-2.00006-2

is essential to use some low-cost devices. However, it is extremely challenging to create such a vision system with precise localization, as well as motion control. To tackle this challenge, we developed a real-time vision system with which a reliable grasping algorithm combining machine vision, robotic kinematics and motor control technology is presented. After the target is captured by the arm camera, the arm camera keeps tracking the target while the arm keeps stretching until the end effector reaches the target. However, if the target is not captured by the arm camera, the arm will make a move to help the arm camera capture the target under the guidance of the head camera. This algorithm is implemented on two robot systems: one with a fixed base and another with a mobile base. The results demonstrate the feasibility and efficiency of the algorithm and system we developed, and our study shows the significance of developing a service robot in a modern household environment.

2.5.1 Introduction

With the aging population and the rising cost of health care, assistant agents are gradually playing important roles in the household environment. To the increasing demand of home assistance, the household assistant robot appears to be a perfect solution, especially for robots that can automatically manipulate objects within everyday settings. Benefiting from the development of all kinds of related technologies, robots are gradually stepping into our home environments. For example, the vacuum cleaning robot can help with lifting dirt,[2] the robotic dog[3] and the robotic dinosaur[4] can entertain us. Although these robots recently became available in the market and are already effectively performing their missions, it is still too early for them to practically satisfy even the most common needs in the home, such as cleaning, dishwashing, ironing and moving heavy objects [1], because the domain of the home is a world away from the laboratory, space or battlefield and is hardly in accordance with most of the assumptions and requirements of these domains. Briefly speaking, the home is not designed to accommodate robots, nor should it be. Rather, if robots are to be used in a household environment, they need to be designed to "artfully integrate" with the structure and practices of the home [2,3].

In hopes of having a robot working at home and artfully integrating with the environment, people developed many sophisticated robots with complex architectures and control strategies. They can be generally divided into two categories: bimanual robots and bipedal robots. Generally, a bimanual robot usually consists of a torso and a pair of arms with dexterous hands [4]. For example, UMass's Dexter [5] has a pair of commercial seven degrees of freedom (DOF) arms with three-fingered four DOF hands. MIT's Domo [6] has two arms consisting of a series elastic actuators, providing inherent compliance for safe

---

[2] iRobot Corporation. Roomba. http://store.irobot.com/category/index.jsp?categoryId=3334619&cp=2804605.

[3] Sony Corporation. Aibo. http://support.sony-europe.com/aibo/.

[4] Innvo Labs. Pleo. http://www.pleoworld.com/Home.aspx.

interactions with objects in the surroundings. NASA's Robonaut [7–10], the first humanoid robot used in space, closely intimates the kinematics of human arms and hands. Some platforms integrate bimanual manipulators with mobile bases. For example, Willow Garage's PR2,[5] the most popular open source robot platform in the world, has two seven DOF arms and employs a chassis with four steered and driven casters providing omnidirectional performance to traverse in the environment. KIT's ARMAR [11] has a pair of seven DOF arms and a four-wheeled mobile base, and TUM's Rosie[6] has a pair of seven DOF KUKA lightweight arms equipped with torque sensors in all joints and a KUKA omnidirectional mobile base. DLR's Rollin' Justin [12,13] has two compliant controlled seven DOF light weight arms linked to a movable and spring born wheeled base by a four DOF torso.

The bipedal robot mainly focuses on bipedal locomotion, because legs seem to be better suited than wheels in human environments. For example, Honda's ASIMO [14–16] has the capability of bipedal walking and five finger grasping, which allows it to operate freely in the human-living space and could be helpful to humans in the future. Kwada Industries' HRP series, including HRP1 [17], HRP2 [18], HRP3 [19], HRP4C,[7] and HRP4[8] also can walk using two legs and perform skillful tasks using their two hands. Other examples are Sony's QRIO,[9] Waseda University Tokyo's WABIAN2R [20,21], and the Toyota Partner Robot.[10] All of these platforms have brought significant advances in the bipedal locomotion and mechanical design, but had limited impact in the area of autonomous mobile manipulation [4]. The above research works suggests that, for household assistant robots, the capability of manipulation and mobility are the keys to making them practically functional in our home.

For the purpose of bringing robots into a household environment sooner, some other robots are mainly designed for performing mobile manipulation tasks with reasonable costs. In the early stages, these robots are primarily being used as applied heavy, large and power-demand commercial manipulators. For example, Stanford Assistant Mobile Manipulator (SAMM) [22] consists of a holonomic Nomadic XR4000 base and a PUMA 560 robotic arm equipped with parallel-jaw grippers. Later, with the emerging of light weight commercial arms, such as Exact Dynamics' iArm[11] and Barrett's WAM arm,[12] both of which can be flexibly teleoperated by users to perform many kinds of manipulation tasks, the mobile manipulation platforms have gradually become lighter and smaller. For

[5] Willow Garage Corp. PR2. http://www.willowgarage.com/pages/pr2/overview.
[6] TUM. Rosie. http://ias.cs.tum.edu/research-areas/robots/tum-robot.
[7] AIST. HRP-4C. http://www.aist.go.jp/aist e/latest research/2009/20090513/20090513.html.
[8] AIST. HRP-4. http://www.aist.go.jp/aist e/latest research/2010/20101108/20101108.html.
[9] Sony. Qrio. http://www.sony.net/SonyInfo/QRIO/.
[10] Tokyo TOYOTA MOTOR CORPORATION. Toyota partner robot. http://www.toyota.co.jp/en/special/robot/.
[11] Exact Dynamics. iArm. http://www.exactdynamics.nl/site/?page=iarm.
[12] Barrett Technology Inc. WAM arm. http://www.barrett.com/robot/products-arm.htm.

example, CMU's Herb [23] automatically manipulates its robotic arm guided by the visual system and chessboards placed in the environment, and its mobility comes from a two-wheeled mobile base. Herb can efficiently perform manipulation tasks, but the use of marking chessboards slightly limits its applications in the household environment. Another example is the Stanford AI Robot (STAIR) [24] which has a five DOF arm with parallel-jaw grippers mounted on a modified Segway RMP[13] working in a tractor mode. Other examples can be found at Obro University (PANDI-1[14]), the Center for Autonomous Systems at the Royal Institute of Technology and the Robotics and Automation Laboratory at Michigan State University.

Accurate and sophisticated capabilities of manipulation and mobility rely on highly-delicate mechanisms, highly-sensitive sensors, highly-complex computations, and highly-accurate actuators, none of which can be obtained at a low price. However, in order to eventually come into the household, robots should not only provide useful service with limited user interventions but also be available at a reasonable price. As illustrated in this table, these robots are too expensive to enter our home right now, even in a long period of time, according to the current speed of the development of electronic devices. For further reducing the building cost, our system is constructed using very common and low-cost off-the-shelf devices. For example, the cameras applied in our system just cost $40 for each and the DC motors cost $300 for each. However, these low-cost devices can only provide limited precision. The resolution of the head camera and arm camera are 640×480 and 320×240 pixels respectively, and the motor's control precision is 1.0°. Clearly, these performances are much lower than the devices used in most of the robots mentioned above. It is difficult to accurately localize the target and then directly control the joint motors. For instance, as the resolution of the arm camera is 320×240 pixels, considering that the estimation of the target position on the image plane has at least an error of ± 3 pixels, the localization of the target bears an error of at least ± 3 cm. Similarly, as the accuracy of the motors are only $0.5^\circ$for each, the accuracy of the gripper's position has an error of ± 3 cm.

In order to overcome the low-precision deficiency and provide efficient grasping capability, a new algorithm is presented to solve the grasping problem with no need for marking chessboards. The main idea is to take a balance between rough and accurate localizations. This algorithm proceeds as follows: After the target is captured by the arm camera, the arm camera keeps tracking the target while the arm keeps stretching until the end effecter reaches the target. If the target is not captured by the arm camera, the head camera with a wider field-of-view is used to localize the target and thus guiding the arm to move and help the camera capture the target. If the target is not captured in the head camera either, the whole robot rotates to look for the target.

---

[13] Segway Inc. Segway RMP. http://rmp.segway.com/.

[14] AASS Intelligent Control Lab. PANDI-1. http://www.aass.oru.se/Research/Control/pandi1.html.

For systems that aim at home use and elderly caring, it is very important to find a convenient and low-cost way for the user to command the system. As the cellphone has already become a ubiquitous personal property and can function in almost every human-living space, it is selected to become the remote controller of our system. There is already some work existing about controlling robots through a cellphone. For instance, the authors build a hoist trolley robot which communicates its state signal, voice, and video information interactively with the user's cellphone and the robot's movement pattern can be remotely controlled by the user simultaneously [25]. But, due to the feedback delay, transmitting video and voice data is not very practical. So in our system, a cellphone text message based command is applied to command the robot and the experiment showed that a text message based command is efficient and convenient for elderly people to use.

This chapter is organized as follows: The system architecture is described in Section 2.5.2. The two subproblems that combine together to solve the whole grasping problem are described in Section 2.5.3 and their solutions are given in Sections 2.5.4 and 2.5.5, respectively. In Section 2.5.6, a series of representative experiments and their results are illustrated. Finally, the conclusion and the direction of our future work are discussed in Section 2.5.7.

2.5.2 System Architecture

As illustrated in Figure 1, the mobile manipulation system consists of seven components:

1. A wide-angle camera in head position, denoted as "head camera."
2. A narrow-angle camera mounted on the arm, denoted as "arm camera."
3. A multiple degree of freedom robotic arm, denoted as "arm."
4. A two-fingered gripper with a pressure sensor in palm, denoted as "hand."
5. A PC, functions as "brain."

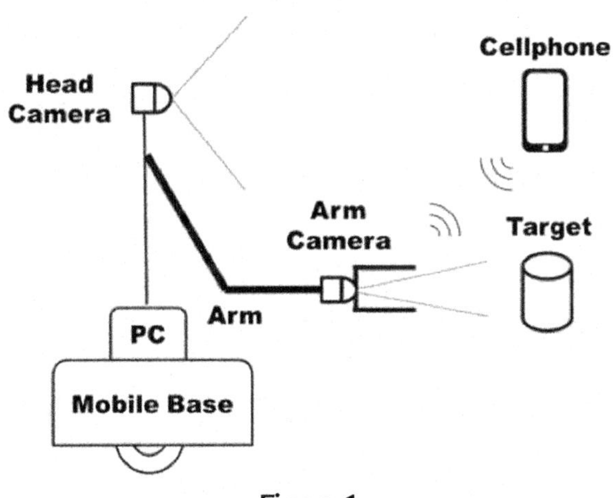

Figure 1
System architecture.

6. A wheeled mobile base, denoted as "chassis."
7. A cellphone as remote controller.

Under this chapter's context, the variation range of the joint position is defined as $[0,2\pi]$ and counter-clockwise is defined as the positive rotating direction.

The cellphone works in a standard global system of a mobile (GSM) communication network commercially maintained by China Mobile Corporation. Inside the robot, a GSM module connected to the PC is applied to parse the text message sent by the user and then invoke the corresponding action of the robot.

2.5.3 Grasping Algorithm

In our system shown in Figure 1, the arm camera is mounted in the center of the gripper. Then if the target has been captured by the arm camera, it means that the gripper is pointing at the target. What should we do next? Here is our idea: if the arm keeps stretching while the gripper keeps pointing at the target, i.e., the arm camera keeps tracking the target, it will finally end up with two different results. The first one is that the gripper touches the position of the target and the second one is that the arm reaches its longest length before the gripper touches the target because the target is too far away. Getting the first result means that the gripper is ready to grasp the target and the second result means that the target is currently unreachable for the system. Both of them are meaningful, so this idea is feasible for grasping.

But as the field of view (FOV) of the arm camera is very narrow (about 40°), it is easy for the arm camera to lose track of the target, in addition, in the beginning of grasping, the target is not captured by the arm camera either. Under these two situations, what should the system do? As the FOV of the head camera is very wide (about 160°), it is more likely and easier for the head camera to find the target. Therefore an intuitive idea is to localize the target using the head camera and guide the arm to move to a position under which the arm camera can capture the target.

Based on these considerations, the whole grasping problem is split into these two subproblems:

- **Subproblem 1**: Suppose that the target object is captured by the arm camera, i.e., in the FOV of the arm camera. How should the arm move to get it?
- **Subproblem 2**: If the target is not in the FOV of the arm camera, by using the head camera, how should the arm make a move to have the target come into the FOV of the arm camera?

As long as these two subproblems are solved, the target can be grasped by the arm if it is reachable, otherwise the target is determined as unreachable, as shown in Figure 2.

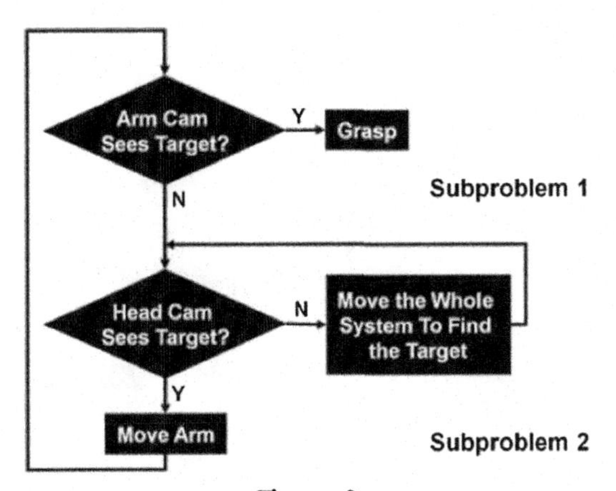

Figure 2
Grasping algorithm.

2.5.4 Solving Subproblem 1

Solving Subproblem 1 deals with two issues: how to recognize the target and how to stretch the arm while the arm camera keeps tracking the target.

2.5.4.1 Target Recognition

Fetching a coke out of a fridge is a common and classic task in a household environment. So this task is chosen as the examination of the performance of our system. Under the circumstance of our experiments, a coke mainly looks bright red and the background, i.e., inside of fridge, mainly looks diffuse silver. Thus the coke appears as high saturation and high intension while the background appears as low saturation and low intension. It can be inferred that a color based method should perform a good recognition of the coke. In hue-saturation-value (HSV) color space, estimation areas of the target are extracted by the following rules:

1. Hue value lies in the red region, i.e., Hue $\in$ [0, 10] $\cup$ [170, 180].
2. Saturation value is greater than half of the maximum, i.e., Saturation $\in$ [128, 255].
3. Value is greater than a quarter of the maximum, i.e., Value $\in$ [64, 255].

After doing that, all the areas that possibly belong to the coke are extracted. Then these areas are taken as a binary threshold to form blobs. The biggest blob region is taken as the estimation of the coke. In order to repress noises, the estimations are filtered by a three-frame median filter, that is:

$$\tilde{E}^i = \text{Median}\left(E^{i-2}, E^{i-1}, E^i\right), \tag{1}$$

Figure 3
Target recognition using the arm camera.

where E^i is the most possible blob estimation of ith frame, and $\tilde{E}^i$ is the final estimation of ith frame. The process of coke recognition is shown in Figure 3.

2.5.4.2 Arm Stretching and Target Tracking

After the coke is recognized by the arm camera, what the system needs to do is to stretch the arm while continuing to track the target in the center of the FOV. The configuration shown in Figure 4 is defined as the initial position, where, all the joint angles are 0.

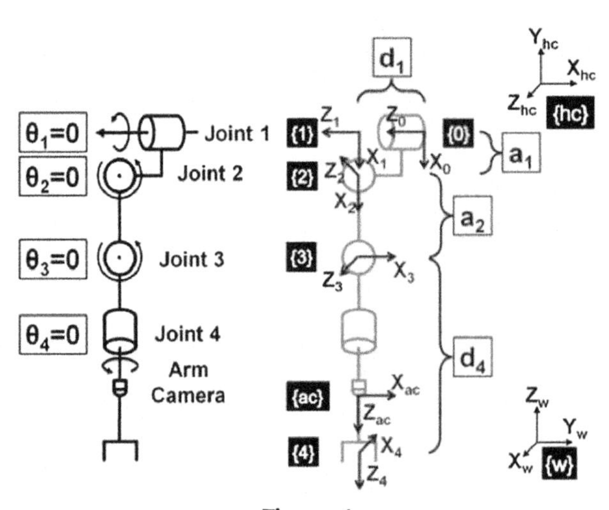

Figure 4
Initial configuration and attached frames.

2.5.4.2.1 *Calculating* $\Delta\theta_1$

First, the target needs to be tracked in the vertical center of the arm camera's FOV. As shown in Figure 5, if $\Delta\theta_1$ is denoted as the rotation amount that joint 1 needs to rotate to make the target lie in the vertical center of the FOV, based on simple geometry, it can be inferred that:

$$\Delta\theta_1 = \theta_1^i - \theta_1^{i-1} \approx -{}^{ac}y_t / {}^1L_t, \tag{2}$$

$$^1L_t = {}^1L_{ac}(\theta_2, \theta_3) + {}^{ac}z_t, \tag{3}$$

where T is the center point of the surface of the target, $^{ac}y_t$ and $^{ac}z_t$ are the y-axis and z-axis coordinates of point T under frame {ac} and ^1L_t is the distance between T and frame {1}'s origin point O_1. Moreover, $^1L_{ac}(\theta_2, \theta_3)$ is the distance between frame {ac}'s origin O_{ac} and frame {1}'s origin O_1 and can be determined only by θ_2 and θ_3:

$$^1L_{ac} = \frac{a_2 \sin(\theta_3)}{\sin(\theta_2 - \theta_3)} + \left(d_4 - \frac{a_2 \sin(\theta_2)}{\sin(\theta_2 - \theta_3)}\right)\cos(\theta_3 - \theta_2). \tag{4}$$

The calculation of $^{ac}y_t$ and $^{ac}z_t$ will be described in the next section.

2.5.4.2.2 *Calculating* $\Delta\theta_3$

Second, the target needs to be kept in the horizontal middle of the arm camera's FOV. Based on Figure 6, $\Delta\theta_3$, denoted as the rotation amount that joint 3 needs to rotate to make the target lie in the horizontal center of the FOV, can be determined as:

$$\Delta\theta_3 = \theta_3^i - \theta_3^{i-1} \approx -{}^{ac}x_t / {}^3L_t, \tag{5}$$

$$^3L_t = {}^3L_{ac} + {}^{ac}z_t, \tag{6}$$

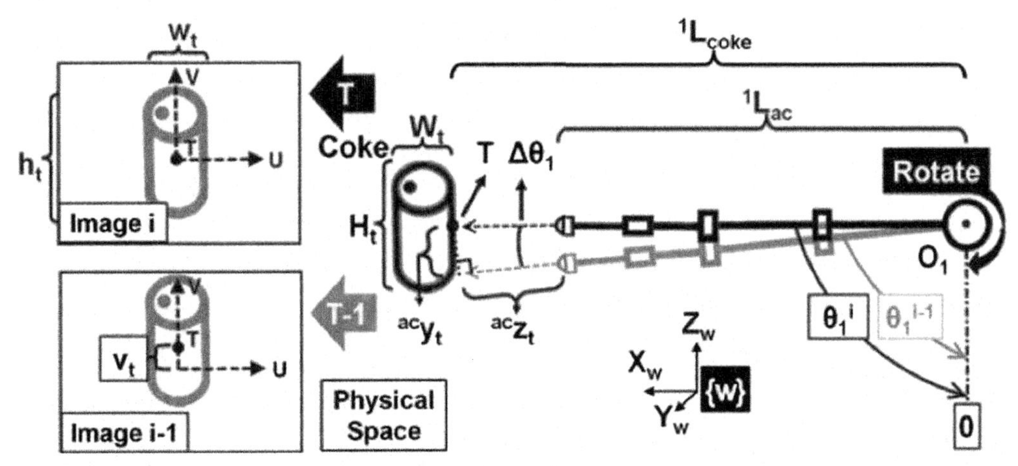

Figure 5
Rotate joint 1 to track coke. (Note: T is the center of the coke's front face; O_1 is the origin of joint 1; {ac} is the frame attached to the arm camera.)

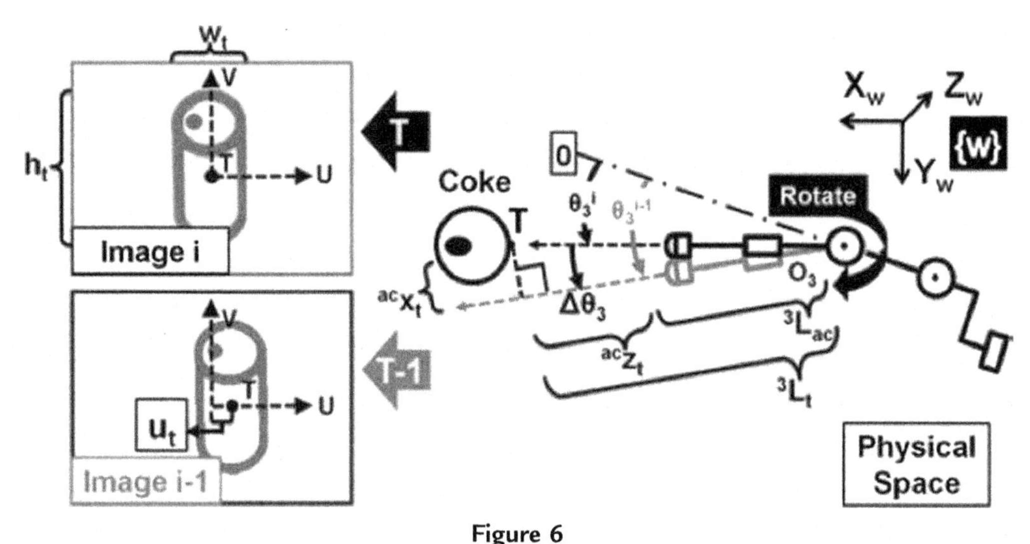

Figure 6
Rotate joint 3 to track coke. (Note: O_3 is the origin of joint 3.)

where ${}^3L_t$ is the distance between T and frame {3}'s origin O_3 and ${}^3L_{ac}$ is a constant that denotes the distance between frame {ac}'s origin O_{ac} and frame {3}'s origin O_3.

2.5.4.2.3 Revising arm position

Stretching the arm means increasing the distance between the arm camera and joint 2. As the length is maximized when θ_2 and θ_3 are both equal to zero. So stretching the arm is equivalent to making θ_2 and θ_3 decreasing to zero: $\theta_2 \rightarrow 0$ and $\theta_3 \rightarrow 0$.

Finally, here is how the arm keeps stretching while still keeping track of the target. Under the precondition of Subproblem 1, the arm camera has seen the target at time $t-1$. The first thing to do is to calculate $\Delta\theta_1$ and $\Delta\theta_3$ and then rotate joint 1 and joint 3 to ensure the target lies in the center of the image plane vertically and horizontally at time t. Then joint 2 rotates a small angle $\Delta\theta_2$ toward zero and joint 3 rotates the same angle but in reverse direction as compensation at time $t+1$. After that, as $\Delta\theta_2$ is small, the target still can be seen by the arm camera but no longer lies in the center of the image plane. So as the situation of time $t-1$, joints 2 and 3 rotate to move the target back into the center of the image plane at time $t+1$. Comparing the situation of time $t-1$ and time $t+1$, it can be seen that the arm is stretched while the target still lies in the center of the image plane. That means Subproblem 1 is solved. The process is shown in Figure 7.

If we keep repeating this process, the arm keeps stretching while still keeping the tracking target in the center of the image plane. It means that the gripper is approaching the target. A pressure sensor is mounted in the root of the gripper (as shown in Figure 1). If the root of the gripper reaches the target, the pressure sensor outputs a positive signal. Then the

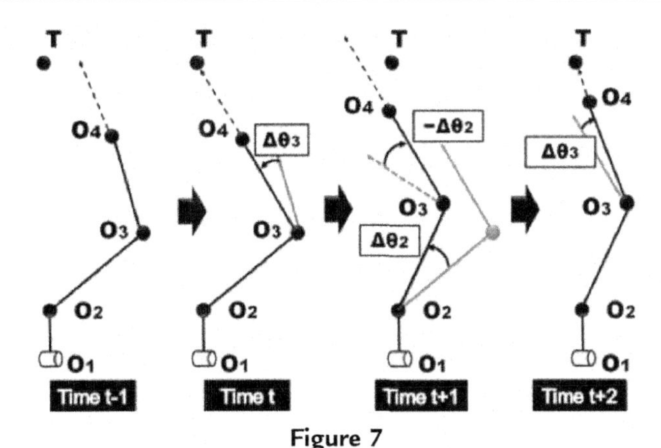

Figure 7
Track target while stretching arm.

gripper is closed to grab the coke. If the arm reaches its maximum length (i.e., $\theta_3 = 0$) before touching the coke, then it means that the target is unreachable right now, and the whole arm system needs to move toward the fridge. That involves the solution of Subproblem 2.

2.5.5 Solving Subproblem 2

After solving Subproblem 1, it is easier to solve Subproblem 2. The main idea is how to configure the arm to make the arm camera capture the target in its FOV. For doing that, a wide-angle camera mounted in the "head" position is applied to localize the target and indicate how the arm should move to the right posture. As the FOV of head camera is very large, even if the arm camera loses track of the target, head camera still has the target in sight. Consequently, the head camera can localize the target and calculate the objective posture that the arm should take to make the arm camera see the target again.

2.5.5.1 Target Localization

The method used to recognize the target using the head camera is the same as the one used by the arm camera in the last section. After recognizing the target, optical geometry is used to localize the target. In our system, the image plane of the head camera is perpendicular to the horizontal plane. As the fridge stands straight and the beverages in the fridge either stand up or lie down, then the principal directions of all the targets are parallel to the image plane. As shown in Figure 8, we have:

$$^{\mathrm{hc}}\mathrm{y}_t/v_t = {}^{\mathrm{hc}}z_t/f_{hc} = H_t/h_t. \tag{7}$$

Similarly to the arm camera, {hc} is the coordinate system attached to the head camera. Y_{hc} and Z_{hc} are the y-axis and z-axis of the frame {hc}, respectively. $^{\mathrm{hc}}\mathrm{y}_t$ and $^{\mathrm{hc}}z_t$ are the

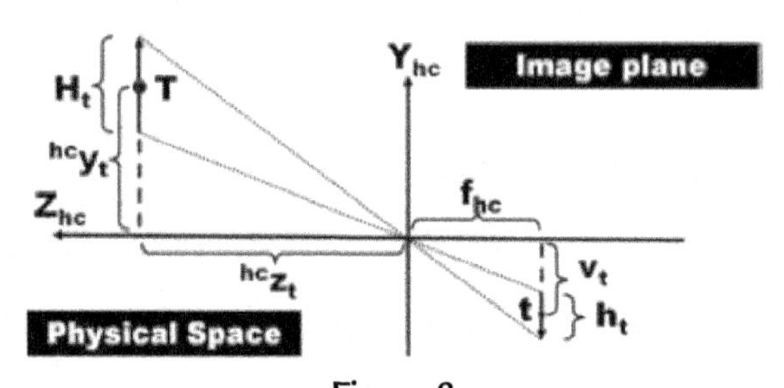

Figure 8
Target localization.

y-axis and z-axis coordinates of the T under frame {hc}. f_{hc} is a constant related to the head camera and v_t is the v-axis coordinate of T in the image plane of the head camera. H_t and h_t are the height of the target in physical space and image plane, respectively.

Similarly, the x-axis coordinate of T under frame {hc} is:

$$^{\mathrm{hc}}x_t/u_t = {}^{\mathrm{hc}}z_t/f_{hc} = W_t/w_t. \tag{8}$$

Where u_t is the u-axis coordinate of T in the image plane of the head camera. W_t and w_t are the height of the target in the physical space and image plane, respectively.

Then coordinates of T in frame {hc} are:

$$^{\mathrm{hc}}P_t = {}^{\mathrm{hc}}(x_t y_t z_t)^T = (u_t v_t f_{hc})^T \cdot \left(H_t/h_t + W_t/w_t\right)/2. \tag{9}$$

2.5.5.2 Calculate Objective Joint Angles

Then the coordinates of T under frame {0} are:

$$^0P_t = {}^0T_{hc}{}^{\mathrm{hc}}P_t, \tag{10}$$

where $^0T_{\mathrm{hc}}$ is the transformation matrix from {0} to {hc}.

As shown in Figure 9, suppose that the arm camera has tracked T in the center of FOV, then O_1, O_2 and T are co-linear. So the first objective joint angle θ_1 is determined as:

$$\theta_1 = \arctan\left({}^0y_t/{}^0x_t\right). \tag{11}$$

As shown in Figure 4, the transformation matrix between frame {0} and {1} $^0T_1$ is only related to θ_1 and the transformation matrix between frame {1} and {2} $^1T_2$ are constant. So if joint 1 rotates to θ_1, the coordinates of T in frame 2 are:

$$^2P_t = {}^2(x_t y_t z_t)^T = {}^2T_1{}^1T_0{}^0P_t, \tag{12}$$

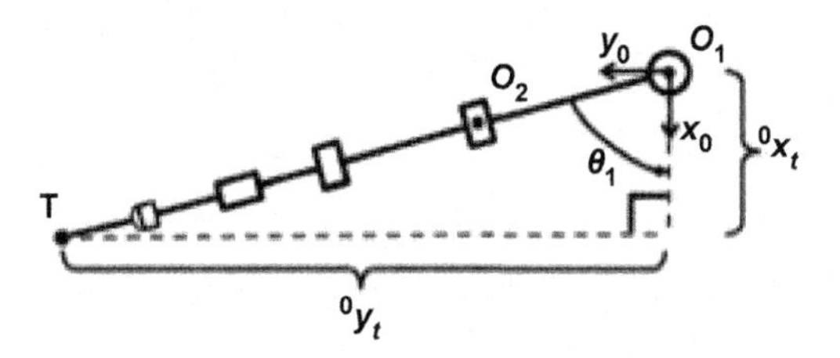

Figure 9
Calculate objective θ_1.

where O_4 is the center of the root of gripper, so the objective θ_2 and θ_3 should make O_4 reach the position of T. As shown in Figure 10, we have:

$$\theta_2 = \beta - \alpha, \tag{13}$$

$$\theta_3 = \pi - \arccos\left(\frac{d_4^2 + a_2^2 - s^2}{2d_4a_4}\right), \tag{14}$$

$$\alpha = \arctan\left(\frac{^2y_t}{^2x_t}\right), \tag{15}$$

$$\beta = \arccos\left(\frac{s^2 + a_2^2 - d_4^2}{2sa_2}\right), \tag{16}$$

$$s = \sqrt{^2x_t^2 + {}^2y_t^2}. \tag{17}$$

If any of these three angles has no reasonable solution, then the target is considered to be unreachable for the system and the arm will not take any action. Otherwise, a trajectory is generated in joint space. The effector is moving toward itd destination, once the object is seen by the arm camera, the solution of Subproblem 1 is applied to deal with the task remaining. Thus, the whole grasping problem is solved.

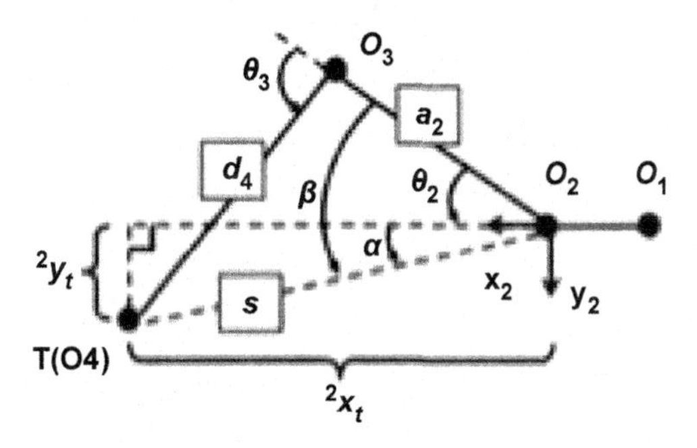

Figure 10
Calculate objective θ_2 and θ_3.

2.5.6 Experiments

2.5.6.1 Fixed Base Manipulation

To test the performance of the grasping algorithm, a fixed-based manipulation system illustrated in Figure 11 is implemented. As shown in Figure 12, the target is placed in three situations: The first one is that the target is seen by the arm camera, the second one is that the target is only seen by the head camera and the third one is that the target is seen by the head camera but is unreachable. As the FOV of head camera is so big and the FOV of the arm camera is so small that if the target is seen by the arm camera it is also seen by the head camera.

This experiment was performed 30 times, 10 times for each situation. The coke in each experiment is placed in a random position. The results and the time-consumption are shown in Table 1.

The results showed that our system grabbed the target efficiently and quickly if the target is reachable and correctly recognized unreachability if the target is unreachable, in spite of a few failures. These failures only occur when the target is in the vicinity of the boundary between reachable space and unreachable space of the system. Under that circumstance, the localization error of the head camera may have led to a wrong estimation of the reachability of the target. Figure 13 shows an image sequence of one of the successful experiments with a target in the reachable space.

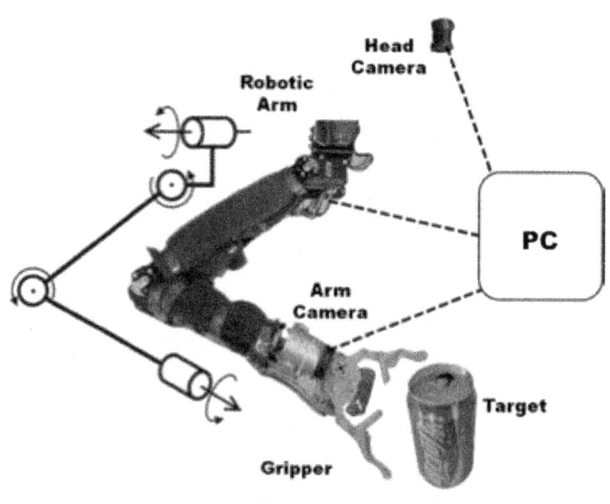

Figure 11
System architecture.

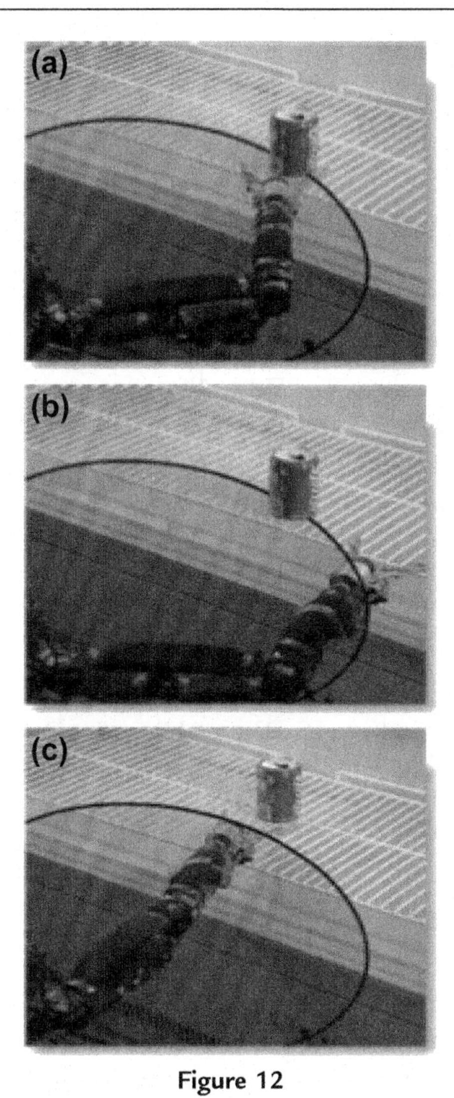

Figure 12
Three kinds of situations. (a) Target seen by arm camera, (b) target only seen by head camera, (c) target is unreachable.

Table 1: Results of 30 times experiments of the stationary arm

Target State	Seen by Arm Camera	Only Seen by Head Camera	Unreachable
Success rate	9/10	9/10	9/10
Avg. time consuming	5.3 s	10.1 s	3.2 s

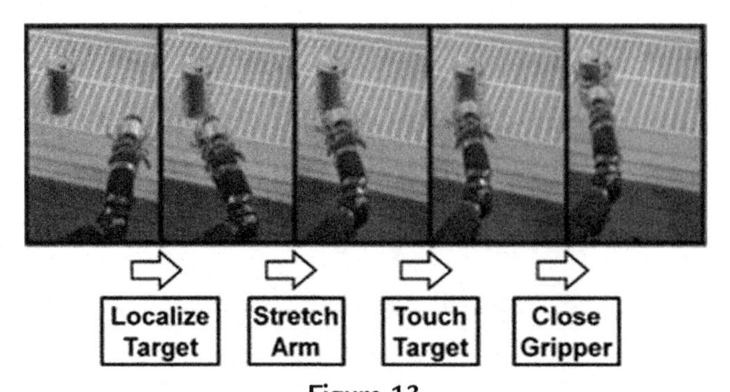

Figure 13
Image sequence of grasping a coke with the stationary arm.

2.5.6.2 Mobile Manipulation and Cellphone Control

To further evaluate this algorithm when combined with mobility. This algorithm is implemented on a wheel-based mobile bimanual robot built by our lab, as shown in Figure 14.

This robot has two four DOF arms, one wide-angle head camera, one narrow-angle arm camera below each front arm and one P3-DXn mobile robot as the mobile base. As both arms of this robot have the same joint structure as with the stationary arm, the same

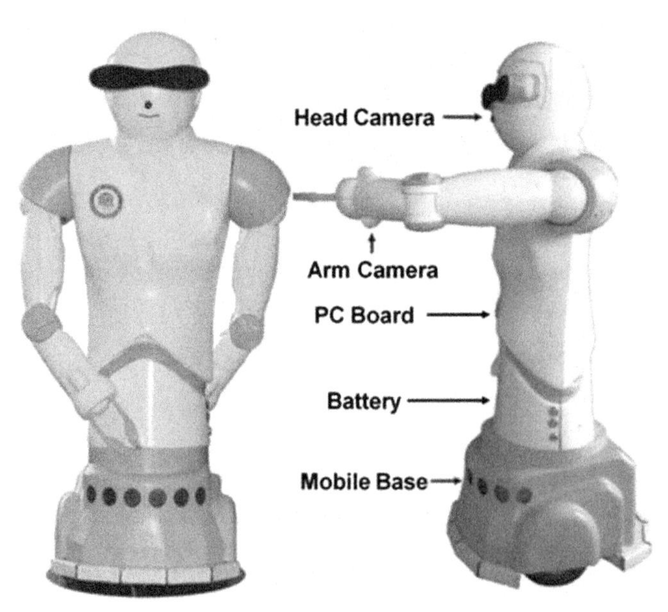

Figure 14
Wheel-based bimanual robot.

algorithm and program used in the stationary arm can be easily applied to both arms with only one extra consideration: dealing with the freedoms brought by the mobile base. The control strategy of the mobile base is to move the whole body toward the center of the target until the target comes into the work space of the arm while the head camera keeps tracking the target.

Actually for this mobile robot, there is a Subproblem 3. As the head camera is not panoramic, it cannot capture the target all the time. Then Subproblem 3 is how to move the chassis to make the head camera find the target. In our experiments, the target is placed in the fridge which is stationary, so the fridge is localized using indoor navigation technology [26]. The potential-field based navigation technology [27] is used to localize the fridge in this chapter. After the fridge is localized, the whole body moves to face the front of the fridge, making the FOV of the head camera cover the inside of the fridge.

Similarly, the right arm of this mobile-based bimanual robot is required to perform the same coke-grasping experiment for 30 times, 10 times for the three situations described before. The results and the time-consumption are shown in Table 2. Every trail starts after the robot has received a text message based command sent by the user's cellphone, as shown in Figure 15.

Table 2: Results of 30 times experiments of the mobile arm

Target State	Seen by Arm Camera	Only Seen by Head Camera	Unreachable
Success rate	10/10	10/10	10/10
Avg. command delay	5.1 s	4.8 s	5.9 s
Avg. time consuming	4.4 s	9.3 s	2.1 s

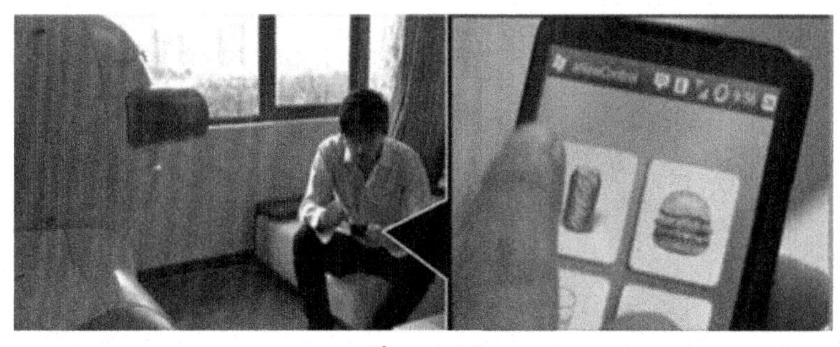

Figure 15
Cellphone commanding.

The results showed that even with uncertainty brought by a mobile base, the success rate of grasping is improved, time-consumption is reduced and the command delay is acceptable. This result is easy to explain: this mobile-based robot has much better equipment, like sensors, cameras and motors. Figure 16 shows an image sequence of one of the successful experiments with a target in the reachable space.

These two results show that our new system works fine with either low performance devices, or better performance devices.

2.5.7 Conclusion and Future Work

In this chapter, a control concept about commanding a robot based on a cellphone text message device, aiming to provide convenience for the elderly and an algorithm is presented to achieve efficient grasping ability without the need for marking chessboards. The experimental results clearly show that our system performs coke-grasping tasks

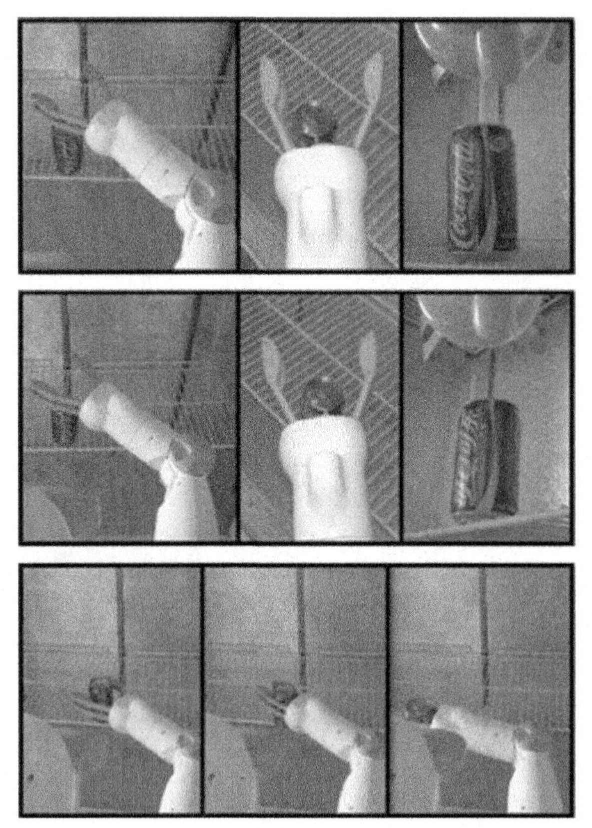

Figure 16
Image sequence of grasping a coke with the mobile arm.

successfully and efficiently under cellphone commands. Although the usage of low-cost devices inevitably brings low-precision localization and control to our system, the results of these experiments clearly illustrated that our low-cost manipulation system overcame these disadvantages, and took a balance between low-complexity rough localization and high-complexity fine localization at the same time.

After solving the grasping problem using performance-limited devices, in the future we will work on the next step of the manipulation problem, which is solving problems utilizing low-cost off-the-shelf devices as well. That will further enlarge the application field of our manipulation system.

Acknowledgments

The author would like to thank Yong Yang, Keke Wang, Ruiqing Fu, Jianquan Sun, Ke Xu, Liping Ouyang, Guogang Xiong, Peng Han, and Hong Li for their highly constructive suggestions and substantial help.

The work described in this chapter is partially supported by the Guangdong-Hongkong Technology Cooperation Funding (2011A091200001), Shenzhen Fundamental Research Program (JC201005270365A), Shenzhen Key Lab for Computer Vision and Pattern Recognition (CXB201104220032A), and Guangdong Innovative Research Team Program (201001D0104648280).

References

[1] C. Ray, F. Mondada, R. Siegwart, What do people expect from robots?, in: IEEE/RSJ Int. Conf. Intelligent Robots and Systems, 2008. IROS, IEEE, 2001, pp. 3816–3821.

[2] L. Suchman, Located accountabilities in technology production, Scand. J. Inform. Syst. 14 (2) (2002) 7.

[3] A.S. Taylor, L. Swan, Artful systems in the home, in: Proc. SIGCHI Conf. Human Factors in Computing Systems, ACM, 2005, pp. 641–650.

[4] D. Katz, E. Horrell, Y. Yang, B. Burns, T. Buckley, A. Grishkan, V. Zhylkovskyy, O. Brock, E. Learned-Miller, The UMass mobile manipulator UMan: an experimental platform for autonomous mobile manipulation, in: Workshop on Manipulation in Human Environments at Robotics: Science and Systems, Citeseer, 2006.

[5] L.M. Garcia, A.A.F. Oliveira, R.A. Grupen, et al., Tracing patterns and attention: humanoid robot cognition, IEEE Intell. Syst. Appl. 15 (4) (2000) 70–77.

[6] A. Edsinger-Gonzales, Massachusetts Inst of Tech Cambridge Artificial Intelligence Lab, Design of a Compliant and Force Sensing Hand for a Humanoid Robot, Citeseer, 2005.

[7] R.O. Ambrose, H. Aldridge, R.S. Askew, R.R. Burridge, W. Bluethmann, M. Diftler, C. Lovchik, D. Magruder, F. Rehnmark, Robonaut: NASA's space humanoid, IEEE Intell. Syst. Appl. 15 (4) (2000) 57–63.

[8] W. Bluethmann, R. Ambrose, M. Diftler, S. Askew, E. Huber, M. Goza, F. Rehnmark, C. Lovchik, D. Magruder, Robonaut: a robot designed to work with humans in space, Auton. Robots 14 (2) (2003) 179–197.

[9] M.A. Diftler, R.O. Ambrose, Robonaut: a robotic astronaut assistant, in: Proc. 6th International Symposium on Artificial Intelligence, Robotics & Automation in Space: i-SAIRAS 2001, Canadian Space Agency, St-Hubert, Quebec, Canada, June 18-22, (2001).

[10] M.A. Diftler, C.J. Culbert, R.O. Ambrose, R. Platt Jr., W.J. Bluethmann, Evolution of the NASA/DARPA robonaut control system, in: Proc. ICRA'03. IEEE Int. Conf. Robotics and Automation, 2003, vol. 2, IEEE, 2003, pp. 2543–2548.

[11] B. Dillman, P. Steinhaus, ARMAR II — a learning and cooperative multimodal humanoid robot system, Int. J. Humanoid Rob. 1 (1) (2004) 143−155.
[12] M. Fuchs, C. Borst, P.R. Giordano, A. Baumann, E. Kraemer, J. Langwald, R. Gruber, N. Seitz, G. Plank, K. Kunze, et al., Rollin'justindesign considerations and realization of a mobile platform for a humanoid upper body, in: IEEE Int. Conf. Robotics and Automation, 2009. ICRA'09, IEEE, 2009, pp. 4131−4137.
[13] C. Borst, T. Wimbock, F. Schmidt, M. Fuchs, B. Brunner, F. Zacharias, P.R. Giordano, R. Konietschke, W. Sepp, S. Fuchs, et al., Rollin'justin-mobile platform with variable base, in: IEEE Int. Conf. Robotics and Automation, 2009. ICRA'09, IEEE, 2009, pp. 1597−1598.
[14] K. Hirai, Current and future perspective of honda humamoid robot, in: Proceedings of the 1997 IEEE/RSJ Int. Conf. Intelligent Robots and Systems, 1997. IROS'97, vol. 2, IEEE, 1998, pp. 500−508.
[15] K. Hirai, M. Hirose, Y. Haikawa, T. Takenaka, The development of honda humanoid robot, in: Proc. 1998 IEEE Int. Conf. Robotics and Automation, 1998, vol. 2, IEEE, 1998, pp. 1321−1326.
[16] M. Hirose, K. Ogawa, Honda humanoid robots development, Philos. Trans. R. Soc. A Math. Phys. Eng. Sci. 365 (1850) (2007) 11.
[17] K. Kaneko, F. Kanehiro, S. Kajita, K. Yokoyama, K. Akachi, T. Kawasaki, S. Ota, T. Isozumi, Design of prototype humanoid robotics platform for HRP, in: IEEE/RSJ Int. Conf. Intelligent Robots and Systems, 2002, vol. 3, IEEE, 2002, pp. 2431−2436.
[18] K. Kaneko, F. Kanehiro, S. Kajita, H. Hirukawa, T. Kawasaki, M. Hirata, K. Akachi, T. Isozumi, Humanoid robot HRP-2, in: Proc. ICRA'04. 2004 IEEE Int. Conf. Robotics and Automation, 2004, vol. 2, IEEE, 2004, pp. 1083−1090.
[19] K. Akachi, K. Kaneko, N. Kanehira, S. Ota, G. Miyamori, M. Hirata, S. Kajita, F. Kanehiro, Development of humanoid robot HRP-3P, in: 5th IEEE-RAS Int. Conf. Humanoid Robots, 2005, IEEE, 2005, pp. 50−55.
[20] S. Hashimoto, S. Narita, H. Kasahara, K. Shirai, T. Kobayashi, A. Takanishi, S. Sugano, J. Yamaguchi, H. Sawada, H. Takanobu, et al., Humanoid robots in Waseda University Hadaly-2 and Wabian, Auton. Robots 12 (1) (2002) 25−38.
[21] Y. Ogura, H. Aikawa, K. Shimomura, A. Morishima, H. Lim, A. Takanishi, Development of a new humanoid robot Wabian-2, in: Proc. 2006 IEEE Int. Conf. Robotics and Automation, ICRA, IEEE, 2006, pp. 76−81.
[22] R. Holmberg, O. Khatib, Development and control of a holonomic mobile robot for mobile manipulation tasks, Int. J. Rob. Res. 19 (11) (2000) 1066.
[23] S.S. Srinivasa, D. Ferguson, C.J. Helfrich, D. Berenson, A. Collet, R. Diankov, G. Gallagher, G. Hollinger, J. Kuffner, M.V. Weghe, HERB: a home exploring robotic butler, Auton. Robots 28 (1) (2010) 5−20.
[24] M. Quigley, E. Berger, A.Y. Ng, et al., Stair: Hardware and Software Architecture, AAAI 2007 Robotics Workshop, Vancouver, BC, 2007.
[25] A. Yagi, M. Sakai, K. Kashiwabara, K. Matsumiya, K. Masamune, T. Dohi, Telecare system by hoist-trolley robot arm from multi controllers and cellphone controller, in: World Congress on Medical Physics and Biomedical Engineering 2006, Springer, 2007, pp. 2872−2875.
[26] G.N. DeSouza, A.C. Kak, Vision for mobile robot navigation: a survey, IEEE Trans. Pattern Anal. Mach. Intell. 24 (2) (2002) 237−267.
[27] Y. Koren, J. Borenstein, Potential field methods and their inherent limitations for mobile robot navigation, in: Proc. 1991 IEEE Int. Conf. Robotics and Automation, IEEE, 1991, pp. 1398−1404.

PART 3

Mapping and Navigation

CHAPTER 3.1

The State of the Art in Mapping and Navigation for Household Service

Long Han[1], Borislav Dzodzo[1], Huihuan Qian[1,2], Yangsheng Xu[1,2]
[1]The Chinese University of Hong Kong, Hong Kong, China; [2]Shenzhen Institutes of Advanced Technology, Chinese Academy of Sciences, Shenzhen, China

Chapter Outline

Robust indoor robot navigation is generally achieved by first creating a map of the environment and then pinpointing within the environment. Because the starting and goal positions of a service robot tend to change, path planning is also necessary. To ensure safe navigation, the household robot needs to be able to sense local obstacles and navigate around them.

3.1.1 Map Building and Localization

A wide range of information has been used for mapping and localization, and robot platforms have successfully been localized with Wi-Fi and other wireless signals, laser

Household Service Robotics. http://dx.doi.org/10.1016/B978-0-12-800881-2.00007-4

scanners, camera images, special markers, and many other techniques, which are reviewed below.

3.1.1.1 Localization with Wi-Fi and Other Networked Radio Communications

Wireless radio signal positioning can be achieved in a number of ways. Today, Wi-Fi signals are available in almost all homes. Cisco Systems [1], as well as Progri [2], provide an overview of techniques that can achieve Wi-Fi positioning. Techniques such as finding the cell of the signal origin, time of arrival, time difference of arrival, received signal strength, angle of arrival, and location patterning are discussed in greater detail here. The cell of the signal origin is an approximate localization method that determines the position of a client by taking note of which Wi-Fi router has the clearest connection and which happens to be connected to the client. The time of arrival technique requires a number of routers and the client to cooperatively measure the time that the signal took to travel from the client to the routers or vice versa. Time difference of arrival can measure the position of any Wi-Fi device as long as a sufficient number of time-synced Wi-Fi routers at known positions detect the signal. Time-of-arrival methods usually require modified hardware and a mostly obstruction-free communication path. If the environment is simple, the information on received signal strength may be sufficient for localization. Signal strength is attenuated by distance; however, signal obstructions, reflections, noise, and other challenges deteriorate the performance of this algorithm in more complex environments. The angle of arrival, as the name implies, allows routers with adequate hardware to identify the angle of the client Wi-Fi source emission, and the actual position in space can also be determined if several routers at known positions combine their readings. A specially equipped Wi-Fi client could also measure its relative angle to routers and thereby infer its position. Location patterning does not take into account geometric relationships between the router and the client; rather, access point (AP) identification signals from many routers are collected at each location on the map. In complex settings, router signal strengths vary greatly with distance because they depend on the nature of the communication paths. The environment may change to a certain extent, and this alters the unique signal strengths at each position. Various position pattern algorithms have been developed that deal with the unique challenges of this approach. An off-the-shelf client can be used to take measurements of router signal strengths from various positions. Cisco and other major wireless router manufacturers provide equipment that employs these techniques to position the Wi-Fi client. It is also possible to develop independent Wi-Fi localization solutions, especially if location patterning is used. Wi-Fi localization methods have been applied to Bluetooth, XBee, television signals, radio signals, cell phone signals, and many other means of radio communication.

3.1.1.2 Ultra Wideband Localization

As demonstrated by Krishnan et al. [3], Segura et al. [4], and Pivato et al. [5], ultra wideband communication differs from most other radio communication systems and has been used very effectively for robot mapping and localization. Instead of using a single frequency at any one time, a very broad range of frequencies is used at the same time to enable communication. To avoid the issues of interference and power consumption, ultra wideband emissions are extremely brief in time and most radio equipment either does not detect the signal or at best treats it as commonly encountered background noise. A signal that is composed of a very wide spectrum of frequencies is far less likely to be blocked by obstructions made of materials that would stop any particular range of frequencies. This property, which is unique to ultra wideband communications, enables far more accurate time-of-flight calculations even in cluttered environments. Ultra wideband is a relatively novel means of communication and signaling because it requires far more complicated means of transmission and reception compared with classical radio communications. Because ultra wideband communications are also difficult to detect and are technically challenging to create, they were first used in the expensive and secret world of military radio communications. Until recently, commercial sets were almost impossible to acquire for home service robot navigation experiments, but the march of time and continued technology development have allowed research in this field.

3.1.1.3 Radio Frequency Identification Localization

Wi-Fi and ultra wideband offer their own unique advantages, but Radio Frequency Identification (RFID) is yet another well-known and inexpensive radio wave technology that has been successfully applied to indoor robot positioning and navigation by Ma et al. [6] and many others. In studies by Lin et al. [7] and Enriquez et al. [8], the RFID tags were embedded into floors or ceiling tiles and were used in a dense grid, as a means of global positioning, or sporadically, as landmarks. RFID tags were also used to provide robots with information about surrounding objects that may obstruct their path. Because RFID readers and tags tend to be directional, it is possible to localize to some extent using the angle of arrival as well as the tag ID. It should also be noted that RFID tags can store information and can also be reprogrammed. Navigation and map building algorithms were devised to take advantage of this unique RFID feature. As shown by Johansson and Saffiotti [9], mapping and navigating robots can leave up-to-date information on their activities in the tags that surround them and this information can be used for navigation by other robots that are new to the environment.

3.1.1.4 North-Star Active Infrared Localization

In addition to radio navigation, many other navigation systems are available based on visible, infrared, and ultraviolet light. One unique positioning system, known as North-Star and researched by Gutmann et al. [10], uses modulated infrared light. More specifically, a small box with infrared directional light-emitting diodes projects a couple of dots on the ceiling. The level of luminescence of each dot is modulated at different frequencies. The robot uses three or four photodiodes generally pointed toward the ceiling but tilted and rotated away from each other. With this inexpensive setup, it is possible to derive the position of the robot by measuring the light intensity of each dot frequency and comparing the readings from all of the photodiodes.

3.1.1.4.1 Two-dimensional planar code localization

As shown by Kabuka [11], Becker et al. [12], and many others [13–17], it is also possible to use special two-dimensional (2D) black and white codes to mark locations. The most widely known 2D codes are quick-response codes but many other 2D patterns exist. Ota et al. [18], as well as Kiva Systems [19], placed 2D codes on floors, walls, and ceilings. The codes can be visible, ultraviolet, infrared, or even fluorescent as demonstrated in part by Huh et al. [20]. The codes can be placed on objects as well as at locations of interest. Lin et al. [7] demonstrated that the maps of such systems consist of positions of 2D codes.

3.1.1.4.2 Laser-ranging localization

Rotating laser rangers and their low noise, highly accurate results have been used to create a map and pinpoint locations therein. Laser rangers can vary from the industrial SICK safety-rated rangers to shorter range and more consumer-oriented Hokuyo rangers, down to the very cheapest short-range laser-ranging modules such as those used in Neato robotic vacuum cleaners. As explained by Grisetti et al. [21], laser-ranging sensors can build maps through Rao-Blackwellization particle filtering [21]. Kurt-Yavuz and Yavuz [22] compared many ways to use this method, of which EKF, UKF, FastSLAM 2.0, and UKF-based FastSLAM are some of the most successful (Figure 1). Maps are generally only 2D; however, Biber et al. [23] showed that 3D maps can be constructed under certain circumstances in which the spinning laser rangers are properly positioned or even actively actuated.

3.1.1.5 Camera-Based Localization

Stereo camera systems are yet another line of sensors that have come down in price through Moore's law. Today, we can achieve real-time processing of vast data flows that come from these sensors. Stereo camera vision usually requires two cameras to derive depth as well as color information from the scene. Seitz et al. [24], among others, also

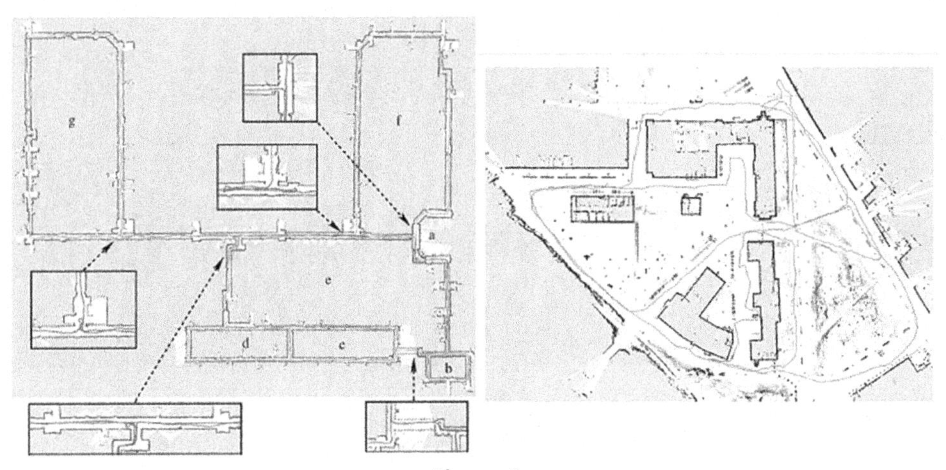

Figure 1
Indoor and outdoor results of Rao-Blackwellization-based SLAM.

used more than two cameras. Some stereo camera systems rely on static or dynamic projection systems and use only one camera to derive depth. Kinect, a popular low-price sensor, is an example of such a system as shown by Biswas and Veloso [25]. Prior to Kinect, only a few companies such as Point Gray produced high-quality, but also highly priced, stereo vision systems. Many other manufacturers are now releasing their own stereo cameras, such as the Sony Playstation 4 stereo camera or the Leap touchless interface controller. At the time of this writing, these newer post-Kinect sensors were not properly integrated into domestic robotic platforms; however, because of their low cost and very attractive capabilities, it is inevitable that their integration will begin shortly. Most frequently, the data from stereo sensors are used to build dense point clouds, as demonstrated by Lee et al. [26]. A point cloud is a collection of 3D positions that the sensor detects, and sometimes researchers such as Rusu et al. [27] augment the point cloud with color information. Point cloud maps are dense 3D representations of the mapped area. The most common technique for localization within the point cloud is to calculate the best fitting overlap of newly sensed information and the information already in the map.

Instead of point clouds, very unique features that the stereo camera picks up from both imagers are sometimes stored in a special and far less dense cloud where the uncertainty of each feature position is also tracked. Rao-Blackwellization can be used in this case to create a SLAM result. The advancements in robot vision have even allowed the development of single-camera mapping and localization, whereby only one camera with no supplementary data can be used to build a navigation map consisting of a cloud of features (Figure 2) [28–30].

Figure 2
Monocular image used for segmenting and positioning major structures in the scene.

3.1.1.6 Time-of-Flight Camera Localization

Cameras, such as SwissRanger, can also sense the depth of the scene and are also used to build point clouds. As explained by Yuan et al. [31], each pixel in this camera detects the time that it takes for a flash of light from the camera to travel to a given position in the scene and then back to the related pixel on the imager array. The new Kinect sensor for the Xbox One is a promising lower-cost implementation for the time-of-flight camera. Huhle et al. [32] and others [27] demonstrated that, although these cameras generate distance data that are similar to those of the stereo cameras, they can be integrated with a color camera to generate dense color and distance measurements for point cloud mapping and localization.

3.1.1.7 Low-Cost Sensor Localization

Zhang et al. [33] have shown that robots with only contact or near-field sensors can, under certain circumstances, create very adequate 2D maps of the environment and even localize within acquired maps. These navigational techniques are of particular interest to robotic vacuum cleaner manufacturers, as the study by Park et al. [34] demonstrates.

3.1.2 Navigation, Path Planning, and Obstacle Avoidance

Navigational maps have been built with one or many of the techniques described, together with a great variety of additional sensor data such as odometry, inertial measurement, and many other supplemental data sources. Although mapping and localization are serious challenges that household robots usually have to satisfy adequately, obstacle detection and avoidance are another domain that requires satisfactory solutions.

Most of the current household robots are lightweight and slow, which greatly enhances safety and simplifies sensing. The slow speed and light weight allow most of these robots to detect obstacles using inexpensive contact sensors or very close range infrared sensors. Many, if not most, of the depth-perception sensors could actually be used to detect obstacles. The indoor environment presents unique challenges, such as completely blank white walls that cannot be detected by some sensors, such as stereo cameras, or the interference of overlapping active sensors from multiple household robots. Moreover, obstructions must be sensed in real time and detected throughout the entire region where they may be encountered by the robot.

3.1.2.1 Structured Light Obstacle Avoidance

Wei et al. [35] and others [36,37] have demonstrated the possibility of detecting deviations from the ground plane inexpensively using simple structured light vision approaches. In cases in which the floor is textured, these can be detected by stereo cameras with a much simpler algorithm and simplified calibration. Stereo cameras could also use structured light, which is provided by the robot platform, or even an interfering robot platform, especially in map building and detection. However, in the case of obstacle detection, the robot may not have the option to coordinate its actions to avoid interference. Active light-aided stereo cameras would not be hampered by large, blank, featureless surfaces.

3.1.2.2 Linear and Planar Distance Sensors and Obstacle Avoidance

Nieuwenhuisen et al. [38] successfully used ultrasonic sensors and rotating laser rangers as obstruction sensors on paper; however, in practice the narrow field of view that these sensors provide is insufficient to safely operate in complex, unstructured, and physically interconnected household environments.

3.1.2.3 Optic Flow, Ground-Plane Segmentation, and Feature-Based Obstacle Avoidance

Soria et al. [39] and Liau et al. [40], among others, achieved the avoidance of obstacles in experiments by measuring optic flow. Roh et al. [41], Wang et al. [42], and Lin et al. [43] showed that, if the texture or the pattern of the floor is known to the robot, it would allow for ground-plane segmentation from the horizon and thereby infer the presence or absence of obstacles between the floor and the camera. The detection of known features was used to sense obstacle-free paths between the camera and the feature, and Jia et al. [44] even used RFID tags to detect obstacle-free paths.

3.1.2.4 Optimization of Path Planning for Obstacle Avoidance

Under certain circumstances, dynamic obstacle avoidance may be impossible owing to obstructions. Kim et al. [45] have already managed to identify such conditions during map

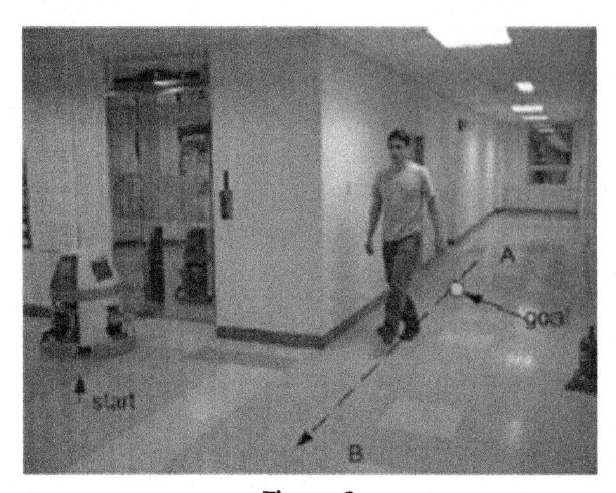

Figure 3
The robot has planned the slower path to maintain adequate sensing capacity and ensure obstacle avoidance.

building and automatically adjusted any path planning to take into consideration the limited view and the necessary deceleration in such locations (Figure 3).

3.1.2.5 Optimization of Robot Behavior and Interaction for Obstacle Avoidance

Another active field of research in navigation is the challenge of crossing an environment populated with dynamic obstacles, such as humans or other automated devices, with which communication is impossible. Some mobile robots simply stop moving and attempt to communicate their intention or outwait the dynamically positioned obstacle [38]. Henry et al. [46] made attempts to predict the path of the dynamic obstacle, whereas others study the behavior of humans under the assumption that they have already optimized their travel paths. Most path planning approaches use the A* algorithm for global as well as local planning and dynamic replanning. El Halawany et al. [47] and Pradhan et al. [48], among others, demonstrated how successful path planning algorithms model dynamic obstacles with repulsive force fields and goal positions with attractive force fields.

3.1.3 Summary of Case Studies

Zhang and Song [49] demonstrated that the detection and tracking of vertical lines combined with inertial measurement can localize a robot platform faster and more reliably than Harris and SLAM feature-based systems (Figure 4).

Sgorbissa and Zaccaria [50] presented a path planning algorithm called Roaming Trail, which provides a degree of freedom in the planning and execution of following a path.

Figure 4
Identified vertical lines and their positions on the map.

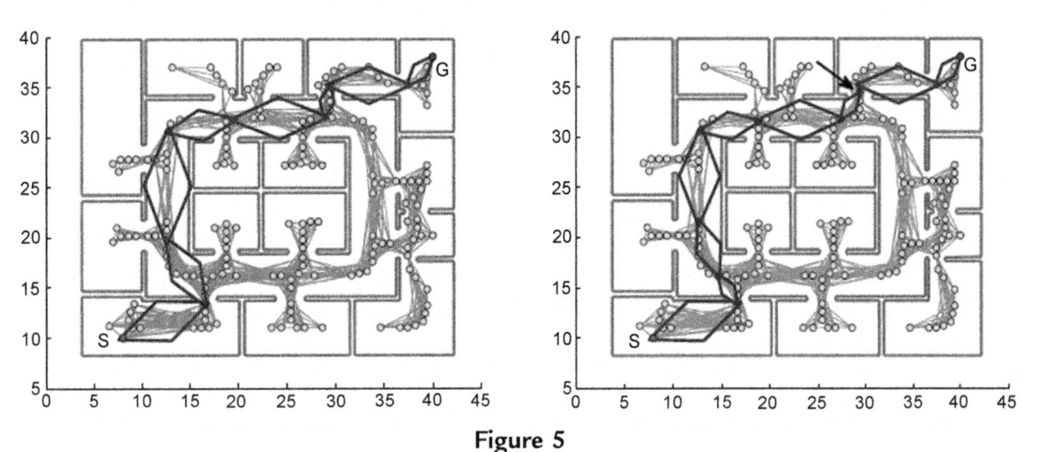

Figure 5
Roaming trial path planner initialized with different parameters.

This approach is particularly well suited to robots that navigate spaces that are populated by dynamic obstacles and are often crossed by humans (Figure 5).

Choi et al. [51] combined an inertial measurement unit and odometry with a low-frame-rate monocamera using an optimized Extended Kalman Filter SLAM. The result was a system that had a rapid initialization and high-quality tracking due to the sophisticated use of feature information even when only partial feature information was known.

Lam et al. [52] included models of human navigation and cooperative navigation in robot path planning. The robots in this study optimized movement and navigation goals to maximize the sense of comfort and security of a human passerby (Figure 6).

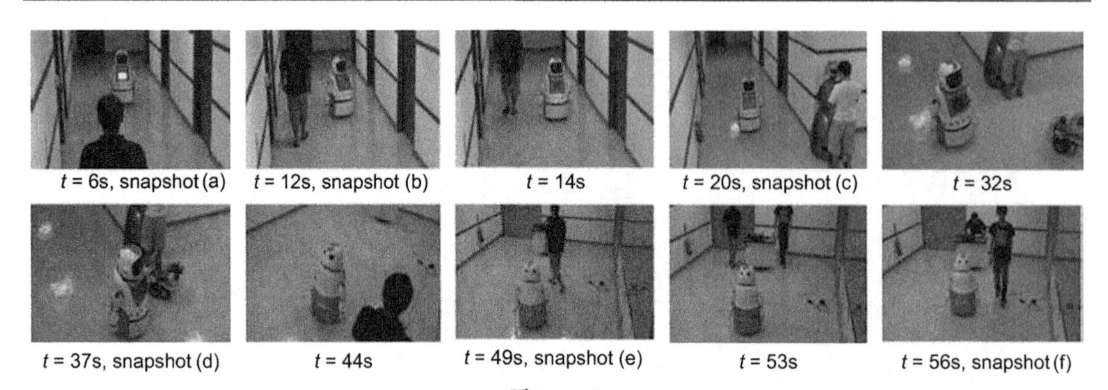

Figure 6
Robot Julia interacting naturally with a human passerby.

References

[1] Cisco Systems, Wi-Fi Location-based Services 4.1 Design Guide, 2008. http://www.cisco.com/en/US/docs/solutions/Enterprise/Mobility/wifich2.html.
[2] I.F. Progri, Wireless-enabled GPS indoor geolocation system, in: IEEE International Conference on Position Location and Navigation Symposium (PLANS), IEEE, 2010, pp. 526–538.
[3] S. Krishnan, P. Sharma, Z. Guoping, O.H. Woon, A UWB based localization system for indoor robot navigation, in: IEEE International Conference on Ultra-wideband 2007, IEEE, 2007, pp. 77–82.
[4] M. Segura, V. Mut, C. Sisterna, Ultra wideband indoor navigation system, IET Radar, Sonar and Navig. 6 (5) (2012) 402–411.
[5] P. Pivato, S. Dalpez, D. Macii, Performance evaluation of chirp spread spectrum ranging for indoor embedded navigation systems, in: SIES, 2012, pp. 307–310.
[6] Y. Ma, S. Kim, D. Oh, Y. Cho, A study on development of home mess-cleanup robot MCBOT, in: IEEE/ASME International Conference on Advanced Intelligent Mechatronics 2008, IEEE, 2008, pp. 114–119.
[7] W. Lin, S. Jia, T. Abe, K. Takase, Localization of mobile robot based on id tag and web camera, in: IEEE Conference on Robotics, Automation and Mechatronics 2004, vol. 2, IEEE, 2004, pp. 851–856.
[8] G. Enriquez, S. Park, S. Hashimoto, Wireless sensor network and RFID sensor fusion for mobile robots navigation, in: IEEE International Conference on Robotics and Biomimetics (ROBIO) 2010, IEEE, 2010, pp. 1752–1756.
[9] R. Johansson, A. Saffiotti, Navigating by stigmergy: a realization on an RFID floor for minimalistic robots, in: IEEE International Conference on Robotics and Automation 2009, ICRA'09, IEEE, 2009, pp. 245–252.
[10] J.-S. Gutmann, P. Fong, L. Chiu, M.E. Munich, Challenges of designing a low-cost indoor localization system using active beacons, in: IEEE International Conference on Technologies for Practical Robot Applications (TePRA) 2013, IEEE, 2013, pp. 1–6.
[11] M. Kabuka, A. Arenas, Position verification of a mobile robot using standard pattern, IEEE J. Rob. Auton. 3 (6) (1987) 505–516.
[12] C. Becker, J. Salas, K. Tokusei, J.-C. Latombe, Reliable navigation using landmarks, in: Proceedings of IEEE International Conference on Robotics and Automation 1995, vol. 1, IEEE, 1995, pp. 401–406.
[13] A. Takahashi, I. Ishii, H. Makino, M. Nakashizuka, A method of measuring marker position/orientation for VR interface by monocular image processing, Electron. Commun. Japan (Part III: Fundam. Electron. Sci.) 80 (3) (1997) 1–12.
[14] C.-J. Wu, W.-H. Tsai, Location estimation for indoor autonomous vehicle navigation by omni-directional vision using circular landmarks on ceilings, Rob. Auton. Syst. 57 (5) (2009) 546–555.

[15] H-T. Jiang, G-H. Tian, Y-H. Xue, R-K. Li, Design, recognition, localization and application of a new artificial landmark, J. Shandong Univ. (Eng. Sci.) 2 (022) (2011).

[16] G. Lin, X. Chen, A robot indoor position and orientation method based on 2d barcode landmark, J. Comput. 6 (6) (2011) 1191–1197.

[17] H. Kobayashi, A new proposal for self-localization of mobile robot by self-contained 2d barcode landmark, in: Proceedings of SICE Annual Conference (SICE) 2012, IEEE, 2012, pp. 2080–2083.

[18] J. Ota, M. Yamamoto, K. Ikeda, Y. Aiyama, T. Arai, Environmental support method for mobile robots using visual marks with memory storage, in: Proceedings of IEEE International Conference on Robotics and Automation 1999, vol. 4, IEEE, 1999, pp. 2976–2981.

[19] Kiva Systems, Kiva Mapping Localization and Navigation System Overview. http://www.kivasystems.com/solutions/system-overview.

[20] J. Huh, W.S Chung, S.Y. Nam, W.K. Chung, Mobile robot exploration in indoor environment using topological structure with invisible barcodes, ETRI J. 29 (2) (2007) 189–200.

[21] G. Grisetti, C. Stachniss, W. Burgard, Improved techniques for grid mapping with Rao-blackwellized particle filters, IEEE Trans. Rob. 23 (1) (2007) 34–46.

[22] Z. Kurt-Yavuz, S. Yavuz, A comparison of EKF, UKF, Fastslam2. 0, and UKF-based Fastslam algorithms, in: IEEE 16th International Conference on Intelligent Engineering Systems (INES) 2012, IEEE, 2012, pp. 37–43.

[23] P. Biber, H. Andreasson, T. Duckett, A. Schilling, 3D modeling of indoor environments by a mobile robot with a laser scanner and panoramic camera, in: Proceedings of IEEE/RSJ International Conference on Intelligent Robots and Systems 2004 (IROS 2004), vol. 4, IEEE, 2004, pp. 3430–3435.

[24] S.M. Seitz, B. Curless, J. Diebel, D. Scharstein, R. Szeliski, A comparison and evaluation of multi-view stereo reconstruction algorithms, in: IEEE Computer Society Conference on Computer Vision and Pattern Recognition 2006, vol. 1, IEEE, 2006, pp. 519–528.

[25] J. Biswas, M. Veloso, Depth camera based indoor mobile robot localization and navigation, in: IEEE International Conference on Robotics and Automation (ICRA) 2012, IEEE, 2012, pp. 1697–1702.

[26] S. Lee, D. Jang, E. Kim, S. Hong, J.H. Han, A real-time 3D workspace modeling with stereo camera, in: IEEE/RSJ International Conference on Intelligent Robots and Systems 2005 (IROS 2005), IEEE, 2005, pp. 2140–2147.

[27] R.B. Rusu, Z.C. Marton, N. Blodow, M. Dolha, M. Beetz, Towards 3D point cloud based object maps for household environments, Rob. Auton. Syst. 56 (11) (2008) 927–941.

[28] T. Adachi, K. Kondo, S. Kobashi, Y. Hata, Self-location estimation of a moving camera using the map of feature points and edges of environment, in: World Automation Congress 2006, WAC'06, IEEE, 2006, pp. 1–6.

[29] J. Courbon, Y. Mezouar, L. Eck, P. Martinet, Efficient visual memory based navigation of indoor robot with a wide-field of view camera, in: 10th International Conference on Control, Automation, Robotics and Vision 2008, ICARCV 2008, IEEE, 2008, pp. 268–273.

[30] F. Schaffalitzky, A. Zisserman, Viewpoint invariant texture matching and wide baseline stereo, in: Proceedings of Eighth IEEE International Conference on Computer Vision 2001, ICCV 2001, vol. 2, IEEE, 2001, pp. 636–643.

[31] F. Yuan, A. Swadzba, R. Philippsen, O. Engin, M. Hanheide, S. Wachsmuth, Laser-based navigation enhanced with 3D time-of-flight data, in: IEEE International Conference on Robotics and Automation 2009, ICRA'09, IEEE, 2009, pp. 2844–2850.

[32] B. Huhle, S. Fleck, A. Schilling, Integrating 3D time-of-flight camera data and high resolution images for 3DTV applications, in: 3DTV Conference 2007, IEEE, 2007, pp. 1–4.

[33] Y. Zhang, J. Liu, G. Hoffmann, M. Quilling, K. Payne, P. Bose, A. Zimdars, Real-time indoor mapping for mobile robots with limited sensing, in: IEEE 7th International Conference on Mobile Adhoc and Sensor Systems (MASS) 2010, IEEE, 2010, pp. 636–641.

[34] S.-H. Park, Y.-H. Choi, S.-H. Baek, T.-K. Lee, S.e-Y. Oh, Feature localization using neural networks for cleaning robots with ultrasonic sensors, in: International Conference on Control, Automation and Systems 2007, ICCAS'07, IEEE, 2007, pp. 449–453.

[35] B. Wei, J. Gao, K. Li, Y. Fan, X. Gao, B. Gao, Indoor mobile robot obstacle detection based on linear structured light vision system, in: IEEE International Conference on Robotics and Biomimetics 2008, ROBIO 2008, IEEE, 2009, pp. 834–839.
[36] G. Fu, P. Corradi, A. Menciassi, P. Dario, An integrated triangulation laser scanner for obstacle detection of miniature mobile robots in indoor environment, IEEE/ASME Trans. Mechatronics 16 (4) (2011) 778–783.
[37] K. Konolige, Projected texture stereo, in: IEEE International Conference on Robotics and Automation (ICRA) 2010, IEEE, 2010, pp. 148–155.
[38] M. Nieuwenhuisen, J. Stückler, S. Behnke, Intuitive multimodal interaction for domestic service robots, in: 41st International Symposium on and 6th German Conference on Robotics (ISR) 2010 (ROBOTIK), VDE, 2010, pp. 1–8.
[39] C.M. Soria, R. Carelli, M. Sarcinelli-Filho, Using panoramic images and optical flow to avoid obstacles in mobile robot navigation, in: IEEE International Symposium on Industrial Electronics 2006, vol. 4, IEEE, 2006, pp. 2902–2907.
[40] Y.S. Liau, Q. Zhang, Y. Li, S.S. Ge, Nonmetric navigation for mobile robot using optical flow, in: IEEE/RSJ International Conference on Intelligent Robots and Systems (IROS) 2012, IEEE, 2012, pp. 4953–4958.
[41] K.S. Roh, W.H. Lee, I.S. Kweon, Obstacle detection and self-localization without camera calibration using projective invariants, in: Proceedings of the 1997 IEEE/RSJ International Conference on Intelligent Robots and Systems, IROS'97, vol. 2, IEEE, 1997, pp. 1030–1035.
[42] Y. Wang, S. Fang, Y. Cao, H. Sun, Image-based exploration obstacle avoidance for mobile robot, in: Control and Decision Conference 2009, CCDC'09, IEEE, 2009, pp. 3019–3023 (in Chinese).
[43] C.-H. Lin, S.-Y. Jiang, Y.-J. Pu, K.-T. Song, Robust ground plane detection for obstacle avoidance of mobile robots using a monocular camera, in: IEEE/RSJ International Conference on Intelligent Robots and Systems (IROS) 2010, IEEE, 2010, pp. 3706–3711.
[44] S. Jia, J. Sheng, D. Chugo, K. Takase, Obstacle localization for a nonholonomic mobile robot based on RFID technology, in: SICE Annual Conference 2007, IEEE, 2007, pp. 270–273.
[45] S. Kim, W. Chung, C. Bae Moon, J.-B. Song, Safe navigation of a mobile robot using the visibility information, in: IEEE International Conference on Robotics and Automation 2007, IEEE, 2007, pp. 1304–1309.
[46] P. Henry, C. Vollmer, B. Ferris, D. Fox, Learning to navigate through crowded environments, in: IEEE International Conference on Robotics and Automation (ICRA) 2010, IEEE, 2010, pp. 981–986.
[47] B.M. El Halawany, H.M. Abdel-Kader, A. TagEldeen, A.E. Elsayed, Z.B. Nossair, Modified A* algorithm for safer mobile robot navigation, in: Proceedings of International Conference on Modelling, Identification and Control (ICMIC) 2013, IEEE, 2013, pp. 74–78.
[48] N. Pradhan, T. Burg, S. Birchfield, U. Hasirci, Indoor Navigation for Mobile Robots Using Predictive Fields, American Control Conference (ACC), June 17–19, 2013, pp. 3237–3241.
[49] J. Zhang, D. Song, Error aware monocular visual odometry using vertical line pairs for small robots in urban areas, in: AAAI, 2010.
[50] A. Sgorbissa, R. Zaccaria, Planning and obstacle avoidance in mobile robotics, Rob. Auton. Syst. 60 (4) (2012) 628–638.
[51] K. Choi, J. Park, Y.-H. Kim, H.-K. Lee, Monocular SLAM with undelayed initialization for an indoor robot, Rob. Auton. Syst. 60 (6) (2012) 841–851.
[52] C.-P. Lam, C.-T. Chou, K.-H. Chiang, Li-C. Fu, Human-centered robot navigation towards a harmoniously human-robot coexisting environment, IEEE Trans. Rob. 27 (1) (2011) 99–112.

CHAPTER 3.2

An Error-Aware Incremental Planar Motion Estimation Method Using Paired Vertical Lines for Small Robots in Urban Areas[1]

Ji Zhang[1], Dezhen Song[2]

[1]The Robotics Institute, Carnegie Mellon University, Pittsburgh, PA, USA; [2]Computer Science and Engineering Department, Texas A&M University, College Station, TX, USA

Chapter Outline

[1] This work was supported in part by the National Science Foundation under Grant IIS-1318638.

Household Service Robotics. http://dx.doi.org/10.1016/B978-0-12-800881-2.00008-6

We report on our development of a monocular visual odometry system based on vertical lines such as vertical edges of buildings and poles in urban areas. Because vertical lines are easy to extract, insensitive to lighting conditions/shadows, and sensitive to robot movements on a ground plane, they are excellent landmarks. We derive an incremental visual odometry method based on vertical line pairs. We analyze how errors are introduced and propagated in the continuous odometry process by deriving the closed-form representation of a covariance matrix. We formulate the minimum variance ego-motion estimation problem and present two different algorithms. The two algorithms have been extensively tested in physical experiments. The error-aware odometry method has also been compared with two popular methods and consistently outperforms these two counterparts in robustness, speed, and accuracy. The relative errors of the odometry are less than 2% in physical experiments.

3.2.1 Introduction

When a small robot travels in an urban area, tall buildings along the roadside form a deep valley and often block GPS signals. Wheel encoders and low-cost inertial measurement units cannot provide enough accuracy for ego-motion estimation. Visual odometry becomes an important supplemental motion estimation method for the robot. Although capable of providing motion estimation for all six degrees of freedom, existing visual odometry methods require extensive computation and cannot be trivially scaled down to be implemented on a low-power computation platform. We are interested in designing a light-weight planar motion estimation scheme for these low-power platforms.

Urban environments often offer a rich set of structured features. As illustrated in Figure 1(a), building edges and poles are common features in urban areas. These vertical lines are insensitive to lighting conditions and shadows. They are parallel to one another and to the gravity direction. Extracting parallel lines using the gravity direction as a reference can be done quickly and accurately for a low-power computation platform. Moreover, vertical lines are sensitive to robot motion on the ground plane. Hence vertical lines are natural choices for landmarks.

Here we present a visual odometry method based on paired vertical lines for a robot equipped with a single camera. We first show that a single pair of vertical lines can provide a minimal solution for estimating the robot ego-motion up to similarity. Because there often exists multiple vertical edges in urban scenes (Figure 1(b)), there are multiple vertical line pairs. Different choices of the vertical line pairs affect the ego-motion estimation accuracy. We analyze how errors are introduced and propagated in the continuous odometry process by deriving the recursive and closed form representation of the error covariance matrix. We formulate the minimum variance ego-motion estimation problem and present an algorithm that outputs weights for different vertical line pairs. The resulting visual odometry method is tested in physical experiments and compared with two existing methods that are based on point features and line features, respectively. Our result

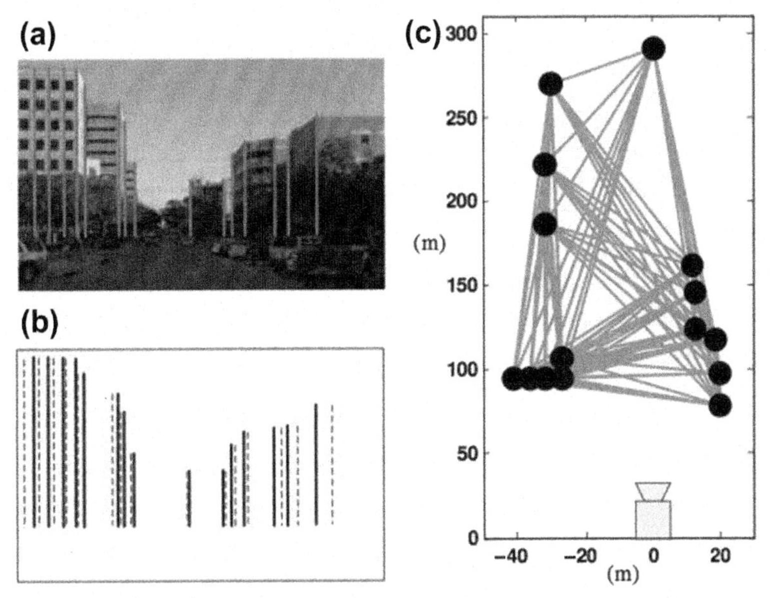

Figure 1
An illustration of monocular visual odometry using vertical line pairs. (a) An image frame taken by the robot with vertical lines highlighted in orange. (b) Corresponding vertical lines in two consecutive frames after the robot has moved forward along the optical axis direction of the camera by 10 m. (c) A top view of the vertical edges in (a) (black dots) and potential choices of pairs (edges between black dots). (For interpretation of the color in this figure legend, the reader is referred to the online version of this book.)

outperforms the two counterparts in robustness, accuracy, and speed. The relative error of our method is less than 2% in conducted experiments.

The rest of this chapter is organized as follows. First, we review related studies in Section 3.2.2. We formulate the vertical line pair-based ego-motion estimation problem in Section 3.2.3. Modeling and analysis of ego-motion estimation for a single vertical line pair are presented in Section 3.2.4. Building on this result, we present the variance minimization method to aggregate motion estimation results from multiple vertical line pairs in Section 3.2.5. We summarize the two resulting algorithms in Section 3.2.6. The algorithms are extensively tested in physical experiments in Section 3.2.7 before we conclude the chapter in Section 3.2.8.

3.2.2 Related Studies

Visual odometry [1] utilizes images taken from an onboard camera(s) to estimate robot motion. It can be viewed as a supplementary method when GPS signals are challenged. Visual odometry has many successful applications including aerial vehicles [2], underwater vehicles [3], legged robots [4], and ground mobile robots [5,6]. Visual odometry is closely related to simultaneous localization and mapping (SLAM) [7] and can be viewed as a building block for visual SLAM [8–16]. A better visual odometry method can certainly increase SLAM performance.

Although our study is based on a regular pinhole camera system that follows a minimalist design to save power usage, visual odometry and visual SLAM can be performed with different sensor combinations such as omnidirectional cameras, stereo vision systems, and laser range finders.

An omnidirectional camera has been a popular choice for odometry applications [17]. Scaramuzza and Siegwart proposed a real-time visual odometry algorithm for estimating vehicle ego-motion using the omnidirectional camera [1,18]. Also using an omnidirectional camera, Wongphati et al. proposed a fast indoor SLAM method [19]. New vertical line detection and matching methods have been developed for omnidirectional cameras [20,21]. In our system, we use a regular camera because the regular camera distributes pixels into space more evenly than an omnidirectional camera and hence can achieve better accuracy.

A very popular sensor in visual odometry is the stereo vision system [22,23]. Nister et al. developed a visual odometry system to estimate the motion of a stereo head or a single camera on a ground vehicle [24,25]. The stereo-vision-based visual odometry used on Mars rovers is a well-known example [26–28]. Laser range finders [29,30] and sonars [31] that provide proximity data can also be used in assisting visual odometry or SLAM. Inspired by the fact that a human can perform odometry with a single eye, we focused on monocular-vision-based approaches.

A different way of classifying visual odometry and visual SLAM is based on what kind of features or landmarks have been used. Point features, such as Harris corners, scale-invariant feature transformation points [32], speed-up robust feature points [33], or center surround extremas feature points [34], are the most popular approaches in visual odometry [24] because they are readily available and well developed in computer vision literature. However, point features usually contain a large amount of noisy data and must be combined with filtering methods such as random sample consensus [35,36] to allow correct correspondence across frames. Such combinations usually result in very high computation costs.

Moreover, a point feature mathematically is a singularity in the feature space and sometimes might not have an actual geometric meaning, which could lead to problems if serving as a landmark because we are unsure how robust such singularity would be under different lighting/shadow conditions. Humans do not view a scene as isolated points and are still capable of performing odometry tasks. Lines are often used by humans in estimating distance. Easy to be extracted [37], lines are inherently robust and insensitive to lighting conditions or shadows. Because of this characteristic, line features see many successful applications in visual SLAM [9,38–42], structure from motion [43,44], indoor localization [45], scene analysis [46], and camera orientation estimation [47,48].

Vertical lines are a special class of lines and widely exist in urban environments. Earth gravity forces us to construct buildings with vertical edges. They are inherently parallel to each other and dramatically reduce the feature extraction difficulties [49]. Moreover, they are very sensitive to robot motion on the ground. All of these properties make vertical lines perfect for visual odometry applications. Building on existing studies, we aim to develop a new systematic method to utilize those advantages, which can dramatically reduce the computation cost of odometry without sacrificing accuracy.

Our group is interested in developing algorithms for vision-based navigation [50,51] and motion estimation [52]. This chapter extends our two previous conference papers [53,54] with a comprehensive and complete approach and includes more experimental results.

3.2.3 Problem Definition

We want to estimate robot motion on a horizontal plane. The robot periodically takes frames to estimate its ego-motion in each step. To set up this ego-motion estimation problem, we begin with assumptions.

3.2.3.1 Assumptions

1. We assume that the initial step of the robot motion is known as a reference. This is the requirement for a monocular vision system. Otherwise the ego-motion estimation is only up to similarity.
2. We assume that the vertical lines, such as poles and building vertical edges, are stationary.
3. We assume that the camera lens distortion is removed by calibration and the camera follows the pinhole camera model with square pixels and a zero skew factor. If not, we can use intrinsic parameters from precalibration to correct the discrepancies.
4. For simplicity, we assume the camera image planes are perpendicular to the horizontal plane and parallel to each other. If not, we can use homography matrices [36] constructed from vanishing points [47] to rotate the image planes to satisfy the condition.

3.2.3.2 Notations and Coordinate Systems

In this chapter, all coordinate systems are right-hand systems (RHSs). For camera coordinate systems (CCSs), we define the z-axis as the optical axis of the camera and let the y-axis point upward toward the sky. The optical axis is always parallel to the x–z plane, which is horizontal. The corresponding image coordinate system (ICS) is defined on the image plane parallel to the x–y plane of the CCS, with its u-axis and v-axis parallel to the x-axis and y-axis, respectively. The optical axis intersects the ICS at its origin on the image plane. To maintain RHS, the x-axis of the CCS and its corresponding u-axis in the ICS must point left (see Figure 2).

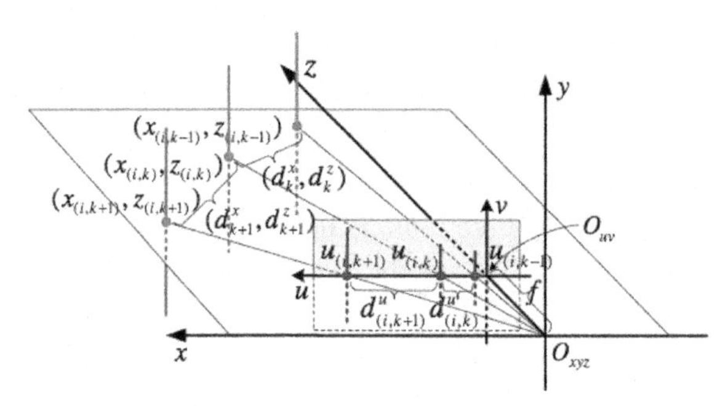

Figure 2
Superimposed CCSs x–y–z and ICSs u–v for the vertical line i over frames $k-1$, k and $k+1$.

Because image planes are perpendicular to the horizontal plane, and parallel to each other, the corresponding CCSs are iso-oriented during computation. Therefore, the robot ego-motion on the horizontal plane in different CCSs is equivalent to the displacement of vertical lines in a fixed CCS in the opposite direction. The x–y–z coordinate in Figure 2 illustrates the superimposed CCSs for three consecutive frames $k-1$, k, and $k+1$, respectively. At time k, $k \in N^+$, let $(x_{(i,k-1)}, z_{(i,k-1)})$, $(x_{(i,k)}, z_{(i,k)})$, and $(x_{(i,k+1)}, z_{(i,k+1)})$ be the (x,z) coordinate of the intersection between the corresponding vertical line i and the x–z plane for frames $k-1$, k, and $k+1$, respectively. Let (d^x_k, d^z_k) be the vertical line i's displacement from frame $k-1$ to k, and we have $d^x_k = x_{(i,k)} - x_{(i,k-1)}, d^z_k = z_{(i,k)} - z_{(i,k-1)}$.

The u–v coordinate in Figure 2 shows the corresponding superimposed ICSs for frames $k-1$, k, and $k+1$. Let $u_{(i,k-1)}$, $u_{(i,k)}$, and $u_{(i,k+1)}$ be the u coordinate of the intersections between the vertical line i and the u-axis in frames $k-1$, k, and $k+1$, respectively. Let $d^u_{(i,k)}$ be the vertical line i's displacement in the ICS from frame $k-1$ to k; we then have $d^u_{(i,k)} = u_{(i,k)} - u_{(i,k-1)}$. With the above notations and coordinate systems defined, we will describe our task.

3.2.3.3 Problem Description

Define n as the number of corresponding vertical lines in three consecutive frames $k-1$, k, and $k+1$. Define $I = \{1, 2, \ldots, n\}$ as the index set of the lines. Let $u_i = [u_{(i,k-1)}, u_{(i,k)}, u_{(i,k+1)}]^T$ be vertical line i's u coordinate in frames $k-1$, k, and $k+1$. Given the robot displacement in the previous step, $d_k = [d^x_k, d^z_k]^T$, we can calculate the displacement of the current step $d_{k+1} = [d^x_{k+1}, d^z_{k+1}]^T$ using the corresponding vertical line positions in the three images, $u_{1:n} = \{u_i, i \in I\}$, as follows

$$d_{k+1} = F(d_k, u_{1:n}) \tag{1}$$

where function $F(\cdot)$ will be determined later in the chapter.

Equation (1) provides a recursive format for us to estimate the robot ego-motion that is represented by d_{k+1}. However, in each step of the calculation, errors are brought into the system. We do not know the actual values of d_k and u_i, which are defined as d_k^* and u_i^*, respectively. d_k and u_i are measurements of d_k^* and u_i^*, respectively. As a convention in this chapter, we use the starred notation a^* to indicate the true value of variable a and define the error value e^a of a as $e^a = a^* - a$. Hence we have $e_k^d = d_k^* - d_k$, $e_{k+1}^d = d_{k+1}^* - d_{k+1}$ and $e_i^u = u_i^* - u_i$, where $e_i^u = [e_{(i,k-1)}^u, e_{(i,k)}^u, e_{(i,k+1)}^u]^T$ describes the measurement error from the line segment extraction for line i in frames $k-1$, k, and $k+1$.

Define Σ_k^d and Σ_{k+1}^d as the covariance matrices for e_k^d and e_{k+1}^d, respectively. At time k, Σ_k^d is known from the previous step. Σ_{k+1}^d is influenced by the errors from the previous step, namely, Σ_k^d, and the new measurement error e_i^u. For measurement error e_i^u, we assume that each vertical line follows an independent and identical Gaussian distribution with zero mean and a variance of σ_u^2. The covariance matrix Σ^u of e_i^u is a diagonal matrix, $\Sigma^u = \text{diag}(\sigma_u^2, \sigma_u^2, \sigma_u^2)$.

To measure how Σ_{k+1}^d changes, we use its trace $\sigma_{k+1}^2 = \text{Tr}[\Sigma_{k+1}^d]$ as a metric. Hence our incremental error-aware motion estimation problem becomes:

Definition 1: Given d_k with Σ_k^d and new measurements ($u_i, i \in I$) with Σ^u, derive $\boldsymbol{F}(\cdot)$ and Σ_{k+1}^d while minimizing σ_{k+1}^2 with respect to design options.

There are two design options: a minimal solution using a single vertical line pair and a multiple vertical line pair-based solution. We begin with the minimal solution.

3.2.4 Deriving a Minimum Solution with a Single Vertical Line Pair

With two equations for two unknowns, a pair of vertical lines can offer us a minimum solution for the ego-motion estimation. The minimum solution is a foundation for multiple vertical line-based solutions. The minimum solution can also help us understand how factors, such as locations of vertical lines and relative positions of the lines, affect the solution quality.

Let u_i and u_j be the input pair of vertical lines, where $i \in I$, $j \in I$, and $i \neq j$. Then Eqn (1) can be rewritten as

$$d_{k+1} = F_s\left(d_k, u_i, u_j\right) \tag{2}$$

where $F_s(\cdot)$ is the motion estimation function for the minimum solution.

3.2.4.1 Deriving $F_s(\cdot)$

Define f as the camera focal length in units of camera pixel width. Because the camera has square pixels and a zero skew factor, we can reduce the camera to the simple pinhole camera model to obtain the following relationship between $(x_{(l,k)}, z_{(l,k)})$ and $u_{(l,k)}, l = i,j$:

$$u_{(l,k)} = \frac{fx_{(l,k)}}{z_{(l,k)}}, l = i, j \tag{3}$$

Combining Eqn (3) with $x_{(i,k-1)} = x_{(i,k)} - d_k^x, x_{(i,k+1)} = x_{(i,k)} + d_{k+1}^x, z_{(i,k-1)} = z_{(i,k)} - d_k^z$, and $z_{(i,k+1)} = z_{(i,k)} + d_{k+1}^z$, we have

$$u_{(i,k-1)} = \frac{fx_{(i,k-1)}}{z_{(i,k-1)}} = \frac{f\left(x_{(i,k)} - d_k^x\right)}{z_{(i,k)} - d_k^z}, \tag{4}$$

$$u_{(i,k)} = \frac{fx_{(i,k)}}{z_{(i,k)}}, \tag{5}$$

$$u_{(i,k+1)} = \frac{fx_{(i,k+1)}}{z_{(i,k+1)}} = \frac{f\left(x_{(i,k)} + d_{k+1}^x\right)}{z_{(i,k)} + d_{k+1}^z}. \tag{6}$$

Combining Eqns (4–6) to eliminate $x_{(i,k)}$ and $z_{(i,k)}$, we have

$$d_{k+1}^x + a_i d_{k+1}^z = b_i, \tag{7}$$

where $a_i = -\frac{u_{(i,k+1)}}{f}$, $b_i = \frac{u_{(i,k+1)} - u_{(i,k)}}{u_{(i,k)} - u_{(i,k-1)}}\left(d_k^x - \frac{u_{(i,k-1)}}{f} d_k^z\right)$.

Similarly, we have the following for vertical line j,

$$d_{k+1}^x + a_j d_{k+1}^z = b_j, \tag{8}$$

where $a_j = -\frac{u_{(j,k+1)}}{f}$, $b_j = \frac{u_{(j,k+1)} - u_{(j,k)}}{u_{(j,k)} - u_{(j,k-1)}}\left(d_k^x - \frac{u_{(j,k-1)}}{f} d_k^z\right)$.

Combining Eqns (7) and (8), we have the $F_s(\cdot)$ function:

$$\mathrm{d}_{k+1} = F_s(d_k, u_i, u_j) = \mathrm{M}_{k+1}^{-1}\mathrm{M}_k d_k \tag{9}$$

where

$$\mathrm{M}_k = \begin{bmatrix} f\left(u_{(i,k+1)} - u_{(i,k)}\right) & -u_{(i,k-1)}\left(u_{(i,k+1)} - u_{(i,k)}\right) \\ f\left(u_{(j,k+1)} - u_{(j,k)}\right) & -u_{(j,k-1)}\left(u_{(j,k+1)} - u_{(j,k)}\right) \end{bmatrix},$$

$$\mathrm{M}_{k+1} = \begin{bmatrix} f\left(u_{(i,k)} - u_{(i,k-1)}\right) & -u_{(i,k+1)}\left(u_{(i,k)} - u_{(i,k-1)}\right) \\ f\left(u_{(j,k)} - u_{(j,k-1)}\right) & -u_{(j,k+1)}\left(u_{(j,k)} - u_{(j,k-1)}\right) \end{bmatrix}.$$

3.2.4.2 Computing the Jacobian Matrices

When errors are brought into the system, Eqn (2) becomes

$$d_{k+1} + \mathrm{e}_{k+1}^{d} = F_s\left(d_k + \mathrm{e}_k^{d}, \mathrm{u}_i + \mathrm{e}_i^{u}, \mathrm{u}_j + \mathrm{e}_j^{u}\right). \tag{10}$$

Because we are interested in how errors propagate, we want to derive the following relationship from Eqn (10):

$$\mathrm{e}_{k+1}^{d} = G\left(\mathrm{e}_k^{d}, \mathrm{e}_i^{u}, \mathrm{e}_j^{u}\right). \tag{11}$$

When errors are small, function $\boldsymbol{G}$ can be approximated by a linear expression,

$$\mathrm{e}_{k+1}^{d} = P_{(i,j)}\mathrm{e}_k^{d} + Q_{(i,j)}\mathrm{e}_i^{u} + Q_{(j,i)}\mathrm{e}_j^{u} \tag{12}$$

where $P_{(i,j)} = \partial F_s/\partial \mathrm{e}_k^{d}$, $Q_{(i,j)} = \partial F_s/\partial \mathrm{e}_i^{u}$, and $Q_{(j,i)} = \partial F_s/\partial \mathrm{e}_j^{u}$ are the Jacobian matrices. Note that $Q_{(i,j)}$ and $Q_{(j,i)}$ are for the vertical lines i and j, respectively.

Obtaining the Jacobian matrices is necessary for studying how errors propagate. It is possible to take an algebraic approach. However, it is more intuitive to use a geometric approach, which helps one to understand the error propagation process.

Figure 3 illustrates the geometric approach. Let l_i be the line described by Eqn (7), which intersects with the d^x axis at b_i with angle α_i. Recalling that a_i is defined in Eqn (7), we have $\tan\alpha_i = -1/a_i = f/u_{(i,k+1)}$. Similarly, let l_j be the line described by Eqn (8), which intersects with the d^x-axis at b_j with angle α_j. Also, we have $\tan\alpha_j = 1/a_j = -f/u_{(j,k+1)}$. l_i and l_j intersect at point A, which is the robot displacement d_{k+1}.

Let e_i^{α}, e_i^{b} be the parameter errors of α_i,b_i, respectively. Owing to the existence of e_i^{b}, l_i shifts to l_i', where l_i' is a line parallel to l_i. Owing to the existence of e_i^{α}, l_i' shifts to l_i'', where l_i'' is a line that intersects with l_i' on x. Let e_j^{α} and e_j^{b} be the parameter errors of α_j and b_j, respectively. Similarly, we have lines l_j' and l_j''. Accordingly, the intersection

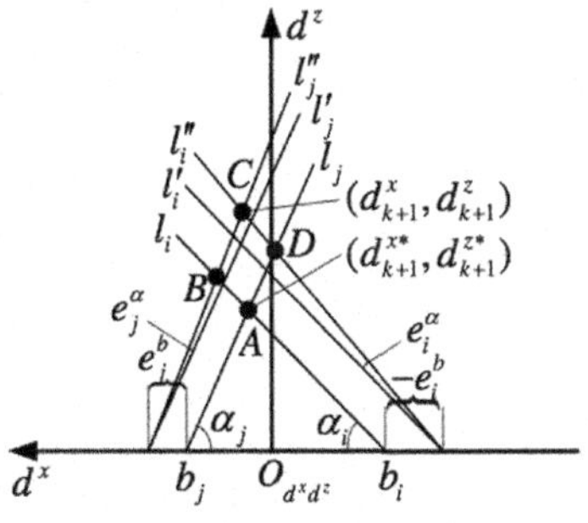

Figure 3
Computing the Jacobian matrices using a geometric approach. Point A at $(d_{k+1}^{x*}, d_{k+1}^{z*})$ is the unknown true location of the displacement.

between l_i and l_j becomes that of l_i'' and l_j'', located at point C, which is the estimated displacement d_{k+1}. The difference between C and A is the robot ego-motion estimation error e_{k+1}^d.

Let B be the intersection between l_i and l_j'' and D be the intersection between l_i'' and l_j. Because e_i^α and e_j^α are both very small, we can approximate $ABCD$ as a parallelogram. Thus, we have

$$e_{k+1}^x = |\overline{AB}|\cos\alpha_i - |\overline{AD}|\cos\alpha_j, \tag{13}$$

$$e_{k+1}^z = |\overline{AB}|\sin\alpha_i - |\overline{AD}|\sin\alpha_j. \tag{14}$$

From the geometric relationship, we have

$$|\overline{AD}| = -\frac{e_i^b \sin(\alpha_i)}{\sin(\alpha_i + \alpha_j)} + \frac{e_i^\alpha (b_j - b_i)\sin(\alpha_j)}{\sin^2(\alpha_i + \alpha_j)} \tag{15}$$

Let e_i^α be the parameter error of a_i in Eqn (7); because $a_1 = -1/\tan\alpha_i$, we have $e_i^a = e_i^\alpha/\sin^2\alpha_i$. At the same time, applying $\tan\alpha_i = f/u_{(i,k+1)}$ and $\tan\alpha_j = -f/u_{(j,k+1)}$, Eqn (15) becomes

$$|\overline{AD}| = \eta/\sin\alpha_j \tag{16}$$

where $\eta = \frac{-fe_i^b}{u_{(i,k+1)} - u_{(j,k+1)}} + \frac{f^2 e_i^a (b_j - b_i)}{(u_{(i,k+1)} - u_{(j,k+1)})^2}$. Similarly, we have

$$|\overline{AB}| = \mu/\sin\alpha_i \tag{17}$$

where $\mu = \frac{fe_j^b}{u_{(i,k+1)} - u_{(j,k+1)}} + \frac{f^2 e_j^a (b_j - b_i)}{(u_{(i,k+1)} - u_{(j,k+1)})^2}$. Substituting Eqns (16) and (17) into Eqns (13) and (14), and using $\tan\alpha_i = f/u_{(i,k+1)}$ and $\tan\alpha_j = -f/u_{(j,k+1)}$, we have

$$e_{k+1}^x = \mu u_{(i,k+1)}\big/f + \eta u_{(j,k+1)}\big/f \tag{18}$$

$$e_{k+1}^z = \mu + \eta. \tag{19}$$

Recalling a_i in Eqn (7), we have

$$e_i^a = -e_{(i,k+1)}^u\big/f. \tag{20}$$

From Eqns (4) and (5), we have $u_{(i,k)} - u_{(i,k-1)} = (fd_k^x - u_{(i,k-1)}d_k^z)/z_{(i,k)}$. Recalling that $d_{(i,k)}^u$ is the u displacement of vertical line i in the superimposed ICS from frame $k-1$ to k, we have $d_{(i,k)}^u = u_{(i,k)} - u_{(i,k-1)}$. After deriving b_i in Eqn (7) and substituting the above equations, we have

$$e_i^b = e^u_{(i,k+1)}\frac{z_{(i,k)}}{f} - e^u_{(i,k)}\frac{z_{(i,k)}}{f}\left(1+\frac{d^u_{(i,k+1)}}{d^u_{(i,k)}}\right) + e^u_{(i,k-1)}\frac{z_{(i,k-1)}}{f}\frac{d^u_{(i,k+1)}}{d^u_{(i,k)}}\bigg) \\ + \left(e_k^x - \frac{u_{(i,k-1)}}{f}e_k^z\right)\frac{d^u_{(i,k+1)}}{d^u_{(i,k)}}\bigg) \tag{21}$$

Substituting Eqns (20) and (21), b_i, and b_j into the expression of η in Eqn (16), and applying the same substitution that we used in Eqn (21), we have the expression of η and, similarly, the expression of μ in Eqn (17). Then, we substitute η and μ into Eqns (18) and (19). Finally, we can obtain the Jacobian matrices shown in Eqns (22) and (23) as the following:

$$P_{(i,j)} = \frac{1}{u_{(j,k+1)} - u_{(i,k+1)}} \\ = \begin{bmatrix} \frac{d^u_{(i,k+1)}}{d^u_{(i,k)}}u_{(j,k+1)} - \frac{d^u_{(j,k+1)}}{d^u_{(j,k)}}u_{(i,k+1)} & -\frac{d^u_{(i,k+1)}}{d^u_{(i,k)}f}u_{(j,k+1)}u_{(i,k-1)} - \frac{d^u_{(j,k+1)}}{d^u_{(j,k)}f}u_{(i,k+1)}u_{(j,k-1)} \\ \frac{d^u_{(i,k+1)}}{d^u_{(i,k)}}f - \frac{d^u_{(j,k+1)}}{d^u_{(j,k)}}f & -\frac{d^u_{(i,k+1)}}{d^u_{(i,k)}}u_{(i,k-1)} - \frac{d^u_{(j,k+1)}}{d^u_{(j,k)}}u_{(j,k-1)} \end{bmatrix}, \tag{22}$$

$$Q_{(i,j)} = \frac{1}{u_{(j,k+1)} - u_{(i,k+1)}} \\ = \begin{bmatrix} \frac{d^u_{(i,k+1)}}{d^u_{(i,k)}}u_{(j,k+1)}z_{(i,k-1)} & -\left(1+\frac{d^u_{(i,k+1)}}{d^u_{(i,k)}}\right)u_{(j,k+1)}z_{(i,k)} & u_{(j,k+1)}z_{(i,k-1)} \\ \frac{d^u_{(i,k+1)}}{d^u_{(i,k)}}fz_{(i,k-1)} & -\left(1+\frac{d^u_{(i,k+1)}}{d^u_{(i,k)}}\right)fz_{(i,k)} & fz_{(i,k+1)} \end{bmatrix}. \tag{23}$$

3.2.4.3 Sensitivity Analysis

With the Jacobian matrices ready, we can analyze how errors are introduced and propagated over the computation. The first analysis we conduct is to study which dimension of the ego-motion estimation error e^d_{k+1} is more suspectable to the error introduced by line detection. In this case, matrix $Q_{(i,j)}$ is scrutinized. We have the following result:

Theorem 1: Let $Q^{gh}_{(i,j)}$ be the (g,h)-th entry of $Q_{(i,j)}$. If the camera horizontal field of view (HFOV) is $\leq 50°$, then $|Q^{1h}_{(i,j)}/Q^{2h}_{(i,j)}| \leq 0.46, \;\; h = 1, 2, 3.$

Proof: From Eqn (23), we have

$$Q^{1h}_{(i,j)} \Big/ Q^{2h}_{(i,j)} = u_{(j,k+1)}, \quad h = 1, 2, 3. \tag{24}$$

Because the HFOV$\leq$ 50°, we have

$$-\tan 25^\circ \leq u_{(j,k+1)} \leq \tan 25^\circ. \tag{25}$$

Combining Eqn (25) with Eqn (3), we have

$$-0.46 \leq \frac{x_{(j,k+1)}}{z_{(j,k+1)}} \leq 0.46. \tag{26}$$

Thus,

$$\left| Q^{1h}_{(i,j)} \Big/ Q^{2h}_{(i,j)} \right| \leq 0.46, j = 1, 2, 3. \tag{27}$$

This theorem indicates that the introduced error in the x direction is smaller than that in the z direction. The result could also be explained by Figure 3, in which point C moves only inside $\angle BAD$. Because the HFOV$\leq$ 50°, angles α_i and α_j are a bounded inside set [65°, 90°]. Hence the quadrilateral $ABCD$ is long in the d^z direction and narrow in the d^x direction. Because a regular camera has an HFOV less than 50°, the conclusion is that the depth error is at least twice more than the lateral error.

Another interesting question is how the ego-motion estimation error e^d_{k+1} relates to the position of the vertical line pair. In other words, if there are many vertical line pairs available in the scene, how do you find the pair that provides the most accurate ego-motion estimation? Define $\delta^u_{k+1} = u_{(i,k+1)} - u_{(j,k+1)}$ as the distance between the two vertical lines in the ICS. Recall that $z_{(i,k+1)}$ is the depth of vertical line i. We have:

Theorem 2: $\partial |Q^{gh}_{(i,j)}| / \partial |\delta^u_{k+1}| \leq 0$, $\partial |Q^{gh}_{(i,j)}| / \partial z_{(i,k+1)} \geq 0, g = 1, 2. h = 1, 2, 3.$

Proof: The first step is to prove $\partial |Q^{2h}_{(i,j)}| / \partial |\delta^u_{k+1}| \leq 0, h = 1, 2, 3$. We prove the inequality for the case of $h = 1$ because other cases can be proved similarly. From Eqn (23), we have

$$Q^{21}_{(i,j)} = -\frac{d^u_{(i,k+1)} z_{(i,k-1)}}{d^u_{(i,k)} \left(u_{(i,k+1)} - u_{(j,k+1)} \right)}. \tag{28}$$

Because $\delta^u_{k+1} = u_{(i,k+1)} - u_{(j,k+1)}$, we have

$$\frac{\partial Q^{21}_{(i,j)}}{\partial \delta^u_{k+1}} = \frac{d^u_{(i,k+1)} z_{(i,k-1)}}{d^u_{(i,k)} {\delta^u_{k+1}}^2} = -\frac{Q^{21}_{(i,j)}}{\delta^u_{k+1}} \tag{29}$$

For the case in which $Q^{21}_{(i,j)} \geq 0$ and $\delta^u_{k+1} > 0$, or the case in which $Q^{21}_{(i,j)} < 0$ and $\delta^u_{k+1} < 0$, we have

$$\partial\left|Q^{21}_{(i,j)}\right| \Big/ \partial\left|\delta^u_{k+1}\right| = -Q^{21}_{(i,j)} \Big/ \delta^u_{k+1} \leq 0 \tag{30}$$

For the case in which $Q^{21}_{(i,j)} \geq 0$ and $\delta^u_{k+1} < 0$, or the case in which $Q^{21}_{(i,j)} < 0$ and $\delta^u_{k+1} > 0$, we have

$$\partial\left|Q^{21}_{(i,j)}\right| \Big/ \partial\left|\delta^u_{k+1}\right| = Q^{21}_{(i,j)} \Big/ \delta^u_{k+1} \leq 0 \tag{31}$$

Thus,

$$\partial\left|Q^{21}_{(i,j)}\right| \Big/ \partial\left|\delta^u_{k+1}\right| \leq 0 \tag{32}$$

The second step is to prove $\partial|Q^{gh}_{(i,j)}|/\partial z_{(i,k+1)} \geq 0, g = 1, 2, \quad h = 1, 2, 3$. Here we prove only the inequality for $g = 1$ and $h = 1$. All other cases with different g and h values can be proved similarly. From Eqn (23), we have

$$Q^{11}_{(i,j)} = -\frac{d^u_{(i,k+1)} u_{(j,k+1)} z_{(i,k-1)}}{d^u_{(i,k)}\left(u_{(i,k+1)} - u_{(j,k+1)}\right)} \tag{33}$$

Substituting $z_{(i,k-1)} = z_{(i,k+1)} - d^z_k - d^z_{k+1}$ into Eqn (33), we have

$$\frac{\partial Q^{11}_{(i,j)}}{\partial z_{(i,k+1)}} = \frac{-d^u_{(i,k+1)} u_{(j,k+1)}}{d^u_{(i,k)}\left(u_{(i,k+1)} - u_{(j,k+1)}\right)} = -\frac{Q^{11}_{(i,j)}}{z_{(i,k-1)}} \tag{34}$$

Because $z_{(i,k-1)} > 0$, when $Q^{11}_{(i,j)} \geq 0$, we have

$$\partial\left|Q^{11}_{(i,j)}\right| \Big/ \partial z_{(i,k+1)} = Q^{11}_{(i,j)} \Big/ z_{(i,k-1)} \geq 0 \tag{35}$$

When $Q^{11}_{(i,j)} < 0$, we have

$$\partial\left|Q^{11}_{(i,j)}\right| \Big/ \partial z_{(i,k+1)} = -Q^{11}_{(i,j)} \Big/ z_{(i,k-1)} > 0 \tag{36}$$

Thus,

$$\partial\left|Q^{11}_{(i,j)}\right| \Big/ \partial z_{(i,k+1)} \geq 0 \tag{37}$$

This theorem indicates that the ego-motion estimation error e^d_{k+1} grows as the depth of the vertical line $z_{(i,k+1)}$ increases. Also the depth error, which is the d^z direction of d_{k+1}, decreases as $|\delta^u_{k+1}|$ increases. From Theorem 1, we know that the depth error dominates the lateral error. Therefore, choosing the vertical line pair with a short depth and a large distance between the two lines can improve the accuracy of the ego-motion estimation.

3.2.5 Error-Aware Ego-Motion Estimation Using Multiple Vertical Line Pairs

There are often multiple vertical lines in the scene. For n vertical lines, there are $n(n-1)/2$ $n(n-1)/2$ pairs. Each pair is capable of providing a minimum solution. The intuition is that we should be able to combine those solutions to yield a motion estimation with minimal error variance. To achieve this, we first define the final motion estimation result as the weighted sum of the minimum solutions from all possible pairs. Plugging Eqn (2) in, the new recursive ego-motion estimation function is

$$d_{k+1} = \sum_{i=1}^{n-1} \sum_{j=i+1}^{n} \omega_{(i,j)} F_s(d_{k,} u_i, u_j) \tag{38}$$

where $\omega_{(i,j)}$ is the weight of vertical line pair (i,j).$\omega_{(i,j)}$ are standardized:

$$\sum_{i=1}^{n-1} \sum_{j=i+1}^{n} \omega_{(i,j)} = 1 \tag{39}$$

$$\omega_{(i,j)} = \omega_{(j,i)} \geq 0, i \in I, j \in I, \text{and } i \neq j. \tag{40}$$

We want to compute a set of $\omega_{(i,j)}$ to minimize σ_{k+1}^2:

$$\left\{\omega_{(i,j)}, \forall i, j \in I, i \neq j\right\} = \arg \min_{\{\omega_{(i,j)}\}} \sigma_{k+1}^2 \tag{41}$$

To solve this optimization problem, we need to derive the closed form of σ_{k+1}^2. Let us begin with deriving the expression of the estimation error e_{k+1}^d and its covariance matrix Σ_{k+1}^d. We know that e_{k+1}^d has two parts,

$$e_{k+1}^d = e_{k+1}^p + e_{k+1}^m \tag{42}$$

where e_{k+1}^p is the estimation error propagated from the previous step e_k^d, and e_{k+1}^m is introduced from the measurement errors of the current step $e_i^u, i \in I$. From Eqns (12) and (38), we have the expressions of e_{k+1}^p and e_{k+1}^m as

$$e_{k+1}^p = \sum_{i=1}^{n-1} \sum_{j=i+1}^{n} \omega_{(i,j)} P_{(i,j)} e_k^d = T e_k^d, \tag{43}$$

where

$$T = \sum_{i=1}^{n-1} \sum_{j=i+1}^{n} \omega_{(i,j)} P_{(i,j)}, \tag{44}$$

and

$$e_{k+1}^{m} = \sum_{i=1}^{n-1} \sum_{j=i+1}^{n} \omega_{(i,j)} \left(Q_{(i,j)} e_{i}^{u} + Q_{(j,i)} e_{j}^{u} \right) = \sum_{i=1}^{n} \left(\sum_{j=1, j \neq i}^{n} \omega_{(i,j)} Q_{(i,j)} \right) e_{i}^{u} = \sum_{i=1}^{n} S_{i} e_{i}^{u} \tag{45}$$

where $S_i = \sum_{j=1, j \neq i}^{n} \omega_{(i,j)} Q_{(i,j)}$. In the above equations, T and S_i are just the Jacobian matrices corresponding to e_k^d and e_i^u, respectively. With the error relationship, we can derive the covariance matrices.

Similar to Eqn (42), the covariance matrix Σ_{k+1}^{d} of the estimation error e_{k+1}^{d} also has two parts because errors propagated from the previous step are independent of the measurement errors in the current step. Hence,

$$\Sigma_{k+1}^{d} = \Sigma_{k+1}^{p} + \Sigma_{k+1}^{m}, \tag{46}$$

where Σ_{k+1}^{p} and Σ_{k+1}^{m} correspond to e_{k+1}^{p} and e_{k+1}^{m}, respectively.

Recall that the covariance matrix Σ_k^d of e_k^d in Eqn (43) is known from the previous step. Recall that the covariance matrix Σ^u of e_i^u in Eqn (44) is a diagonal matrix, $\Sigma^u = \mathrm{diag}(\sigma_u^2, \sigma_u^2, \sigma_u^2)$. Using the covariance matrices with Eqns (43) and (44), we have

$$\Sigma_{k+1}^{p} = T \Sigma_{k}^{d} T^{T}, \tag{47}$$

$$\Sigma_{k+1}^{m} = \sum_{i=1}^{n} S_i \Sigma^{u} S_i^{T} = \sigma_u^2 \sum_{i=1}^{n} S_i S_i^{T}. \tag{48}$$

Therefore, Σ_{k+1}^{p} and its trace can be obtained:

$$\Sigma_{k+1}^{d} = \Sigma_{k+1}^{p} + \Sigma_{k+1}^{m} = T \Sigma_{k}^{d} T^{T} + \sigma_u^2 \sum_{i=1}^{n} S_i S_i^{T}, \tag{49}$$

$$\sigma_{k+1}^{2} = \mathrm{Tr}\left(\Sigma_{k+1}^{d}\right) = \mathrm{Tr}\left(T \Sigma_{k}^{d} T^{T}\right) + \sigma_u^2 \sum_{i=1}^{n} \mathrm{Tr}\left(S_i S_i^{T}\right). \tag{50}$$

With the closed form of σ_{k+1}^{2} derived, we can solve the problem defined in Eqn (41). Let us define vector $w = [\omega_1, \ldots, \omega_{n(n-1)/2}]^T$ with its g-th entry obtained as follows:

$$\omega^{g} = \omega_{(i,j)}, \quad \text{where} \begin{cases} i = 1, \ldots, n-1, \; j = i+1, \ldots, n, \\ g = (i-1)(n - i/2) + j - i. \end{cases} \tag{51}$$

Vector w is our decision vector for the optimization problem in Eqn (41), which can be rewritten as

$$\min_{w} \sigma_{k+1}^{2} = w^{T} A w, \text{subject to:} \; -w \leq 0, \;\text{ and }\; c^{T} w = 1, \tag{52}$$

where $c = 1_{n(n-1)/2 \times 1}$ is a vector with all elements being 1 and A is the $n(n-1)/2 \times n(n-1)/2$ matrix obtained from Eqn (50).

Let us detail how to obtain each entry for A, which actually represents the correlations between the vertical line pairs. A also consists of two parts, $A = A_p + A_m$, where A_p is the error propagation from the previous step and A_m is newly introduced in the current step.

Define A_p^{gh} as the (g,h)-th entry of A_p. Similarly, A_m^{gh} is the (g,h)-th entry of A_m. Then A_p and A_m are obtained from Eqn (50) as follows:

$$\begin{cases} A_p^{gg} = \mathrm{Tr}\left(P_{(i,j)} \sum_k^d P_{(i,j)}^T\right), & i = r, j = m, \\ A_p^{gh} = A_p^{hg} = \mathrm{Tr}\left(P_{(i,j)} \sum_k^d P_{(r,m)}^T\right), & \text{otherwise}, \end{cases} \tag{53}$$

$$\begin{cases} A_m^{gg} = \sigma_u^2(Tr(Q_{(i,j)}Q_{(i,j)}^T) + Tr(Q_{(j,i)}Q_{(j,i)}^T), & \\ & i = r, j = m, \\ A_m^{gh} = A_m^{hg} = \sigma_u^2 \,\mathrm{Tr}\left(Q_{(i,j)}Q_{(r,m)}^T\right), & i = r, j \neq m, \\ A_m^{gh} = A_m^{hg} = \sigma_u^2 \,\mathrm{Tr}\left(Q_{(j,i)}Q_{(m,r)}^T\right), & i \neq r, j = m, \\ A_m^{gh} = A_m^{hg} = \sigma_u^2 \,\mathrm{Tr}\left(Q_{(j,i)}Q_{(r,m)}^T\right), & j = r, j \neq m, \\ A_m^{gh} = A_m^{hg} = 0, & \text{otherwise}, \end{cases} \tag{54}$$

where

$$\begin{cases} i = 1, \ldots, n-1, \ j = i+1, \ldots, n, \\ r = i, \ldots, n-1, \ m = j, \ldots, n, \\ g = (i-1)(n-i/2) + j - i, \\ h = (r-1)(n-r/2) + m - r. \end{cases}$$

Each diagonal entry of A is exactly the estimation error variance of each single vertical line pair. Defining it as $\sigma_{(i,j)}^2$, we have

$$\sigma_{(i,j)}^2 = A^{gg} = \mathrm{Tr}\left(P_{(i,j)}\Sigma_k^d P_{(i,j)}^T\right) + \sigma_u^2\left(\mathrm{Tr}\left(Q_{(i,j)}Q_{(i,j)}^T\right) + \mathrm{Tr}\left(P_{(j,i)}Q_{(j,i)}^T\right)\right) \tag{55}$$

For the vertical line pair (i,j), this can be simply proved by degenerating Eqn (50) into the case containing only two vertical lines. Because A is a positive definite, the feasible set in the optimization problem in Eqn (52) is convex; hence, this problem is a quadratic convex optimization problem. For such a problem, it is well studied and has various solving methods [55]. Here, we use the well-known interior-point method (IPM) [56] to solve it.

3.2.6 Algorithms

The above analysis implies two different algorithms: best single pair (BSP) and minimum variance ego-motion estimation (MVEE) with multiple pairs. The BSP selects the best pair from multiple vertical line pairs using the results from the sensitivity analysis in Section 3.2.4.3. The MVEE uses the weights computed from solving the optimization problem (Eqn (41)).

The two algorithms share a common structure as indicated in Algorithm 1. Note that the function $T(n)$ in Algorithm 1 is the complexity of either the BSP subroutine in Algorithm 2 or the MVEE subroutine in Algorithm 3. For simplicity, we adopt the approximations that $z_{(i,k-1)} \approx z_{(i,k)}$ and $z_{(i,k+1)} \approx z_{(i,k)}$ in the algorithm (see lines 4–9). Because $d^z_{(i,k)} \ll z_{(i,k)}$ and $d^z_{(i,k+1)} \ll z_{(i,k)}$ for consecutive image frames, this approximation is reasonable. At line 29, we call either the BSP subroutine in Algorithm 2 or the MVEE subroutine in Algorithm 3 to obtain the ego-motion estimation results. It is not difficult to find that

Theorem 3: The computation times for the BSP algorithm and the MVEE algorithm are $O(n^4)$ and $O(n^6)$, respectively.

Algorithm 1: BSP and MVEE Algorithms

Line	Statement	Cost
1	**input** : $\mathbf{d}_k, \mathbf{u}_i, i \in I, f, \sigma^2_u, \Sigma^{\mathbf{d}}_k$	
2	**output** : $\mathbf{d}_{k+1}, \Sigma^{\mathbf{d}}_{k+1}$	
3	**begin**	
4	**for** $i = 1$ **to** n **do**	$O(n)$
5	$d^u_{(i,k+1)} = u_{(i,k+1)} - u_{(i,k)}$;	$O(1)$
6	$d^u_{(i,k)} = u_{(i,k)} - u_{(i,k-1)}$;	$O(1)$
7	$z_{(i,k)} = \frac{(d^z_k - u_{(i,k-1)} d^x_k)}{(u_{(i,k)} - u_{(i,k-1)})}$;	$O(1)$
8	$z_{(i,k-1)}, z_{(i,k+1)} = z_{(i,k)}$;	$O(1)$
9	**end**	
10	**for** $i = 1$ **to** $n-1$ **do**	$O(n)$
11	**for** $j = i+1$ **to** n **do**	$O(n)$
12	Calculate $P_{(i,j)}, Q_{(i,j)}, Q_{(j,i)}$ based on (22) and (23);	$O(1)$
13	**end**	
14	**end**	
15	**for** $i = 1$ **to** $n-1$ **do**	$O(n)$
16	**for** $j = i+1$ **to** n **do**	$O(n)$
17	$g = (i-1)(n-i/2)+j-i$;	$O(1)$
18	$w^g = w_{(i,j)}$, (51);	$O(1)$
19	**for** $r = i$ **to** $n-1$ **do**	$O(n)$
20	**for** $m = j$ **to** n **do**	$O(n)$
21	$h = (r-1)(n-r/2)+m-r$;	$O(1)$
22	Calculate A^{gh}_p using $P_{(i,j)}, P_{(r,m)}$ based on (53);	$O(1)$
23	Calculate A^{gh}_m using $Q_{(i,j)}, Q_{(j,i)}, Q_{(r,m)}, Q_{(m,r)}$ based on (54);	$O(1)$
24	$A^{gh} = A^{gh}_p + A^{gh}_m$;	$O(1)$
25	**end**	
26	**end**	
27	**end**	
28	**end**	
29	Call BSP subroutine or MVEE subroutine;	$T(n)$
30	Return $\mathbf{d}_{k+1}, \Sigma^{\mathbf{d}}_{k+1}$;	
31	**end**	

At first glance, the computation complexity seems to be high. However, there are $O(n^2)$ pairs to start with. The dominating computation in MVEE is from the use of the IPM [56] to get w, which takes $O(n^6)$ time for the worst case in our IPM implementation, which apparently can be improved. However, the speed is not a concern because n is the number of vertical lines and usually no more than 20. The problem size is still small and our testing results have also confirmed that.

Algorithm 2: BSP Subroutine

```
1  input : d_k, u_i, i ∈ I, f, σ_u^2, Σ_k^d, A
2  output : d_{k+1}, Σ_{k+1}^d
3  begin
4      for i = 1 to n − 1 do                                          O(n)
5          for j = i + 1 to n do                                      O(n)
6              g = (i − 1)(n − i/2) + j − i;                          O(1)
7              Record the maximum A^{gg} and the
               corresponding i, j as i*, j*;                          O(1)
8          end
9      end
10     Calculate T based on (43) for (i*, j*);                        O(1)
11     Calculate Σ_{k+1}^p based on (47);                             O(1)
12     Calculate S_{i*}, S_{j*} based on (44) for (i*, j*);           O(1)
13     Calculate Σ_{k+1}^m based on (48) for (i*, j*);                O(1)
14     Calculate Σ_{k+1}^d based on (46);                             O(1)
15     Compute d_{k+1} based on (9) for (i*, j*);                     O(1)
16     Return d_{k+1}, Σ_{k+1}^d;
17 end
```

Algorithm 3: MVEE Subroutine

```
1  input : d_k, u_i, i ∈ I, f, σ_u^2, Σ_k^d, w, A
2  output : d_{k+1}, Σ_{k+1}^d
3  begin
4      Calculate w in (52) using IPM;                                 O(n^6)
5      for i = 1 to n − 1 do                                          O(n)
6          for j = i + 1 to n do                                      O(n)
7              g = (i − 1)(n − i/2) + j − i;                          O(1)
8              w_{(i,j)} = w^g, (51);                                 O(1)
9          end
10     end
11     Calculate T based on (43);                                     O(n^2)
12     Calculate Σ_{k+1}^p based on (47);                             O(1)
13     for i = 1 to n do                                              O(n)
14         Calculate S_i based on (44);                               O(n)
15     end
16     Calculate Σ_{k+1}^m based on (48);                             O(n)
17     Calculate Σ_{k+1}^d based on (46);                             O(1)
18     for i = 1 to n − 1 do                                          O(n)
19         for j = i + 1 to n do                                      O(n)
20             Calculate F(d_k, u_i, u_j) based on (9);               O(1)
21         end
22     end
23     Calculate d_{k+1} based on (38);                               O(n^2)
24     Return d_{k+1}, Σ_{k+1}^d;
25 end
```

3.2.7 Experiments

3.2.7.1 Experiment Setup

The algorithms are implemented on a Compaq V3000 laptop PC with an Intel 1.6 GHz dual core CPU and 1.0G RAM and programmed in the MatLab environment. We use a Sony DSC-F828 camera mounted on a robot in the experiment as shown in Figure 4(a). The camera's HFOV is set to 50° with a resolution of 640 × 480 pixels. The robot was custom-made in our lab. The robot measures 50 × 47 × 50 cm^3 in size. It has two front drive wheels and one rear cast wheel and uses a typical differential driving structure. The robot can travel at a maximum speed of 50 cm/s.

We define a relative error metric ε for comparison purposes. Let $d_k^{x^*}$ and $d_k^{z^*}$ be the true displacements (i.e., *ground truth*) of the robot in the x and z directions at step k, respectively, which are obtained using a tape measure in our experiments. Recall that the corresponding outputs of visual odometry are d_k^x and d_k^z. ε is defined as

$$\varepsilon = \sqrt{\varepsilon_x^2 + \varepsilon_z^2} \tag{56}$$

where ε_x and ε_z are relative errors in the x and z directions, respectively,

$$\varepsilon_x = \frac{\left|\sum_k d_k^x - \sum_k d_k^{x^*}\right|}{\sum_k \sqrt{\left(d_k^{x^*}\right)^2 + \left(d_k^{z^*}\right)^2}}, \varepsilon_z = \frac{\left|\sum_k d_k^z - \sum_k d_k^{z^*}\right|}{\sum_k \sqrt{\left(d_k^{x^*}\right)^2 + \left(d_k^{z^*}\right)^2}}.$$

This metric describes the ratio of the ego-motion estimation error in comparison to the overall distance traveled.

During the experiments, we employ Gioi et al.'s method to extract the line segments from the images [57]. The vertical lines are found using an inclination angle threshold [49] and vanishing points. Then, we employ the vanishing point method [47] for vertical and horizontal lines to construct homographies that project images into the iso-oriented ICSs with their u–v planes parallel to the vertical lines, which allows us to align the ICSs at frames $k-1$, k, and $k+1$ for step $k+1$. The correspondence between lines in the adjacent frames is found by directly matching pixels of the vertical stripes at the neighboring region of the vertical lines.

3.2.7.2 Validating the Minimum Solution and the BSP Algorithm

We first validate the minimum solution, and its sensitivity analysis results in the physical experiment. The experiment site is in front of a building on the Texas A&M University campus (see Figure 4(b)). We use eight pairs of vertical lines on the frontal plane of the building. In Figure 4(b), two lines with the same number belong to the same pair. The

(a) 

(b) 

Figure 4

(a) The camera and the robot used in the experiment. (b) Experiment site from the robot's view for the minimum solution. We use the vertical lines on the frontal plane of a building as highlighted in yellow. The vertical lines are numbered in pairs. (For interpretation of the color/colour in this figure legend, the reader is referred to the online version of this book.)

relative distance between the two lines in each pair is defined as δ. During the experiment, the camera has to face the building's frontal plane to obtain edge position readings. Hence, the z direction is the direction perpendicular to the frontal plane and the x direction is the direction parallel to the frontal plane.

During each trial, the robot moves along a straight line with 11 incremental steps and a step length of 0.5 m. The robot takes images at each of the 12 positions introduced by the 11-step movement. Recall that the robot displacement of the first step is given as a reference. For the subsequent 10 steps, we compute the robot ego-motion using Eqn (9) and compare it with the ground truth. Each trial is the average of the outcome of the 10 steps.

Combinations of four different experimental conditions are tested in different trials:

C1: Two different robot headings including x and z directions.
C2: Eight different relative distance settings δ between the vertical line pair.
C3: Eight different depths of vertical line pair z. The initial positions of the robot with respect to the building frontal plane are from eight different depth settings ranging from 35 to 70 m with 5-m intervals.
C4: Camera rotation versus no camera rotation. For cases without camera rotation, we adjust the camera pan and tilt in the experiment to force CCSs to be iso-oriented. For cases with camera rotation, we introduce CCSs with $\pm 10°$ orientational differences.

Therefore, we conducted a total of 256 trials in the experiments.

Figure 5 illustrates the experimental results. As shown in Figure 5(a), regardless of the robot's moving direction, the depth direction estimation error ε_z is always over two times as large as the lateral direction error ε_x, which confirms Theorem 1.

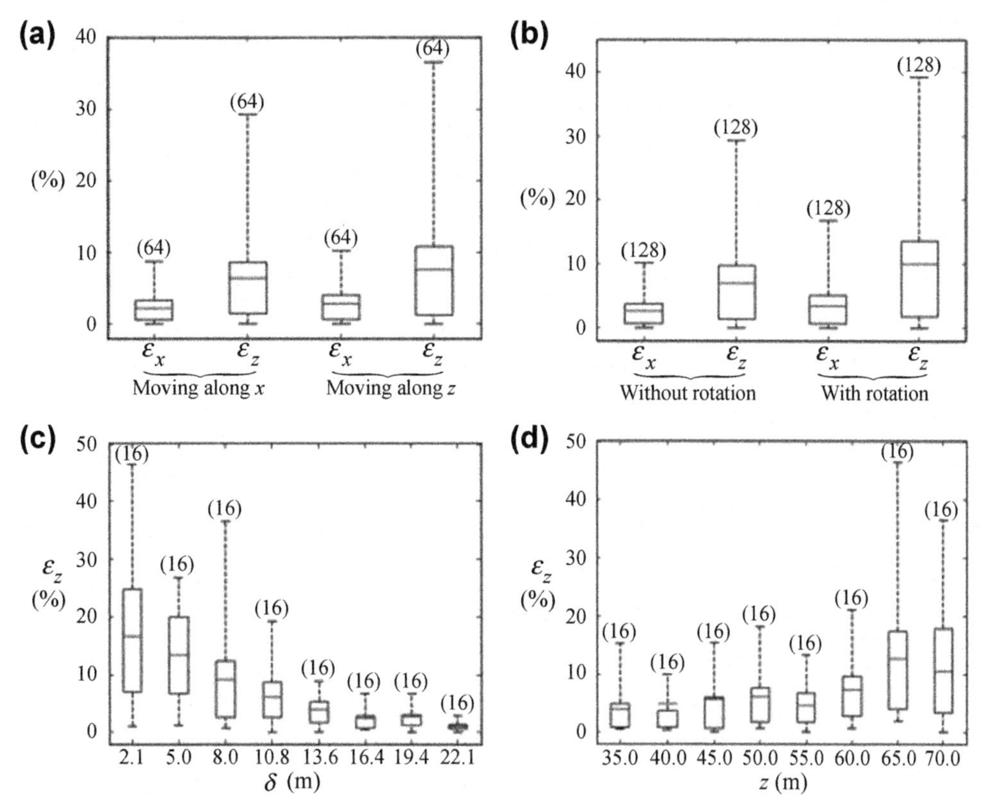

Figure 5

Statistical experimental results for the minimum solution. Note that the red lines are the mean values, the blue boxes represent the population range from the 25th to the 75th percentile, and the black dashed intervals indicate the data range. Numbers inside parentheses are the numbers of trials used to compute the statistics. (a) ε_z versus ε_x. (b) Camera rotation versus no camera rotation. (c) ε_z versus δ. (d) ε_z versus z. The number in the parenthesis is the number of trails used to compute the statistics.

Because the depth error ε_z is the dominating error, we compare only ε_z. Figure 5(c) illustrates how ε_z changes with respect to different δ settings. It is clear that as δ increases, ε_z decreases. Figure 5(d) illustrates how ε_z changes as z changes. There is a trend for ε_z to decrease as z decreases, although the trend is not clear when z is relatively small because factors other than z dominate the error. These results confirm Theorem 2.

Additionally, Figure 5(b) illustrates how camera rotation influences ε_x and ε_z. It is clear that there is no significant difference between the two cases for either ε_x or ε_z. The results show that assuming CCSs are iso-oriented in the analysis is reasonable.

3.2.7.3 Validating the MVEE Algorithm

3.2.7.3.1 A sample case in simulation

For MVEE validation, we first present a sample case to see how the weights are assigned by the MVEE algorithm for a typical vertical line distribution. We want to know which pairs are more important than others. We set up a scenario in which eight vertical lines are symmetrically located along each side of a road as illustrated in Figure 6(a). This is to simulate the urban case in which buildings are evenly distributed on both sides of a road.

Originally, the robot is located at ($x = 0$, $z = 0$). Then the robot moves two steps by traveling 1 m at a time in the z direction. The first step is given as an initial reference and the MVEE algorithm is executed at the end of the second step.

Figure 6(a) highlights the more heavily weighted pairs through a darker edge. The pairs with weights that are less than 1% of the maximum weight make very little contribution to the final ego-motion estimation and are not drawn. It is clear that the vertical line pair that is closest to the camera has the heaviest weight, which is expected according to our results in the minimum solution sensitivity analysis.

Figure 6(b) illustrates ordered weights by their values. It is clear that only a small part of the vertical line pairs (20%) make contributions with more than 1% of the maximum weighted pair. This result suggests that it is not necessary to track all edges if the computation power is limited. In fact, the best pair actually contributes over 70% to the final result, which indicates that BSP is a viable choice for cases in which computation power is extremely limited.

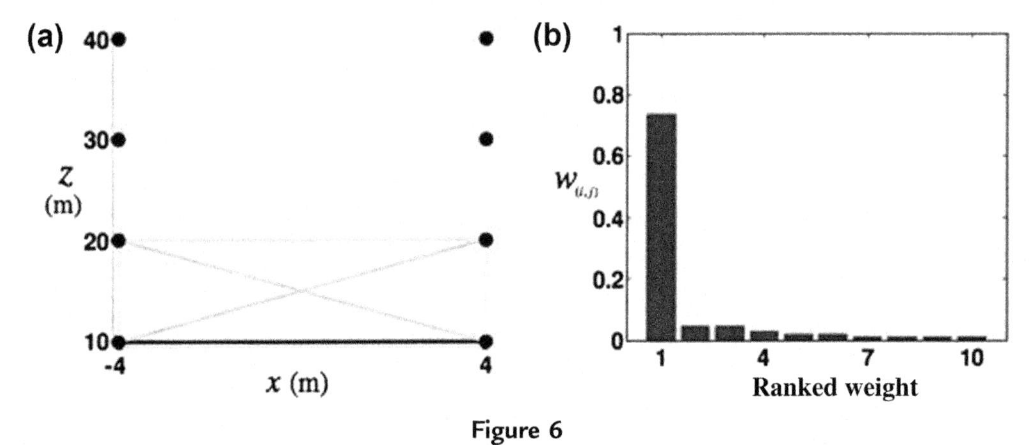

Figure 6
(a) A top view of vertical lines and corresponding weights for pairs for the sample case. The black dots are vertical lines. The resulting weights for each vertical line pair are presented as edges in gray scale. A darker edge means a heavier weight. (b) Weight distribution in decreasing order corresponds to the edges in (a).

3.2.7.3.2 A comparison of pair aggregation methods in physical experiments

The MVEE aggregates motion estimation results from multiple vertical line pairs using the variance minimization method. Here we compare MVEE with the results from BSP and a simple equal weighting for all (EWA) pairs. The experiment site is shown in Figure 7(a). In each trial, the robot moves 31 steps along a zigzagging poly line with a step length of 1 m for odd steps and 0.5 m for even steps (Figure 8(a)). We also repeat the experiment with three different camera resolutions: 640×480, 1280×960, and 2560×1920 pixels. With 10 trials for each resolution, we have a total of 30 trials.

The experimental results of the three different pair aggregation methods are shown in Figure 8. With a camera resolution of 640×480 pixels, Figure 8(a) presents the sample ego-motion estimation results in the form of robot trajectories for one trial. Because EWA is much worse than either BPS or MVEE, it has to be shown on a bigger scale in the small thumbnail at the lower left corner of the figure. The comparison between BPS and MVEE is shown both in the format of the robot trajectories in Figure 8(a) and in $\overline{\varepsilon}$ over the steps in Figure 8(b). Each $\overline{\varepsilon}$ in Figure 8(b) is an average of ε over the trials with all camera resolutions at the same step number. Unsurprisingly, MVEE consistently outperforms BPS at all resolution settings (Figure 8(c)).

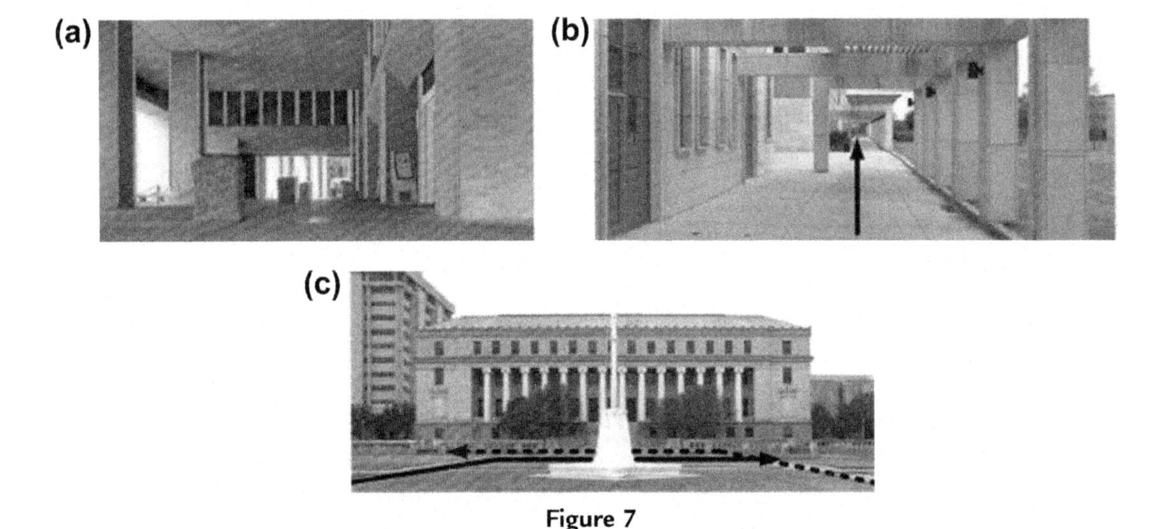

Figure 7
(a) Experiment site 1 from the robot's view with vertical edges highlighted in green. (b) and (c) Experiment sites 2 and 3 with robot trajectories highlighted in black. (For interpretation of the color in this figure legend, the reader is referred to the online version of this book.)

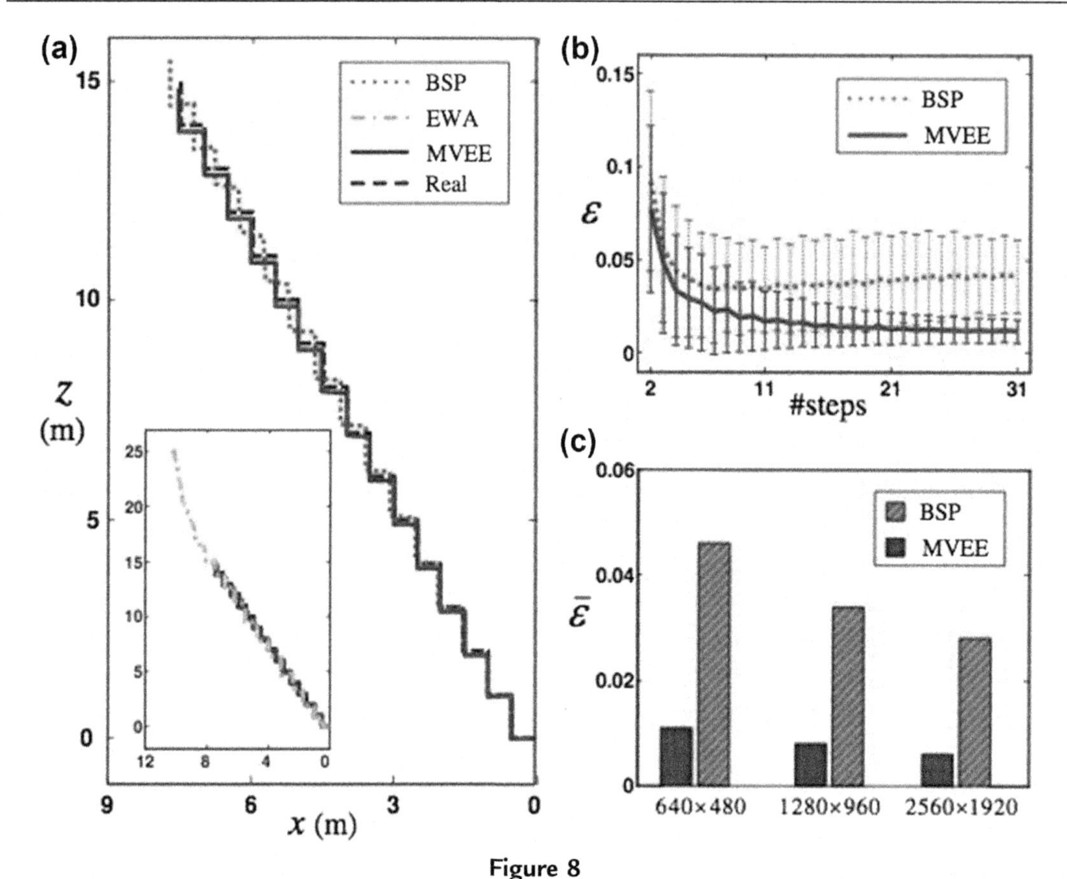

Figure 8
(a) A comparison of robot trajectories from BSP, MVEE, and EWA with ground truth (dashed black poly line). (b) Mean relative error $\bar{\varepsilon}$ and its standard deviation over #steps for both BSP and MVEE. (c) $\bar{\varepsilon}$ versus camera resolution.

3.2.7.4 Comparison of MVEE with Existing Point and Line-Based Odometry Methods

We compare MVEE with two popular ego-motion estimation methods in physical experiments:

- Nister [24]: This method is selected because it is a representative point-feature-based method. The method employs Harris corner points as landmarks. This method supports both monocular and stereo configurations. We use its monocular configuration in the experiments.
- L&L [9]: This method is selected because it is a representative line-feature-based method. The method is a monocular-vision-based SLAM method using general line segments as landmarks. We turn off the loop closing for visual odometry comparison purposes.

Both methods estimate 3D robot movements. Because our method is 2D, we compare only the odometry results on the $x{-}z$ ground plane.

We ran tests at three experiment sites (Figure 7) for all three methods. At each site, the robot moved along a planned trajectory for a certain number of steps. The details about each site are described below:

- Site 1: The same 31 steps performed in Figure 8(a) for the site in Figure 7(a).
- Site 2: The robot moves toward the depth direction for 51 steps with a step length of 1 m (Figure 7(b)).
- Site 3: The robot has two trajectories as indicated by the black solid and dashed lines, respectively. Each trajectory has 31 steps along the depth direction followed by 20 steps along the lateral direction with a step length of 1 m (Figure 7(c)).

We ran the robot for 10 trials at each site (for site 3, each trajectory took five trials), which leads to a total of 30 trials.

The experimental results of the three methods are shown in Figure 9. Figure 9(a) presents a representative sample trial of estimated trajectory comparisons at site 1. Figure 9(b) compares the mean values of ε for the three methods at each site. It is clear that MVEE outperforms its two counterparts in estimation accuracy.

Table 1 compares feature quality and computation speed for the three methods. Each row in Table 1 is the average of the 30 trials. It is obvious that the two line-feature-based methods, MVEE and L&L, outperform the point-feature-based Nister method, which conforms to our expectation. MVEE is slightly faster than L&L owing to its smaller input

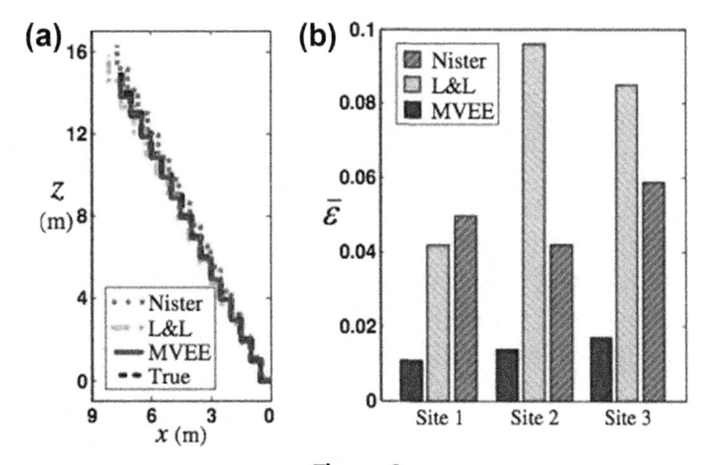

Figure 9

Physical experimental results. (a) A comparison of robot trajectories from the three methods with the ground truth (dashed black poly line). (b) A comparison of $\bar{\varepsilon}$ values for the three methods at each experiment site.

Table 1: Feature quality and computation speed comparison

Method	Feature			Speed	
	Total	Inliers	Ratio	Time	Factor
Nister	3425	245	7%	15.2 s	6.6x
L&L	122	41	34%	3.4 s	1.5x
MVEE	59	25	42%	2.3 s	1.0x

sets because vertical lines are a subset of general lines. Note that all implementations are in MatLab and the speed should be much faster if converted to C++ but the factors should remain the same.

For feature quality, it is clear that the Nister method employs many more features than MVEE and L&L, whereas its inliers/total features ratio is the lowest. In contrast, MVEE has the smallest number of features with the highest inlier ratio. This indicates that MVEE is more robust than the other two methods. Overall, MVEE outperforms the other two methods in robustness, accuracy, and speed.

3.2.8 Conclusion and Future Work

We have discussed our development of an incremental error-aware monocular visual odometry method that utilizes vertical edges of buildings in urban areas. We derived how to estimate the robot ego-motion using vertical line pairs. To improve the accuracy, we analyzed how errors are introduced and propagated in the continuous odometry process by deriving the recursive and closed form representation of the error covariance matrix. We formulated the minimum variance ego-motion estimation problem and presented two algorithms. The resulting visual odometry methods were extensively tested in physical experiments. The proposed odometry method was compared with two popular existing methods and consistently outperformed the two counterparts in speed, robustness, and accuracy.

In the future, we will extend this approach by exploring various combinations of geometric features such as vertical planes, horizontal lines, and points with geometric meanings (e.g., intersections between lines and planes) in visual odometry. We will also look into methods using texture features in combination with geometric features.

Acknowledgments

We acknowledge the insightful thoughts from Y. Xu, A. Perera, and S. Oh in Kitware. We also thank W. Li, Y. Lu, M. Hielsberg, J. Lee, Z. Gui, M. Hirami, S. Mun, S. Jacob, and P. Peelen for their input and contributions to the Networked Robots Lab at Texas A&M University.

References

[1] D. Scaramuzza, R. Siegwart, Appearance-guided monocular omni-directional visual odometry for outdoor ground vehicles, IEEE T. Robot. 24 (5) (October 2008) 1015–1026.

[2] O. Amidi, T. Kanade, J. Miller, Vision-based autonomous helicopter research at carnegie mellon robotics institute 1991-1997, in: American Helicopter Society International Conference, Heli, Japan, April 1998.

[3] R. Marks, H. Wang, M. Lee, S. Rock, Automatic visual station keeping of an underwater robot, in: Oceans Engineering for Today's Technology and Tomorrow's Preservation, vol. 2, Brest, France, September 1994, pp. 137–142.

[4] R. Ozawa, Y. Takaoka, Y. Kida, et al., Using visual odometry to create 3D maps for online footstep planning, in: 2005 IEEE International Conference on Systems, Man and Cybernetics, vol. 3, October 2005, pp. 2643–2648.

[5] P. Corke, D. Strelow, S. Singh, Omnidirectional visual odometry for a planetary rover, in: 2004 IEEE/RSJ International Conference on Intelligent Robots and Systems, Sendai, Japan, September 2004, pp. 4007–4012.

[6] M. Agrawal, K. Konolige, Rough terrain visual odometry, in: International Conference on Advanced Robotics, August 2007.

[7] S. Thrun, W. Burgard, D. Fox, Probabilistic Robotics, The MIT Press, Cambridge, Massachusetts, 2005.

[8] A. Davison, L. Reid, N. Molton, O. Stasse, Monoslam: real-time single camera slam, IEEE T. Pattern Anal. 29 (6) (June 2007) 1052–1067.

[9] T. Lemaire, S. Lacroix, Monocular-vision based slam using line segments, in: IEEE International Conference on Robotics and Automation (ICRA), Roma, Italy, May 2007.

[10] B. Steder, G. Grisetti, C. Stachniss, Visual slam for flying vehicles, IEEE T. Robot. 24 (5) (2008) 1088–1093.

[11] T. Marks, Gamma-slam: visual slam in unstructured environments using variane grid maps, J. Field Robot. 26 (1) (2009) 26–51.

[12] K. Konolige, M. Agrawal, Frameslam: from bondle adjustment to real-time visual mapping, IEEE T. Robot. 24 (5) (2008) 1066–1077.

[13] J. Civera, A. Davison, J. Montiel, Inverse depth parametrization for monocular slam, IEEE T. Robot. 24 (5) (2008) 932–945.

[14] J. Civera, D. Bueno, A. Davison, J. Montiel, Camera self-calibraction for sequential bayesian structure form motion, in: IEEE International Conference on Robotics and Automation (ICRA), Kobe, Japan, May 2009.

[15] R. Sim, P. Elinas, M. Griffin, J. Little, Vision-based SLAM using the rao-blackwellised particle filter, in: International Joint Conference on Artificial Intelligence, Edinburgh, Scotland, July 2005.

[16] G. Klein, D. Murray, Parallel tracking amd mapping for small ar workspaces, in: International Symposium on Mixed and Augmented Reality (ISMAR), Nara, Japan, November 2007.

[17] T. Lemaire, S. Lacroix, Slam with panoramic vision, J. Field Robot. 24 (1/2) (2007) 91–111.

[18] D. Scaramuzza, R. Siegwart, Monocular omnidirectional visual odometry for outdoor ground vehicles, in: Computer Vision Systems, vol. 5008, Springer-Berlin, 2008, pp. 5206–5215.

[19] M. Wongphati, N. Niparnan, A. Sudsang, Bearing only fastslam using vertical line information from an omnidirectional camera, in: IEEE International Conference on Robotics and Biomimetics, Bangkok, Thailand, February 2009.

[20] D. Scaramuzza, R. Seigwart, A robust descriptor for tracking vertical lines in omnidirectional images and its use in mobile robotics, Int. J. Robot. Res. 28 (2) (2009) 149–171.

[21] G. Caron, E. Mouaddib, Vertical line matching for omnidirectional stereovision images, in: IEEE International Conference on Robotics and Automation (ICRA), Kobe, Japan, May 2009.

[22] J. Sola, A. Monin, M. Devy, T. Vidal-Calleja, Fusing monocular information in multicamera slam, IEEE T. Robot. 24 (5) (2008) 958–968.

[23] L. Paz, P. Pinies, J. Tardos, Large-scale 6-dof slam with stereo-in-hand, IEEE T. Robot. 24 (5) (2008) 946–957.

[24] D. Nister, O. Naroditsky, J. Bergen, Visual odometry for ground vechicle applications, J. Field Robot. 23 (1) (January 2006) 3–20.

[25] D. Nister, O. Naroditsky, J. Bergen, Visual odometry, in: IEEE Computer Society Conference on Computer Vision and Pattern Recognition, vol. 1, June 2004, pp. 652–659.
[26] M. Maimone, Y. Cheng, L. Matthies, Two years of visual odometry on the mars exploration rovers, J. Field Robot. 24 (2) (March 2007) 169–186.
[27] Y. Cheng, M. Maimone, L. Matthies, Visual odometry on the mars exploration rovers, in: 2005 IEEE International Conference on Systems, Man and Cybernetics, vol. 1, October 2005, pp. 903–910.
[28] D. Helmick, Y. Cheng, D. Clouse, Path following using visual odometry for a mars rover in high-slip environments, in: 2004 IEEE Aerospace Conference, vol. 2, March 2004, pp. 772–789.
[29] D. Cobzas, H. Zhang, M. Jagersand, Image-based localization with depth-enhanced image map, in: IEEE International Conference on Robotics and Automation (ICRA), September 2003. Taipei, Taiwan.
[30] W. Zhou, J. Miro, G. Dissanayake, Information-efficient 3-d visual slam for unstructured demains, IEEE T. Robot. 24 (5) (2008) 1078–1087.
[31] S. Li, T. Kanbara, A. Hayashi, Making a local map of indoor environments by swiveling a camera and a sonar, in: International Conference on Intelligent Robots and Systems (IROS), Kyongju, Korea, October 1999.
[32] D. Lowe, Distinctive image features from scale-invariant keypoints, Int. J. Comput. Vision 60 (4) (November 2004) 91–110.
[33] H. Bay, A. Ess, T. Tuytelaars, L. Gool, Surf: speeded up robust features, Comput. Vis. Image Und. (CVIU) 110 (3) (2008) 346–359.
[34] M. Agrawal, K. Konolige, M. Blas, Sensure: center surround extremas for realtime feature detection and matching, in: The 10th European Conference on Computer Vision, Marseille, France, October 2008.
[35] M.A. Fischler, R.C. Bolles, Random sample consensus: a paradigm for model fitting with applications to image analysis and automated cartography, Commun. ACM 24 (6) (June 1981) 381–395.
[36] R. Hartley, A. Zisserman, Multiple View Geometry in Computer Vision, second ed., Cambridge University Press, 2004.
[37] D. Ziou, S. Tabbone, Edge detection techniques—an overview, Int. J. Pattern Recogn. Image Anal. 8 (4) (1998) 537–559.
[38] P. Smith, I. Reid, A. Davison, Real-time monocular slam with straight lines, in: The 17th British Machine Vision Conference, Edinburgh, UK, September 2006.
[39] Y. Choi, T. Lee, S. Oh, A line feature based slam with low grad range sensors using geometric constrains and active exploration for mobile robot, Auton. Robot 24 (2008) 13–27.
[40] M. Dailey, M. Parnichkun, Landmark-based simultaneous localization and mapping with stereo vision, in: The 4th Asian Conference on Industrial Automation and Robotics, Bangkok, Thailand, May 2005.
[41] A. Gee, W. Mayol-Cuevas, Real-time model-based SLAM using line segments, Adv. Visual Comput. 4292 (2006) 354–363.
[42] A. Martignoni, W. Smart, Localizing while mapping: a segment approach, in: The AAAI Conference on Artificial Intelligence, Edmonton, Alberta, Canada, July 2002.
[43] C. Taylor, D. Kriegman, Structure and motion from line segments in multiple images, T. Pattern Anal. 17 (11) (1995) 1021–1032.
[44] J. Montiel, J. Tardoh, L. Montano, Structure and motion from straight line segments, Pattern Recogn. 32 (2000) 1295–1307.
[45] M. Kim, S. Lee, K. Lee, Self-localization of mobile robot with single camera in corridor environment, in: IEEE International Symposium on Industrial Electronics, Pusan, Korea, June 2001.
[46] Y. Sakamoto, M. Aoki, Street model with multiple movable panels for pedestrian environment analysis, in: IEEE Intelligent Vehicles Symposium, Parma, Italy, June 2004.
[47] A. Gallagher, Using vanishing points to correct camera rotation in images, in: The 2nd Canadian Conference on Computer and Robot Vision, Victoria, BC, Canada, May 2005.
[48] J. Guerrero, R. Martinez-Cantin, C. Sagues, Visual map-less navigation bsed on homographies, J. Field Robot. 22 (10) (2005) 569–581.
[49] J. Zhou, B. Li, Exploiting vertical lines in vision-based navigation for mobile robot platforms, in: IEEE International Conference on Acoustics, Speech and Signal Processing (ICASSP), April 2007, pp. I-465–I-468. Honolulu, Hawaii, USA.

[50] D. Song, H. Lee, J. Yi, A. Levandowski, Vision-based motion planning for an autonomous motorcycle on ill-structured roads, Auton. Robot. 23 (3) (October 2007) 197–212.
[51] D. Song, H. Lee, J. Yi, On the analysis of the depth error on the road plane for monocular vision-based robot navigation, in: International Workshop on Algorithmic Foundations of Robotics (WAFR), Guanajuato, Mexico, December 2008.
[52] J. Yi, H. Wang, J. Zhang, D. Song, S. Jayasuriya, J. Liu, Modeling and analysis of skid-steered mobile robots with applications to low-cost inertial measurement unit-based motion estimation, IEEE T. Robot. 25 (5) (October 2009) 1087–1097.
[53] J. Zhang, D. Song, On the error analysis of vertical line pair-based monocular visual odometry in urban area, in: International Conference on Intelligent Robots and Systems (IROS), St. Louis, MO, USA, October 2009.
[54] J. Zhang, D. Song, Error aware monocular visual odometry using vertical line pairs for small robots in urban areas, in: Special Track on Physically Grounded AI (PGAI), the Twenty-fourth AAAI Conference on Artifical Intelligence (AAAI-10), Atlanta, Georgia, USA, July 2010.
[55] J. Nocedal, S.J. Wright, Numerical Optimization, second ed., Springer-Verlag, Berlin, New York, 2006.
[56] S. Boyd, L. Vandenberghe, Convex Optimization, Cambridge University Press, 2006.
[57] R. Gioi, J. Jakubowicz, J. Morel, G. Randall, Lsd: a line segment detector, IEEE T. Pattern Anal. 32 (4) (December 2008).

CHAPTER 3.3

Planning and Obstacle Avoidance in Mobile Robotics[1]

Antonio Sgorbissa, Renato Zaccaria
DIST University of Genova, Genova, Italy

Chapter Outline

This chapter focuses on the navigation subsystem of a mobile robot which operates in human environments to carry out different tasks, such as transporting waste in hospitals or escorting people in exhibitions. The chapter describes a hybrid approach (Roaming Trails), which integrates a priori knowledge of the environment with local perceptions in order to carry out the assigned tasks efficiently and safely: that is, by guaranteeing that the robot can never be trapped in deadlocks even when operating within a partially unknown dynamic environment. This chapter includes a discussion about the properties of the approach, as well as experimental results recorded during real-world experiments.

3.3.1 Introduction

In the last twenty years, scientists have foreseen a close future in which Service Mobile Robots will be able to operate within human populated environments to carry out different tasks, such as surveillance [1,2], transportation of heavy objects [3–5], or escorting people

[1] Reprinted from Robotics and Autonomous Systems, vol. 60, Antonio Sgorbissa and Renato Zaccaria, Planning and obstacle avoidance in mobile robotics, pp. 628–638, 2013, with permission from Elsevier.

Household Service Robotics. http://dx.doi.org/10.1016/B978-0-12-800881-2.00009-8

in exhibitions and museums [6]. Unfortunately, and in spite of the huge number of studies on this topic, very few existing systems are able to work continuously over a long period of time without performance degradation, i.e., the "six months between missteps" that Moravec advocated in 1999 [7]. It is commonly accepted that one of the causes of failure is that human populated environments cannot be completely known a priori since they are highly dynamic: this has a dramatic effect both on self-localization (i.e., it makes it difficult to recognize and match features against environmental models) and on navigation (i.e., it is possible that a planned path is no more available or temporarily obstructed).

From a philosophical perspective, navigation in a dynamic environment has a straightforward solution: even if the path has been planned on the basis of the available a priori knowledge of the environment, the robot must also be free to make decisions in real-time on the basis of its current perceptions. However, how much should the robot be free to deviate from the planned path, is a question for which the answer is neither simple nor unique: in a metaphorical sense, it resembles the never-ending theological dispute between Free Will and Predestination.

From a technical perspective, Artificial Potential Fields (APFs) and similar force field-based models [8–10] have often represented a good solution to achieve a fast and reactive response to a dynamically changing environment; however, it has been widely demonstrated that they suffer from unavoidable drawbacks [11]. In particular, since the law of motion of the robot is basically determined by descending the gradient of the potential field generated by the goal and the obstacles, it is very likely for the robot to get trapped into a local minimum. The problem can be solved by invoking a planner to search for alternative paths; however, in the presence of moving obstacles and sensor noise, significant deviations from the original path can lead to a deadlock configuration from which it is even harder to escape. This is further complicated by the fact that the robot is subject to geometric and kinematic constraints which puts severe limitations on its motion capabilities. As an extreme example, consider that the approach used by commercial Automated Guided Vehicles (AGVs) to avoid obstacles in real-world situations (e.g., in a corridor crowded with people) is often not to avoid them at all[2]: experience tells that, if the vehicle slows down or stops, warns people, and waits, it has more probability of success than trying to find a path to the goal by other means.

By recalling the dialectic between Free Will and Predestination, this chapter metaphorically proposes a solution (Roaming Trails) which is well exemplified by Roger Hestons epitaph in Edgar Lee Masters' Spoon River: "OH many times did Ernest Hyde and I/Argue about the freedom of the will./My favorite metaphor was Pricketts cow/Roped out to grass, and free you know as far/As the length of the rope". Specifically, the chapter describes a novel two-layered

[2] See the description of a successful case study in www.swisslog.com/hcs-agv-memorialhermann.pdf.

approach to mobile robot navigation that integrates two navigation methods: an off-line method, which generates a reference path on the basis of an a priori map, and an online method, which adapts the generated path to avoid static and moving obstacles, but only to a given extent (i.e., "as the length of the rope allows"). With respect to solutions proposed in the literature [12–20], the major contribution of the approach consists in the formalism adopted to represent a path: instead of a sequence of way points, Motor Schemata, or Elastic Strips, a path is represented as a Roaming Trail, i.e., a chain of diamond-shaped or elliptic areas whose boundaries define the maximum allowed distance for the robot to deviate from its path when generating the online trajectory. Roaming Trails, as shown in the following, are able by construction to prevent the robot from being trapped into deadlocks caused by static obstacles (i.e., obstacles depicted in the a priori map), as well as by moving and removable obstacles (i.e., obstacles not depicted in the a priori map). Smooth motion in presence of geometric and kinematic constraints is guaranteed by the control law adopted for navigation.

Section 3.3.2 describes Related work. Section 3.3.3 describes in detail a prototypical two-level multi-agent navigation architecture, by focusing on the details of the agents involved. Section 3.3.4 describes how the same architecture can be modified to implement the Roaming Trail approach. Section 3.3.5 describes experimental results. Conclusions follow.

3.3.2 Related Work

Some of the most successful approaches to obstacle avoidance [8,12] rely on the idea of computing artificial repulsive forces which are exerted on the robot by surrounding obstacles, and an attractive force exerted by the goal. All of these forces are then summed up to produce a resulting force vector that is used to control the motion of the robot. APF and similar approaches have well known problems [11], among which the presence of local minima in the field (e.g., in correspondence of narrow passages). In addition, if occupancy grids [21] are used to store sensor data, APF can produce oscillations in the resulting velocity vector (and hence on the robot path) due to the finite resolution of the grid. The problem is well described in [9], which argues that the latter two problems are both due to the fact that a huge amount of sensor data are summarized in just one force vector, and proposes the Vector Field Histogram (VFH) as a solution. In the VFH, the concept of "resulting force" is substituted with the concepts of "valleys" toward which the robot is allowed to navigate. VFH has been applied both in indoor and outdoor robotic applications, and improvements are described in [22].

In spite of the many important differences, the approaches above share three drawbacks:

1. they are local methods, and therefore not able to deal with complex obstacle configurations encountered by the robot during navigation, e.g., narrow passages, cluttered areas, or a cul-de-sac in which global planning capabilities or recovery strategies would be required;

2. they face obstacle avoidance assuming an ideal point-like robot, therefore requiring to transform the velocity vector into commands to actuators in a separate phase: this is not a trivial task whenever it is desirable to produce a smooth navigation trajectory in the presence of geometric and kinematic constraints;
3. they do not put any upper bound on the maximum allowed deviation from the path while avoiding obstacles, thus possibly producing an undesirable behavior in crowded environments, where it would be more efficient to simply stop, warn people, and wait.

To overcome problem 1, a class of approaches try to combine force field methods with traditional planning algorithms, by asking a symbolic planner to dynamically generate a sequence of local force fields which are free of minima (e.g., Motor Schemata [12] or Navigation Templates [13]), thus being able to guide the robot in different situations and areas of the environment while avoiding obstacles in real-time. Research in the field of the so-called "hybrid cognitive architectures", where the term "hybrid" refers to the concurrency of deliberative planning and reactive behaviors, has been very active up to one decade ago, producing notable examples such as AuRA, 3T, ATLANTIS [15,23]. Solutions based on the VHF method have been proposed as well. In [18] the enhanced method VFH* is presented, which verifies that a particular candidate direction guides the robot around an obstacle, by using the A* search algorithm and appropriate cost and heuristic functions. In [24] obstacle avoidance for an electric wheelchair is semi-automatically supported by the minimum vector field histogram (MVFH) method, whereas global planning is guaranteed by the manual intervention of the user. Obstacle avoidance in very dense, complex and cluttered scenarios has been considered in [25], and integrated with planning in [26]. The idea is that there are situations where it is more suitable to direct the motion toward a given zone of the space (that ameliorates the situation to reach the goal latter), rather than directly toward the goal itself. The algorithm finds sub-goals based on the obstacle structure, and then associate a motion restriction to each obstacle to compute the most promising motion direction. The major limitation of the class of approaches above is that they require performing planning in run-time, depending on the configuration of obstacles encountered during navigation. Moreover, they usually do not deal with problems 2 and 3.

Problem two is explicitly considered in another class of approaches. In [16] the Curvature–Velocity Method (CVM) is presented, that formulates obstacle avoidance as a problem of constrained optimization in velocity space. Physical limitations (velocities and accelerations) and the configuration of obstacles place constraints on the translational and rotational velocities of the robot. The robot chooses velocity commands that satisfy all the constraints and maximize an objective function that trades off speed, safety and goal-directedness. A variant of the approach is proposed in [27]. In a similar spirit [17], presents the Dynamic Window approach (DW), which relies on the idea of performing a local search for admissible velocities which allow the robot to avoid obstacles while meeting kinematic constraints: in order to reduce computational complexity, the search is performed within a dynamic window which is centered around the current velocities of the robot in the velocity space, and only circular curvatures are considered. A theoretical treatment of convergence properties of the algorithm is proposed in [28]. Recently [29], has proposed the Forbidden Velocity Map, a generalization of the Dynamic Window concept that considers obstacle and robot shape, velocity and dynamics, to deal with navigation in unpredictable and cluttered scenarios. CVM and DW are able to produce smooth trajectories while dealing with kinematic constraints; however, they still have the problem of local minima (a solution is proposed in [30] by introducing a planning stage in DW). Obstacle avoidance is solved as a non-linear feedback control problem in [20]: path following is achieved by controlling explicitly the rate of progression of a "virtual target" to be tracked along the path [31,32], and obstacle avoidance relies on the deformable virtual zone principle, that defines a safety zone around the vehicle, in which the presence of an obstacle drives the vehicle's reaction. However, as stated by various authors, the combination of path following with a reactive obstacle avoidance strategy has a natural limitation coming from the situation where both controllers yield antagonist system reactions. This situation leads to a local minimum, where a heuristic switch between controllers is necessary. Again, problem three is not considered.

Problems 1–3, are faced in an integrated fashion in [14,19], where a path is planned off-line, and then the whole path is ideally deformed in real-time on the basis of forces exerted by surrounding obstacles. In particular, the initial path is augmented by a set of paths homotopic to it, represented implicitly by a volume of free space in the work space which describes the maximum allowed deviation from the path. During execution, reactive control algorithms are used to select a valid path from the set of homotopic paths, using proximity to the environment in a sense very similar to [8]. The concept of path deformation is considered also in [33], in which the current path is described as a mapping from an interval of real numbers into the configuration space of the robot, and iteratively deformed in order to get away from obstacles and satisfy the nonholonomic constraints of the robot. The approach has been shown to work with complex nonholomic systems (e.g., a trailer) with complex shapes. Approaches based on path deformation avoid local minima since the path connection is preserved during deformation, and they put boundaries on the maximum allowed deviation from the original path. However, they have higher computational and memory requirements than the approach proposed in this chapter, because they require the robot either to memorize a set of alternative paths, or to compute the path deformation in run-time.

3.3.3 Navigation Architecture

A prototypical two-level navigation architecture is introduced: many different choices are possible in the design process and, consequently, many different models have been presented in literature. However, it seems that the following considerations are valid for most existing approaches. The described architecture has been implemented on real robots whose behavior has been evaluated through experiments, thus providing a starting point to incrementally implement the Roaming Trail approach above it. In the following, we ignore all the problems related to position uncertainty, by assuming that a sufficiently accurate self-localization subsystem is available, such as the laser-based localization system described in [34].

As usual, both a deliberative and a reactive component are present in the architecture, each component being implemented as a multi-agent system exploiting the ETHNOS real-time programming environment [35]: only agents related to navigation are shown in Figure 1. Agents (rounded squares) communicate by posting messages (rectangles) on a distributed message board or by accessing shared representations (shaded rectangles). Reactive agents (labeled with an "R") are responsible for trajectory generation and other simple reactive behaviors. Deliberative agents (labeled with a "D") are responsible of generating sequences of actions in order to accomplish a given mission and monitor the state of advancement of the plan.

3.3.3.1 Deliberative Agents

Spatial Reasoner has the purpose of building a graph-like representation of the environment referred to as the roadmap $\mathfrak{R}$: the nodes of $\mathfrak{R}$ represent significant locations in the

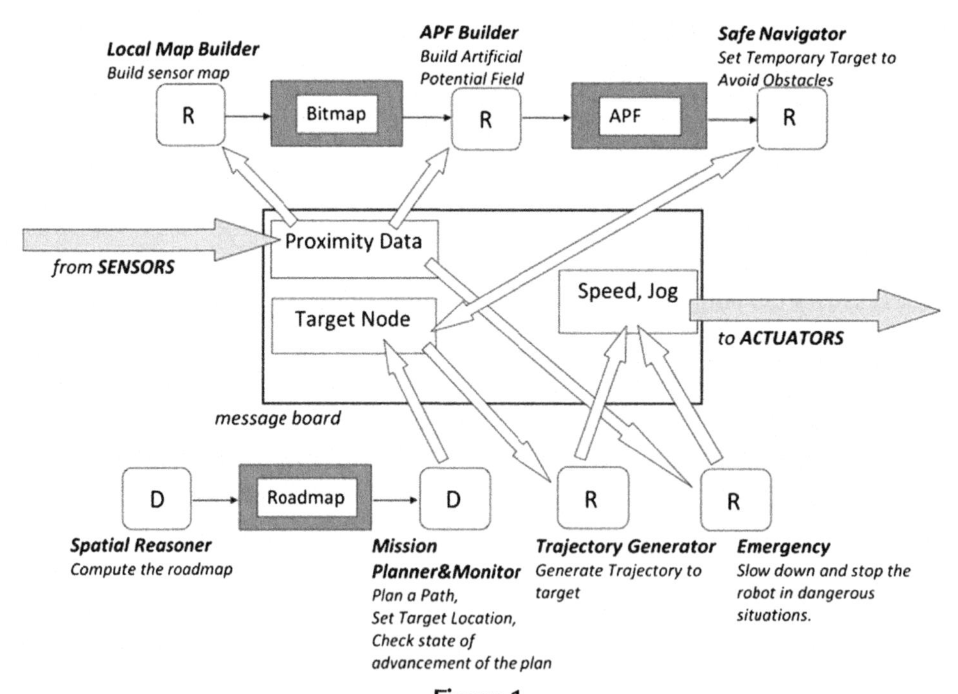

Figure 1
Multi-agent navigation system.

environment (i.e., the center of a room, a T-junction, a door, or the middle of a corridor), whereas links store information about the connectivity of these places. The roadmap is built starting from an a priori geometrical map $\mathcal{M} = \{s_i\}, i = 1...S$, which represents the workspace, where s_i is an oriented segment defined by an ordered pair of points. Ordered segments allow for defining not only the boundary between obstacles and the free space, but also the subspace which must be interpreted as free: for instance, if a segment goes from A to B, one can assume that the free space lies on the right half plane. Obstacles are represented through closed polygons, and the free space is assumed to be finite, i.e., it exists in a closed polygon made with segments of $\mathcal{M}$ such that no free space exists outside the polygon.

The approach adopted to extract information about the topology of the free space relies on a popular method based on the Voronoi diagram [36], described in Algorithm 1 (see, among the others, an alternative approach proposed in [37]).

Algorithm 1 is self-explaining. It is worth spending a few additional words about step 1, which requires solving a typical problem of computational geometry, i.e., checking if a cell $m_i \in G$ lies inside or outside a polygon defining the boundary between obstacles and the free space. Since all polygons are closed and made of oriented segments, it can be solved by ideally tracing an arbitrary line passing through the center of m_i, and by finding the closest segment s_j which intersects the line (if the line does not intersect any segment, m_i does not lie in the free space, since the latter is assumed to be finite). Then, it is possible to classify m_i as free space or not free space depending on where the center of m_i is located with respect to s_j.

Algorithm 1 Compute Roadmap $\Re$
Require: M Ensure: G , $\Re$

(1) The work space is sampled with arbitrary resolution, producing a grid $G = \{m_i\}, i = 1...M$, where each cell m_i is either labeled as free space or not free space, using the information in M;
(2) each cell of the grid m_i labeled as free space produces a Voronoi node if and only if its center is (approximately) equidistant from the three closest segments $s_j, s_k, s_l \in M$;
(3) the resulting nodes $n_i, i = 1...N$, are added to the roadmap $\Re$ and labeled with the position of the corresponding cell m_j;
(4) for every couple of nodes $n_i, n_j, i \neq j$, an undirected link is established in $\Re$ if and only if it is possible to draw a straight line which connects their corresponding positions without intersecting any segment in M.

Mission Planner & Monitor is a goal-based agent responsible of plan selection and adaptation, allowing the robot to plan and execute high level tasks such as “go to office A”, “call elevator” or “transport waste”. These tasks, depending on the current robot context, may be decomposed into many, possibly concurrent, subtasks such as "localize, go to the door, open door" and finally into primitive actions such as "go to configuration(x_i,y_i,θ_i). In the following, we focus on navigation tasks: when Mission Planner & Monitor receives a mission (i.e., a location in the environment to be reached) it executes Algorithm 2.

Algorithm 2 Compute Path$P$ Require: M , $\Re$
Ensure: P

(1) A couple of nodes n_s and n_g, corresponding to the start and goal positions, are added to $\Re$;
(2) for every node $n_i \in \Re$, an undirected link is established between n_i and n_s – respectively, n_g – if and only if it is possible to draw a straight line which connects the positions of n_i and n_s – respectively, n_g –without intersecting any segment in M ;
(3) the resulting roadmap is searched for the shortest path between n_s and n_g using the A* algorithm, thus producing a sequence of nodes $P = (n_1 \ldots n_t)$ in $\Re$ to be visited in sequence, where $n_1 = n_s$ is the first node (corresponding to the robot current position) and $n_t = n_g$ is the final destination.

At any time, Mission Planner & Monitor knows the location of the next node n_i to be reached; it posts this information to the message board for Trajectory Generator and the other reactive agents which are responsible for the robot motion, and periodically checks the state of advancement of the plan.

3.3.3.2 Reactive Agents

From the point of view of reactive agents, the problem to be solved is point-to-point navigation in an unknown environment. In fact the a priori map $\mathcal{M}$ is used exclusively to build the roadmap $\Re$ and to consequently choose the nodes $\mathcal{N}$ to be reached in sequence: since we want the robot to deal with a dynamic environment where unpredictable things can happen, the robot must be able to find a way to the current target node n_i by relying on its local perceptions. The agents involved in the process are the following:

Trajectory Generator retrieves from the message board the last posted node n_i, i.e., the next location to be visited in the environment. Then, it is capable of generating and executing smooth trajectories from the robot's current configuration $q = (x_r, y_r, \theta_r)$ to a target configuration (x_i, y_i, θ_i) relying on a closed-loop control function, where the target orientation θ_i can be manually specified or automatically computed, e.g., to minimize the path's curvature. In the current implementation, the control function is a biologically inspired motion generator called ξ-model [38]. As already stated, the smoothness of the trajectory is a fundamental characteristic, whenever we ask robots to operate in the real world: mobile platforms for real world applications (such as transportation of heavy loads) have severe geometric and kinematic constraints, to reduce both the stress on their mechanical parts and errors in dead reckoning (e.g., nonholomic or quasiholonomic geometries are preferred to holomic ones). To deal with these constraints, Trajectory Generator is able to generate and execute a smooth trajectory in closed-loop, by producing a sequence of speed and jog values (i.e., linear and angular velocity) which are transformed into velocity commands to be issued to the rear motors of a differentially driven vehicle. However, Trajectory Generator cannot deal with obstacles in the robot path.

Local Map Builder, APF Builder and Safe Navigator are the agents responsible of obstacle avoidance. In Figure 1 two shared representations (indicated as shaded rectangles) can be observed: the bitmap, an ecocentrical statistical dynamic description of the environment which is periodically updated through sensor readings [21,39], and the APF, based on the bitmap and on direct sensor information [10]. Map Builder and APF Builder update these representations to maintain consistency with the real world on the basis of sensor data. Safe Navigator periodically executes a virtual navigation in the APF (continuously aligned with the real world), thus determining a segment of safe trajectory which allows the robot to successfully avoid obstacles while heading to the goal. To achieve this, Safe Navigator reads the target node n_i produced by Mission Planner & Monitor and uses it to compute a new target node n_i' as the endpoint of the generated safe trajectory: n_i' is then posted to the message board and becomes available to Trajectory Generator for smooth obstacle avoidance. Figure 2 should help to clarify this concept: when in position q', Mission Planner & Monitor sets n_1 as the current node; Safe Navigator performs a virtual navigation in the APF and moves the current target to n_i'; Trajectory Generator computes a smooth trajectory to n_i' and the robot starts moving toward it. However, n_i' is never reached by the robot, since at the next time step (when the robot is in q'') Safe Navigator computes a new target n_i'' which is fed to Trajectory Generator: this procedure is iterated until the real target node n_1 is eventually reached. The approach guarantees, at the same time, smoothness (as a consequence of the control law implemented by Trajectory Generator) and safety (since Safe Navigator reactively compute a free path in the APF), and share some similarities with other virtual target approaches [20,31,32].

Emergency Handler intervenes whenever an unforeseen collision is imminent. Since the actual trajectory followed by the robot differs from the virtual trajectories periodically generated in the APF, it is possible for the robot to collide with some obstacles in the environment. Emergency Handler prevents this from happening: on the basis of raw

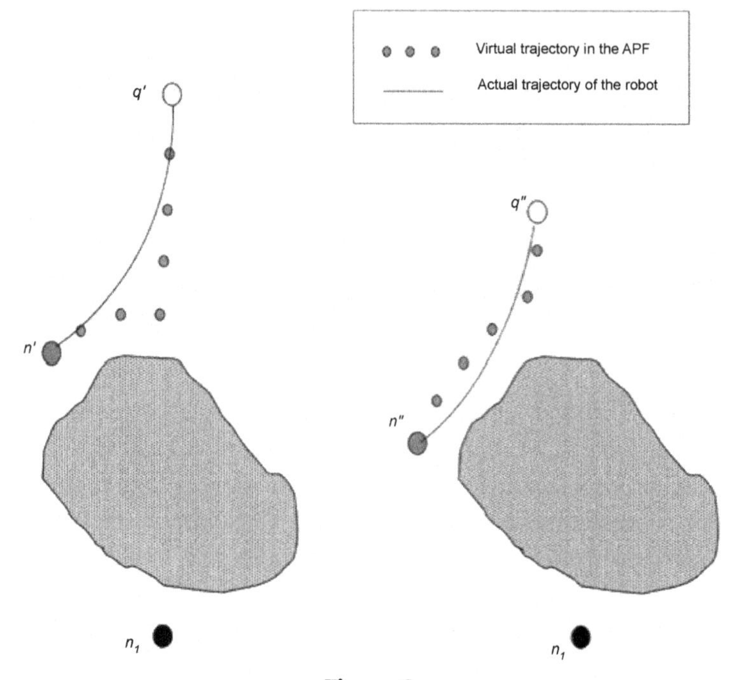

Figure 2
Virtual navigation and real trajectory.

proximity data, in critical conditions it suddenly switches off the smooth navigation mechanism, and slows down or even stops the motors. Next, it forces the robot to turn in place until a free passage is found, before switching to the normal navigation mode again. This allows the robot to correctly and safely avoid obstacles even when the environment is highly dynamic, and a person or another robot is quickly approaching without allowing the robot to maneuver around it. Clearly, the resulting trajectory will no longer be smooth.

In spite of its good properties, the navigation system described in this Section has the typical drawbacks of systems relying on local perceptions and navigation strategies (see Section 2). First, since obstacle avoidance ultimately relies on APF, we have to deal with local minima in the potential field, which can prevent the robot from reaching the target node if an appropriate escape strategy has not been implemented. Second, the navigation system does not provide a policy to determine how much the robot should be allowed to deviate from the original path, and an unlucky configuration of obstacles (or a nasty person) could cause the robot to move very far from its original path in the attempt to search for an alternative solution. Roaming Trails offer a solution to deal with these problems in an integrated fashion.

3.3.4 Roaming Trails

To begin with, obstacles in human environments are categorized into three classes, defined as follows:

1. *Static objects*: this class includes walls and furniture that are guaranteed to stay still during the robot mission, and therefore can be depicted in the a priori map $\mathscr{M}$.
2. *Moving objects*: this class includes people, as well as objects carried around by people (e.g., suitcases and bags) that can be temporarily abandoned and lie on the robot's path, but are not depicted in the a priori map $\mathscr{M}$.
3. *Removable objects*: this class includes objects (usually furniture, e.g., chairs) that are static in most cases, but can be occasionally displaced and lie on the robot's path. As well as moving objects, removable objects are not depicted in the a priori map.

The Roaming Trail approach is able to deal very efficiently with environments in which most obstacles encountered by the robot during motion are of type 1 and 2. In addition, it is able to manage situations in which the robot encounters an object of type 3, even if less efficiently.

To achieve this, the underlying idea is the following: instead of planning the path on the basis of the topology of the roadmap $\Re$, a more complex representation is extracted from the a priori map $\mathscr{M}$ and used for path planning and navigation. An example is shown in Figure 3: some of the roadmap nodes are connected by means of diamond-shaped areas $\Re\mathscr{T}_{ij}$, which—taken together—constitute a Roaming Trail leading from the start to the goal. Specifically.

1. each area $\Re\mathscr{T}_{ij}$ is defined by the triplet (n_i, n_j, σ), where n_i, n_j are two roadmap nodes, and σ is an additional parameter which determines the dimensions of the area;
2. Roaming Trails are drawn in such a way that the intersection of each area $\Re\mathscr{T}_{ij}$ with the free space in $\mathscr{M}$ is always a convex area.

During navigation, the robot is allowed to deviate from its path to avoid obstacles on the basis of reactive navigation strategies, but it is never allowed to exit from the area $\Re\mathscr{T}_{ij}$ which connects the current target node n_j with the previous target noden_i. Thus, since the robot is constrained to move within a convex area which includes the location of the target node, in presence of static obstacles it is guaranteed to reach the target by following a straight line. If we consider also moving obstacles (whose location is obviously not shown in the a priori map), the robot has two choices: avoiding them or simply stopping and waiting for them to move on. In both case, it can happen that the robot is temporarily unable to find a way to the goal, or even trapped in a deadlock caused by moving obstacles. However, if humans are collaborative as they are expected to be, they will let the robot pass: more important, it cannot happen that the robot, while trying to avoid a

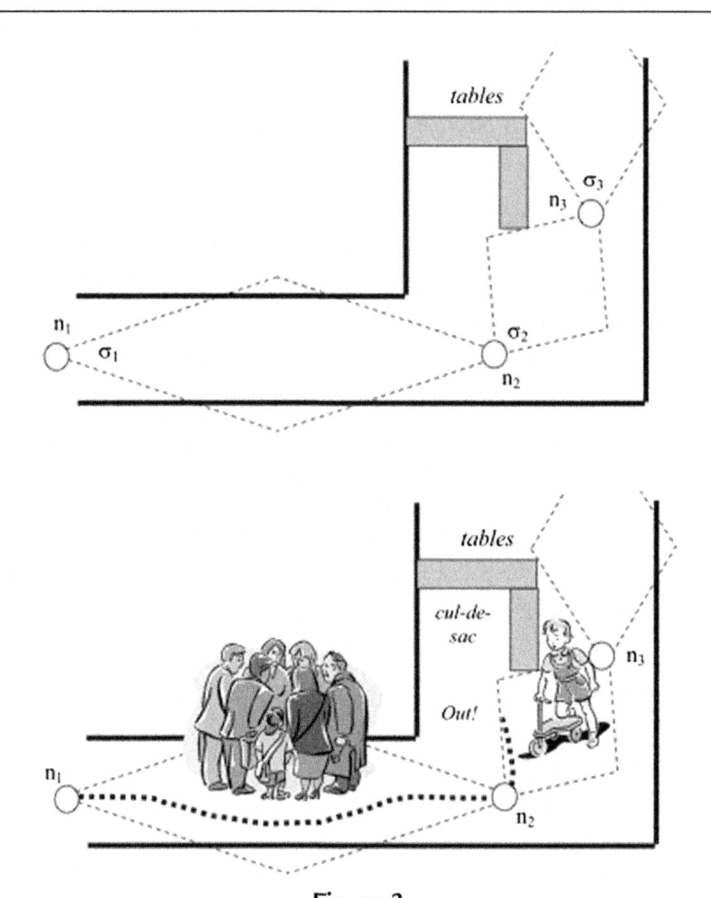

Figure 3
Top: apriori map and RoamingTrails. Bottom: trajectory of the robot (dotted line) in presence of people.

moving obstacle, ends its path in a concavity formed by an unlucky configuration of static obstacles (e.g., a cul-de-sac formed by the left wall and the two tables in Figure 3): in fact, the robot is stopped as soon as it reaches the boundaries of the Roaming Trail. One could argue that the system relies on the fact that humans are collaborative and, sooner or later, let the robot pass. However, this is a limitation of every system programmed not to hit people to meet a safety requirement: a single non-collaborative human can stop the robot for an arbitrary time, if the former is faster than the latter, whichever algorithm is adopted for navigation.

The presence of removable objects poses major limitations: in fact, even when the robot remains within the boundaries of Roaming Trails, it is always possible that it gets trapped into a concavity formed by removable objects (e.g., chairs or waste baskets) whose position is different from time to time, and hence are not depicted in the a priori map. Obviously, waiting for a chair to move is not considered to be a winning strategy, unless there is a human in the neighborhood that quickly understands the situation and decides to directly intervene to help the robot! One could argue that, in this case, it would be more efficient to let the robot wander outside Roaming Trails, instead of waiting for the intervention of *a deus ex machina*. Of course, after some time that the robot is still, this can be done. However, we still claim that—in many realistic situations—Roaming Trails reveals to be a more efficient and safe strategy to merge a priori knowledge and planning with sensor-based reactive methods.

It should also be noted that diamonds are not necessarily the best shape. In fact, this choice is rather arbitrary, and different shapes could be considered as well, e.g., ellipses, rectangles, or generalized cones. This means also that known methods in literature for motion planning could be adapted to work in a similar manner (e.g., methods for exact or approximate cell decomposition [36]). We now show in detail how the prototypical architecture introduced in the previous Section can be adapted to implement these concepts.

3.3.4.1 Deliberative Agents

Spatial Reasoner behaves as described in Algorithm 1, by building the roadmap $\Re$ in four steps. However, step four is now modified to meet the constraints placed by Roaming Trails, i.e., for every couple of nodes n_i, n_j, $i \neq j$, an undirected edge is established in $\Re$ if and only if the intersection between the area $\Re\mathcal{T}_{ij} = (n_i, n_j, \sigma)$ and the free space in $\mathcal{M}$ is a convex area.

An analogous modification is made to step two of Algorithm 2, required by Mission Planner & Monitor to connect the start and the goal node to $\Re$. Moreover, instead of searching for the shortest path, step three of Algorithm two can be modified to search for the path with the lowest number of nodes: less nodes obviously corresponds to less and bigger areas $\Re\mathcal{T}_{ij}$, which give the robot more freedom to deviate from the nominal path to avoid moving and removable obstacles. As in the previous case, Mission Planner & Monitor communicates the location of the next node to be reached to Trajectory Generator and to the other reactive agents which are responsible for the robot motion, and periodically checks the current state of advancement of the plan.

3.3.4.2 Reactive Agents

The reactive agents Trajectory Generator, Local Map Builder, APF Builder and Safe Navigator behave as in Section 3.3.3.2. On the opposite, Emergency Handler is different, since it now implements a set of rules to inhibit the behavior of other agents not only when a collision is imminent, but also when the robot is heading toward an area of the workspace in which it could be trapped in a deadlock. Specifically, Emergency Handler intervenes if:

1. the robot, when traveling from n_i to n_j, reaches the boundaries of an area $\Re\mathcal{T}_{ij} = (n_i, n_j, \sigma)$, which can happen when the robot is trying to avoid a moving or a removable obstacle;
2. the robot gets very close to an obstacle, which can happen either because the robot is trapped in a local minimum or because it has not enough space to maneuver around it.

If either condition is verified, Emergency Handler slows down and stops the robot, and turns it toward the next node n_j. Next, if the path is free, it allows the robot to move again; otherwise, it forces the robot to wait until all the moving obstacles have moved on.[3] Remember that, by construction, n_j is reachable from every position within $\Re\mathcal{T}_{ij} = (n_i, n_j, \sigma)$, since the robot motion is always constrained within a convex area. As a consequence, if an obstacle intersects the straight line connecting the robot position and the target, it necessarily is a moving or a removable obstacle which is not depicted in the a priori map.

3.3.5 Experimental Results

Figures 4–9 show how Roaming Trails are built. First, the workspace is sampled at an arbitrary resolution R, obtaining a grid $\mathcal{G}$ with M cells (Algorithm 1, step one). Next, every cell m_i which is labeled as free space is searched for a Voronoi node, producing $N \ll M$ nodes (steps two and three). Notice that, depending on the threshold chosen when checking if m_i is equidistant from the three closest segments, it can happen that a number of neighboring nodes are found. To avoid this, experiments are performed with the additional constraint that, whenever a Voronoi node is found in a cell m_i, the algorithm does not search for additional nodes within a square of side P (arbitrarily chosen) centered in m_i: as a consequence, the minimum distance between nodes cannot be lower than $(P - 1)/2$. Finally, couples of nodes n_i, n_j are connected in $\Re$ whenever it is possible to draw an area $\Re\mathcal{T}_{ij} = (n_i, n_j, \sigma)$ whose intersection with the free space is convex (step four). When the planner is asked to plan a path between a start S and a goal G, it updates the roadmap by adding the start and goal nodes (Algorithm 2, steps one and two), and computes a chain of areas $\Re\mathcal{T}_{ij}$ with proper characteristics (step 3).

It is worth spending a few words about time complexity. Step one of Algorithm 1 has O(SM) complexity: for every cell of the grid $m_i \in \mathcal{G}, i = 1...M$, it is necessary to consider all segments of the map $s_j \in \mathcal{M}, j = 1...S$, to check whether the cell lies in the free space

[3] Optionally, if the robot waits for too long, rule one can be disabled, thus letting the robot free to wander at its own risk.

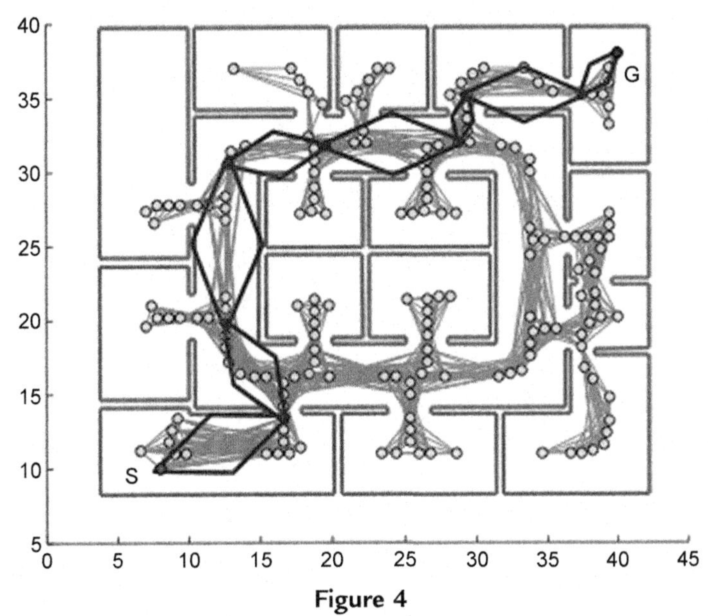

Figure 4
Diamond shapes, $R = 5$ cells per meter, $P = 9$, solution with the least number of links.

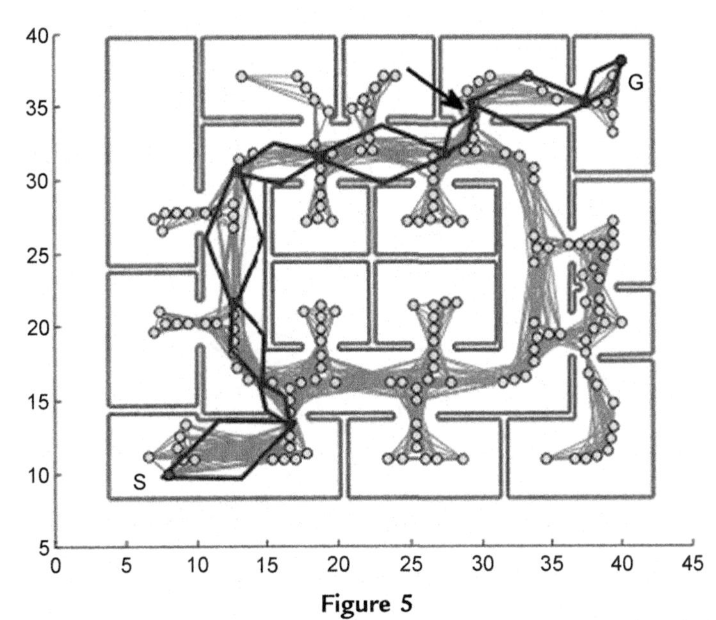

Figure 5
Diamond shapes, $R = 5$ cells per meter, $P = 9$, shortest path solution.

or not. Finding Voronoi nodes in steps 2, 3 has O(SM) complexity: there are at most M cells in the grid which are labeled as free space, and for every cell m_i it is necessary to compute the three closest segments $s_j, s_k, s_l \in \mathscr{M}$. Adding links in step 4, given that N nodes have been produced in the previous steps, has $O(SN^2)$ complexity: for every couple

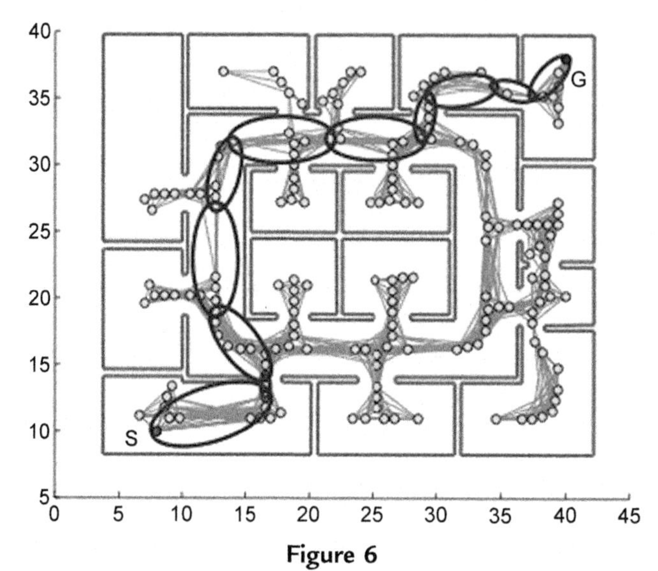

Figure 6
Ellipses, $R = 5$ cells per meter, $P = 9$, solution with the least number of links.

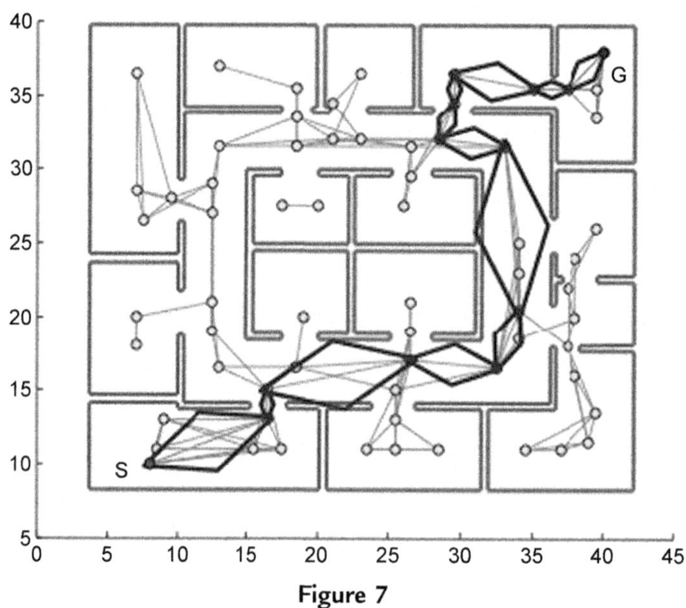

Figure 7
Diamond shapes, $R = 2$ cells per meter, $P = 9$, solution with the least number of links.

of nodes n_i, n_j, $i \neq j$, it is necessary to check whether the intersection between $\Re\mathcal{T}_{ij} = (n_i, n_j, \sigma)$ and the free space in $\mathcal{M}$ is convex, which-once again-requires considering each segment $s_l \in \mathcal{M}, l = 1...S$. Updating the roadmap in step two of Algorithm 1 has $O(SN)$ complexity, since it is necessary to check whether the start and the goal can be linked or

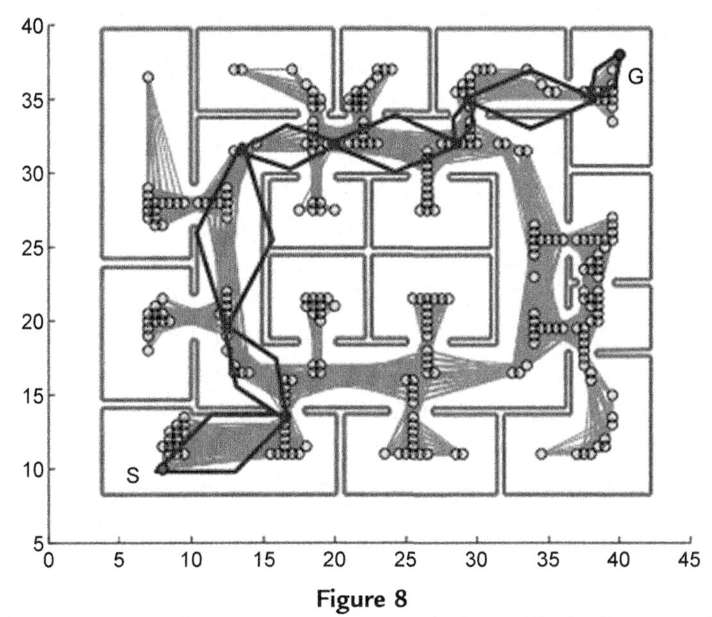

Figure 8
Diamond shapes, $R = 2$ cells per meter, $P = 0$, solution with the least number of links.

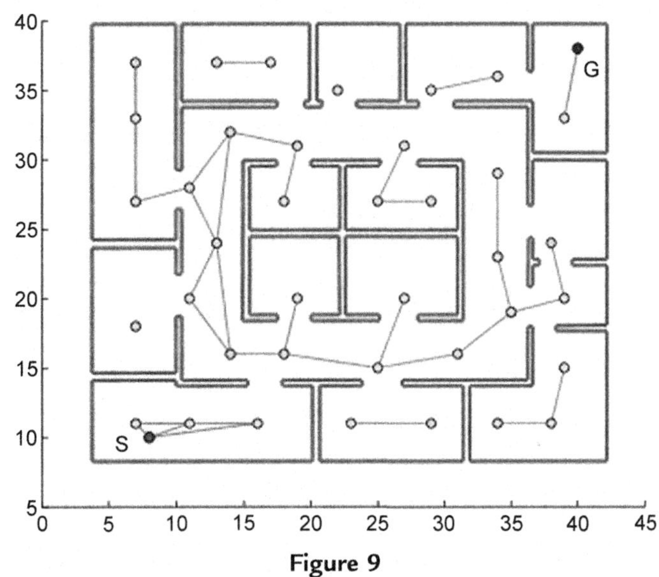

Figure 9
Diamond shapes, $R = 1$ cells per meter, $P = 9$, no solution found.

not with any of the pre-existing N nodes, whereas the complexity of planning in step three is the complexity of the A* algorithm, which—in the worst case—is exponential in the length of the solution, and is known to be optimum among search algorithms. In general, the roadmap in Algorithm one can be computed off-line, since it does not depend on a particular navigation problem to be solved. On the opposite, a new path must be computed whenever the robot is assigned a new mission, and therefore the computational complexity of Algorithm two is an issue which deserves attention.

Experiments have been performed on a 2.66 GHz Intel(R) Xeon(R) CPU.[4] The problem in Figure 4 has been solved by setting $R = 5$ cells per meter and $P = 9$, and by asking the planner to find the solution with the least number of links (which turns out to be eight). The time required to build $\mathcal{G}$ by classifying each of the M cells as free space or not free space is $t(\mathcal{G}) \approx 1.58$ s; the time required to build $\Re$ is $t(\Re) \approx 1.61$ s; the time required to update R and plan a path $\mathcal{P}$ is $t(\mathcal{P}) \approx 0.02$ s. The solution in Figure 5 has been found by asking the planner to search for the shortest path, instead of the path with the least number of links. The resulting path consists of 10 links, corresponding to a smaller area $\Re\mathcal{T}_{ij}$ than the previous case: notice, for instance, the very small area which connects the 8th and the 9th nodes in the path (pointed by arrow). The solution in Figure 6 uses a different shape for Roaming Trails, i.e., ellipses instead of diamonds: the robot is allowed to deviate more from the planned path, especially in the proximity of nodes, but a bigger number of links is required. The time required to find a solution is approximately the same in all cases.

The complexity analysis tells that, if required to reduce computational cost, (1) the resolution R of the grid can be decreased, which consequently decreases the maximum number M of free space cells, and (2) the maximum number of nodes N in the roadmap can be reduced by increasing P, which determines the minimum distance between roadmap nodes.

Figure 7 shows the same problem as in Figure 4, with a lower resolution $R = 2$, and $P = 9$: the times required are $t(\mathcal{G}) \approx 0.26$ s, $t(\Re) \approx 0.18$ s, and $t(\mathcal{P}) \approx 0.006$ s. Figure 8 shows the solution found by setting $R = 2$ and $P = 0$, i.e., without any constraints on the distance among nodes: the times required are $t(\mathcal{G}) \approx 0.26$ s, $t(\Re) \approx 2.36$ s, and $t(\mathcal{P}) \approx 0.038$ s. Figure 9 shows that, by setting $R = 1$ and $P = 9$, no solution can be found, since the resulting roadmap is made of disconnected components: the times required are $t(\mathcal{G}) \approx 0.068$, $t(\Re) \approx 0.042$ s, and $t(\mathcal{P}) \approx 0.004$ s.

In order to quantitatively evaluate how the number of nodes and the computational times vary depending on R and P, a set of experiments have been performed in a very large environment with 10×10 rooms: each room is approximately 6×6 m, and it is connected to neighboring rooms. Table 1 summarizes the results corresponding to the problem in Figure 10 (roadmap nodes and links are not shown): rows are ordered according to the increasing number of nodes N, and report both the solution $\mathcal{P}_1$ with the shortest path $l(\mathcal{P}_1)$, and the solution $\mathcal{P}_2$ with the least number of links $l(\mathcal{P}_2)$. It can be noted that, in this very large environment, a large number of roadmap nodes are produced,

[4] The C++ implementation of Roaming Trails which has been used for experiments is freely available at the address www.robotics.laboratorium.dist.unige.it.

Table 1: Computational time in environments with 100 rooms

R	P	N	$t(g)(s)$	$t(R)$	$t(p_1)(s)$	$t(p_2)$	$l(p_1)$	$l(p_2)$
1	11	129	2.0	4.8 s	0.15	0.15 s	n.a.	n.a
1	9	182	2.0	9.5 s	0.21	0.22 s	n.a.	n.a
1	7	265	2.0	20.6 s	0.31	0.31 s	n.a.	n.a
2	11	360	7.9	38.6 s	0.43	0.43 s	n.a.	n.a
2	9	475	7.9	66.2 s	0.5 7	0.59 s	107.9	26
1	5	496	2.0	71.9 s	0.59	1.34 s	n.a.	n.a
2	7	755	7.9	3 min	0.91	0.92 s	97.8	27
5	11	1074	48.9	6 min	1.28	1.34 s	88.3	19
5	9	1609	49.1	12 min	2.10	2.04 s	86.3	19
1	3	1619	2.0	13 min	1.9	2.1 s	83.8	22
2	5	1667	7.9	14 min	2.0	2.18 s	87.2	22
5	7	2426	48.8	29 min	2.9	3.5 s	85.4	19
5	5	5039	48.9	2 h	6.2	12.7 s	83.6	19
2	3	5835	7.9	3 h	7.3	15.9 s	84.1	19
5	3	14 436	48.9	17 h	22.2	3 min	83.4	19

and the computational times $t(\mathcal{G})$, $t(\Re)$, and $t(\mathcal{P})$ increase accordingly: specifically, the time $t(\Re)$ increases with the square of N, up to $t(\Re) \approx 17$ h in the last row of the Table. However, it is worth noting that $t(\Re)$ corresponds to steps 2, 3, 4 of Algorithm one which can be performed off-line, whereas the time $t(\mathcal{P})$ corresponding to online planning is much lower. Also, even if it can happen that a solution cannot be found when the number of nodes is too small (label n.a. in rows one to four and six), increasing the number of nodes above a given threshold affects only marginally the length of the solution: this is

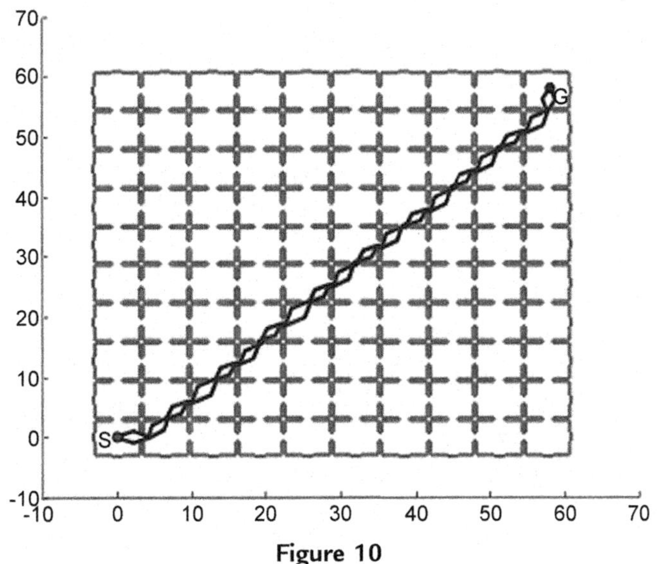

Figure 10
Diamond shapes, $R = 5$ cells per meter, $P = 9$, least cost solution.

particularly evident by inspecting the length $l(\mathscr{P}_2)$ of the solution with the least number of links. The solution shown in Figure 10 corresponds to the 9th row of the table, with $t(\mathscr{G}) \approx 49.1$ s, $t(\Re) \approx 12$ m, $t(\mathscr{P}_1) \approx 2.10$ s, and $t(\mathscr{P}_2) \approx 2.04$ s.

The approach has been extensively tested, for more than a decade, on all the robots of the Robotics Laboratory at DIST Università di Genova: primarily the autonomous robot Staffetta, a car-like platform with two rear motors and a front active steering wheel (a design choice which improves stability and reduces slippage) that we have designed for autonomous transportation within hospitals. Staffetta has a payload of about 120 kg and it is able to move at a maximum speed of about 1 m/s, it is equipped with proximity sensors for obstacle detection/avoidance and touch sensors for collision recovery, and it relies on a laser-based localization system to periodically correct its position in the environment. Figure 11 shows still images from a video showing smooth obstacle avoidance thanks to

Figure 11
Staffetta (chassis only) smoothly avoiding a static obstacle.

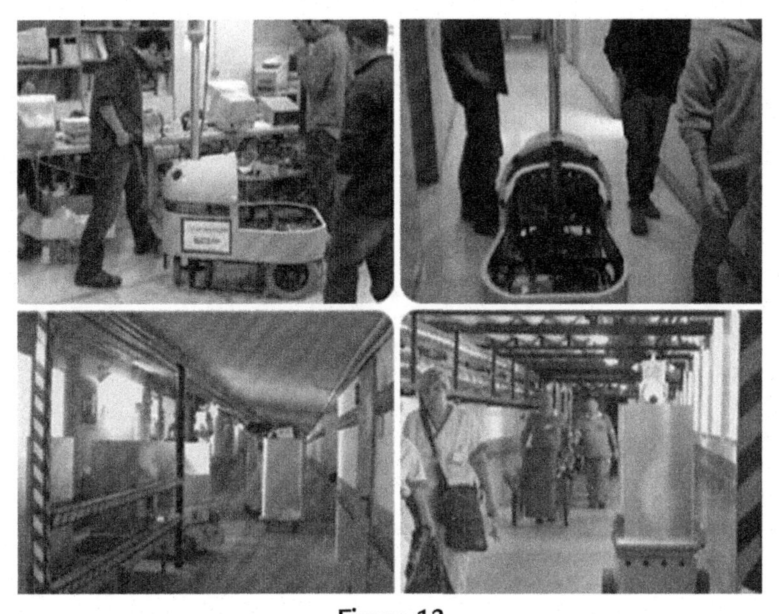

Figure 12
Top: Staffetta at the 2nd floor of DIST. Bottom: Merry Porter™ at Polyclinic of Modena, loading waste (left) and moving along a crowded corridor (right).

the joint action of Trajectory Generator, Local Map Builder, APF Builder and Safe Navigator. When the front steering wheel is actively controlled, smooth motion capabilities help preserve kinematic coherency between the front and the rear wheels. Consider the following non-smooth motion behavior, very frequent in reactive navigation approaches: the robot is moving on a straight line, slows down in front of an obstacle, and starts searching for a safe direction of motion. In order to preserve kinematic coherency, every time the signs of the angular speed changes while the linear velocity has a small value above zero, an almost 180° rotation of the front steering wheel is required. Under these conditions, smooth motion is required to prevent the robot from spending all its time adjusting the orientation of the front wheel.

Experiments have been performed at the 2nd floor of DIST Università di Genova (Figure 12 on the top), at the Gaslini Hospital of Genoa and in public exhibitions, such as the Tmed 2001 exhibition (Magazzini Del Cotone, Porto Antico di Genova, October 2001). Moreover, thanks to the lessons learned with Staffetta, a second generation of robots have been developed (Merry Porter™, Figure 12 on the bottom), and are currently being set-up within the Polyclinic of Modena for autonomous waste transportation.[5]

[5] Staffetta and the Merry Porter™ are commercialized by the company Genova Robot srl, www.genovarobot.com.

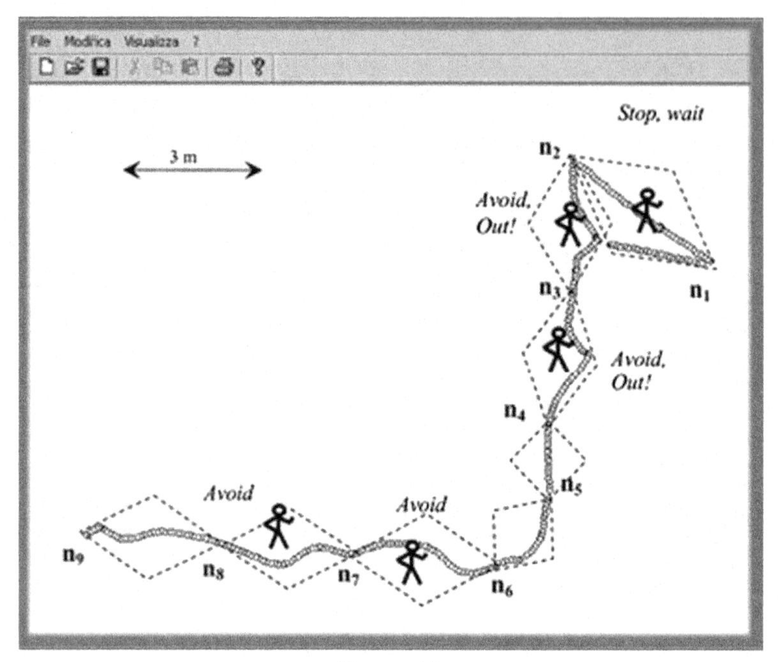

Figure 13
Trajectory of the robot during one mission.

All the agents described in the chapter are executed concurrently on board the robot, allowing for merging global planning with a reactive, yet smooth and safe motion behavior which takes into account the available a priori knowledge. Figure 13 shows the trajectory followed by the robot while executing one mission: in some cases the robot has to stop and wait for persons to move on (labels "Stop" and "Wait"), whereas in some other cases it manages to avoid them (label "Avoid"). Finally, it can happen that the robot reaches the Roaming Trail boundaries while trying to avoid a person on its path (label "Out!"). There are no rough turns in the robot trajectory, except for the following cases: (1) a target node has been reached, (2) Emergency Handler intervenes either because the robot has reached the boundaries of the Roaming Trail, or an obstacle is too close to be smoothly avoided.

The experiences gained in one decade of experiments with TRC Labmate, Staffetta, and Merry Porter™ allows us to draw the following qualitative conclusions concerning the usage of Roaming Trails in crowded indoor environments: robots are able to smoothly avoid approximately half of the obstacles; the remaining obstacles are dealt with by stopping, warning, and waiting for people to move on or to remove objects. Moreover, thanks to the presence of Roaming Trails—and given that an accurate localization system is available [34]—robots never reach deadlock situations from which it is difficult to recover, yielding approximately a 100% success in navigation missions.

3.3.6 Conclusions

Experience shows that, when working within human environments, most existing systems are unable to work continuously with no failure or performance degradation. This is mainly due to the fact that human environments cannot be completely known a priory since they are highly dynamic, causing a dramatic effect both on self-localization and on navigation. It is not a case that the few examples of autonomous mobile robots which successfully hit the market have been cleaning robots and lawnmowers, i.e., robots in which superior navigation and self-localization capabilities are not required in the first place: since the task is simply to maximize the coverage of a given area in a statistical sense, a pre-planned path is not required at all, and navigation can be dealt with through a set of heuristic rules plus random noise, allowing the robot to change direction of motion when required. As an aside, notice also that the same approach would not work with other coverage tasks, such as painting or humanitarian de-mining: even if the task is similar, painting and humanitarian de-mining cannot be statistically faced: a 95% cleaned floor can be reasonably defined "a cleaned floor", whereas a 95% de-mined field or a 95% painted surface are definitely NOT a "de-mined field" or a "painted surface".

Then, in most robotics applications, superior navigation capabilities are required to plan and follow a path. The chapter describes a two-level cognitive architecture which faces the problem by integrating, as usual, deliberative and reactive activities: the approach is original in that it introduces the concept of Roaming Trails. Instead of a path to the goal, the planner output is a Roaming Trail, i.e., a chain of diamond-shaped or elliptic areas which define the maximum allowed distance for the robot to deviate from its path to the goal. During mission execution, local navigation algorithms are allowed to reactively compute or simply update the robot trajectory, but always within the boundaries of the planned Roaming Trail. The approach has been tested for more than one decade on the robots TRC Labmate, Staffetta, and Merry Porter™ in many different indoor environments, significantly improving the performance of the navigation system.

References

[1] F. Capezio, F. Mastrogiovanni, A. Sgorbissa, R. Zaccaria, The ANSER project: airport nonstop surveillance expert robot, in: IEEE/RSJ International Conference on Intelligent Robots and Systems, 2007. IROS 2007, 2007, pp. 991–996.

[2] R. Vidal, O. Shakernia, H. Kim, D. Shim, S. Sastry, Probabilistic pursuit-evasion games: theory, implementation, and experimental evaluation, IEEE Trans. Rob. Autom. 18 (5) (2002) 662–669.

[3] A. Sgorbissa, R. Zaccaria, The artificial ecosystem: a distributed approach to service robotics, in: 2004 IEEE International Conference on Robotics and Automation, 2004. Proceedings. ICRA '04, vol. 4, 2004, pp. 3531–3536.

[4] S. Berman, Y. Edan, M. Jamshidi, Navigation of decentralized autonomous automatic guided vehicles in material handling, IEEE Trans. Rob. Autom. 19 (4) (2003) 743–749.

[5] A. Yamashita, T. Arai, J. Ota, H. Asama, Motion planning of multiple mobile robots for cooperative manipulation and transportation, IEEE Trans. Rob. Autom. 19 (2) (2003) 223–237.
[6] R. Siegwart, K.O. Arras, S. Bouabdallah, D. Burnier, G. Froidevaux, X. Greppin, et al., Robox at expo.02: a large-scale installation of personal robots, Rob. Autom. Syst. 42 (3–4) (2003) 203–222.
[7] H. Moravec, Rise of the Robots, Scientific American, 1999, pp. 124–135.
[8] O. Khatib, Real-time obstacle avoidance for manipulators and mobile robots, Int. J. Rob. Res. 5 (1) (1986) 90–98.
[9] J. Borenstein, Y. Koren, The vector field histogram-fast obstacle avoidance for mobile robots, IEEE Trans. Rob. Autom. 7 (3) (1991) 278–288.
[10] M. Piaggio, A. Sgorbissa, AI-CART: an algorithm to incrementally calculate artificial potential fields in real-time, in: 1999 IEEE International Symposium on Computational Intelligence in Robotics and Automation, 1999. CIRA '99. Proceedings, 1999, pp. 238–243.
[11] Y. Koren, J. Borenstein, Potential field methods and their inherent limitations for mobile robot navigation, in: 1991 IEEE International Conference on Robotics and Automation, 1991. Proceedings, vol. 2, 1991, pp. 1398–1404.
[12] R.C. Arkin, Motor schema based mobile robot navigation, Int. J. Rob. Res. 8 (4) (1989) 92–112.
[13] M. Slack, Navigation templates: mediating qualitative guidance and quantitative control in mobile robots, IEEE Trans. Syst. Man Cybern. 23 (2) (1993) 452–466.
[14] S. Quinlan, O. Khatib, Elastic bands: connecting path planning and control, in: 1993 IEEE International Conference on Robotics and Automation, 1993. Proceedings, vol. 2, 1993, pp. 802–807.
[15] R.P. Bonasso, D. Kortenkamp, D.P. Miller, M. Slack, Experiences with an architecture for intelligent, reactive agents, J. Exp. Theor. Artif. Intell. 9 (1995) 237–256.
[16] R. Simmons, The curvature-velocity method for local obstacle avoidance, in: 1996 IEEE International Conference on Robotics and Automation, 1996. Proceedings, vol. 4, 1996, pp. 3375–3382.
[17] D. Fox, W. Burgard, S. Thrun, The dynamic window approach to collision avoidance, Rob. Autom. Mag. IEEE 4 (1) (1997) 23–33.
[18] I. Ulrich, J. Borenstein, VFH: local obstacle avoidance with look-ahead verification, in: IEEE International Conference on Robotics and Automation, 2000. Proceedings. ICRA '00, vol. 3, 2000, pp. 2505–2511.
[19] O. Brock, O. Khatib, Elastic strips: a framework for motion generation in human environments, Int. J. Rob. Res. 21 (12) (2002) 1031–1052.
[20] L. Lapierre, R. Zapata, P. Lepinay, Simultaneous path following and obstacle avoidance control of a unicycle-type robot, in: 2007 IEEE International Conference on Robotics and Automation, 2007, pp. 2617–2622.
[21] A. Elfes, Using occupancy grids for mobile robot perception and navigation, Computer 22 (6) (1989) 46–57.
[22] I. Ulrich, J. Borenstein, VFH+: reliable obstacle avoidance for fast mobile robots, in: 1998 IEEE International Conference on Robotics and Automation, 1998. Proceedings, vol. 2, 1998, pp. 1572–1577.
[23] R.C. Arkin, T. Balch, Aura: principles and practice in review, J. Exp. Theor. Artif. Intell. 9 (1997) 175–189.
[24] R. Kurozumi, T. Yamamoto, Implementation of an obstacle avoidance support system using adaptive and learning schemes on electric wheelchairs, in: 2005 IEEE/RSJ International Conference on Intelligent Robots and Systems, 2005. (IROS 2005), 2005, pp. 1108–1113.
[25] J. Minguez, The obstacle-restriction method for robot obstacle avoidance in difficult environments, in: 2005 IEEE/RSJ International Conference on Intelligent Robots and Systems, 2005. (IROS 2005), 2005, pp. 2284–2290.
[26] J. Minguez, Integration of planning and reactive obstacle avoidance in autonomous sensor-based navigation, in: 2005 IEEE/RSJ International Conference on Intelligent Robots and Systems, 2005. (IROS 2005), 2005, pp. 2486–2492.
[27] F. Zhang, A. O'Connor, D. Luebke, P. Krishnaprasad, Experimental study of curvature-based control laws for obstacle avoidance, in: 2004 IEEE International Conference on Robotics and Automation, 2004. Proceedings. ICRA '04, vol. 4, 2004, pp. 3849–3854.

[28] P. Ogren, N. Leonard, A convergent dynamic window approach to obstacle avoidance, IEEE Trans. Rob. 21 (2) (2005) 188–195.
[29] B. Damas, J. Santos-Victor, Avoiding moving obstacles: the forbidden velocity map, in: IEEE/RSJ International Conference on Intelligent Robots and Systems, 2009. IROS 2009, 2009, pp. 4393–4398.
[30] K. Arras, J. Persson, N. Tomatis, R. Siegwart, Real-time obstacle avoidance for polygonal robots with a reduced dynamic window, in: IEEE International Conference on Robotics and Automation, 2002. Proceedings. ICRA '02, vol. 3, 2002, pp. 3050–3055.
[31] M. Aicardi, G. Casalino, A. Bicchi, A. Balestrino, Closed loop steering of unicycle like vehicles via lyapunov techniques, Rob. Autom. Mag., IEEE 2 (1) (1995) 27–35.
[32] D. Soetanto, L. Lapierre, A. Pascoal, Adaptive, non-singular path-following control of dynamic wheeled robots, in: 42nd IEEE Conference on Decision and Control, 2003. Proceedings, vol. 2, 2003, pp. 1765–1770.
[33] F. Lamiraux, D. Bonnafous, O. Lefebvre, Reactive path deformation for nonholonomic mobile robots, IEEE Trans. Rob. 20 (6) (2004) 967–977.
[34] F. Mastrogiovanni, A. Sgorbissa, R. Zaccaria, Designing a system for map-based localization in dynamic environments, in: T. Arai, R. Pfeifer, T.R. Balch, H. Yokoi (Eds.), IAS, IOS Press, 2006, pp. 173–180.
[35] M. Piaggio, A. Sgorbissa, R. Zaccaria, Pre-emptive versus non-pre-emptive real time scheduling in intelligent mobile robotics, J. Exp. Theor. Artif. Intell. 12 (2) (2000) 235–245.
[36] J.-C. Latombe, Robot Motion Planning, Kluwer Academic Publishers, Norwell, MA, USA, 1991.
[37] K. Sugihara, Approximation of generalized voronoi diagrams by ordinary voronoi diagrams, CVGIP: Graph. Model. Image Process 55 (1993) 522–531.
[38] P. Morasso, V. Sanguineti, Computational maps and target fields for reaching movements, in: Self-organization, Computational Maps, and Motor Control, Advances in Psychology, vol. 119, North-Holland, 1997, pp. 507–546.
[39] M. Piaggio, A. Sgorbissa, G. Vercelli, R. Zaccaria, Fusion of sensor data in a dynamic representation, in: Proceedings of the First Euromicro Workshop on Advanced Mobile Robot, 1996, 1996, pp. 10–16.

CHAPTER 3.4

Monocular SLAM with Undelayed Initialization for an Indoor Robot[1]

Kiwan Choi, Jiyoung Park, Yeon-Ho Kim, Hyoung-Ki Lee

Micro Systems Laboratory, Samsung Advanced Institute of Technology, Samsung Electronics Inc., Yongin-Si, Gyeonggi-Do, South Korea

Chapter Outline

This chapter presents a new feature initialization method for monocular EKF SLAM (Extended Kalman Filter Simultaneous Localization and Mapping) which utilizes a 3D measurement model in the camera frame rather than a 2D pixel coordinate in the image plane. The key idea is to consider a camera as a range and bearing sensor, of which the

---

[1] Reprinted from Robotics and Autonomous Systems, Vol. 60, Kiwan Choi, Jiyoung Park, Yeon-Ho Kim, Hyoung-Ki Lee, Monocular SLAM with Undelayed Initialization for an Indoor Robot, pp. 841–851, 2013, with permission from Elsevier.

Household Service Robotics. http://dx.doi.org/10.1016/B978-0-12-800881-2.00010-4

range information contains numerous uncertainties. 2D pixel coordinates of measurement are converted to 3D points in the camera frame with an assumed depth. The element of the measurement noise covariance corresponding to the depth of the feature is set to a very high value. It is then shown that the proposed measurement model has very little linearization error, which can be critical for the EKF performance. Furthermore, this chapter proposes an EKF SLAM system that combines odometry, a low-cost gyro, and low frame rate (1−2 Hz) monocular vision. A low frame rate is crucial for reducing the price of the processor. This system combination is cost-effective enough to be commercialized for a real vacuum cleaning application. Simulations and experimental results show the efficacy of the proposed method with computational efficiency in indoor environments.

3.4.1 Introduction

Samsung Electronics has been developing vacuum cleaning robots for the home environment that generate a cleaning path with minimal overlap of the traveled region. To do this, the robots require the capability of localization and map building. For the localization capability, our robots are equipped with a camera for external position measurements. They execute a visual SLAM algorithm based on an EKF. In home environments, the need for the visual features for an SLAM algorithm does not permit robots to travel under a bed, for example. Thus, our robot has a dead-reckoning system of two wheel encoders and a low-cost gyro for relative position calculations when the visual features are not available. A calibrated microelectromechanical systems (MEMS) gyro gives a good orientation estimate for a few minutes and very accurate relative position information in the frame of the Kalman filter [1,2]. Thanks to this dead-reckoning system, the required frame rate of images for visual feature tracking can be reduced to 1 or 2 Hz while previous monocular SLAMs with only one camera need a high frame rate of more than 20 Hz. The sensor combination of a low-cost MEMS gyro and a low frame rate camera leads to a practical system for vacuum cleaning applications.

Davison [3] presented the feasibility of a real-time monocular SLAM with a standard low-cost camera using an EKF framework. In his work, the camera obtains only bearing information, meaning that multiple observations from different points of view must be processed to achieve accurate depth information. Davison used an extra particle filter to estimate the feature depths. However, this method causes a delay in feature initialization. Bailey [4] also proposed a delayed initialization method for a bearing-only SLAM. This method used a Gaussian approximation approach similar to EKF. Costa et al. [5] presented another delayed initialization method which calculated the initial uncertainty of feature locations. Their method projected measurements from the sensor space onto the Cartesian world frame and assumed that their depth directional uncertainties had large values. Their method was also a delayed method. Such delayed initialization schemes have the drawback that new features, held outside the main probabilistic state, are not able to contribute to the estimation of the camera position until finally included in the map. Further, features that retain low parallax over many frames (those very far from the camera or close to the motion epipole) are usually rejected completely because they never pass the test for inclusion.

Kwok and Dissanayake [6] presented a multiple hypothesis approach that outlined their undelayed feature initialization scheme. Each landmark is initialized in the form of multiple hypotheses with different depths distributed along the direction of the bearing measurement. Using subsequent measurements, the depth of the landmark is estimated in an EKF framework and hypothesized landmarks with different depths are tested by a sequential probability ratio test. Invalid hypotheses are removed from the system state. Sola et al. [7] improved computational efficiency of this approach by approximating a Gaussian sum filter.

Eade and Drummond [8] estimated the inverse depths of features rather than their depths. The inverse depth concept has an advantage in terms of linearity, allowing linear Kalman filtering to operate successfully. While this approach utilizes a FastSLAM-type particle filter, Montiel et al. [9,10] applied this concept to EKF SLAM. Using an inverse depth entry in the state vector and a modified measurement model allows unified modeling and processing within the standard EKF without a special initialization process. Measurement equations have little linearization error for distant features, and the estimation uncertainty is accurately modeled as the Gaussian in inverse depth. A feature is represented by the state vector of six dimensions at the first observation—three for the camera position, two for the direction of the ray and one for the reciprocal depth. This is an over-parameterization compared to three dimensions for Euclidean *XYZ* coding and hence increases the computational cost.

This chapter presents a new method for initializing landmarks directly from the first measurement without any delay. Our approach updates the measurement in the 3D camera frame. This is in contrast to previous methods that utilize 2D image coordinates in the image plane. The proposed method was inspired by the measurement model of 3D cameras (such as the SR4000 by Mesa Imaging) and by 3D range sensors that simultaneously measure the range and bearing. A camera is considered as a 3D range sensor but with a considerable amount of depth uncertainty. When a feature is initially measured, the depth value of the feature in the 3D coordinates is set to an arbitrary value with a sufficiently large covariance. It should be noted that the arbitrary value has little effect on the EKF performance due to its large covariance. Another advantage of the proposed scheme is its high degree of linearity. A more linear measurement equation leads to better performance from the EKF. Our approach is much simpler and more efficient than earlier approaches that use multiple hypotheses or an inverse depth technique.

This chapter is organized as follows. Section 3.4.2 presents our EKF framework with the new measurement update model. Section 3.4.3 explains the undelayed initialization scheme using our EKF framework and introduces the methods adopted to implement SLAM. Section 3.4.4 shows the simulation and experimental results. Section 3.4.5 closes the chapter with a conclusion.

3.4.2 EKF Framework

3.4.2.1 Coordinate Frames

For a clear explanation, we will start with a definition of the Cartesian coordinate frames (see Figure 1). The letters *w*, *b*, and *c* denote the world frame, the body frame, and the camera frame, respectively. The origin of the camera frame is the focal point of the camera and that of the body frame is the center of the two driving wheels. The body frame and the camera frame are in the same rigid body.

3.4.2.2 State Vector Definition

A robot moving on the irregular floor of a home environment involves attitude changes. Thus, a 6-DOF (degree of freedom) model is used despite the fact that the robot moves with 3-DOF most of the time.

The state vector to represent the position and orientation of the robot in the world frame is given by

$$\mathrm{x}_v = \begin{bmatrix} \mathrm{x}_{vp} \\ \Psi \end{bmatrix} = \begin{bmatrix} x & y & z & \theta & \varphi & \psi \end{bmatrix}^T \tag{1}$$

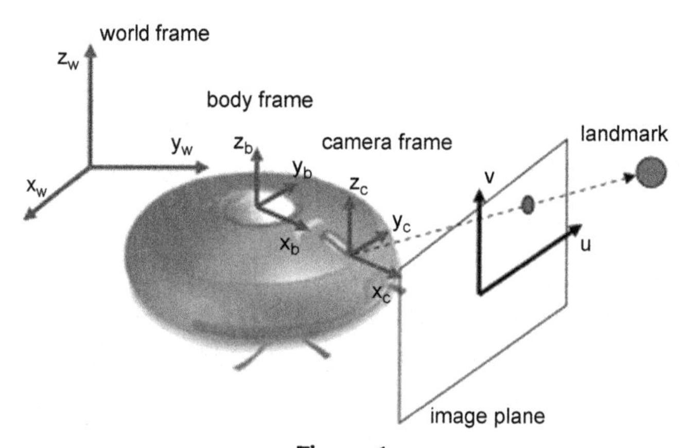

Figure 1
Coordinate frames.

where θ, φ, and ψ are the roll, pitch, and yaw of the Euler angles respectively; x_{vp} is the position vector of the origin of the camera frame and ψ is the orientation vector of the camera. It should be noted that the robot position and orientation are represented by the camera frame rather than the body frame, which can reduce the level of complexity. Because the pitch angle of our robot is always between $-\pi/2$ and $\pi/2$, the Euler angles are more computationally efficient than the quaternion which is commonly adopted for visual SLAM algorithms to use a hand-held camera. Because a dead-reckoning system can provide good prediction information, a velocity term doesn't appear in the state vector. The state vector includes the 3D position vector **y** of the landmark in the world frame, as do the conventional SLAM frameworks. Thus, the state vector **x** and its error covariance matrix **P** for our EKF framework are given as

$$x = \begin{bmatrix} x_v \\ y_1 \\ \vdots \\ y_n \end{bmatrix}, \quad P = \begin{bmatrix} P_x & P_{xy_1} & \cdots & P_{xy_n} \\ & P_{y_1} & & P_{y_1 y_n} \\ \vdots & & \ddots & \vdots \\ P_{y_n x} & & \cdots & P_{y_n} \end{bmatrix} \tag{2}$$

where n is the number of registered landmarks.

3.4.2.3 Prediction Model

Our dead-reckoning system consists of a yaw gyroscope and two wheel encoders. It can calculate a translation in the *XY*-plane of the body frame and the rotation around the *z*-axis of the body frame. Given that it cannot provide any information about the translation in the *Z* direction and the roll and pitch rotations, they are assumed to remain unchanged. Instead, the uncertainty in these values will increase as the robot moves. Feature locations are also assumed to be stationary. Thus the prediction model at step k can be given as

$$x_{vp,k|k-1} = x_{vp,k-1|k-1} + C_c^w \Delta x^c \tag{3}$$

$$\psi_{k|k-1} = \psi_{,k-1|k-1} + \begin{bmatrix} 1 & \sin\theta\tan\varphi & \cos\theta\tan\varphi \\ 0 & \cos\theta & -\sin\theta \\ 0 & \sin\theta\sec\varphi & \cos\theta\sec\varphi \end{bmatrix} \Delta\psi^c \tag{4}$$

$$y_{i,k|k-1} = y_{i,k-1|k-1} \tag{5}$$

where C_c^w is the transformation matrix from the camera frame to the world frame [11]. Δx^c and $\Delta\psi^c$ are the incremental changes in the position vector and the orientation vector in the camera frame respectively. They are simply calculated from the incremental changes in the body frame given by the dead-reckoning system. Equation (3.4.4) is a simple forward difference representation of the partial differential equation which relates the Euler angle to the body rates. It is acceptable if $\Delta\psi^c$ is small enough and if the roll and pitch angles change within small values. EKF predicts the state vector **x** and its covariance matrix **P**, as follows:

$$\mathrm{x}_{v,k|k-1} = \mathrm{f}\left(\mathrm{x}_{v,k-1|k-1}, \mathrm{u}_k\right) \tag{6}$$

$$\mathrm{P}_{\mathrm{x},k|k-1} = \frac{\partial \mathrm{f}}{\partial \mathrm{x}_v}\mathrm{P}_{\mathrm{x},k-1|k-1}\frac{\partial \mathrm{f}^T}{\partial \mathrm{x}_v} + \frac{\partial \mathrm{f}}{\partial \mathrm{u}}Q\frac{\partial \mathrm{f}^T}{\partial \mathrm{u}} \tag{7}$$

$$\mathrm{P}_{\mathrm{xy},k|k-1} = \frac{\partial \mathrm{f}}{\partial \mathrm{x}_v}\mathrm{P}_{\mathrm{xy},k-1|k-1} \tag{8}$$

where

$$\mathrm{u}_k = \begin{bmatrix} \Delta \mathrm{x}^c \\ \Delta\psi^c \end{bmatrix}_k \tag{9}$$

and Q is the covariance matrix of **u**.

3.4.2.4 Measurement Model

In the conventional monocular SLAM, observations are tracked feature points in the image plane. Naturally, all in the case of the previous monocular SLAM, update the measurements in the image plane. This means that the measurement vector **z** is expressed in the image plane. In contrast, all components of the state vector **x** are in the world frame. To perform EKF, the feature positions of the state vector must inevitably be transformed into those in the image plane. The upper block in Figure 2 shows this process. Particularly, the transformation from the camera frame to the image plane can cause a critical linearization error due to the large uncertainty of the depth.

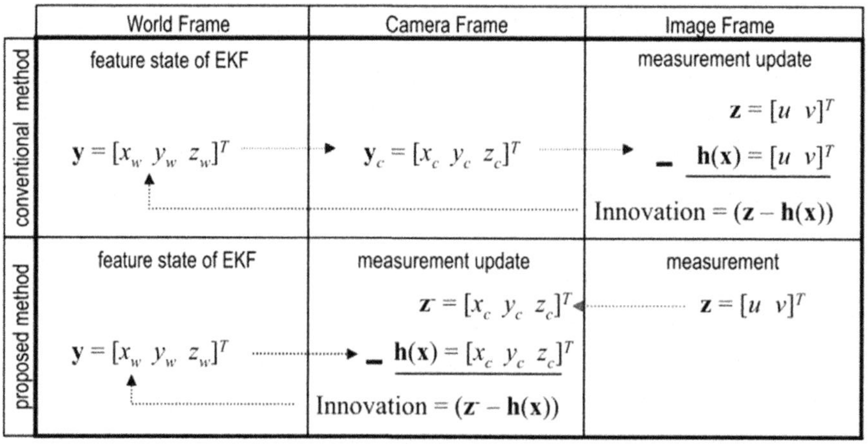

	World Frame	Camera Frame	Image Frame
conventional method	feature state of EKF $\mathbf{y} = [x_w\ y_w\ z_w]^T$	$\mathbf{y}_c = [x_c\ y_c\ z_c]^T$	measurement update $\mathbf{z} = [u\ v]^T$ $-\ \mathbf{h}(\mathbf{x}) = [u\ v]^T$ Innovation = $(\mathbf{z} - \mathbf{h}(\mathbf{x}))$
proposed method	feature state of EKF $\mathbf{y} = [x_w\ y_w\ z_w]^T$	measurement update $\mathbf{z}' = [x_c\ y_c\ z_c]^T$ $-\ \mathbf{h}(\mathbf{x}) = [x_c\ y_c\ z_c]^T$ Innovation = $(\mathbf{z}' - \mathbf{h}(\mathbf{x}))$	measurement $\mathbf{z} = [u\ v]^T$

Figure 2
Measurement model.

To overcome this, we propose a different approach in which the measurement is represented in the camera frame rather than in the image plane. In Figure 2, y_C is the 3D position vector of the landmark in the camera frame. As shown in the lower block in Figure 2, this technique fundamentally removes the linearization error induced by transformation from the camera frame to the image plane. This is described in greater detail in the linearization error analysis in the next section.

Our measurement and its covariance matrix are

$$z = \begin{bmatrix} v_1^c \\ \vdots \\ v_m^c \end{bmatrix}, \quad R = \begin{bmatrix} R_{v1} & & 0 \\ & \ddots & \\ 0 & & R_{vm} \end{bmatrix} \tag{10}$$

where m is the number of observed features, v^c is the feature position in the camera frame, and R_v is the error covariance matrix of v^c. The observation vector **z** requires the 3D coordinates of the feature, referring to Figure 3. This involves the problem of how to convert the 2D information of the camera measurement to 3D information.

Here, we explain how to obtain v^c from the observed value of **v**. The 3×4 camera projection matrix which describes the mapping of a camera from 3D points in the camera frame to 2D points in the image plane is written as

$$C = K_C[M|t] \tag{11}$$

where **M** is the 3×3 rotation matrix, K_C is the camera calibration matrix and **t** is the translation vector which is actually the zero vector because the origin of the camera frame is the focal point. The directional vector c_t in Figure 3 can be obtained by the following equation:

$$c_t = \begin{bmatrix} c_{t1} \\ c_{t2} \\ c_{t3} \end{bmatrix} = (K_C M)^{-1} \begin{bmatrix} v \\ 1 \end{bmatrix} \tag{12}$$

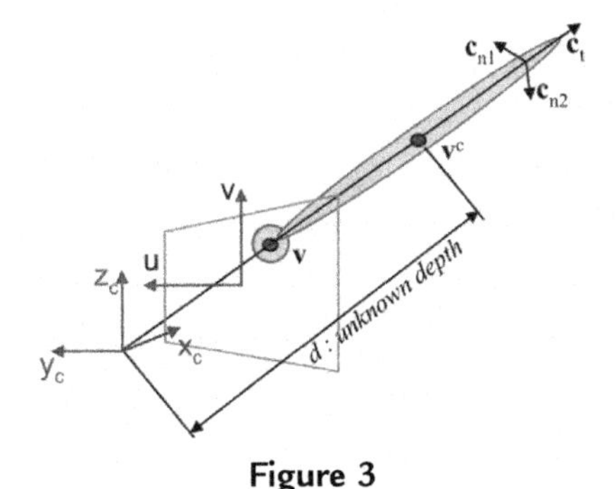

Figure 3
Landmark initialization.

In addition, v^c can be obtained by

$$v^c = d\frac{c_t}{|c_t|} \tag{13}$$

where d is an unknown depth. Here, c_t is in the camera frame. If prior information about the depth is available, d can be estimated with it. Otherwise, d is set to an arbitrary value. For example, when we observe it for the first or second time, there is no information. Therefore, it can be assumed to be the maximum distance of the workspace or another proper value. Instead, the covariance matrix R_v will be used through careful consideration.

To determine R_v, an error covariance of the feature in the (c_t, c_{n1}, c_{n2}) frame is introduced as

$$R_{v0} = \begin{bmatrix} \varepsilon_{ct}^2 & 0 & 0 \\ 0 & \varepsilon_{cn1}^2 & 0 \\ 0 & 0 & \varepsilon_{cn2}^2 \end{bmatrix} \tag{14}$$

where ε_{ct}, ε_{cn1}, and ε_{cn2} are the standard deviations of the error in the c_t, c_{n1}, and c_{n2} directions, respectively (Figure 3). c_t, c_{n1}, and c_{n2} are orthogonal to each other, and c_{n1} and c_{n2} can be chosen in any direction.

The most important point is that the value of ε_{ct} must be set large enough to reflect the uncertainty of the depth. The ideal value of ε_{ct} is an infinite value, but this is not feasible for numerical computation. In our system, a value over 10^5 cm occasionally caused problems during the process of inverting a matrix, and SLAM resulted in failure. It is recommended that

$$3000 \text{ cm} \leq \varepsilon_{ct} \leq 10^5 \text{ cm}. \tag{15}$$

The values above were obtained empirically. They serve simply to provide a guideline when selecting a proper value of ε_{ct}.

Because the value of ε_{ct} cannot actually be infinite, there is one problem that must be addressed. The same observed images can lead to erroneous results. For example, when a robot does not move, it observes nearly identical images and tracks the same points that have the same errors. Given that the Kalman filter was induced with the assumption that there is no correlation between the measurements and predicted states, these measurements are against this assumption. This can lead to an underestimation of the covariance, which means the situation when the estimated covariance is smaller than the true one. This problem also applies to conventional methods, but it is more serious in our approach as it has the additional effect of decreasing the depth of uncertainty incorrectly. To avoid this, a measurement should not be updated when the robot does not move.

The values of ε_{cn1} and ε_{cn2} depend approximately on the tracking error, calibration error, and the depth such that

$$\varepsilon_{cn} \approx \sqrt{\varepsilon_{\mathrm{trk}}^2 + \varepsilon_{\mathrm{cal}}^2}\,\frac{d}{f_p} \tag{16}$$

where $\varepsilon_{\mathrm{trk}}$ is the standard deviation of the image feature tracking error, $\varepsilon_{\mathrm{cal}}$ is the standard deviation of the camera calibration error, and f_p is the focal length in pixel units.

After determining R_{v0}, R_v is determined by solving the following equation.

$$\mathrm{R}_v = \mathrm{M}_{ct}\mathrm{R}_{v0}\mathrm{M}_{ct}^T \tag{17}$$

where M_{ct} is the rotation matrix that makes the $\boldsymbol{x}$-axis parallel to the c_t-direction. M_{ct} can be calculated from c_t as follows:

$$\mathrm{M}_{ct} = \begin{bmatrix} \cos\psi_{ct} & -\sin\psi_{ct} & 0 \\ \sin\psi_{ct} & \cos\psi_{ct} & 0 \\ 0 & 0 & 1 \end{bmatrix} \begin{bmatrix} \cos\varphi_{ct} & 0 & \sin\varphi_{ct} \\ 0 & 1 & 0 \\ -\sin\varphi_{ct} & 0 & \cos\varphi_{ct} \end{bmatrix} \tag{18}$$

where ψ_{ct} and φ_{ct} are defined as follows

$$\cos\psi_{ct} = \frac{c_{t1}}{\sqrt{c_{t1}^2 + c_{t2}^2}}, \quad \sin\psi_{ct} = \frac{c_{t2}}{\sqrt{c_{t1}^2 + c_{t2}^2}} \tag{19}$$

$$\cos\varphi_{ct} = \frac{\sqrt{c_{t1}^2 + c_{t2}^2}}{\sqrt{c_{t1}^2 + c_{t2}^2 + c_{t3}^2}}, \quad \sin\varphi_{ct} = \frac{-c_{t3}}{\sqrt{c_{t1}^2 + c_{t2}^2 + c_{t3}^2}} \tag{20}$$

In the next step, it is necessary to transform the features of the state vector to the camera frame. The measurement model of the j-th feature is given as

$$\mathrm{h}_j(\mathrm{x}_v, \mathrm{y}_j) = \mathrm{M}_w^c(\mathrm{y}_j - \mathrm{x}_{vp}) \tag{21}$$

where M_w^c is the rotation matrix from the world frame to the camera frame. The Jacobian matrix H_j of function h_j is given as

$$\mathrm{H}_j = \left[-\mathrm{M}_w^c \middle| \delta\mathrm{M}_{w,j}^c \middle| \mathrm{M}_w^c\right] \tag{22}$$

$$\delta\mathrm{M}_{w,j}^c = \left[\frac{\partial\mathrm{M}_w^c}{\partial\theta}\Delta_j \middle| \frac{\partial\mathrm{M}_w^c}{\partial\varphi}\Delta_j \middle| \frac{\partial\mathrm{M}_w^c}{\partial\psi}\Delta_j\right] \tag{23}$$

where

$$\Delta_j = \mathrm{y}_j - \mathrm{x}_{vp}. \tag{24}$$

At this point, we can obtain h and H by arranging h_j and H_j. The measurement update equations are well known, as follows:

$$x_{k|k} = x_{k|k-1} + K\left(z_k - h\left(x_{k|k-1}\right)\right) \tag{25}$$

$$P_{k|k} = \left(I - KH\right)P_{k|k-1} \tag{26}$$

$$K = P_{k|k}H^T\left(HP_{k|k}H^T + R_k\right)^{-1}. \tag{27}$$

3.4.2.5 Linearization Error Analysis

EKF SLAM has a drawback in that linearization errors can make a system unstable. The trigonometric functions and the perspective projection are the main factors that cause this. Equations (3) and (4) in the prediction process and Eqn (21) in the measurement process have nonlinearities due to the trigonometric functions of the robot orientation. When the orientation error is small, the trigonometric linearization error does not cause much trouble. A precise dead-reckoning system or a high update rate can be helpful to keep the orientation error small. Our dead-reckoning system is designed well enough to give about a 1.5° of yaw angle error after 10 rotations [1]. Furthermore, multi-map approaches [12,13] can reduce these trigonometric nonlinearities, as they maintain small orientation errors in local maps. The iterated Kalman filter framework [14] can also decrease the overall nonlinearities by applying an optimization theory to EKF, but this increases the computation time in keeping with the increased number of iterations.

Especially with regard to feature initialization, the nonlinearity of the perspective projection is more critical because the depth directional uncertainty is very large. Therefore, we focus on this aspect in this chapter. Our method is compared with two previous methods: one is the conventional method to use a measurement model expressed in the image plane, which includes most of the delayed feature initialization of SLAM and the multiple hypothesis method [6,7] for undelayed feature initialization. The other is the inverse depth method [8–10].

Figure 4 shows measurement models and their Jacobians. To simplify this analysis, function h in Eqn (25) is decomposed into two functions:

$$h(x_v, y) = h_z(y_c) = h_z(h_y(x_v, y)) = h_z \circ h_y(x_v, y). \tag{28}$$

h_z is introduced to transform y_c into the measurement $\mathbf{z}$ and h_y is a function to satisfy $y_c = h_y(x_v, y)$.

	conventional method	inverse depth method	proposed method
feature's state	$\mathbf{y}=[x_w\ y_w\ z_w]^T$	$\mathbf{y}=[x_\rho\ y_\rho\ z_\rho\ \theta_\rho\ \varphi_\rho\ \rho]^T$ $=[\mathbf{x}_\rho \mid \theta_\rho\ \varphi_\rho\ \rho]^T$	$\mathbf{y}=[x_w\ y_w\ z_w]^T$
measurement	$\mathbf{z}=[u\ v]^T$	$\mathbf{z}=[u\ v]^T$	$\mathbf{z}=[x_c\ y_c\ z_c]^T$
$\mathbf{y}_c=\mathbf{h}_y(\mathbf{x}_v,\mathbf{y})$	$\mathbf{y}_c=\mathbf{M}_w^c(\mathbf{y}-\mathbf{x}_{vp})$	$\mathbf{y}_c=\mathbf{M}_w^c(\mathbf{x}_\rho+\frac{1}{\rho}\mathbf{m}-\mathbf{x}_{vp})$	$\mathbf{y}_c=\mathbf{M}_w^c(\mathbf{y}-\mathbf{x}_{vp})$
$\mathbf{h}=\mathbf{h}_z(\mathbf{y}_c)$	$\mathbf{h}=\begin{bmatrix}u\\v\end{bmatrix}=\begin{bmatrix}u_0+f_u\frac{y_c}{x_c}\\v_0+f_v\frac{z_c}{x_c}\end{bmatrix}$	$\mathbf{h}=\begin{bmatrix}u\\v\end{bmatrix}=\begin{bmatrix}u_0+f_u\frac{y_c}{x_c}\\v_0+f_v\frac{z_c}{x_c}\end{bmatrix}$	$\mathbf{h}=\begin{bmatrix}x_c\\y_c\\z_c\end{bmatrix}=\mathbf{y}_c$
Jacobian of $\mathbf{h}_y$ $\mathbf{J}_{hy}(\mathbf{x}_{vp},\mathbf{y})=\frac{\partial\mathbf{y}_c}{\partial(\mathbf{x}_{vp},\mathbf{y})}$	$\mathbf{J}_{hy}(\mathbf{x}_{vp},\mathbf{y})=\left[-\mathbf{M}_w^c \mid \mathbf{M}_w^c\right]$	$\mathbf{J}_{hy}(\mathbf{x}_{vp},\mathbf{x}_\rho,\rho)$ $=\left[-\mathbf{M}_w^c \mid \mathbf{M}_w^c \mid -\frac{1}{\rho^2}\mathbf{M}_w^c\mathbf{m}\right]$	$\mathbf{J}_{hy}(\mathbf{x}_{vp},\mathbf{y})=\left[-\mathbf{M}_w^c \mid \mathbf{M}_w^c\right]$
Jacobian of $\mathbf{h}_z$ $\mathbf{J}_{hz}(\mathbf{y}_c)=\frac{\partial\mathbf{h}}{\partial\mathbf{y}_c}$	$\mathbf{J}_{hz}(\mathbf{y}_c)=\begin{bmatrix}-\frac{f_u y_c}{x_c^2} & \frac{f_u}{x_c} & 0\\-\frac{f_v z_c}{x_c^2} & 0 & \frac{f_v}{x_c}\end{bmatrix}$	$\mathbf{J}_{hz}(\mathbf{y}_c)=\begin{bmatrix}-\frac{f_u y_c}{x_c^2} & \frac{f_u}{x_c} & 0\\-\frac{f_v z_c}{x_c^2} & 0 & \frac{f_v}{x_c}\end{bmatrix}$	$\mathbf{J}_{hz}(\mathbf{y}_c)=\left[\mathbf{I}_{3\times3}\right]$

Figure 4
Extended Kalman Filter measurement model comparison.

C in (11) is assumed to be a simple pinhole camera model based on our coordinate system defined in Figure 1:

$$\mathrm{C}=\begin{bmatrix}0 & f_u & 0 & 0\\0 & 0 & f_v & 0\\1 & 0 & 0 & 0\end{bmatrix} \tag{29}$$

where f_u and f_v are the focal length in pixel units with respect to each of the values of u and v in the image plane. For the inverse depth method in Figure 4, x_ρ, θ_ρ, φ_ρ, and ρ are the position of the camera frame, the feature's azimuth, the elevation and the inverse depth w. r. t. and the world frame at the first observation of a feature, respectively. $\mathbf{m}$ is the unit depth directional vector. (Refer to Ref. [10] for more details.)

To analyze nonlinearity for only the perspective projection, the following assumption is made:

Assumption 1. The robot's orientation has no error.

This implies that ψ and M_w^c are assumed to be constant; thus, only x_{vp} and $\mathbf{y}$ in the state are variables. Using the chain rule, the Jacobian matrix of $\mathbf{h}$ can be derived as

$$\mathrm{J}_h(\mathrm{x}_{vp},\mathrm{y})=\mathrm{J}_{hz}(\mathrm{y}_c)\mathrm{J}_{hy}(\mathrm{x}_{vp},\mathrm{y}). \tag{30}$$

If all elements in the Jacobian matrix are constant, it is a perfectly linear function.

First, we analyze our method. The Jacobian of **h** based on the proposed method is

$$\mathrm{J}_h^p(\mathrm{x}_{vp}, \mathrm{y}) = \left[-\mathrm{M}_w^c \middle| \mathrm{M}_w^c\right]. \tag{31}$$

All its elements are obviously constant. Consequently, it is clear that our measurement model has no perspective linearization error under Assumption 1.

Second, the Jacobian of **h** based on the conventional method is

$$\mathrm{J}_h^c(\mathrm{x}_{vp}, \mathrm{y}) = \begin{bmatrix} -\dfrac{f_u y_c}{x_c^2} & \dfrac{f_u}{x_c} & 0 \\ -\dfrac{f_v z_c}{x_c^2} & 0 & \dfrac{f_v}{x_c} \end{bmatrix} \left[-\mathrm{M}_w^c \middle| \mathrm{M}_w^c\right]. \tag{32}$$

Its elements include highly nonlinear functions of x_c. When features are first observed and x_c-axis is parallel to the depth direction, x_c has significantly high uncertainty. This will lead to the divergence of the EKF SLAM. To reduce the depth directional uncertainty, the multiple hypothesis approach divides one feature with large uncertainty into several features with much less uncertainty.

Lastly, we will obtain the Jacobian of **h** of the inverse depth method. Among the six parameters of the inverse depth method, the others except ρ have much smaller uncertainties than ρ. For this reason, $\partial \mathrm{h}/\partial \theta_\rho$ and $\partial \mathrm{h}/\partial \varphi_\rho$ are neglected here. Then, the Jacobian of **h** based on the inverse depth method is

$$\mathrm{J}_h^\rho(\mathrm{x}_{vp}, \mathrm{x}_\rho, \rho) = \begin{bmatrix} -\dfrac{f_u y_c}{x_c^2} & \dfrac{f_u}{x_c} & 0 \\ -\dfrac{f_v z_c}{x_c^2} & 0 & \dfrac{f_v}{x_c} \end{bmatrix} \left[-\mathrm{M}_w^c \middle| \mathrm{M}_w^c \middle| -\frac{1}{\rho^2}\mathrm{M}_w^c \mathrm{m}\right] \tag{33}$$

This appears more nonlinear than Eqn (32). To examine the linearization error of the Jacobian with respect to ρ, we neglect $\partial \mathrm{h}/\partial \mathrm{x}_{vp}$ and $\partial \mathrm{h}/\partial \mathrm{x}_\rho$ by introducing the following assumption.

Assumption 2. ρ is the only variable in the Jacobian.

For simple explanation, the following definitions are introduced.

$$\mathrm{m}_c = \begin{bmatrix} m_{cx} \\ m_{cy} \\ m_{cz} \end{bmatrix} = \mathrm{M}_w^c \mathrm{m}, \quad \mathrm{x}_{0c} = \begin{bmatrix} x_{0c} \\ y_{0c} \\ z_{0c} \end{bmatrix} = \mathrm{M}_w^c(\mathrm{x}_\rho - \mathrm{x}_{vp}) \tag{34}$$

Using Eqn (34), y_c can be derived as

$$\mathrm{y}_c = \begin{bmatrix} x_c \\ y_c \\ z_c \end{bmatrix} = \mathrm{M}_w^c\left(\mathrm{x}_\rho + \frac{1}{\rho}\mathrm{m} - \mathrm{x}_{vp}\right) = \mathrm{x}_{0c} + \frac{1}{\rho}\mathrm{m}_c. \tag{35}$$

With Assumption 2, the Jacobian can be determined by

$$J_h^\rho(\rho) = \begin{bmatrix} -\frac{f_u y_c}{x_c^2} & \frac{f_u}{x_c} & 0 \\ -\frac{f_v z_c}{x_c^2} & 0 & \frac{f_v}{x_c} \end{bmatrix}\left[-\frac{\mathrm{m}_c}{\rho^2}\right] = \begin{bmatrix} \frac{f_u y_c}{x_c^2\rho^2} m_{cx} - \frac{f_u}{x_c\rho^2} m_{cy} \\ \frac{f_v z_c}{x_c^2} m_{cx} - \frac{f_v}{x_c\rho^2} m_{cz} \end{bmatrix} \tag{36}$$

Finally, inserting Eqn (35) into Eqn (36), the Jacobian with respect to ρ_d becomes

$$\mathrm{J}_h^\rho(\rho) = \begin{bmatrix} f_u \frac{m_{cx}y_{0c} - m_{cy}x_{0c}}{(x_{0c}\rho + m_{cx})^2} \\ f_v \frac{m_{cx}z_{0c} - m_{cz}x_{0c}}{(x_{0c}\rho + m_{cx})^2} \end{bmatrix} \tag{37}$$

This still appears to be nonlinear. However, it should be noted that if the dimensionless term $x_{0c}\rho$ is much smaller than m_{cx}, Eqn (37) becomes a linear function, as follows.

Assumption 3. $x_{0c}\rho \ll m_{cx}$.

This assumption leads to

$$\mathrm{J}_h^\rho(\rho) \approx \begin{bmatrix} f_u \frac{m_{cx}y_{0c} - m_{cy}x_{0c}}{m_{cx}^2} \\ f_v \frac{m_{cx}z_{0c} - m_{cz}x_{0c}}{m_{cx}^2} \end{bmatrix} \tag{38}$$

This shows that $\mathrm{J}_h^\rho(\rho)$ is approximately independent of ρ and constant under Assumptions 1–3. The validity of Assumption three needs to be checked. The minimum value of m_{cx} is 0.5 with a 120° field of view, for example: ρ becomes very small when a feature is very distant. Thus, the inverse depth method will perform well with distant features. In addition, x_{0c} is zero if the camera is moving only in the directions of the y_c- and z_c-axes. In contrast, moving in the direction of the x_c-axis will increase x_{0c}. Thus, the inverse depth method will perform better in case of lateral movement compared to forward and backward movements.

In conclusion, the conventional method has too much non-linearity to initialize a feature without a delay. The multiple hypothesis method divides the nonlinearity by the number of hypotheses, making an undelayed initialization possible. The inverse depth method can be considered as nearly linear under Assumptions 1–3. The proposed method is perfectly linear only under Assumption 1. This explains why the proposed EKF SLAM does not diverge in spite of such a large error covariance. These analyses will be verified with the simulation in Section 3.4.4.3.

3.4.3 Implementation of SLAM

In the following section, we briefly introduce the implementation of the proposed EKF-based SLAM.

3.4.3.1 Undelayed Feature Initialization

In the previous section, we proposed a new approach to update the measurement in EKF. Our feature initialization concept works along the same lines but with one additional step. The other processes are identical to how the measurement update operates. The additional step is transformation from the camera frame to the world frame. Overall, the newly observed feature $\mathbf{v}$ will be transformed to v^c and then to $\mathbf{y}$ for the initialization. v^c can be obtained from $\mathbf{v}$ using Eqns (12) and (13). Its covariance matrix R_v can also be determined by Eqns (14)–(20). Subsequently, v^c and R_v are transformed to y and P_y, as follows:

$$y = x_{vp} + M_c^w v^c \tag{39}$$

$$P_y = T_c^w P_x T_c^{wT} + M_c^w R_v M_c^{wT} \tag{40}$$

where M_c^w is the rotation matrix from the camera frame to the world frame, and

$$T_c^w = \left[I_{3\times 3} \middle| \delta M_c^w \right] \tag{41}$$

$$\delta M_c^w = \left[\frac{\partial M_c^w}{\partial \theta} v^c \middle| \frac{\partial M_c^w}{\partial \varphi} v^c \middle| \frac{\partial M_c^w}{\partial \psi} v^c \right]. \tag{42}$$

After obtaining y and P_y, they are added to the state vector $\mathbf{x}$. Its covariance matrix $\mathbf{P}$ is updated as follows:

$$P_{y_j} = P_y \tag{43}$$

$$P_{y_j x} = T_c^w P_x, \quad P_{x y_j} = P_{y_j x}^T \tag{44}$$

$$P_{y_j y_i} = T_c^w P_{x y_i}, \quad P_{y_i y_j} = P_{y_j y_i}^T \tag{45}$$

where j is the index of the feature to add, and i is one of the other feature indices except j.

This initialization method is much simpler than any other method. For example, the inverse depth method [9,10] needs an additional step in which a feature's state vector (of six dimensions) is reduced to one simple position vector (of three dimensions) after the feature's uncertainty has become small enough. But the proposed method does not require such a post-processing step. We think this simplicity is the most important advantage of the proposed method. It is the measurement update in the camera frame that enables such a simple initialization.

From a different point of view, the proposed method does not appear to accurately express the error distribution. Its ideal shape is conic as shown in Figure 5. However, the shape expressed by a covariance matrix in the Cartesian coordinate is essentially an ellipse. Multiple hypothesis approaches use several hypotheses to make the shape as conical a shape as possible. The inverse depth approach also initializes a feature with conically shaped uncertainties using six parameters. Ours, in contrast, has an oval shape instead of a conical shape. This misrepresentation of the error distribution can induce problems such as underestimation or overestimation. These underestimated and overestimated regions are shown in Figure 5. The underestimation region becomes smaller as d increases while the opposite is true for the overestimation region. The underestimation may be worse than the overestimation because the former can somewhat fail SLAM, whereas the latter mainly increases the convergence time and causes a wider search region. Thus, we recommend that d is given the maximum distance of the workspace if there is no prior information. On the other hand, this implies that our method does not work well for features at infinity. Therefore, our method is recommended for use in indoor environments.

3.4.3.2 Feature Extraction

Image features should be persistent and reliable to perform long-term tracking. There are many methods that can be used to select salient image patches from an image. Although

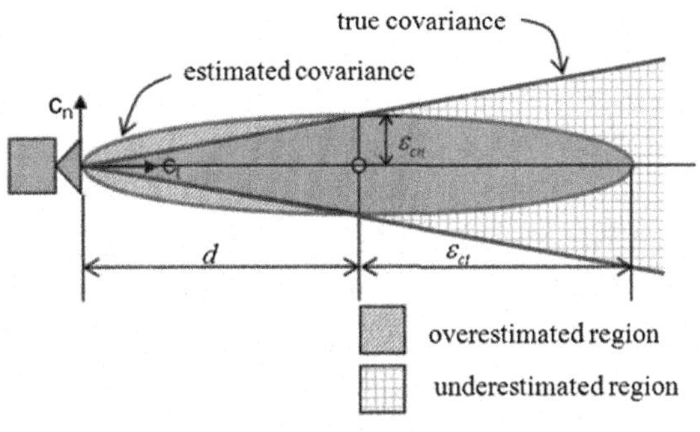

Figure 5
Covariance analysis.

the Scale-Invariant Feature Transform (SIFT)-based algorithm is one of the most commonly used methods at present, it could not be chosen here as it requires more computational power than our target system (ARM processor) can supply. Therefore, we adopted the Harris corner detector [15] due to its high computation efficiency. The number of features in one image is maintained between 5 and 10.

3.4.3.3 Search Region

To track the features efficiently, we set the search region on the image. By reducing the search region, we can reduce the time necessary for the tracking procedure. However, if the region is too small, feature tracking failure occurs more frequently. The search region is expressed as the estimated position of the feature in the image plane and its error covariance matrix.

3.4.3.4 Feature Tracking

As a tracking method, we basically utilize the Lucas–Kanade method [16] with only 2-DOF translation using patches of 15×15 pixels. The initial points are selected as Harris corners inside a search region. The Lucas–Kanade method finds the point that has the minimum Sum of Squared Difference (SSD) value. SSD is not robust against illumination changes. Thus, we modify the Lucas–Kanade method to find the minimum point of the zero-mean SSD and validate it with Normalized Cross Correlation, which makes data association much more robust.

Cleaning robots move along a cleaning path in the manner of a seed sowing machine. This consists of moving forward followed by a 90° rotation. Moving forward involves feature scale changes, which leads to a long-term tracking failure. Therefore, we adopted a feature-scale-prediction scheme. It is similar to predictive multi-resolution descriptors [17] which utilize the descriptors of the images of different resolutions. However, we directly utilize a scale-transformed patch image with an estimated scale without using descriptors.

3.4.3.5 Removing Outliers

The above tracking method gives robust matching pairs. However, because moving objects or repeated patterns can exist, matching errors, also known as outliers, are inevitable and induce critical errors in the EKF SLAM. To remove these, there are two popular methods: Random Sample Consensus (RANSAC) [18] and Joint Compatibility Branch and Bound (JCBB) [19,20]. In home environments, the number of features is occasionally small. JCBB can better deal with a smaller number of matching pairs using the error covariance of the features compared to RANSAC. Moreover, as our tracking method gives only one candidate or nothing for each feature, we use the simplified version of JCBB [20].

However, if there are less than three inliers, JCBB does not work well either. In this case, EKF should skip updating with the current frame and go to the next frame.

3.4.4 Simulation and Experiment

We performed a number of simulations and experiments to validate our approach. The simulations focused on looking into how exactly the positions of the landmarks are calculated in comparison with the previous methods. Additionally, the overall performance of SLAM was validated through actual experiments.

3.4.4.1 Simulation in Ideal Situation

This simulation is intended to show how accurately our method can estimate the location of landmarks in an ideal situation. Thus, the camera is assumed to be ideal with infinite resolutions, no noise, and no lens distortion. In addition, feature tracking is assumed to be perfect. Referring to Figure 6(a), the robot is located in the initial position and a landmark is observed initially. The observed landmark is initialized to $y_{j,0}$ with an assumed depth. Then, the robot moves forward 15 cm and the landmark position is updated to $y_{j,1}$(Figure 6(b)), which is compared to the true position $y_{j,\mathrm{true}}$. Figure 7 shows the position difference between $y_{j,1}$ and $y_{j,\mathrm{true}}$. Ten landmarks are located approximately 10 m away in front of the robot. $y_{j,\mathrm{true}}$ is marked with blue diamonds and red circles represent$y_{j,1}$. The red ellipses are the 1σ bounds of the error covariances. Figure 7 shows very small errors between the true and the estimated even though EKF updates only one time.

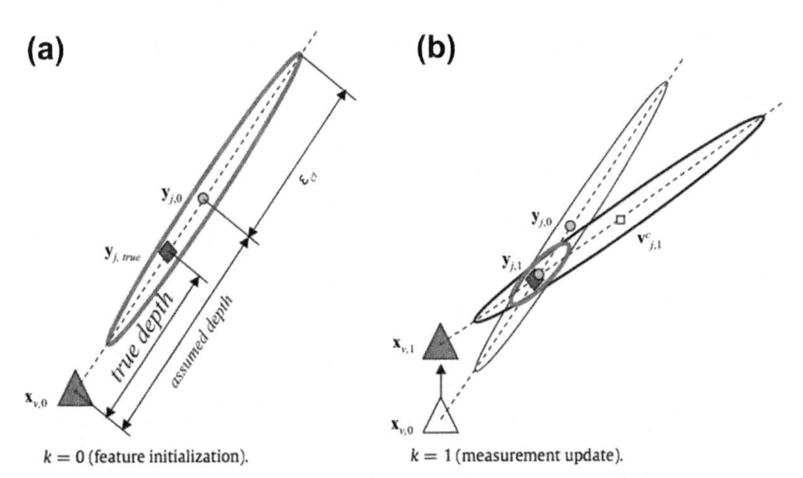

Figure 6
Simulation of one step update of Extended Kalman Filter (diamond: the true position of the landmark, circle: the position that Simultaneous Localization and Mapping estimates, triangle: robot position). (a) k = 0 (feature initialization) and (b) k = 1 (measurement update).

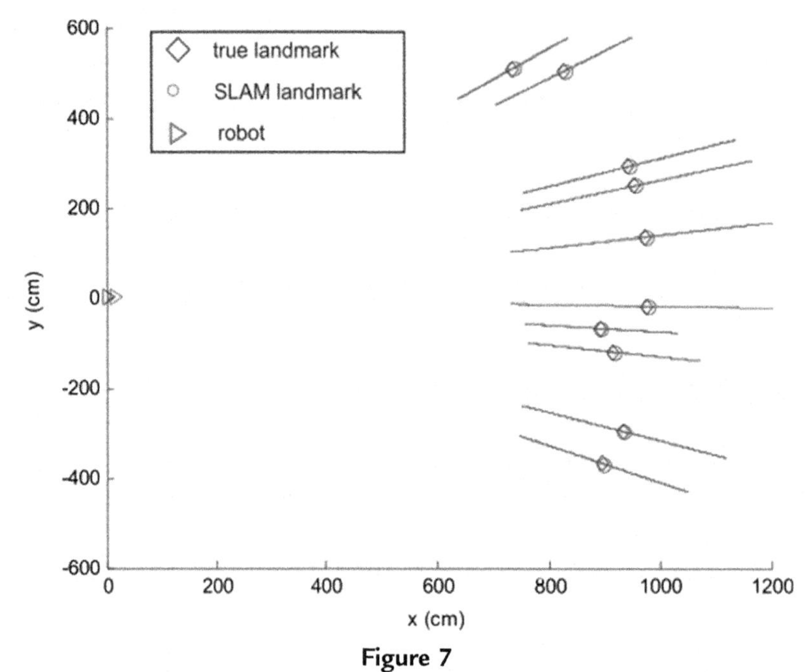

Figure 7
Simulation result in an ideal situation. The true landmark positions, the estimated positions, and the robot location are shown (ε_{ct} = 1e4 cm, assumed depth = 2000 cm).

In Figure 8, the root mean square error of ($y_{j,\text{true}} - y_{j,1}$) for 10 landmarks is shown varying ε_{ct} from 200 to 100,000. Also, the difference between the assumed depth and the true depth is changed from 0 to 30 m. As ε_{ct} is increased, the smaller error of the feature location can be obtained. The result shows that a large value of ε_{ct} can lead to nearly zero errors with only a one step calculation, even though the assumed value of the depth differs greatly from the true one.

3.4.4.2 Simulation with Noise

To simulate the more realistic situation, we adopted two noise sources of feature tracking error and yaw orientation error. Normally distributed random noise was added into the tracked feature point **v** to include feature tracking error. The standard deviation of the noise was set to 0.2 pixels. To include the yaw orientation error, Gaussian noise is added to $\Delta\psi^c$ of the prediction model of Eqn (3.4.4) at every 15 cm movement of the robot. Its standard deviation was set to 0.1°, which was chosen from the experimental data of our yaw gyro sensor. Contrary to the fact that two consecutive frames give a very small error in an ideal situation, 11 frames are used in a noisy situation while the robot moves forward 150 cm. The other simulation parameters were identical to the previous ones of the ideal situation. As shown in Figure 9, ε_{ct} with a value of more than 3000 cm does not

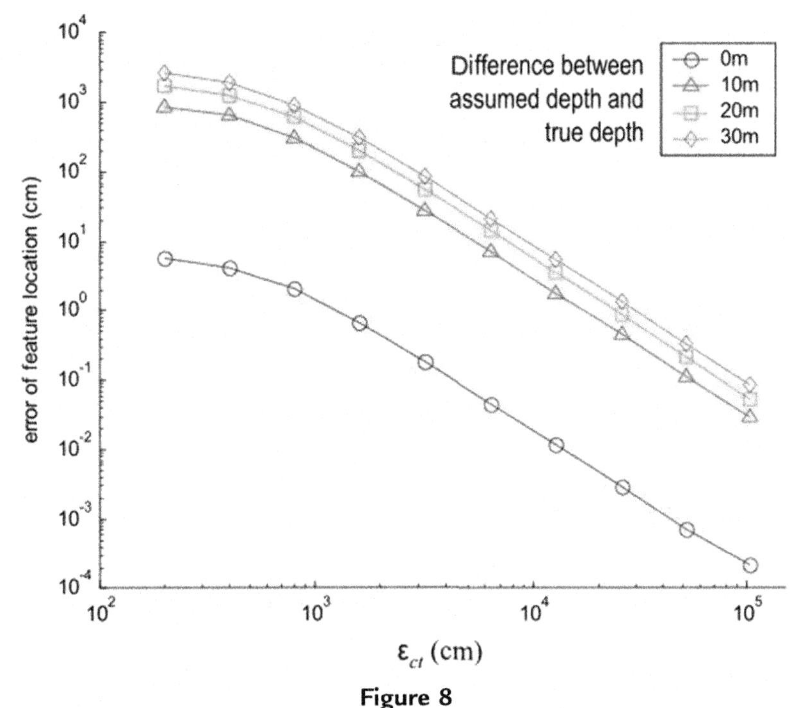

Figure 8
Simulation result in an ideal situation using only two images while the robot moves forward 15 cm.

improve the accuracy any further. From this simulation result, we recommend a value of more than 3000 cm for ε_{ct}. This, of course, depends on the accuracy of the vision system and the size of the environment. On a more accurate vision system in a wider workspace, a larger value than 3000 cm can be chosen.

3.4.4.3 Comparison with Previous Methods

The proposed method was analytically compared to the earlier methods in Section 3.4.2.5, and this is validated through simulations in this section. A feature point is placed from 500 to 1500 cm in front of the initial position of the robot. Its azimuth and elevation are 22° and 15°, respectively. The assumed depth is fixed at 1000 cm. The assumed inverse depth is naturally 1/1000 (1/cm). There is no noise. To confirm the performance difference of the inverse depth method between lateral and forward (or backward) movements, the robot moves forward 30 cm in Figure 10(a) and moves to the left 30 cm laterally without changing the yaw angle in Figure 10(b). As for the simulation parameters of the proposed method, ε_{ct} and ε_{cn} are set to 10^5 and 0.5 cm, respectively. In the inverse depth method, the standard deviation of the inverse depth (ρ) is set to 0.005 (1/cm) as recommended in the earlier studies [10], and the standard deviation of the measurement error in the image

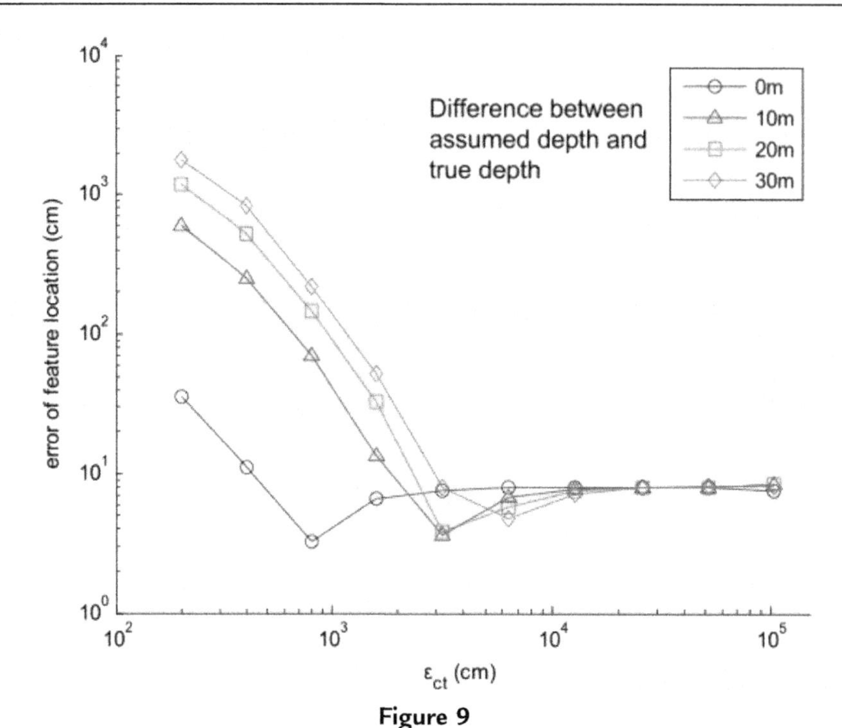

Figure 9
Simulation result with noise using 11 frames while the robot moves forward 150 cm.

plane is set to 0.1 pixels. In the conventional method, a feature is initialized as our method and then updated in the image plane. Figure 10(a) shows that our method has almost no errors even when the true depth of the feature is 1500 cm and the assumed depth is 1000 cm. The conventional method has a very large error in the feature location when the error between the assumed depth and the true depth becomes large. This explains why the multiple hypothesis method was introduced to cover the large distance uncertainty. The multiple hypothesis method can divide the error of the conventional method by the number of hypotheses. According to the graph, the multiple hypothesis method requires approximately five hypotheses until it reaches a similar error as the inverse depth method within the horizontal axis range of the graph. Figure 10(b) shows an interesting result, in which the inverse depth method has no error, similar to the proposed method. This precisely coincides with our analysis in Section 3.4.2.5, which stated that if there is no movement in the direction of the x_c-axis, the inverse depth method is virtually linear. Consequently, considering the linearization error, the proposed method is the best, whereas the inverse depth method appears to be better than the multiple hypothesis method.

Figure 11 shows the comparison between the inverse depth method and the proposed method using experimental data. The robot moves forward 130 cm and the Kalman filter updates 11 times. Eight landmarks are located on the front wall as shown in Figure 13.

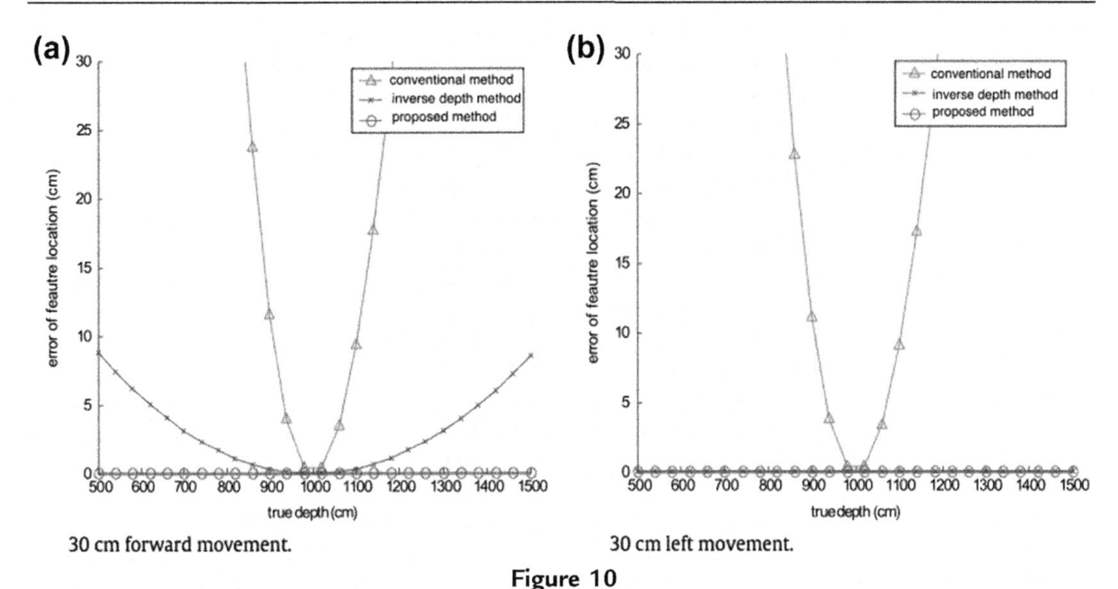

Figure 10

One step estimation for landmark location (assumed depth = 1000 cm). (a) 30 cm forward movement and (b) 30 cm left movement.

The true positions of the landmarks were not measured during the experiment. The converged locations of the landmarks of the two algorithms are the same, so we assume that the converged points are true positions. In the proposed method, ε_{ct} and ε_{cn} are set to 105 cm and 10 cm, respectively and the assumed depth d is 1000 cm. In the inverse depth method, the standard deviation of the inverse depth (ρ) is set to 0.005 (1/cm) and the standard deviation of the measurement error in the image plane is set to 2 pixels. The assumed inverse depth ρ is 1/1000 (1/cm) at the initial time. The average value of

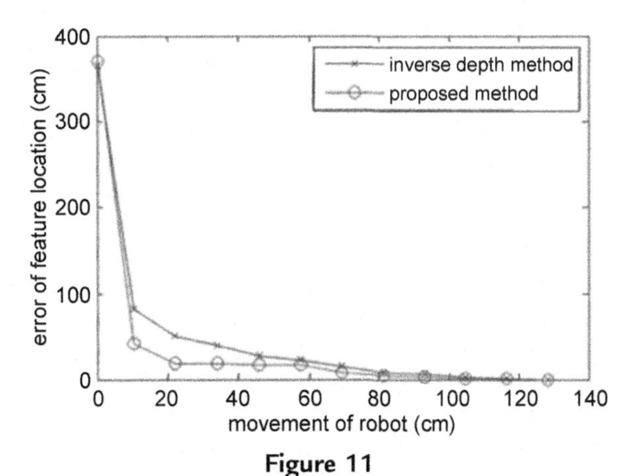

Figure 11

Comparison between the inverse depth method and the proposed method using experimental data.

Figure 12
Robot system used in the experiments.

the feature location error is plotted in Figure 11. Two methods have almost the same error after a few updates of the Kalman filters. It should be noted that the proposed method has advantages on reduced computation time due to the reduced number of states (three). In the inverse depth approach, a feature is represented by the state vector of six dimensions at the first observation—three for the camera position and two for the direction of the ray, and one for the reciprocal depth. This is an over-parameterization compared to dimension three for the Euclidean *XYZ* coding and hence increases the computational cost, although they used three variables after convergence. We just use three variables to represent a feature from the start, which simplifies the overall computation.

3.4.4.4 Experimental Results

We conducted experiments in an indoor environment. The Samsung cleaning robot VC-RE70V, which is commercially available in the market, is used as a robot platform [1]. The robot has one camera of 320×240 resolution that faces upward in order to capture ceiling images. We adjusted the camera orientation to capture front view images as shown in Figure 12. The dead-reckoning system includes two incremental encoders that measure the rotation of each motor, and an Inertial Measurement Unit that provides measures of the robot's linear accelerations and yaw angular rate. Three-axis linear accelerations were used to improve the dead-reckoning performance during the wheel slip [2]. An XV-3500 gyroscope from EPSON was used to measure the yaw angle.

Ground truth data are obtained by a Hawk digital motion tracker system produced by Motion Analysis Inc., which has accuracy of less than 5 mm. A monochrome image sequence is acquired when the robot moves along the cleaning path at a speed of 20 cm/s.

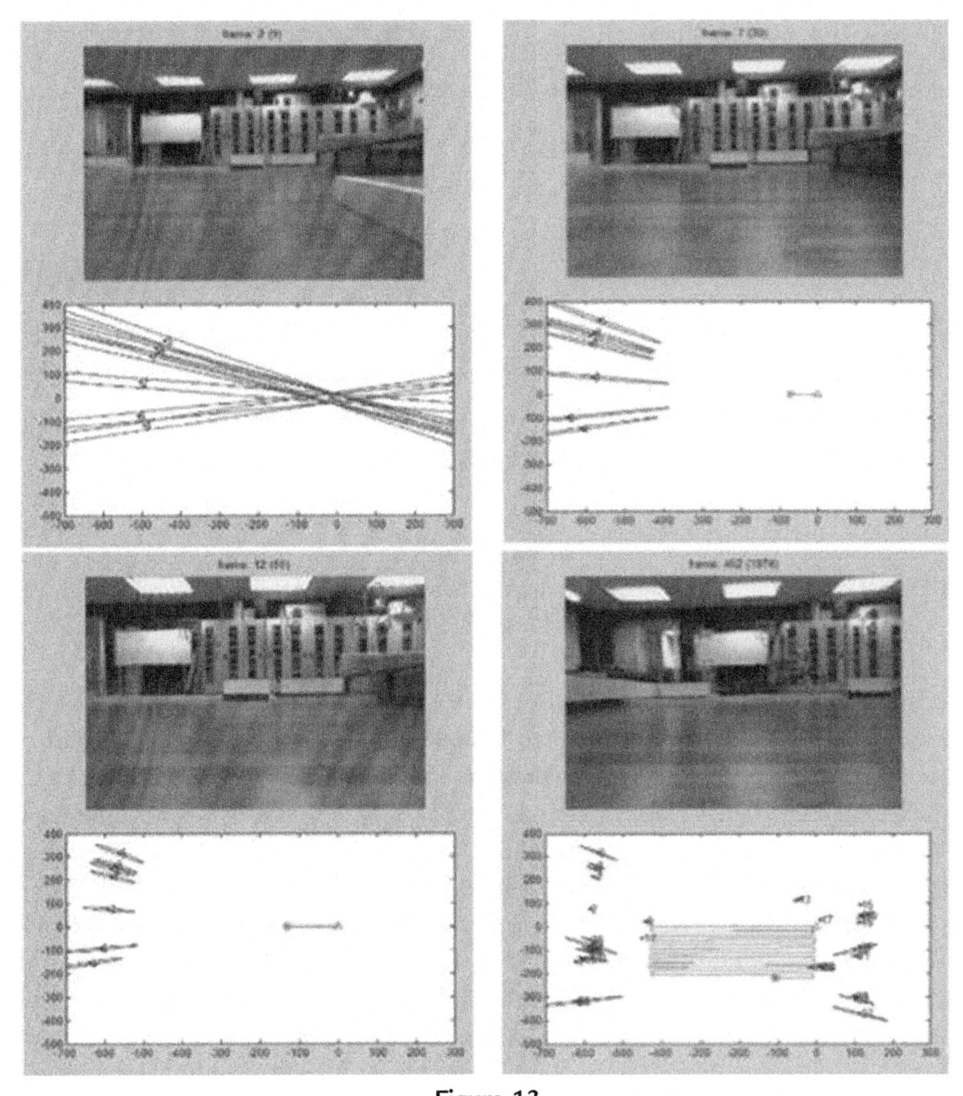

Figure 13
Upper rows: images captured by a camera, lower rows: estimated robot and landmark positions in the *XY*-plane (cm).

At the same time, the data from the dead-reckoning system and the motion tracker system are saved. The sampling rates of the encoder and gyroscope were 20 and 100 Hz, respectively. The frame rate of the camera is only 1–2 frames per second. Our system does not require a high frame rate due to the dead-reckoning system. The SLAM algorithm is run off-line using Matlab.

The results are shown in Figures 13 and 14 and the accompanying video clip entitled slam_result_01.avi is also available. The top left image in Figure 13 shows the seven

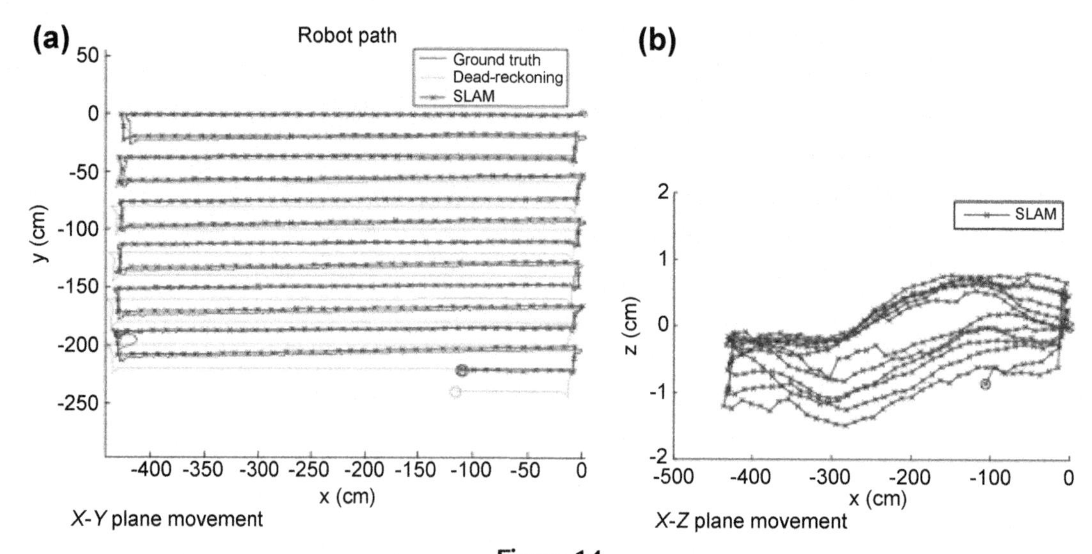

Figure 14
The accuracy of the robot's position estimation. (a) X–Y plane movement and (b) X–Z plane movement.

initialized features in the second frame. The top right image shows the converged features after seven frames of measurement. The red triangle in (0,0) position means the start position of the robot and the red circle represents the current position of the robot. The bottom left image shows the features after 12 frames. These figures show how the features converge while the robot moves forward. The ellipses in blue are the 1σ bounds of the error covariances. The bottom right image shows features in the final frame. Full frames can be seen in the accompanying video, in which the green boxes represent the inliers and red boxes denote the outliers. These boxes show how well JCBB removes outliers. The accuracy of the robot position estimate is compared to that of the dead-reckoning system in Figure 14(a). As shown in the figure, SLAM is more accurate than the dead-reckoning in terms of position estimation. The final position error of SLAM is only 1.4 cm, whereas that of the dead-reckoning is 20.5 cm. Figure 14(b) shows the z-axis robot position estimate. The fluctuation between -1.5 and 1 cm is quite reasonable considering the probable slope of the floor.

3.4.5 Conclusions and Future Work

In this chapter, we presented a 3D measurement model in the camera frame and an undelayed initialization method for monocular EKF SLAM. The 2D pixel coordinates from the camera measurements are converted to 3D points in the camera frame with an assumed arbitrary depth. To account for depth uncertainty, the element of the measurement noise covariance corresponding to the depth of the feature is set to a very large value. This does not cause the divergence of the EKF because the 3D measurement model has very little linearization error. Compared with other undelayed initialization methods, the proposed scheme is very simple to implement and computationally efficient. Furthermore, we presented the efficacy of an EKF SLAM system with a low frame rate camera plus a dead-reckoning system based on odometry and a low-cost MEMS gyro. This system is so practical as to be commercialized for a vacuum cleaning robot.

Although the proposed method cannot deal with features at infinity in the framework of the numerical Kalman filter algorithm, it can be applied for most indoor robot applications. In future works, the authors plan to study the feasibility of extending this algorithm to tackle features at infinity.

Appendix Supplementary Data

Supplementary material related to this chapter can be found online at http://dx.doi.org/10.1016/j.robot.2012.02.002.

References

[1] H. Myung, H.K. Lee, K. Choi, S. Bang, Mobile robot localization with gyroscope and constrained Kalman filter, Int. J. Control Autom. Syst. 8 (3) (2010) 667–676.

[2] H. Lee, K. Choi, J. Park, Y. Kim, S. Bang, Improvement of dead reckoning accuracy of a mobile robot by slip detection and compensation using multiple model approach, Proc. IROS (2008) 1140–1147.

[3] A.J. Davison, Real-time simultaneous localization and mapping with a single camera, in: Proc. International Conference on Computer Vision, 2003.

[4] T. Bailey, Constrained initialization of bearing-only SLAM, in: Proc. of the IEEE International Conference on Robotics and Automation, 2003.

[5] A. Costa, G. Kantor, H. Choset, Bearing-only landmark initialization with unknown data association, in: Proc. of the IEEE International Conference on Robotics & Automation, 2004.

[6] N.M. Kwok, G. Dissanayake, An efficient multiple-hypothesis filter for bearing-only SLAM, in: IEEE/SRJ International Conference on Intelligent Robotics and Systems, Sendai, Japan, 2004.

[7] J. Sola, A. Monin, M. Devy, T. Lemaire, Undelayed initialization in bearing only SLAM, in: IEEE/RSJ International Conference on Intelligent Robots and Systems, 2005.

[8] E. Eade, T. Drummond, Scalable monocular SLAM, in: Proceedings of the IEEE Conference on Computer Vision and Pattern Recognition, 2006.

[9] J.M.M. Montiel, J. Civera, A.J. Davison, Unified inverse depth parameterization for monocular SLAM, in: Proc. Robotics Science and Systems, Philadelphia, USA, 2006.

[10] J. Civera, A.J. Davison, J.M.M. Montiel, Inverse depth parameterization for monocular SLAM, IEEE Trans. Robot. 24 (4) (2008) 932–945.

[11] H.T. David, L.W. John, Strapdown Inertial Navigation Technology, second ed., The Institution of Electrical Engineers, 2004.

[12] J.D. Tardós, J. Neira, P.M. Newman, J.J. Leonard, Robust mapping and localization in indoor environments using sonar data, Int. J. Robot. Res. 21 (3) (2002) 311–330.

[13] L.M. Paz, J.D. Tardós, J. Neira, Divide and conquer: EKF SLAM in O(n), IEEE Trans. Robot. 24 (4) (2008) 1107–1120.

[14] S. Tully, H. Moon, G. Kantor, H. Choset, Iterated filters for bearing-only SLAM, in: IEEE International Conference on Robotics and Automation, 2008.
[15] C. Harris, M. Stephens, A combined corner and edge detector, in: Proceedings of the 4th Alvey Vision Conference, 1988, pp. 147−151.
[16] S. Baker, I. Matthews, Lucas−Kanade 20 years on: a unifying framework part 1, Int. J. Comput. Vision 56 (2) (2004) 221−225.
[17] D. Chekhlov, M. Pupilli, W. Mayol-Cuevas, A. Calway, Real-time and robust monocular SLAM using predictive multi-resolution descriptors, in: 2nd International Symposium on Visual Computing, 2006.
[18] D. Nister, O. Naroditsky, J. Bergen, Visual odometry, in: IEEE Conference on Computer Vision and Pattern Recognition, 2004.
[19] J. Neira, J.D. Tardos, Data association in stochastic mapping using the joint compatibility test, IEEE Trans. Robot. Autom. 17 (5) (2001) 890−897.
[20] L.A. Clemente, A.J. Davision, I.D. Reid, J. Neira, J.D. Tardos, Mapping large loops with a single hand-held camera, in: The Robotics: Science and Systems Conference, US, 2007.

CHAPTER 3.5

Human-Centered Robot Navigation-Towards a Harmoniously Human–Robot Coexisting Environment[1]

Chi-Pang Lam[1], Chen-Tun Chou[1], Kuo-Hung Chiang[1,2], Li-Chen Fu[1,3]

[1]Department of Electrical Engineering, National Taiwan University, Taipei, Taiwan; [2]The New Technology Division, Compal Communications Inc., Taipei, Taiwan; [3]Department of Computer Science and Information Engineering, National Taiwan University, Taipei, Taiwan

Chapter Outline

This chapter was supported by National Science Council, Taiwan, under Grant NSC 97-2218-E-002-015.

Household Service Robotics. http://dx.doi.org/10.1016/B978-0-12-800881-2.00011-6

This chapter proposes a navigation algorithm that considers the states of humans and robots in order to achieve a harmonious coexistence between them. Robot navigation in the presence of humans and other robots is rarely considered in the field of robotics. When navigating through a space filled with humans and robots with different functions, a robot should not only pay attention to obstacle avoidance and goal seeking, it should also take into account whether it interferes with other people or robots. To deal with this problem, we propose several harmonious rules, which guarantee a safe and smooth navigation in a human–robot environment. Based on these rules, a practical navigation method—human-centered sensitive navigation (HCSN)–is proposed. HCSN considers the fact that both humans and robots have sensitive zones, depending on their security regions or on a human's psychological state. We model these zones as various sensitive fields with priori ties, whereby robots will tend to yield socially acceptable movements.

3.5.1 Introduction

Navigation is one of the most fundamental functions of a mobile robot. If robots can smoothly navigate themselves to everywhere in a house, it will be an important milestone to achieve the goal of realizing ubiquitous robotic services in our daily life. Moreover, because robots are going to live or work with human beings, we as robot theorists should pay more attention to robot–human interaction when a robot navigates in a human–robot environment. There are two main issues that should be addressed here: (1) The robot can move autonomously and safely in a human–robot environment in order to complete a specific task; and (2) the robot should behave in both a human- and robot-friendly manner during its movement. In past research, the second issue has hardly been given significant consideration. However, that is exactly what this research primarily investigates. In fact, our study is based on the belief that there should be plausible rules between robots and humans to maintain safe and smooth navigation for all of them, for instance, similar to traffic rules that currently maintain the safety of both drivers and pedestrians. Although we cannot regulate human behaviors and motions, we can achieve the goal by regulating robot behaviors. We believe that these regulation rules should even serve as the most fundamental behaviors to be embedded in all robots' relevant navigation algorithms. As a result, a robot should follow not only a safe and realistic physical path but a socially acceptable path when interacting with humans and other robots in the human–robot environment as well.

Most navigation techniques for mobile robots can be divided into local methods and global ones. In local navigation, researchers mainly deal with the obstacle-avoidance problem. The most popular technique is the "artificial potential fields" method [1], in which the space is modeled as a union of two subspaces, respectively, subjected to attractive or repulsive fields, and the robot navigates according to the resulting fields. This method is simple but has several potential problems, such as being trapped in local minima or failure to pass through small openings [2]. Generally speaking, such a potential-field method is successful in a static environment but is not suitable for a dynamic environment. The "nearness diagram (ND) navigation" method [3] is a reactive navigation approach based on a laser scanner. This method is free of local minima problems and has the advantage of running with low-computational complexity. The "dynamic window approach" [4] considers all the possible velocities attainable by the robot to determine collision-free movement. However, global navigation aims to find an optimal path to the goal position. "Rapidly exploring random trees (RRT)," which has been proposed by Lavalle and Kuffner [5], take samples from the search space and connect valid samples to find a path from the start to the goal point. However, RRT does not consider dynamic objects, especially humans in the environment. Lee et al. proposed a collision-free navigation based on people tracking [6]. Although this algorithm provides a collision-free path, it does not consider whether or not the path will affect the motion of humans or other robots.

Human–robot interaction during the process of a mobile robot's navigation has been widely addressed by several researchers. Alami et al. proposed an idea of designing a human-friendly navigation system [7]. Althaus et al. developed a method for the robot to join a group of people engaged in a conversation in a friendly manner [8]. A human-aware mobile robot motion planner proposed by Sisbot et al. [9] defines a scenario concerning how a robot approaches a human. They consider the non-written rules of human–robot or human–human interactions and integrate these rules into path planning. Topp and Christensen used sample-based joint probabilistic data association filters to track people in order to let a robot follow a specific person [10]. In 2006, Takeshi and Hideki [11] aimed to develop a mobile robot navigation system, which can navigate mobile robots based on the observation of human walking. They observed the moving patterns and trajectories of people in the house and then applied these patterns to improve the mobile robot's navigation therein, i.e., to find the most often used trajectories in order to fit the customs of the human. In 2009, Svenstrup et al. [12] estimate the human pose and uses a "person interest" indicator to generate an artificial potential field, which results in a human-awareness navigation. Sviestins et al. established a hypothesis about how to maintain the relationship between walking speed and relative position while people are walking with robots [13]. Walters et al. adopted a comfort-level device to measure the preference of approaching distance and directions of a robot carrying an object to a person [14]. Hutenrauch et al.

presented a study on special distance and orientation of a robot with respect to a human user during human–robot interaction experiments [15]. In 2006, Pacchierotti et al. [16] described work about the evaluation of the lateral distance for passage in which a robot may pass in a corridor environment. However, these researches mentioned earlier failed to consider the disturbance to humans and other robots when the host robot is moving through them and how the host robot can achieve a socially acceptable navigation.

In this chapter, we propose six harmonious rules that a single robot should obey in order to achieve not only a safe but also a least disturbance motion in a human–robot environment. We will show that these rules are sufficient to guarantee safety and smoothness. Moreover, based on these rules, a practical navigation algorithm, named human-centered sensitive navigation (HCSN), is proposed. Although Sisbot et al. presented algorithms of a motion planner according to humans' positions, field of view, and postures [9], it does not consider velocities of moving people and the interference of people while people are cooperating with robots. Our algorithm not only provides a collision-free navigation in a human–robot environment but also imposes the least disturbance to dynamic people and robots. When taking into account the existence of both humans and other robots, the host robot should be able to know the states of these humans and the other robots around it. To simplify the problem a little bit, here, we assume all the robots have the ability to constantly broadcast their states, including their locations. Since humans are not able to do similar things, the host robot will need to be able to track them persistently. We here adopt the results from the previous work and let human tracking be achieved by multiple particle filters using a laser rangefinder, i.e., the host robot is able to reliably and accurately keep track of a varying number of dynamic objects surrounding it. According to the tracking result and other robots' states, the host robot creates various sensitive fields for each of them, which later serve as the criteria for HCSN. The HCSN is used to handle two problems: (1) Robots and humans should maintain a safety distance, which is varying according to different static and dynamic states of people; and (2) Robots are the least disturbing to both humans and other robots.

This chapter is organized as follows: First, we provide the main characteristics of the harmonious rules in Section 3.5.2. In Section 3.5.3, we describe how we model the sensitive zones of humans and robots. Section 3.5.4 provides the architecture of the overall system. The details of HCSN will be demonstrated in Section 3.5.5. Simulations and experimental results are presented in Sections 3.5.6 and 3.5.7. Finally, we draw a conclusion and provide some further discussions in Section 3.5.8.

3.5.2 Harmonious Rules

Various rules have existed in human society for long time. Among vehicles, traffic rules regulate their behaviors like limiting their velocities, restricting driving directions, or deciding priorities over different vehicles, etc. Although there is no regulation among human motion, in fact, there are some non-written rules among humans. For example, when walking, people tend not to interfere with other people's paths and tend not to disturb other people who are currently working. The research by Hall [17] even proposed social spaces of associated humans and the fact that the area of each space is varying according to the relationships among humans.

Therefore, some rules must be applied to the robots to similarly regulate their behaviors. These rules should consider two major issues, namely, safety and smoothness. Safety ensures a collision-free robot navigating in a human–robot environment, whereas smoothness enables robots not to interfere with humans and one another. The rules are as follows:

1. *Collision-free rule*. It is the most fundamental behavior a robot should have.
2. *Interference-free rule*: The host robot should not enter the personal space of a human and the working space of any other robot unless its task is to approach any of them.
3. *Waiting rule*: Once the host robot enters the personal space of a human carelessly or unwillingly, it has to stop and wait for a threshold time.
4. *Priority rule*: The host robot with low priority should yield to the robot with higher priority when two are both moving.
5. *Intrusion rule*: The host robot intruding other robots' working spaces should leave immediately. The robot whose working space has been intruded should stop working for safety concern.
6. *Human rule*: Humans have the highest priority. Once a robot is serving humans, it only needs to maintain the "collision-free rule" and "interference-free rule."

In addition to the six harmonious rules, robots of the same type may have their internal rules in order to carry out some special mission.

3.5.3 Various Sensitive Fields

We consider the fact that both humans and robots have their sensitive zones, depending either on their security regions or on the psychological feelings of humans, and we then model these zones as various sensitive fields with priorities. These fields will provide criteria to our HCSN.

3.5.3.1 H1: Human-Sensitive Field

A robot entering the human's personal spatial zone will make that human uncomfortable, just like the situation where a stranger enters one's personal spatial zone. Sisbot et al. [9] used cost functions to model the personal spatial zones of stationary people. However, for a moving person, we would like to take more consideration on the influence of his/her velocity upon his/her personal spatial zone rather than only on that of his/her gaze direction. Just imagine that it will be more likely for a robot to enter one's personal spatial zones when that person goes fast. In order to handle this problem, we model the personal spatial zone of a human as a human-sensitive field, which is egg-shaped, i.e., the shape of the field is a combination of a semi-ellipse and a semicircle. As shown in Figure 1(a), the semi-ellipse models the human-sensitive field in front of a person and the semicircle models the field behind the person. A philosophic reason behind this is that a human while walking ahead may prefer to have a longer clear space along his/her way of heading but can accept that an unexpected pop-up robot may get closer to him/her if its approaching direction is within the field of view of the person. More directly speaking, the human-sensitive field in front of a person should be narrow, but long along the sight of view, whereas the range of the field in the back of the person should be equally distant. Moreover, the length of the semi-major axis of the semiellipse should be proportional to the velocity of the person. However, when the person stops moving, the afore-mentioned field will then degenerate into a pure disc. In fact, the length of the semi-minor axis of the egg-shaped field can be a function of the physical states of that person if we can gather sufficient information about that person, such as age and posture, by some kind of human recognition system. For example, children or elders may have

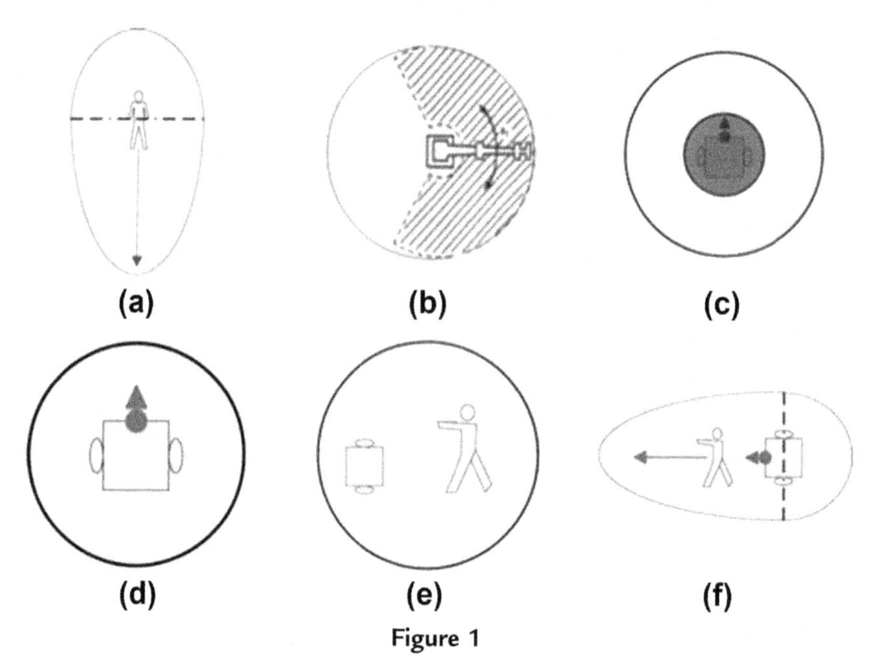

Figure 1

Six kinds of sensitive fields. (a) Human sensitive field ($H1$). (b) Stationary robot working field ($R1$). (c) Movable robot working field ($R2$). (d) Robot normal field ($R3$). (e) Human–robot stationary joint field ($HR1$). (f) Human– robot moving joint field ($HR2$).

longer semi-minor axes than adolescents, and the semi-minor axis of a sitting person may be longer than that of a standing person due to their relatively low mobility. Conceivably, the field $H1$ should have the highest priority among all the sensitive fields because it involves humans.

3.5.3.2 R1: Stationary Robot Working Field

This field models the sensitive field of a stationary robot. It appears in two kinds of situations. The first situation is when the robot is not equipped with mobility but works only at a fixed location, like a manipulator. The second situation is where the robot, although it is able to move, needs to stay at a fixed place when it is working, e.g., when a mobile robot picks something for a human. This sensitive field, as shown in Figure 1(b), is modeled as a round disc, whose radius depends on the working space because some robots require larger working spaces, whereas some others may ask for smaller ones. Here, the priority of this sensitive field $R1$ is set to be the second.

3.5.3.3 R2: Movable Robot Working Field

Some robots can move even though they are working, like robotic vacuum cleaners. Generally, the working space of these types of robots will be smaller than those of $R1$. However, since they are able to choose where to go, their working spaces are moving as well. In this case, the movable robot working field is modeled as a two-layer field, where the inner layer models its current working space, and the outer layer models its possible working space within a short-term future. The shape of such a sensitive field is like a donut, and its priority is set to be the third.

3.5.3.4 R3: Robot Normal Field

In the case where the robot is either being idle or waiting for a human's order, the sensitive zone around the robot is called the robot normal field $R3$. Such a field has the lowest priority, i.e., the fourth rank, among all sensitive fields. This field is also modeled as a round disc with a predefined radius.

3.5.3.5 HR1: Human–Robot Stationary Joint Field

In some situations, the robot is serving a human at a fixed location. As shown in Figure 1(e), the human being served and the serving robot together form a joint field with a disc shape called the human–robot stationary joint field. The center of the disc is at the middle of the robot and the human, and its radius depends on the serving space. Since a human being is involved in the field $HR1$, its priority is the same as that of $H1$.

3.5.3.6 HR2: Human–Robot Moving Joint Field

This situation happens when the robot is following a human or, on the contrary, when the robot is leading a human. The shape of this field, called the human–robot moving joint field, is the same as that of *H*1, which is an egg-shape. With similar reasons, the field *HR*2, as shown in Figure 1(f), has the highest priority as well.

We believe that the six sensitive fields presented earlier cover most of the general situations that a single robot will face. For multiple robots working on a single task, we can simply regard them as a single robot situated at the center of the multi-robot system and then fit it with one of the six sensitive fields.

Besides, we divide the six fields into two different groups: one with solid fields and another with soft fields. The field belonging to the first group is regarded as an obstacle that any other robot cannot intrude, and *H*1, *R*1, *HR*1, *HR*2, and the inner layer of *R*2 all belong to this group. Intrusion into any solid field is not allowed because it violates the "interference-free rule" or "intrusion rule." The soft field belonging to the second group is regarded as a flexible ball with different elasticity, and *R*3 and the outer layer of *R*2 are exactly members of this group. Only when a robot enters the soft field of some other robot, the soft field starts to generate a repulsive force to spring the intruding robot away. If the repulsive force is proportional to the priority, when two robots enter the soft fields of each other, the robot with a lower priority will move further away from its originally established navigation route, whereas the robot with the higher priority will only deviate a little from its original path, and consequently, the "priority rule" is naturally complied with. Table 1 provides a summary of the six sensitive fields.

3.5.4 Human-Centered Sensitive Navigation System Architecture

Based on the six harmonious rules, we design a navigation algorithm named HCSN. Figure 2 shows our overall system architecture. We use pioneer-3DX as our platform, which is equipped with an LMS100 laser sensor serving as the host robot. The inputs of HCSN are the desired goal, the states of the robots and humans, and laser data used to detect them plus obstacles.

As shown in Figure 3, human-sensitive navigation includes sensitive-field sensing, self-situation identification, a motion planner, and a controller. The sensitive-field sensing aims

Table 1: Summary of six sensitive fields

Sensitive Field	Priority	Shape	Solid/Soft Field
H1	1	Egg-shape	Solid
R1	2	Circle	Solid
R2	3	Concentric circle	Inner: solid outer: soft
R3	4	Circle	Soft
HR1	1	Circle	Solid
HR2	1	Egg-shape	Solid

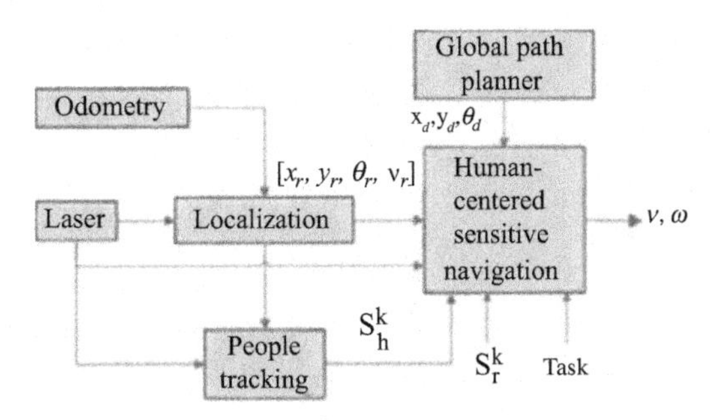

Figure 2
Overall system architecture.

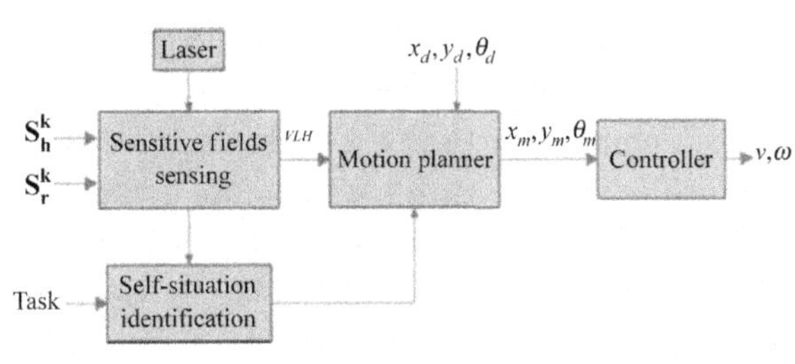

Figure 3
HCSN design.

to sense the existing sensitive fields out of the six kinds, as mentioned earlier resulting from the consideration of the social feeling and safety of people, as well as safety, in particular, of robots. From the previous descriptions, the humans' sensitive fields will be functions of their velocities, their gaze directions, and their status (e.g., children or elder), whereas the robots' sensitive fields will be functions of their states. Next, self-situation identification is intended to verify the current situation of the robot, and its details will be given in Section 3.5.5.2. According to the state of the host robot and sensitive fields of other people and robots, the motion planner designs a dynamic sequence of subgoals so that the controller then generates the corresponding control inputs v and ω. Thus, HCSN will finally lead the host robot to the goal destination, while providing motion causing the least disturbance to humans and other robots.

3.5.5 Human-Centered Sensitive Navigation

3.5.5.1 Communication Signal between Robots

The state of a human is defined as $s_h = [x_h, y_h, \theta_h, v_h]$, including his/her position, heading angle, and velocity, which are the outputs of the human-tracking system. Unlike the human state, we can get much more information about a robot. Here

$$s_r = \left[ID_r, x_r, y_r, \theta_r, v_r, m_r, R_r, R_r^{inn}, H, B_r\right]$$

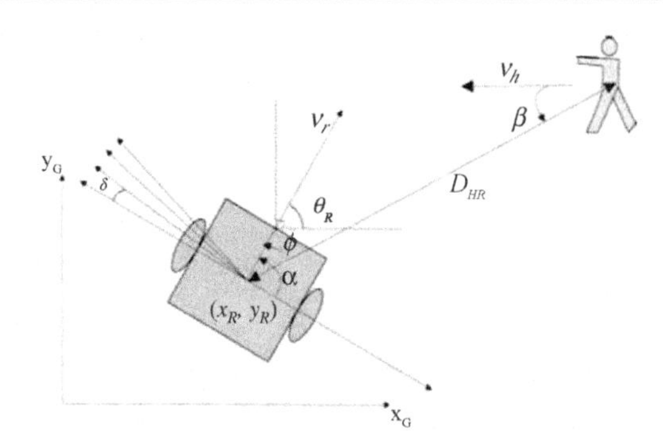

Figure 4
Robot coordination.

is defined as the state of a robot, and we assume that all the robots have the ability to constantly broadcast their states by applying their existing techniques like WiFi or Zigbee. Specifically, ID_r is a robot's identification, and every single robot should have a unique identification just like the media access control address in a computer network. The data vector $[x_r, y_r, \theta_r, v_r]$ is used to describe a robot's posture and velocity; notation $m_r \in \{R1, R2, R3, HR1, HR2\}$ describes the sensitive field the robot deserves, notation R_r is the radius of the sensitive field, notation R_r^{inn}, which is used only when the robot is with the $R2$ field, denotes the radius of the inner layer of $R2$, notation H, which is used only when the robot is with either the $HR1$ or $HR2$ field, denotes the state of human(s), whom the robot is serving, and $B_{rr} = [MoveAway\_ID]$ is a communication signal among robots, and this signal is sent to the robot with "*MoveAway_ID*" if it enters the solid field of the host robot (see Figure 4).

3.5.5.2 Sensitive-Field Sensing

In this chapter, we assume that the robot can gather the state of people near itself by an appropriate human-tracking system, and simultaneously, nearby robots can communicate with one another by broadcasting through a wireless network so that a robot will be able to know the states of other nearby robots. As a result, humans and robots can be divided into six groups, i.e., S_{H1}, S_{R1}, S_{R2}, S_{R3}, S_{HR1} and S_{HR2}, according to their states.

So far from what we have described, overall, there are two kinds of fields: solid and soft fields. From the viewpoint of robots, soft fields are easy to detect because soft fields affect a robot only when the robot enters such fields. A robot only needs to check whether the distance between itself and another relevant robot is less than the radius of the sensitive field of the latter robot. In other words, the soft field of robot j affects robot i if

$$\sqrt{\left(x_{r_i} - x_{r_j}\right)^2 + \left(y_{r_i} - y_{r_j}\right)^2} < R_{r_j}, r_j \in S_{R2} \cup S_{R3}. \tag{1}$$

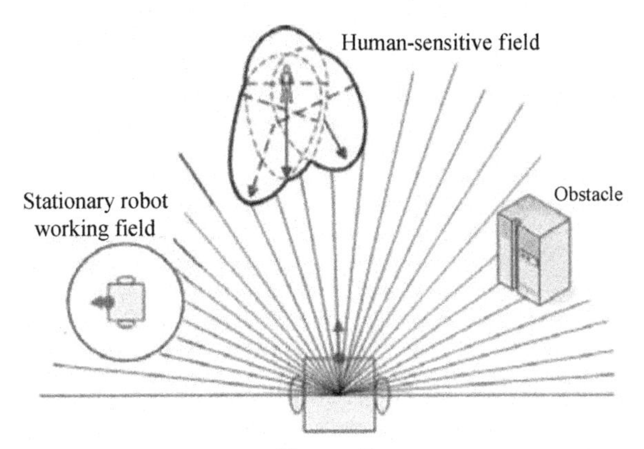

Figure 5
Sensitive fields in a human—robot coexisting environment (here, the human state is uncertain).

However, since solid fields are regarded as a solid obstacle, any robot should be able to "see" those surrounding it through sensing with laser rangefinders. Without loss of generality, in the following discussion, we only consider an egg-shaped field: a combination of a semi-ellipse and a semicircle. A circular field is only a special case of an egg-shaped field when the length of the semi-major axis equals that of the semi-minor axis (see Figure 5). We set the length of the semi-major axis a_{obj_k} by Eqn (2), which is proportional to the velocity of the person in $H1$ or $HR2$

$$a_{\mathrm{obj}_k} = b_{\mathrm{obj}_k} + \lambda v_{\mathrm{obj}_k}, \quad \text{for } \mathrm{obj}_k \in S_{H1} \cup S_{HR2} \tag{2}$$

where v_{obj_k} is the velocity of object k, and λ is a scalar. The length of the semi-minor axis b_{obj_k}, which is the same as the radius of the semicircle, is used to model the general safety zone of a person. As we mentioned earlier, b_{obj_k} is exactly referring to the personal space according to the proxemics theory. In our experiment, we let the length of the semi-minor axis be determined to be 1 m and the length of the semi-major axis be proportional to the velocity of the person, as described by Eqn (2). For the circle-shaped solid field, e.g., the stationary robot working field ($R1$) and human—robot stationary joint field ($HR1$), we set lengths of major and minor axis as follows:

$$\begin{aligned} a_{\mathrm{obj}_k} &= b_{\mathrm{obj}_k} = R_{\mathrm{obj}_k}, \quad \text{for } \mathrm{obj}_k \in S_{R1} \cup S_{HR1} \\ a_{\mathrm{obj}_k} &= b_{\mathrm{obj}_k} = R^{\mathrm{inn}}_{\mathrm{obj}_k}, \quad \text{for } \mathrm{obj}_k \in S_{R2} \end{aligned} \tag{3}$$

Moreover, we define virtual distance as the distance from the host robot to the sensitive fields or obstacles. Since we assume that we can get the positions and the orientations of humans and other robots, we can calculate the virtual distance by simply solving the equations of the semi-ellipses, semicircles, and straight lines (laser beam).

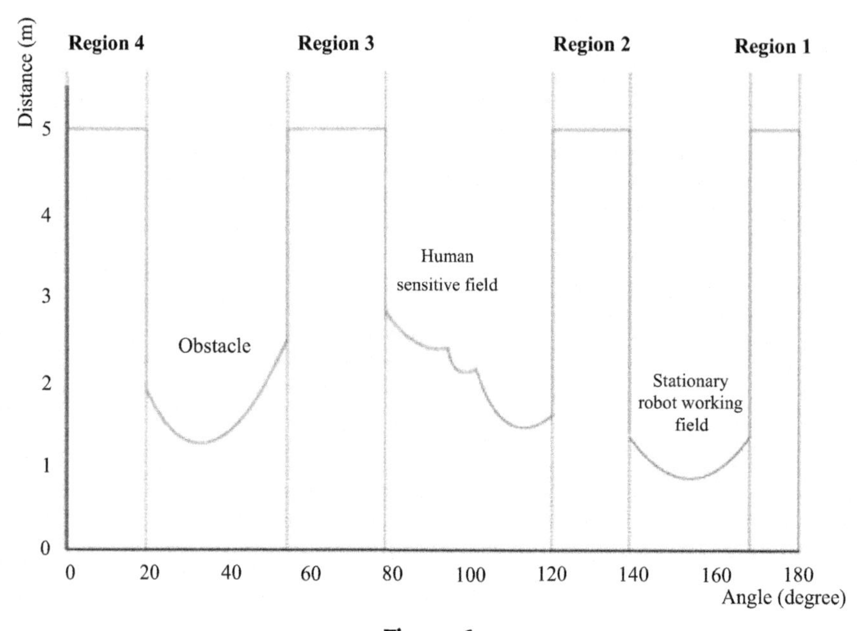

Figure 6
Virtual laser histogram.

By collecting all the virtual distance data, we create a virtual laser histogram (VLH), as shown in Figure 6. In practice, we are only concerned with the virtual distance that is less than, say, 5 m for local navigation purposes. It is noteworthy that this VLH has taken into consideration the sensitive fields of all the observed people. Furthermore, this VLH will change with the dynamic environment, while the robot is navigating through it. Such results serve as input to the motion planner of the robot so that the most efficient way to arrive at the designated goal with collision-free and least-interference movement will be determined.

3.5.5.3 Human Pose Uncertainty

In practical use, we cannot precisely get the people's positions and orientations; therefore, we should handle certain degree of uncertainty about human information. To deal with this problem, we take n_k^{sample} samples from each human's probabilistic distribution obtained from the human-tracking system and combine all the sensitive fields of the samples of each human to generate a complete sensitive field. Therefore, as shown in Figure 5, the resulting sensitive field may not be of a perfect egg-shape or circle after introducing the uncertainty. Thus, the more the uncertainty of the human state there is, the larger the human-sensitive field is.

3.5.5.4 Self-Situation Identification

The term "self-situation" refers to the state of a robot relative to the six kinds of sensitive fields in which the robot itself is currently involved. A robot should always maintain its self-situation in order to provide correct information to other robots and its own motion planner that determines how the robot should move while reacting to the other fields currently under interaction. Figure 7 shows an example of the finite-state machine of the self-situation identification block. A robot will start at a state exhibiting $R3$ field and later will transfer to another state exhibiting an $R1$, $R2$, $HR1$, or $HR2$ field according to its assigned task. As shown in Figure 7, two special states, i.e., "intrusion" and "waiting," are introduced to account for the intermediate stages after the triggering of "intrusion rule" and "waiting rule," respectively. Generally, a robot r_i enters the solid sensitive field of object obj_k in case of the following:

$$\begin{cases} A\left(x_{r_i} - x_{\text{obj}_k}\right)^2 + B\left(y_{r_i} - y_{\text{obj}_k}\right)^2 \\ \quad + C\left(x_{r_i} - x_{\text{obj}_k}\right)\left(y_{r_i} - y_{\text{obj}_k}\right) < 1 \\ \left(\left(x_{r_i} - x_{\text{obj}_k}\right)\cos\theta_{\text{obj}_k} + \left(y_{r_i} - y_{\text{obj}_k}\right)\sin\theta_{\text{obj}_k}\right) > 0 \end{cases}$$

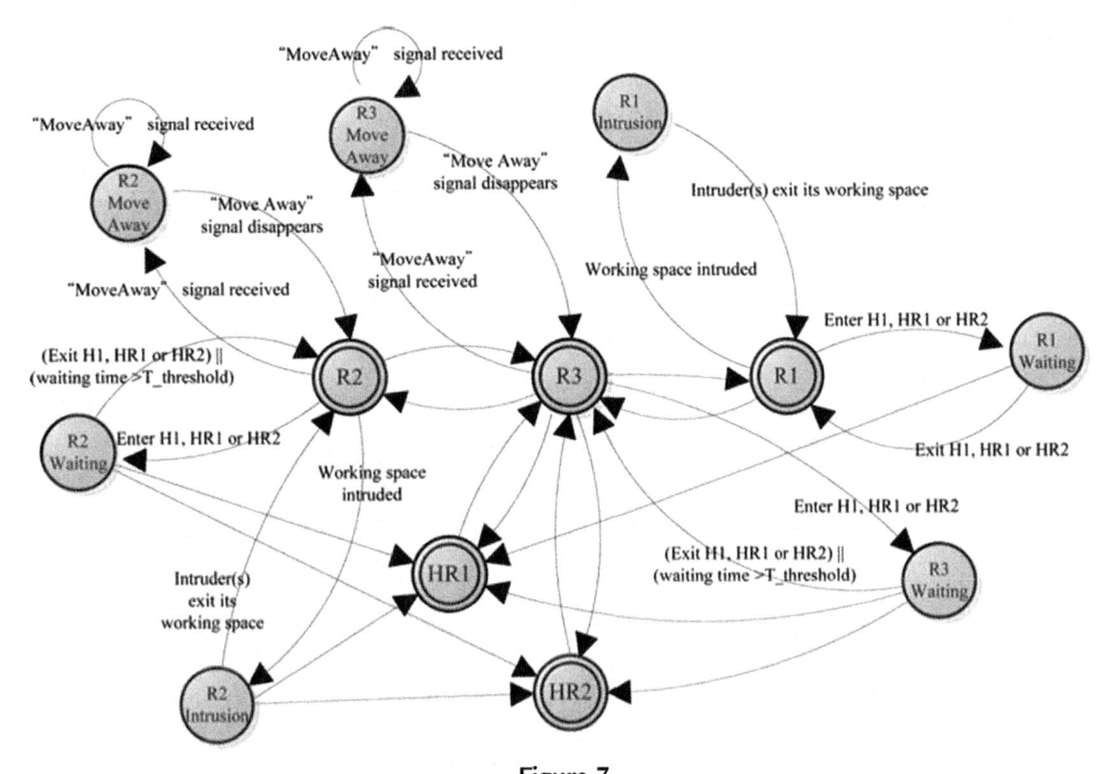

Figure 7
Finite-state machine for "self-situation identification" block.

where

$$A = \frac{\cos^2\theta_{\text{obj}_k}}{a^2_{\text{obj}_k}} + \frac{\sin^2\theta_{\text{obj}_k}}{b^2_{\text{obj}_k}}, \quad B = \frac{\sin^2\theta_{\text{obj}_k}}{a^2_{\text{obj}_k}} + \frac{\cos^2\theta_{\text{obj}_k}}{b^2_{\text{obj}_k}},$$

and

$$C = \left(\frac{1}{a^2_{\text{obj}_k}} - \frac{1}{b^2_{\text{obj}_k}}\right)\sin 2\theta_{\text{obj}_k}$$

which is the condition of entering the semi-ellipse, or

$$\begin{cases} \left(x_{r_i} - x_{\text{obj}_k}\right)^2 + \left(y_{r_i} - y_{\text{obj}_k}\right)^2 - b^2_{\text{obj}_k} < 0 \\ \left(\left(x_{r_i} - x_{\text{obj}_k}\right)\cos\theta_{\text{obj}_k} + \left(y_{r_i} - y_{\text{obj}_k}\right)\sin\theta_{\text{obj}_k}\right) < 0 \end{cases}$$

which is the condition of entering the semicircle, where

$$\text{obj}_k \in S_{H1} \cup S_{HR2} \cup S_{HR1} \cup S_{R1} \cup S_{R2}.$$

However, a robot entering solid fields will be transitioned into the intermediate state, i.e., "intrusion" or "waiting." After the robot is transferred from the state "*R*1" to the state "intrusion" when some other robot enters its solid field, what it does is to send a "MoveAway" signal to the intruding robot. Only robots with field *R*2 or *R*3 state need to react to the "MoveAway" signal because of their lower priorities.

The "waiting" state most likely happens when a human himself or herself approaches the robot so that robot enters the human-sensitive field unwillingly. There are three possible reasons why people will approach a robot.

1. *People pass by the robot only*: In this case, the robot only needs to wait for people to move away from it. As a result, the robot first stops and later returns to its original state after people leave.
2. *People require some services from the robot*: Therefore, the robot waits for the human's command and then directly enters the "*HR*1" state or "*HR*2" state.
3. *People are not aware of the robots nearby*: To deal with this situation, the robot will leave the human-sensitive field by itself if the waiting time exceeds a threshold time.

In fact, Figure 7 shows the most general condition of a robot, but one should beware that some robots may not have all the states as revealed. There can be, in fact, a large variation of state decision as long as the state machine satisfies this general form. For example, a manipulator may only have "*R*3," "*R*1," "waiting," and "intrusion" states. Some robots can also have their internal states inside the main state. For example, within the "*R*3" state, a robot can have internal states, such as "searching human," "surveillance," etc.

3.5.5.5 Motion Planner

The motion planner aims to provide a sequence of robot motions that are subjected to the "collision-free rule," "interference-free rule," and "priority rule," which amounts to providing a solution to the underlying navigation problem. Such a planner is based on two reactive navigation methods, namely, ND navigation [3] and potential-field navigation [2]. Since most indoor environments are dense and complex, ND navigation is good at handling such an environment in real time. As a result, we use "ND" to first find out a suitable free walking area. After this, a potential-field approach is used to establish different potential fields near the free walking area we choose. Apparently, the potential field will be a function of sensitive fields and nearby obstacles. Furthermore, we consider the priorities of different sensitive fields when creating a potential-field function so that a robot with a higher priority will tend to go first.

From sensitive-field sensing, we have obtained the VLH, which is the information about distances of those surrounding the robot, including different sensitive fields and physical obstacles. Here, VLH will replace the raw laser data to find the free walking area from ND. It is noteworthy that when finding the free walking area, the robot itself should consider their working space as well. The radius of working space will replace the original radius of the robot as well.

Referring to Figure 8, let R_{FWA} indicate the free walking area, where s_r and s_l are the start and end sectors of the free walking area, s_{goal} is the goal sector, R is the radius of the robot or the radius of the robot's solid field, d_{max} is the maximum laser distance with which we are concerned, and D_s is the security distance of the robot. All these notations follow the definitions made in [18]. ND navigation divides the robot's behaviors into six

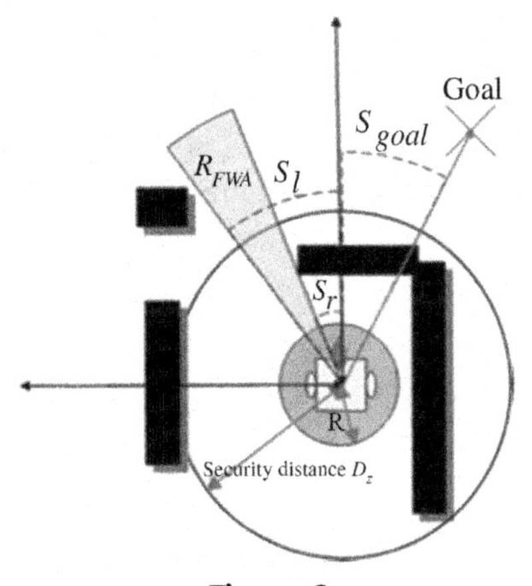

Figure 8
Example of LS2.

scenarios based on the distances of the various obstacle points, namely, HSGR, HSWR, HSNR, LS1, LS2, and LSGR.

1. HSGR: There is high safety, and s_{goal} is in R_{FWA}.
2. HSWR: There is high safety, and R_{FWA} is wide, but s_{goal} is not in R_{FWA}.
3. HSNR: There is high safety, R_{FWA} is narrow, and s_{goal} is not in R_{FWA}.
4. LS1: There is safety, and some distances from VLH are smaller than D_s in one side.
5. LS2: There is safety, and some distances from VLH are smaller than D_s in both sides.
6. LSGR: There is safety, and some distances from VLH are smaller than D_s in one or both sides, but s_{goal} is in R_{FWA}.

After this, a modified direction s_θ for the robot to go along is searched, where s_θ should be a direction that leads the robot to avoid the obstacles and not to interfere with other robots and humans, and to reach the final goal. There are six different strategies to deal with the six different scenarios. For scenarios with HSGR, HSWR, and HSNR, the robot is in a high-safety condition, i.e., the robot does not enter any sensitive field and is not close to any obstacle, and hence, the properly proposed strategies can be very straightforward as shown in the following [18]:

1. HSGR: $s_\theta = s_{\text{goal}}$.
2. HSWR: $\begin{cases} s_\theta = s_r + \arcsin\left(\dfrac{R + D_s}{D_{disc}}\right), & \text{if } s_{\text{goal}} \textit{ is near } s_r \\ s_\theta = s_l - \arcsin\left(\dfrac{R + D_s}{D_{disc}}\right), & \text{if } s_{\text{goal}} \textit{ is near } s_l \end{cases}$ where D_{disc} is the distance to the discontinuity s_r or s_l.
3. HSNR: $s_\theta = \frac{s_r + s_l}{2}$.

However, in scenarios with LS1, LS2, and LSGR, we should take into consideration the obstacles and sensitive fields nearby. The original strategies in [18] only take account of the closest obstacle points in both sides. However, the closest obstacle may not be the most dangerous obstacle. As shown in Figure 8, the closest obstacle is the obstacle at the right-hand side of the robot. However, this obstacle nearly does not affect the safety of the robot because the robot would not take a lateral movement. When the robot continues to go ahead until the obstacle in front of it becomes the closest obstacle, s_θ will change significantly, resulting in a non-smooth motion. As a consequence, the obstacle in front of the robot can be the most dangerous obstacle. Hence, we have to estimate the risk of each obstacle point. The risk measurement rk_i is hereby defined by.

$$\text{rk}_i = \left(\frac{\text{PND}_i}{d_{\max} + 2R}\right)^2 \sqrt{\cos\left(\theta_{\text{obs}(i)}\right)}$$

with

$$\begin{cases} \mathrm{PND}_i = d_{\max} + 2R - D_{\mathrm{VLH}}(i), & \text{if } (D_{\mathrm{VLH}}(i) < d_{\max}) \\ \mathrm{PND}_i = 0, & \text{if } (D_{\mathrm{VLH}}(i) \geq d_{\max}) \end{cases} \tag{4}$$

where $\theta_{\mathrm{obs}(i)} \in [-\pi/2, \pi/2]$ is the angle of the obstacle point from the robot local coordinate, and we only consider the obstacle in front of the robot. The risk measurement increases the weight of the obstacle point, whose direction is close to the facing of the robot.

Furthermore, since "priority rule" has to be satisfied, the priority should be introduced when determining s_θ. The potential-field method provides a good solution to handle it. The attractive potential function takes the form of

$$U_{\mathrm{att}}(q) = \xi \left\| q_{\mathrm{goal}} - q \right\|^m \tag{5}$$

where $q = (x, y)^T$ is the position of the host robot, and m can be 1 or 2 (we let $m = 1$ in our approach). The corresponding attractive force is then given by the negative gradient of the attractive potential

$$F_{\mathrm{att}} = -\nabla U_{\mathrm{att}}(q) = \xi. \tag{6}$$

The direction of F_{att} is determined by.

$$\begin{cases} s_{\mathrm{att}} = \left[\dfrac{s_r + s_l}{2} + \kappa_\theta \left(\mathrm{rk}_{r,\max} - \mathrm{rk}_{l,\max} \right) \right], & \text{if robot is in LS1 or LS2} \\ s_{\mathrm{att}} = s_{\mathrm{goal}}, & \text{if robot is in LSGR} \end{cases}$$

where $\mathrm{rk}_{l,\max}$ and $\mathrm{rk}_{r,\max}$ are the maximum values of the left-hand side and the right-hand side risk measurement, respectively, and κ_θ is the positive constant value that can be tuned experimentally. Here, $\mathrm{rk}_{l,\max}$ and $\mathrm{rk}_{r,\max}$ are used to finely tune the direction of the attractive force. For example, when the left risk measurement is greater than the right risk measurement, s_{att} will be smaller so that the attractive force directs the robot a bit to the right referring to the middle direction of the free walking area.

The repulsive potential function associated with some solid fields and obstacles takes the form of.

$$U_{\mathrm{req}}(q) = \begin{cases} \dfrac{1}{2} \sum \eta_i \left(\dfrac{1}{\left\| q_{\mathrm{obs}} - q \right\|} - \dfrac{1}{D_s} \right)^2, & \text{if } \left\| q_{\mathrm{obs}} - q \right\| \leq D_s \\ 0, & \text{if } \left\| q_{\mathrm{obs}} - q \right\| > D_s \end{cases} \tag{7}$$

where η is a scalar, and to introduce the priority, the value of η depends on the corresponding VLH, i.e., $\eta_{\mathrm{obs}} > \eta_{H1} = \eta_{HR1} = \eta_{HR2} > \eta_{R1} > \eta_{R2}$, where obstacle points have the largest η, and the solid fields of $R2$ have the smallest η. The corresponding repulsive force is given by

$$F_{\text{rep}}^{\text{VLH}}(q) = -\nabla U_{\text{rep}}(q)$$
$$= \begin{cases} \sum \eta_i \left(\frac{1}{\|q_{\text{obs}} - q\|} - \frac{1}{D_s} \right) \frac{1}{\|q_{\text{obs}} - q\|^2} \nabla \|q_{\text{obs}} - q\|, & \text{if } \|q_{\text{obs}} - q\| \le D_s \\ 0, & \text{if } \|q_{\text{obs}} - q\| > D_s \end{cases} \tag{8}$$

The repulsive force associated with the soft field $F_{\text{rep}}^{\text{soft}}$ is rep generated by a heuristic manner. The main concept is that the deeper a host robot enters another robot's soft field, the larger the repulsive force generated to the host robot. $F_{\text{rep}}^{\text{soft}}$ takes the rep form of.

$$F_{\text{rep}}^{\text{soft}} = \begin{cases} \sum_{\text{robot}_i \in R2 \cup R3} k_1 \ln\left(1 + k_2\left(R_{\text{robot}_i} - \left\|q - q_{\text{robot}_i}\right\|\right)\right), & \text{if } \left\|q - q_{\text{robot}_i}\right\| \langle R_{\text{robot}_i} \\ 0, & \text{if } \left\|q - q_{\text{robot}_i}\right\| \rangle R_{\text{robot}_i} \end{cases} \tag{9}$$

where k_1 and k_2 are scalars, and $k_{2,R2} > k_{2,R3}$. The total force is as follows:

$$F_{\text{total}} = F_{\text{att}} + F_{\text{rep}}^{\text{VLH}} + F_{\text{rep}}^{\text{soft}} \tag{10}$$

Finally, s_θ is determined by the direction of F_{total}, i.e., $s_\theta =$ direction of F_{total} in the scenarios with LS1, LS2, or LSGR.

After calculating s_θ, the modified subgoal is set to be.

$$\begin{cases} X_{\text{subgoal}} = D_0 e^{-|s_\theta - \theta_r|} \cos s_\theta \\ Y_{\text{subgoal}} = D_0 e^{-|s_\theta - \theta_r|} \sin s_\theta \end{cases} \tag{11}$$

Where D_0 is a look-ahead distance in order to guide the robot. We will give D_0 a low value when the robot gets into the sensitive field of some other robot whose priority is higher. Moreover, the gain $e^{-|s_\theta - \theta_r|}$ here is used to adjust the exact look-ahead distance, i.e., the larger the turning angle is, the smaller the effective look-ahead distance should be. This is because a large turning angle may mean that the obstacle appears abruptly, and we should reduce our look-ahead distance so that turning is given precedence over forward movement. The modified subgoal will be the input of control law proposed in [19] to achieve stable control.

By introducing the priority into the potential-field function, the higher priority will produce a stronger repulsive potential field and a larger repulsive force as well. As a result, the robot with a lower priority tends not to affect the robot with a higher priority, which satisfies both the "priority rule" and "interference-free rule." Moreover, ND and potential-field methods ensure satisfaction of the "collision-free rule" as well.

3.5.5.6 Complexity Analysis

Here, we will analyze the computational complexity of the HCSN algorithm. In the sensitive-field generator, checking if a robot is in another robots' soft fields takes $O(||S_{R2} \cup S_{R3}||)$ time, and generating solid field takes $O(M \times ||S_{H1} \cup S_{HR2} \cup S_{R1} \cup S_{HR1}||)$ time. Therefore, in the worst case, the sensitive field generator will take $O(M \times N_{\text{obj}})$ time complexity, where M is the total number of laser data of a scan, and N_{obj} is the total number of all the human and robot objects. We have to emphasize that state decision takes constant time O (1). Since a motion planner is based on VLH, the complexity will apparently be O (M). To sum up, the total complexity of the HCSN algorithm is $O(M \times N_{\text{obj}})$ compared with the complexity of ND navigation and potential field, which are both $O(M)$, although HCSN has higher complexity than them, it can still meet the real-time requirement for a limited number of laser data and a limited number of people, and if there are too many people in an area, we can just take an upper bound number of them closest to the robot into account. This way, the robot can still actually achieve a real-time reactive navigation.

3.5.6 Simulations

In this section, we have performed several computer simulations to demonstrate how HCSN achieves a harmonious navigation of a host robot through an environment with a number of humans and robots. We implement the HCSN using the simulation tools of Pioneer-3 DX, MobileSim. A laser rangefinder LMS200 is assumed to be mounted on the Pioneer-3 DX, which uses odometry for localization.

3.5.6.1 Simulation 1

The first scenario is a robot with an $R2$ field (abbreviate $R2$ robot for simplicity) that meets robots with an $R3$ field (abbreviate $R3$ robot for simplicity) in a narrow corridor. As shown in Figure 9(a) and (b), the $R2$ robot goes from location (−3000, −350) to location (−2000, −350), whereas the $R3$ robot goes from location (−3000, −480) to location (−2000, −480), and the affecting radius for them is 350 cm. The $R3$ robot would go much slower because of the repulsive force from the $R2$ robot. On the other hand, the $R2$ robot is affected by a repulsive force from the $R3$ robot as well, but such force is smaller than that exerted on the $R3$ robot. Therefore, we can see that the distance between the $R2$ robot and the upper obstacle will be greater than the distance between the $R3$ robot and the bottom obstacle, i.e., the $R2$ robot is given more free space to move. Moreover, the $R3$

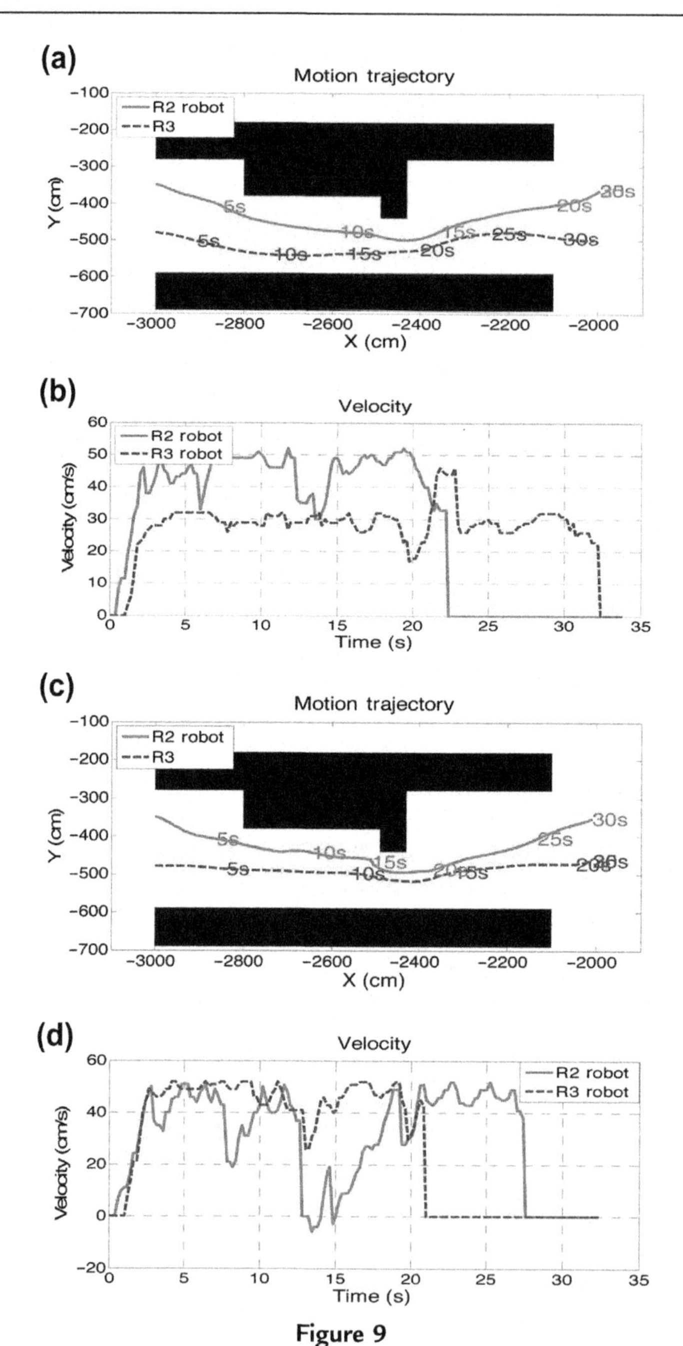

Figure 9

Simulation of R2 robot means R3 robot. (a) Motion trajectory using HCSN. (b) Velocity using HCSN. (c) Motion trajectory using ND. (d) Velocity using normal ND.

robot would reduce its look-ahead distance as well so that its velocity decreases to let the $R2$ robot go first. Once their distance becomes greater than the affecting radius of the $R2$ robot (during the period from 21 to 23 s roughly), the look-ahead distance returns to its original value, which is affected only by the turning angle. By normal (ND) navigation, as shown in Figure 9(c) and (d), the $R3$ robot would almost follow the straight line when reaching the goal because the $R3$ robot can directly "see" the goal. As a result, although there is much free space at the bottom of the corridor, the $R2$ robot would have less free space to go through and has to make a large turn at 14 s to avoid the small protruding obstacle.

3.5.6.2 Simulation 2

The second scenario simulates a situation where a human and a robot move in opposite directions on a corridor. In this scenario, a human is walking from right to left, while the robot is moving from the left side to the right side of the corridor. Figure 10 shows the result of HCSN, whereas Figure 11 is the result of ND navigation without introducing a sensitive field. From Figure 10(a), we can see that at $t = 9$ s, the robot has already started to avoid the human by turning to the upper side of the corridor because it can see the human-sensitive field $H1$, but at about $t = 12$ s, the robot finds that the human is going to the upper side as well. Therefore, the robot readily turns to the lower side. We compare the result of using normal ND navigation, as shown in Figure 11. In Figure 11, the robot is still turning to the upper side of the corridor, although the human has already turned to the upper side. It is because the robot can still see a free walking area between the human and the upper wall, and at about $t = 15$ s, the robot starts to avoid the human from the lower side because at that time, the human has blocked the upper free walking area. However, at the aforementioned critical times, the robot and the human have been very close.

In order to quantify the interference to humans and their safety, we define two indexes: Interference index (II), which measures the interference to a single human, and collision index (CI), which measures the safety of human under consideration.

$$\mathrm{II(t)} = \frac{1 - |\beta(t)|/\pi}{D_{\mathrm{HR}}(t)} \tag{12}$$

$$\mathrm{CI}(t) = H\left(\frac{v_h(t)\cos\beta(t) + v_r(t)\cos\phi(t)}{D_{\mathrm{HR}}(t)}\right)$$

where

$$H(x) = \begin{cases} 0, & \text{if } (x < 0) \\ x, & \text{if } (x \geq 0) \end{cases} \tag{13}$$

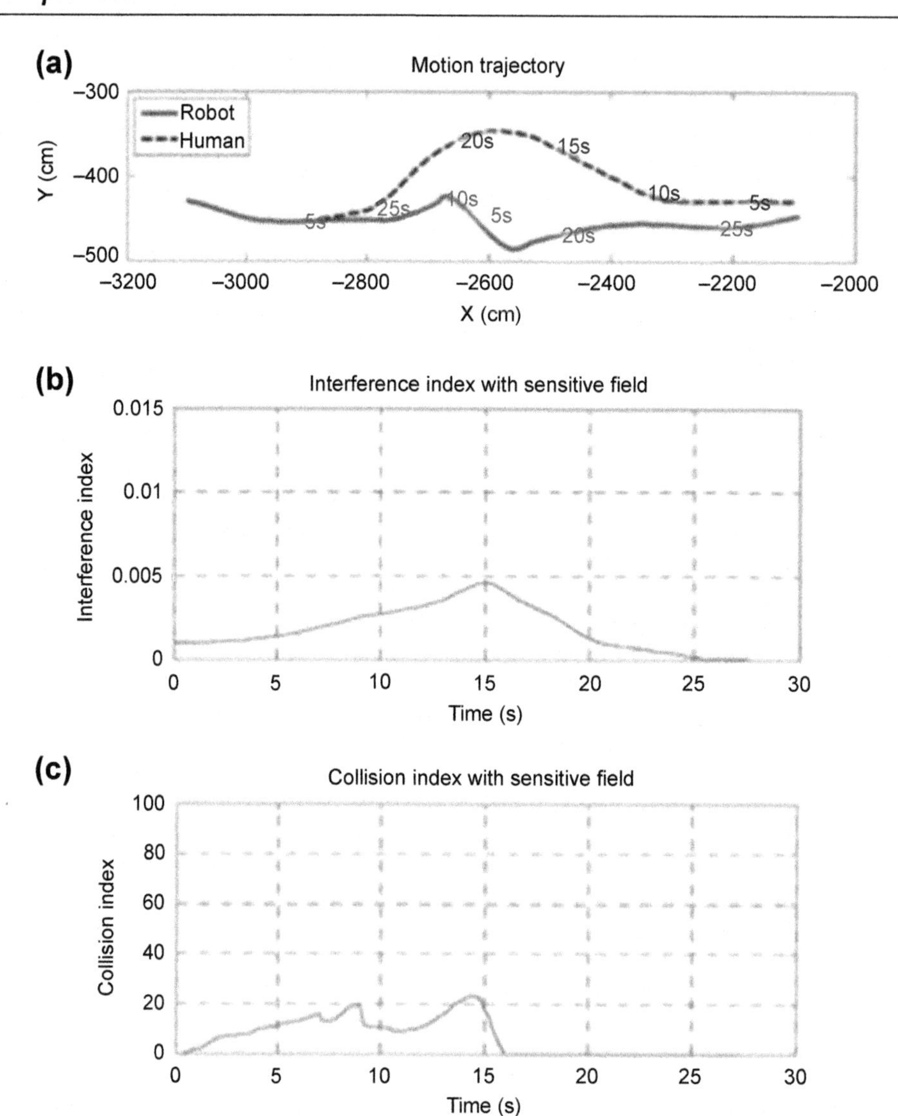

Figure 10

HCSN in the scenario of a human and a robot moving the opposite way on a corridor. (a) Motion trajectory. (b) Interference index. (c) Collision index.

The parameters in II(t) and CI(t) follow Figure 4. The shorter distance between robot and human and the lower value of $|\beta|$ together result in a higher value of II, which means more interference from the robot to human. Moreover, the higher relative velocity between human and robot along the line-of-sight direction and the shorter distance between them then leads to a higher value of CI, which implies a situation that is less safe to the human.

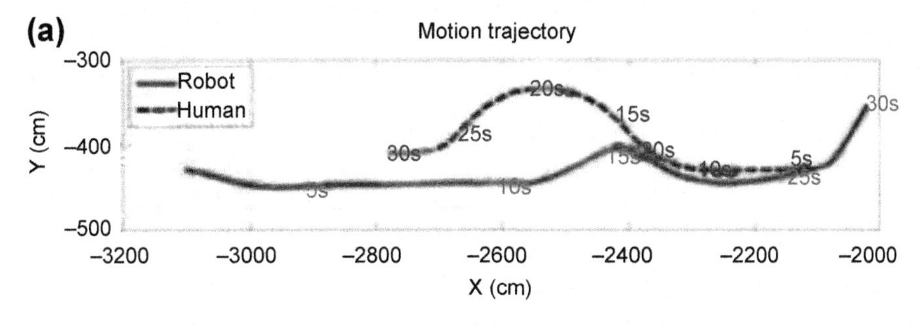

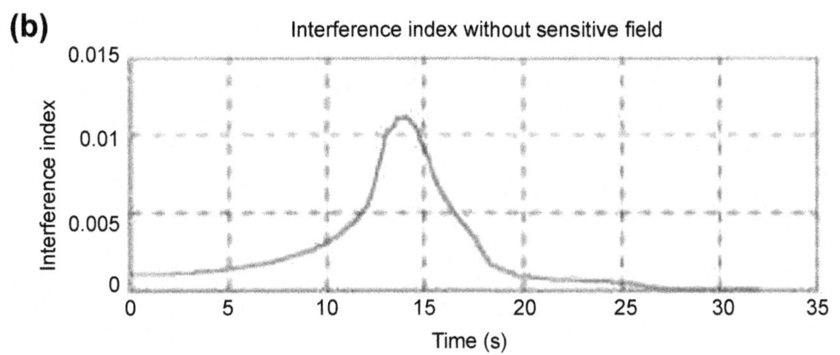

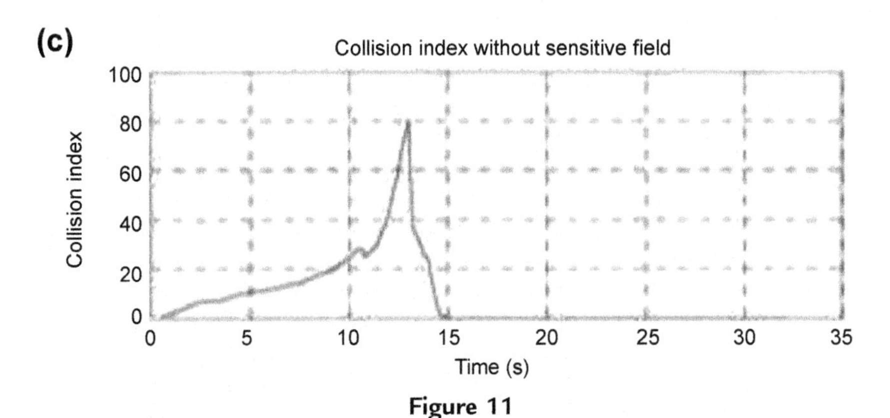

Figure 11
Scenario of human and robot moving the opposite way on a corridor without a sensitive field. (a) Motion trajectory. (b) Interference index. (c) Collision index.

Figure 10(c) and (d) shows the profiles of II(t) and CI(t) of HCSN. The highest values of II(t) and CI(t) happen near $t = 15$ s, which is 0.0046 and 24, respectively. Compared with the result of navigation without taking a sensitive field into consideration, the highest values of II(t) and CI(t) are now 0.011 and 80, respectively, as shown in Figure 11(c) and (d), which clearly indicates that our HCSN performs better.

3.5.6.3 Simulation 3

In the third scenario, a robot needs to avoid two people walking side-by-side toward the robot. Figure 12(a) and (b) shows the robot trajectory by HCSN and the shortest path planner, respectively. We can see that because of the existence of the human-sensitive field ($H1$), the space between the two people is blocked, which leaves the robot without a choice, but the robot has to bypass both of them from one side. However, the normal shortest path planner would plan a path that passes through the middle of the two people. The peak values of II and CI with respect to, say, human1 using our HCSN algorithm are 0.0067 and 33, respectively. Compared with the peak values of II and CI using the shortest path planner, we obtain 0.016 and 130 for the corresponding index values, and apparently, HCSN decreases the interference to humans and the collision risk significantly, although it sacrifices the arrival time.

3.5.7 Experimental Results

We have performed experiments to demonstrate the performance of HCSN. In order to construct a human–robot environment, several robots are used as our experimental

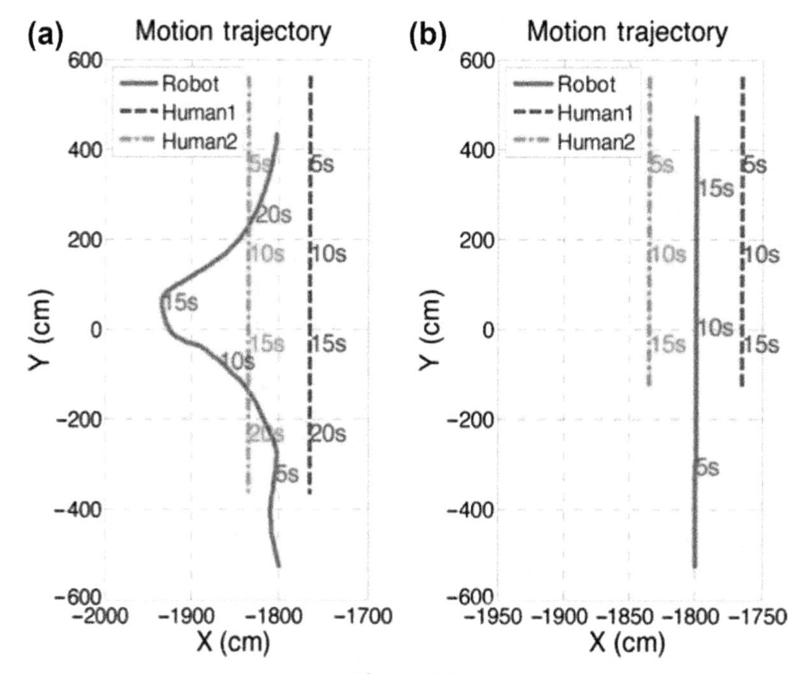

Figure 12
Motion trajectory of robot passing two people walking side-by-side toward the robot by (a) HCSN and (b) shortest path planner.

Figure 13
Experiments platforms. (a) Julia. (b) NTU-1. (c) Pioneer 3-DX.

platforms, including Julia, NTU-1, and Pioneer 3-DX, as shown in Figure 13. All of them are equipped with a laser rangefinder LMS100, whose scanning range is 20 m with 270° of spanning angle. Laptop or personal computers are installed on them to execute the hereby developed algorithm and the control command. In our experiments, the same Monte Carlo localization system is used in all three robots for self-positioning. The human-tracking system refers to [20]. Moreover, the experimental environment is in the MingDa Building of the Department of Electrical Engineering, National Taiwan University.

3.5.7.1 Experiment 1

This experiment compares the practical reaction of HCSN and potential-field-based navigation when a robot meets moving humans. As shown in Figure 14(b), the trajectory of the potential-field-based navigation has only a small change because the potential-field-based method does not react significantly if obstacles are not close enough to the robot. Although a human may sometimes walk close enough to the robot, the short appearance time of the moving human will not make the robot react too much. In HCSN, the awareness of sensitive fields will let the robot react earlier. For example, from 30 to 40 s, the robot changes its path in order to react to the incoming people, achieving a more human-friendly behavior.

3.5.7.2 Experiment 2

In this experiment, Pioneer P3DX acts as an "*R*3" robot that is going to approach the goal (1600, 1650), and Julia is an "*HR*2" robot that is following a certain person. The green dash line is the reference path of P3DX, the circles are P3DX's actual trajectory, and the cross is Julia's trajectory, as shown in Figure 15(a). Because Julia is in the way of P3DX's reference path, the "*R*3" robot is not allowed to enter the joint sensitive field generated by "*HR*2" and the person, as shown in Figure 15(b). As a result, as shown by the trajectories

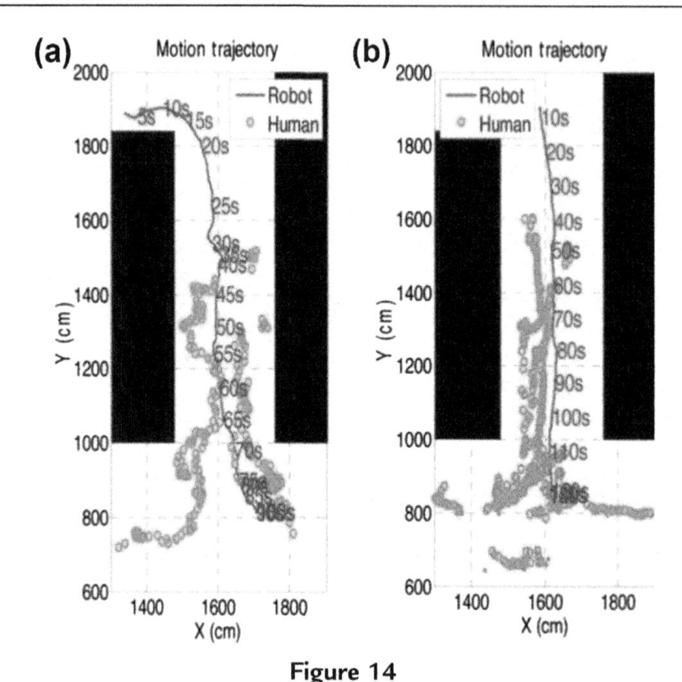

Figure 14
Robot reaction in a common corridor by (a) HCSN and (b) potential-field-base navigation.

in Figure 15(a), the resulting path of P3DX based on HCSN greatly deviates from the sensitive field of Julia, although the reference path of P3DX is not fully occupied by the *HR*2 robot, Julia, and the human it is following.

3.5.7.3 Experiment 3

In the third experiment, as shown in Figure 16, Julia acts as an "*R*2" robot, which is going to navigate from location (1600, 1250) to location (1050, 2100), NTU-1 is an "*HR*1" robot and a human is interacting with it at location (1490, 1870), and Pioneer 3-DX is a "*R*3" robot, which goes from location (1330, 2150) to location (1600, 1300). Note that the moving direction of Pioneer 3-DX along its predefined path is just the opposite of that of Julia's path, and both the maximum speed of Julia and Pioneer is set to be 40 cm/s.

Figure 16 shows the trajectories of Julia, NTU-1, Pioneer 3-DX, and the humans, and Figure 17 shows the snapshots from the viewpoint of Julia. Julia goes in the middle of the corridor (see snapshot (a) in Figure 18) when it discovers a human, and the sensitive field is sensed so that it starts to avoid entering the sensitive field of the human rather than avoiding him in a close distance (see snapshot (b) in Figure 18). After this, it returns to its path (see snapshot (c) in Figure 18) and avoids disturbing NTU-1, which creates a human–robot stationary joint field (*HR*1). At the corner, three robots meet, as shown in

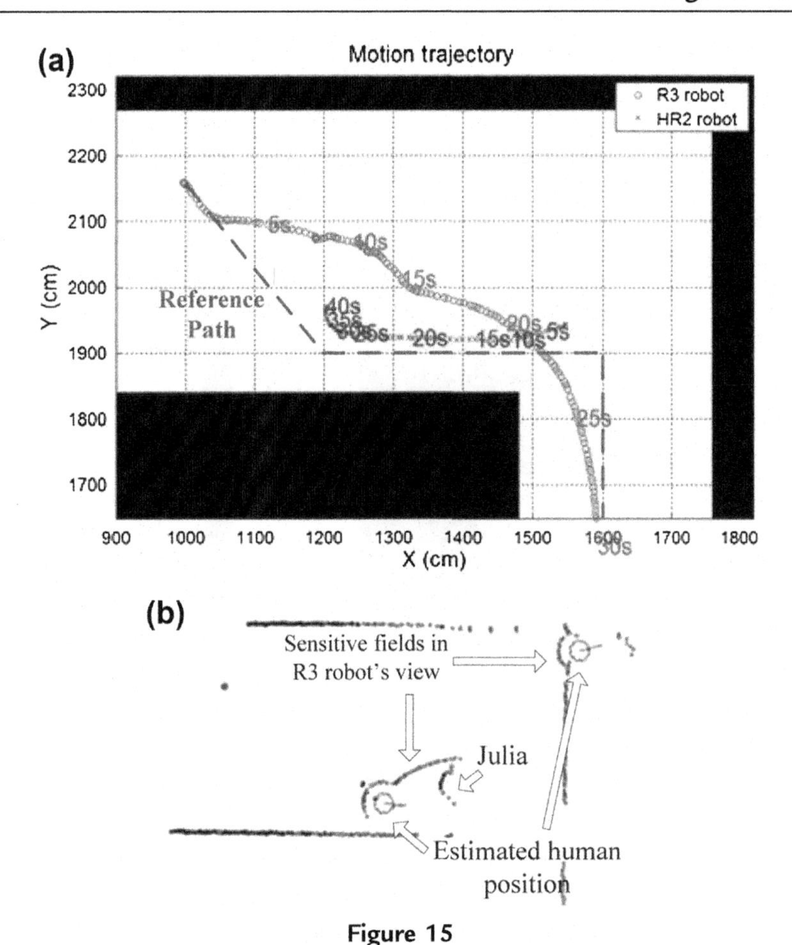

Figure 15
(a) Trajectories of P3DX and Julia robot. (b) Joint sensitive field between Julia robot and the person.

snapshot (d) in Figure 18. Since NTU-1 has the highest priority and *HR*1 is a solid field, both Julia and Pioneer avoid it and try not to interfere with NTU-1. Moreover, the priority of Julia is higher than Pioneer, so Julia is having a relatively larger space, where Pioneer is moving near the *HR*1. To continue, Julia meets two persons and still performs a collision free and least-interference movement (see snapshots (e) and (f) in Figure 18). In the whole process, all the robots using HCSN obeyed the harmonious rules to provide socially acceptable motions.

3.5.7.4 Experiment 4

The second experiment shows the state transitions in the self-situation identification (see Figure 19). Two robots are put in a human–robot environment for about 4 min to show

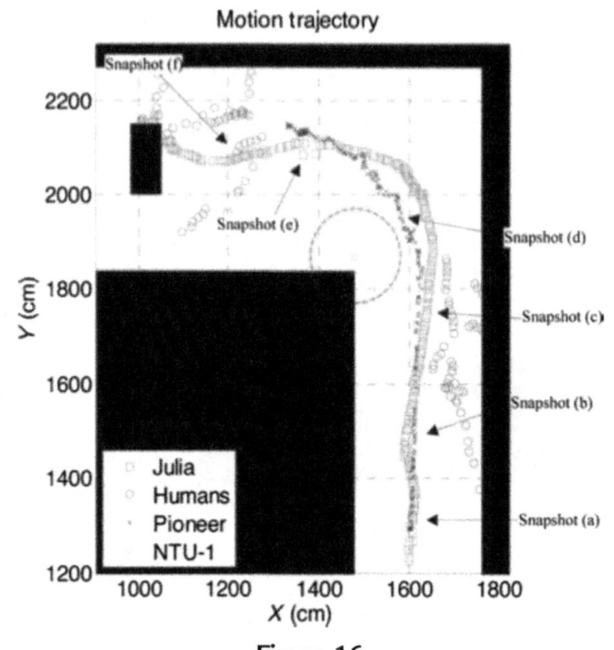

Figure 16
Motion trajectory of robots and humans in experiment 3.

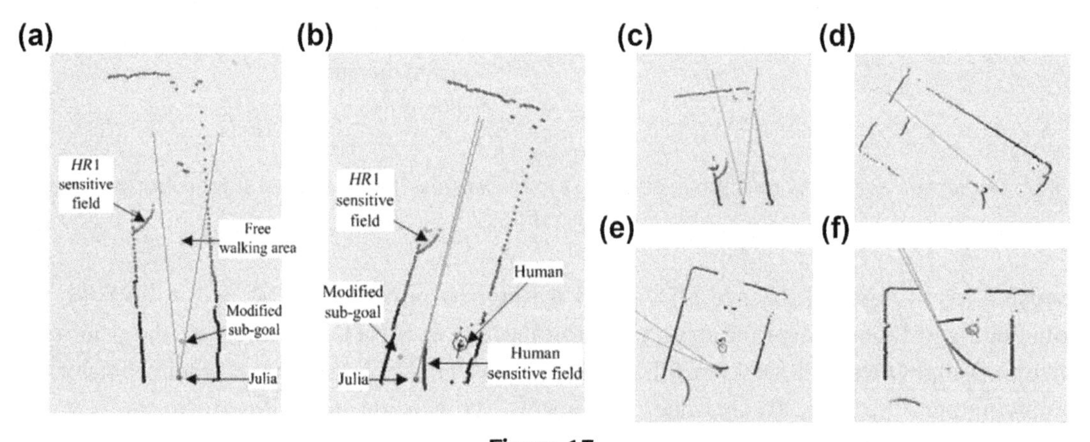

Figure 17
Snapshots from the view point of Julia.

the different behaviors in different situations. In this experiment, Julia is a "*R*2" robot, whose task is to patrol the environment following the path shown in Figure 20. At the beginning, Pioneer is a "*HR*2" robot that leads a human to a particular place, and after Pioneer finishes its task, it becomes a "*R*3" robot that is wandering in the environment and waiting for a human's further command.

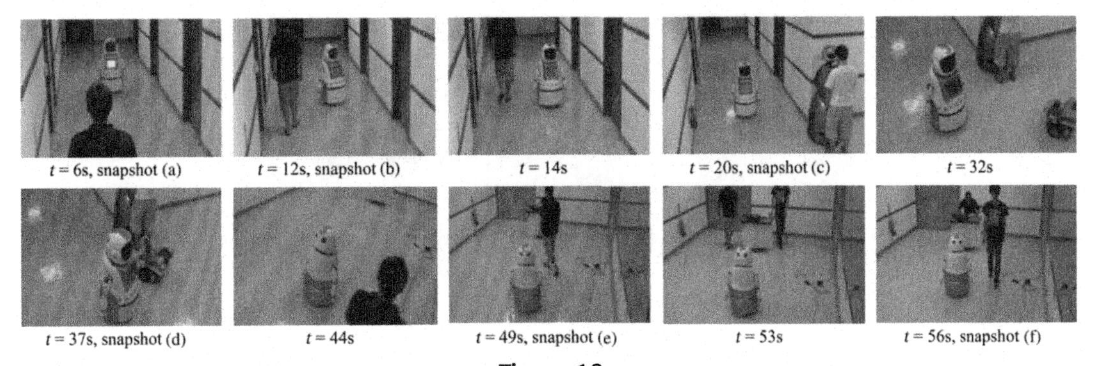

Figure 18
Experiment 3.

As shown in Figure 21, where Pioneer is a "*HR*2" robot, since *HR*2 field is a solid field, the solid field blocks the original patrol path of Julia, which then chooses another path to traverse so that it (Julia) will not affect the motion of *HR*2 robot (pioneer), whose priority is higher than the *R*2 robot.

When someone approaches Julia so that Julia enters the sensitive fields of that person unwillingly, as shown in Figure 22, Julia first enters the "waiting" state for the "waiting rule." However, when this person starts to interact with the robot through the touch panel on Julia, the state of Julia will transfer from state "*R*2" to state "*HR*1," as shown in Figure 23.

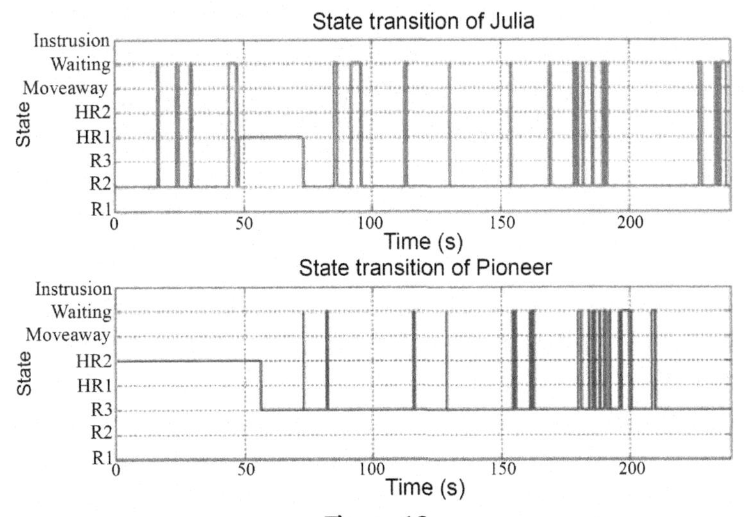

Figure 19
State transitions of Julia and Pioneer.

Figure 20
Patrol path of Julia.

Figure 21
Julia modifies its direction to avoid entering *HR*2 sensitive field.

After this, the two robots and two persons continue to work in this environment for a while. Figure 19 shows the overall state transitions in this experiment.

3.5.8 Conclusion and Future Work

We have proposed a framework that allows robots to harmoniously coexist with humans and other robots. HCSN, which takes account of harmonious rules, personal space of humans, and working space of robots, is proposed. After the robot has successfully tracked people, the sensitive fields are generated. By referring to the sensitive fields in its

Figure 22
Julia enters the human-sensitive field.

Figure 23
Julia is interacting with human.

navigation phase, the robot would be least disturbing to existing humans, behaving more friendly in comparison with the current local navigation approaches. Our approach has been run on different scenarios in both simulations and experiments, and the results showed the feasibility of our human-sensitive navigation.

One of the most common limitations of HCSN is that we have to get the human and robot states to some extent of precision. The inaccuracy of human tracking will adversely affect the performance of HCSN. Therefore, our future work would be taking account of the tracking error when the robot generates the human-sensitive field and the corresponding navigation strategy to these uncertain fields. In addition, it could be a more complicated but interesting problem if we can somehow dynamically estimate the intensions of humans. Their intensions can be another criteria to dynamically adjust the egg-shape field. For example, when a person intends to approach the robot rather than just pass by it, his/her corresponding $H1$ field should become relatively small. Finally, a deadlock may happen when the common passage is not large enough for two or more robots and humans to simultaneously pass through. In order to admit this case, we have to extend our rules or to design a mechanism or protocol to cope with such a problem.

References

[1] O. Khatib, Real-time obstacle avoidance for manipulators and mobile robots, Int. J. Robot. Res. 5 (1986) 90–98.

[2] Y. Koren, J. Borenstein, Potential field methods and their inherent limitations for mobile robot navigation, in: Proc. IEEE Int. Conf. Robot. Autom., 1991, pp. 1398–1404.

[3] M. Javier, L. Montano, Nearness diagram (ND) navigation: collision avoidance in troublesome scenarios, IEEE Trans. Robot. Autom. 20 (1) (February 2004) 45–59.

[4] D. Fox, W. Burgard, S. Thrun, The dynamic window approach to collision avoidance, IEEE Robot. Autom. Mag. 4 (1) (March 1997) 23–33.

[5] S.M. LaValle, J.J. Kuffner Jr., Randomized kinodynamic planning, Int. J. Robot. Res. 20 (2001) 378–400.

[6] J.H. Lee, K. Abe, T. Tsubouchi, R. Ichinose, Y. Hosoda, K. Ohba, Collision-free navigation based on people tracking algorithm with biped walking model, in: Proc. IEEE/RSJ Int. Conf. Intell. Robots Syst., 2008, pp. 2983–2989.

[7] R. Alami, I. Belousov, S. Fleury, M. Herrb, F. Ingrand, J. Minguez, B. Morisset, Diligent: towards a human-friendly navigation system, in: Proc. IEEE/RSJ Int. Conf. Intell. Robot. Syst., 2000, pp. 21–26.

[8] P. Althaus, H. Ishiguro, T. Kanda, T. Miyashita, H.I. Christensen, Navigation for human-robot interaction tasks, in: Proc. IEEE Int. Conf. Robot. Autom., 2004, pp. 1894–1900.

[9] E.A. Sisbot, L.F. Marin-Urias, R. Alami, T. Simeon, A human aware mobile robot motion planner, IEEE Trans. Robot. 23 (5) (October 2007) 874–883.

[10] E.A. Topp, H.I. Christensen, Tracking for following and passing persons, in: Proc. IEEE/RSJ Int. Conf. Intell. Robots Syst., 2005, pp. 2321–2327.

[11] S. Takeshi, H. Hideki, Human observation based mobile robot navigation in intelligent space, in: Proc. IEEE/RSJ Int. Conf. Intell. Robots Syst., 2006, pp. 1044–1049.

[12] M. Svenstrup, S. Tranberg, H.J. Andersen, T. Bak, Pose estimation and adaptive robot behaviour for human-robot interaction, in: Proc. IEEE/RSJ Int. Conf. Intell. Robots Syst., Kobe, Japan, 2009.

[13] E. Sviestins, N. Mitsunaga, T. Kanda, H. Ishiguro, N. Hagita, Speed adaptation for a robot walking with a human, in: Proc. HRI, Arlington, VA, March 8–11, 2007, pp. 349–356.

[14] M. Walters, K.L. Koay, S. Woods, D.S. Syrdal, K. Dautenhahn, Robot to human approaches: preliminary results on comfortable distances and preferences, in: Proc. AAAI Spring Symp. Multidisciplinary Collaborat. Socially Assistive Robot, Stanford University Press, Palo Alto, CA, 2007.

[15] H. Hutenrauch, K.S. Eklundh, A. Green, E.A. Topp, Investigating spatial relationships in human-robot interaction, in: Proc. IEEE/RSJ Int. Conf. Intell. Robots Syst., China, Beijing, October 9–15, 2006.

[16] E. Pacchierotti, H.I. Christensen, P. Jensfelt, Evaluation of passing distance for social robots, in: K. Dautenhahn (Ed.), IEEE workshop robot human interactive communication, September 2006. Hertfordshire, UK.

[17] E.T. Hall, The Hidden Dimension, Anchor, New York, 1966.

[18] J. Minguez, J. Osuna, L. Montano, A "divide and conquer" strategy based on situations to achieve reactive collision avoidance in troublesome scenarios, in: Proc. IEEE Int. Conf. Robot. Autom., 2004, pp. 3855–3862.
[19] Y. Kanayama, Y. Kimura, F. Miyazaki, T. Noguchi, A stable tracking control method for an autonomous mobile robot, in: Proc. IEEE Int. Conf. Robot. Autom., 1990, pp. 384–389.
[20] T. Horiuchi, S. Thompson, S. Kagami, Y. Ehara, Pedestrian tracking from a mobile robot using a laser range finder, in: Proc. IEEE Int. Conf. Syst., Man Cybern., 2007, pp. 931–936.

PART 4

Object Recognition

CHAPTER 4.1

The State of the Art in Object Recognition for Household Services

Guoyuan Liang[1], Xinyu Wu[2], Yangsheng Xu[1,2]

[1]The Chinese University of Hong Kong, Hong Kong SAR, China; [2]Shenzhen Institutes of Advanced Technology, Chinese Academy of Sciences, Shenzhen, China

Chapter Outline

4.1.1 Overview

Household service is one of the most important fields in robot applications. Compared with applications in industry, the household environment is more challenging. First, the household environment is usually unstructured, which makes it more difficult to handle simple tasks such as what the assembly line robot can easily achieve. Second, the object for household services is human, sometimes a small child. For safety reasons, the robot should be very careful when taking actions. Third, the household robot is expected to be cleverer in dealing with complicated situations, which requires flexibility in responding. All these challenges are, to some extent, related to the capability of perceiving various objects, from an old table to a sleeping baby. In this chapter, we discuss the state of the art in object recognition for household service.

In brief, object recognition is identifying certain objects from various kinds of data captured from the real world. The task is extremely difficult because the captured data, no matter what kind of sensor they come from, are subject to more or less ambiguity. For example, the core problem for the visual perception system is that each object in the real world can make numbers of two-dimensional (2D) images when the object's orientation, location, lighting, and background change. On the other hand, objects with different

Household Service Robotics. http://dx.doi.org/10.1016/B978-0-12-800881-2.00012-8

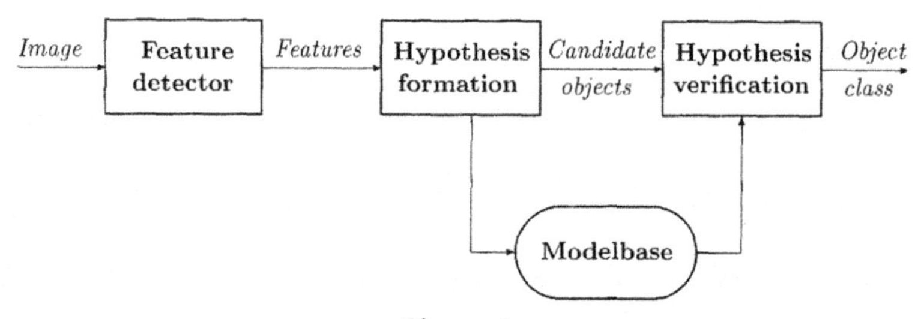

Figure 1
Various components of an object recognition system.

orientations can cast the same 2D image onto the retina because of the uncertainty of the projection from 3D objects to 2D images. This ambiguity readily exists in the computation of object recognition. Ramesh Jain et al. described the general framework for object recognition in their book [1], as shown in Figure 1.

Although it is based on the visual system, actually it is applicable to data acquired by all types of sensors (e.g., range data, audio data, tactile data, etc.). The key part of the system is the model database and feature detector. The former designates what to recognize, the latter, however, indicates by what to make such recognition. The feature detector is the most challenging part because of the complexity of extracting features for different objects, no matter whether it is automatic or semiautomatic.

With the development of mechanical, electronic, and control engineering, researchers have made significant progress in the field of household service robots. In the evolutionary process of the technology for object recognition in a household environment, we note that the new methodologies and techniques developed since 2005 most likely were related to the following aspects.

4.1.1.1 Multimodal Perception

In addition to the camera system, most modern household service robots are equipped with multiple sensors such as a 3D scanner, audio sensor, and tactile sensor. It is more effective to fuse all the information of the object to make a more reliable recognition. Moreover, the fusion of information also opens up a new window for retrieving properties of the object.

Depth data can give accurate descriptions of the object's shape. The depth data scanner used to be too expensive for household applications. With the advent of cheaper depth data sensors, e.g., Kinect, more interest has been focused on shape-based object recognition. Kevin Lai et al. [2] presented an approach for recognizing objects from a 3D point cloud through learning. Figure 2 illustrates the point cloud data from a laser scanner and several 3D models from Google's 3D Warehouse. An important problem for this method is the shortage of labeled training data. Although there are lots of 3D data available on the Web,

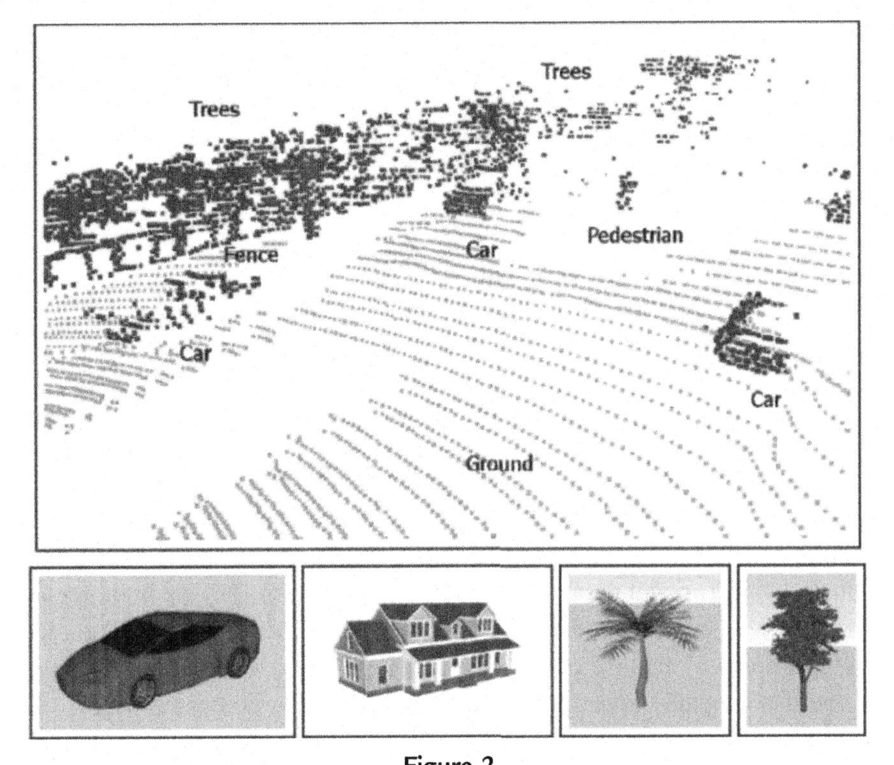

Figure 2
Top: Part of a 3D laser scan taken in an urban environment. Bottom: Four example models from Google's 3D Warehouse.

such as Google's 3D Warehouse, the different characteristics between data collected by the laser scanner installed on a mobile robot and the data from the Web decrease the accuracy of the resulting classifier. To solve this problem, the authors propose a domain adaption approach extended from an image-based classification based on exemplar learning. At the end, they suggest adding visual sensor modalities to improve the accuracy. In a following paper [3], Lai et al. explored a sparse distance learning approach for object recognition combining red, green and blue (RGB) color model and depth information, as shown in Figure 3. They defined a view-to-object distance in which a novel view is compared simultaneously to all views of a previous object. The measure greatly improves the performance of the classification and object recognition and makes it possible to find a sparse solution via Group-Lasso regularization. The proposed technique, instance distance learning, is tested on an RGB-D object data set and its effectiveness is confirmed by comparing with three other state of the art approaches.

Marton et al. [4] presented a categorization and classification approach combining 2D–3D information for multimodal perception. The scenario is set for everyday manipulation tasks in the kitchen environment. The system identifies objects by texture based and 3D

Figure 3
Views of objects from the RGB-D object data set shown as 3D point clouds colored with RGB pixel values.

perception routines, limits the possibilities for identities of objects by categorization modules, and then adds more details until an acceptable solution can be found. The system also maintains an object model database providing information for detection, recognition, and pose estimation as well as for grasping of the objects. The accuracy of the recognition is immensely improved by the use of multiple detectors. This is very important because in some cases one detector may give a totally different judgment compared to the other detector. In some other cases, it is hard to detect objects with only one sensor. The system processing pipeline architecture is illustrated in Figure 4.

Apart from the depth and image sensor, proprioceptive, audio, and tactile sensors are employed in many applications as well, which opens new doors for identifying objects. Sinapov et al. [5] proposed a method for household object recognition using

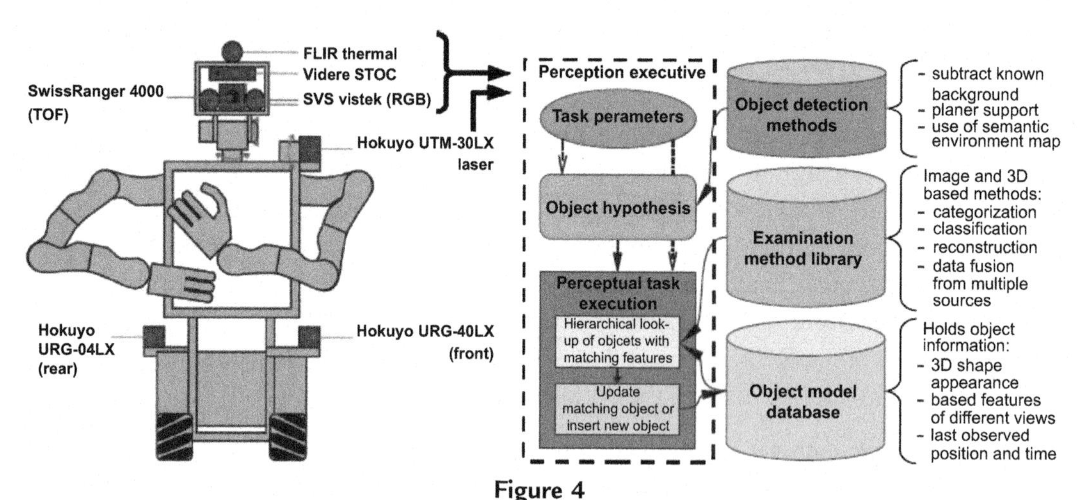

Figure 4
Object processing pipeline architecture: from sensor data to objects.

proprioceptive and audio feedback. The robot records the changes in the data in its proprioceptive and auditory sensor while performing five exploratory behaviors (lift, shake, drop, crush, and push) on 50 household objects (e.g., bottles, cups, balls, toys). The robot then learns a model for each sensory modality using a self-organizing map and eventually is asked to recognize the objects it is manipulating by feeling them and listening. Experiments show that integration from two modalities can achieve better recognition accuracy. Nakamura et al. [6] presented a method for learning novel objects from audiovisual input. The object learning in their mobile manipulation system is implemented using natural speech instruction along with the object segmentation from the background image, as Figure 5 shows. Pezzementi et al. [7] used reading data from a tactile sensor to perform object recognition. The tactile sensor is regarded as an imaging device, and the object representations are learned from these *tactile images* based on mosaics of tactile measurements instead of using a prior geometric model. Their algorithm can not only recognize the objects but locates their positions and evaluates their orientations as well. In another paper [8], Pezzementi extended the topic with exploring unknown objects using tactile-force sensors.

4.1.1.2 Model-Based and Learning-Based Methods

The premise of a successful human–robot interaction most likely depends on how much the robot knows about the service object. In general, we need to build a model database for the objects to be recognized, as illustrated in Figure 1. In a household environment, the objects of interest usually include service objects (people), household objects (cups, bottles, toys, etc.), and environment objects (desks, tables, walls, etc.). Most of them have stable shapes or structures. Therefore, the model-based methods received lots of attention in the recognition system of household robots. Sometimes a predefined geometry or probability model is not good enough for describing the object of interest. Basically the recognition is based on the comparison of object representations or features extracted from

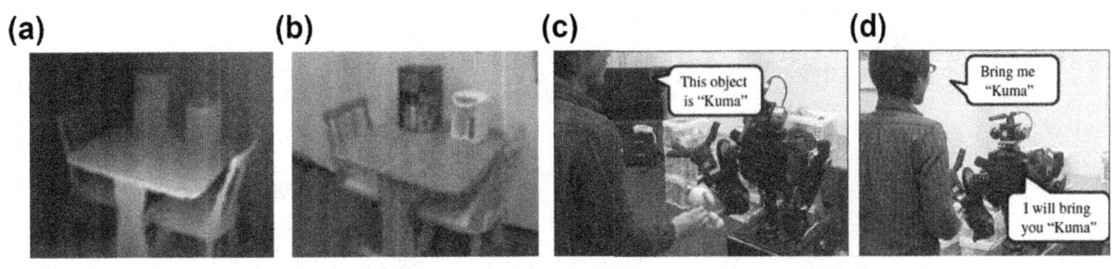

Figure 5

Learning novel objects from audiovisual input. (a) Depth image. (b) Mapped color image. (c) The user teaches the object to the robot. (d) The robot recognizes and utters the out-of-vocabulary word.

objects in the database and the object of interest. Some of the latest research indicates that the manner of feature extraction and selection has a great impact on the recognition results. Therefore, instead of applying a priori knowledge or rules, feature extraction and selection based on learning over a bunch of training samples may be a good option. Automatic feature extraction methods such as deep learning show great potential in this area.

Méndez-Polanco et al. [9] presented an algorithm for detecting and tracking people by using a stereo vision system in dynamic indoor environments. The objects are first segmented from the background using the distance information provided by a stereo camera. Then the locations of the people are roughly estimated according to the proportions of the human body. After that, a contour model is used to calculate the probability of people detection. Finally, by merging the probabilities of the contour model and applying a Bayesian framework, the people are detected. The experimental results are shown in Figure 6. Lin et al. [10] proposed a method for object orientation recognition based on the scale invariant-feature transform (SIFT) and support vector machine (SVM) with a stereo camera system. An affine transform matrix based on SIFT is built as the features, then the SVM is used to classify the object orientation. Collet et al. [11] presented an object recognition and pose registration method from a single image. The 3D model of the object needs to be built and aligned with the real object image in advance. Given a new test image, the local descriptors are matched to the stored models online to give the full pose registration. Rudinac and Jonker tried to use a small number of training samples to find a fast and robust descriptor for multiple view objects. In their paper [12], they picked a combination of invariant color, edges, shapes, and texture descriptors to form the feature vectors and then performed normalization on all the feature vectors from

Figure 6
Experiments performed using a mobile robot in dynamic indoor environments.

the training database to increase the importance of the most dominant feature components and to reduce the less dominant ones.

4.1.1.3 Combination of Interactive and Autonomous Methods

Researchers have already noted that the identification of objects located within a complex background is extremely difficult. No single method can work in every situation. They have figured out that the combination of interactive methods with autonomous methods would provide better performance. Mansur et al. [13] realized that the combination of interactive and autonomous methods could result in better recognition. They proposed several types of interactive recognition methods. Each takes place at the failure of the autonomous methods in different situations. The interactions include interactive instructions by the user and the interactive learning process. Rouanet et al. [14] believed that, in addition to robust machine learning and computer vision algorithm, well-designed human–robot interfaces are crucial for learning efficiency. They designed a system allowing a nonexpert user to give instructions through four alternative human–robot interfaces, which intuitively provide much better training examples to the robot.

4.1.1.4 New Multimodal Data Sets

In earlier days, researchers built their own data sets for their experiments. With the growth of the robot community, the need for public data sets kept growing. Therefore, a lot of public image repositories are available on the Web, such as Google Images and Flickr, as well as object recognition data sets such as Caltech 101, LabelMe, and ImageNet. As the application of multiple sensors in robot recognition systems has expanded, people have acquired deeper insights into and in-depth knowledge on the problems, which has led to the formation of various new multimodal data sets. Lai et al. [15] also introduced a large-scale hierarchical multiview RGB-D (Kinect style) object data set for the task of object recognition. The data set contains 300 objects organized into 51 categories and has been published to enable rapid progress in this technology.

4.1.2 Summary of Case Studies

The remainder of this chapter details case studies on three aspects concerning the building of an object database in the human environment, the object perception algorithm, and a performance evaluation for various approaches.

- New data sets in human environments
  Ciocarlie et al. [16] introduced three data sets for mobile manipulation in human environments. The first set contains a large corpus of robot sensor data collected in typical office environments, which has been annotated with ground-truth information outlining

people in camera images. The second data set consists of 3D models for a number of graspable objects commonly encountered in households and offices. On each of these objects, a large number of grasp points have been marked on a parallel jaw gripper. The third data set contains extensive proprioceptive and ground-truth information regarding the outcome of tasks. Figure 7 shows some examples from this database.

- Vision-based mirrored wall perception
  Lu et al. [17,18] investigated the problem of detecting spurious surfaces in buildings, such as the reflective glassy walls. This is important for collision avoidance when the service robot navigates close to this kind of object. Their approach uses two views from an onboard camera. After deriving geometric constraints for corresponding real—virtual features across two views, they employ a random sample consensus framework and an affine scale-invariant feature transform to develop a robust mirror detection algorithm; the overall detection accuracy rate reaches as high as 91.0%. Figure 8 shows some sample images in their data set.
- Performance evaluation for various object recognition methods
  Ramisa et al. [19] selected three state of the art object recognition methods (the SIFT object recognition algorithm, the Bag of Features, and the Viola and Jones boosted

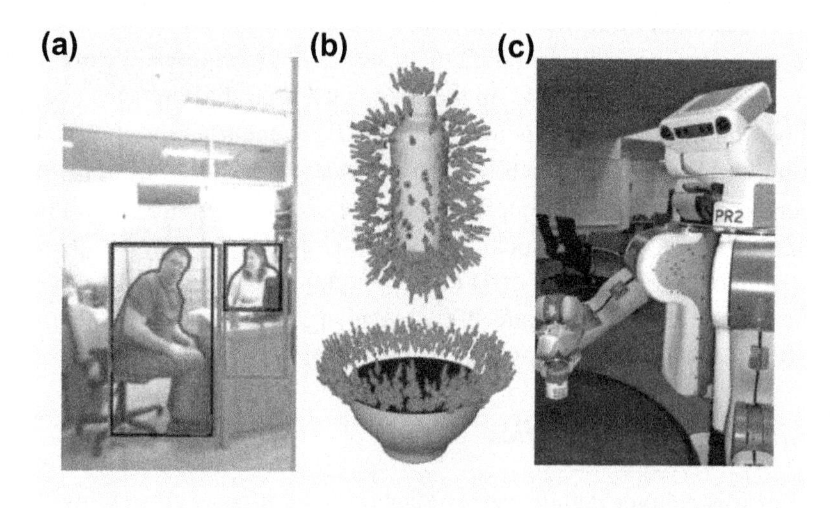

Figure 7
Examples from the data sets. (a) Section of a camera image annotated with people's locations and outlines. (b) 3D models of household objects with grasp-point information (depicted by green arrows) generated in simulation. (c) The PR2 [20] robot executing a grasp while recording visual and proprioceptive information.

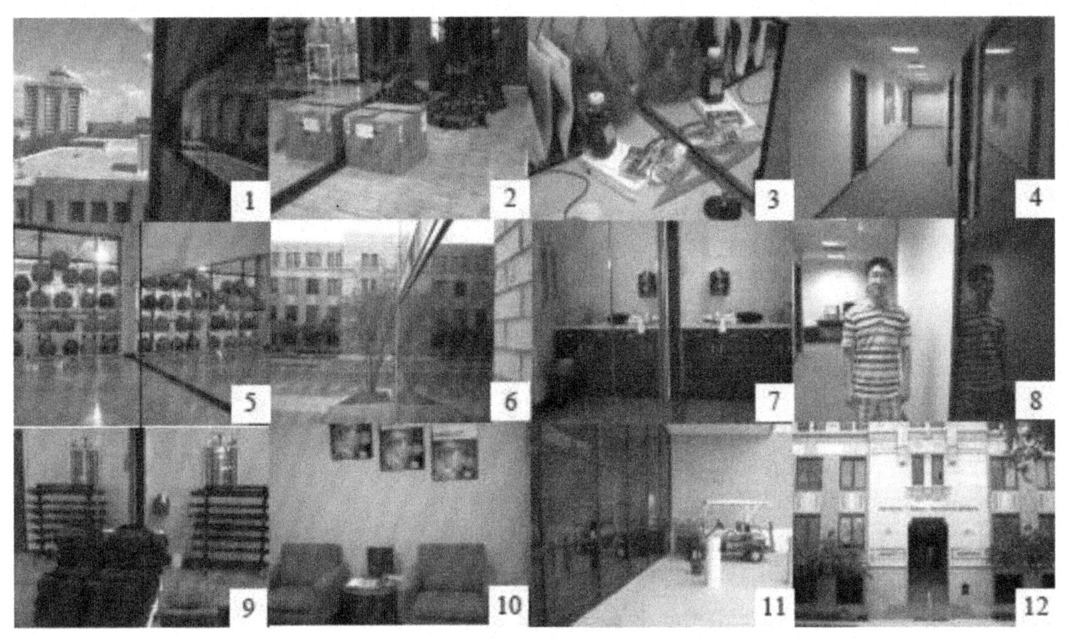

Figure 8
Sample images from the data set of planar mirrored walls.

Table 1: Qualitative summary of results found in experiments

	Scale Invariant-Feature Transform	Vocabulary Tree	Cascade of Simple Classifiers
Detection	Can detect objects under in-plane rotation, scale changes, and small out-of-plane rotations	Must be complemented with a sliding windows approach, a segmentation algorithm, or an interest operator	Is able to determine the most probable bounding box of the object
Pose estimation	Up to an affine transformation	Presence/absence only	Presence/absence only
Classification (intraclass variation and generalization)	No	Yes	Yes
Occlusions	Tolerates it as long as at least three points can be reliably matched (depends on amount of texture)	Showed good tolerance to occlusions	Low tolerance to occlusions
Repetitive patterns	No	Yes	Yes
Minimum training set size	One image	Tens of images	Hundreds or thousands of images

Continued

Table 1: Qualitative summary of results found in experiments—cont'd

	Scale Invariant-Feature Transform	Vocabulary Tree	Cascade of Simple Classifiers
Training set resolution	Video Graphics Array (VGA) resolution is sufficient	Benefits from higher resolution of training data	VGA resolution is sufficient
Trainable online	Easy. Requires clear picture of the object and takes a few seconds to retrain the matching tree	Easy. Requires a few good views of the object and a few seconds to retrain the inverted files	Hard. Requires a large collection of training images and up to a few hours of training for certain objects
Run time	Less than a second per image	Two seconds per image with a segmentation algorithm included	Less than a second per image

cascade of classifiers) and evaluated their performance in 10 aspects including detection, classification, occlusions, repetitive patterns, run time, etc. Table 1 shows the results of the comparison of the three algorithms.

References

[1] R. Jain, R. Kasturi, B.G. Schunck, Machine Vision, McGraw-Hill, Inc, New York, 1995.
[2] K. Lai, D. Fox, Object recognition in 3D point clouds using web data and domain adaptation, Int. J. Robot. Res. 29 (8) (2010) 1019–1037.
[3] K. Lai, L. Bo, X. Ren, D. Fox, Sparse distance learning for object recognition combining RGB and depth information, in: Proc. of IEEE International Conference on Robotics and Automation (ICRA), 2011.
[4] Z.-C. Marton, D. Pangercic, N. Blodow, M. Beetz, Combined 2D–3D categorization and classification for multimodal perception systems, Int. J. Robot. Res. 30 (11) (2011) 1378–1402.
[5] J. Sinapov, T. Bergquist, C. Schenck, U. Ohiri, S. Griffith, A. Stoytchev, Interactive object recognition using proprioceptive and auditory feedback, Int. J. Robot. Res. 30 (10) (2011) 1250–1262.
[6] T. Nakamura, K. Sugiura, T. Nagai, N. Iwahashi, T. Toda, H. Okada, T. Omori, Learning novel objects for extended mobile manipulation, J. Intell. Robot. Syst. 66 (1–2) (2012) 187–204.
[7] Z.A. Pezzementi, C. Reyda, G.D. Hager, Object mapping, recognition, and localization from tactile geometry, in: Proc. of IEEE International Conference on Robotics and Automation (ICRA), 2011, pp. 5942–5948.
[8] Z.A. Pezzementi, E. Plaku, C. Reyda, G.D. Hager, Tactile-object recognition from appearance information, IEEE Trans. Robot. 2 (3) (2011) 473–487.
[9] J.A. Méndez-Polanco, A. Muñoz-Meléndez, E.F. Morales, People detection by a mobile robot using stereo vision in dynamic indoor environments, MICAI, in: A.H. Aguirre, R.M. Borja, C.A.R. García (Eds.), Lecture Notes in Computer Science, vol. 5845, Springer, 2009, pp. 349–359.
[10] C.-Y. Lin, E. Setiawan, Object orientation recognition based on SIFT and SVM by using stereo camera, in: IEEE ROBIO, 2008, pp. 1371–1376.
[11] A. Collet, D. Berenson, S.S. Srinivasa, D. Ferguson, Object recognition and full pose registration from a single image for robotic manipulation, in: IEEE ICRA, 2009, pp. 48–55.
[12] M. Rudinac, P.P. Jonker, A fast and robust descriptor for multiple-view object recognition, in: IEEE ICARCV, 2010, pp. 2166–2171.

[13] Al Mansur, K. Sakata, Y. Kuno, Recognition of household objects by service robots through interactive and autonomous methods, ISVC (2), in: G. Bebis, R.D. Boyle, B. Parvin, D. Koracin, N. Paragios, T.F. Syeda-Mahmood, T. Ju, Z. Liu, S. Coquillart, C. Cruz-Neira, T. Müller, T. Malzbender (Eds.), Lecture Notes in Computer Science, vol. 4842, Springer, 2007, pp. 140–151.

[14] P. Rouanet, P.-Y. Oudeyer, F. Danieau, D. Filliat, The impact of human-robot interfaces on the learning of visual objects, IEEE Trans. Robot. 29 (2) (2013) 525–541.

[15] K. Lai, L. Bo, X. Ren, D. Fox, A large-scale hierarchical multi-view RGB-D object dataset, in: IEEE ICRA, 2011, pp. 1817–1824.

[16] M. Ciocarlie, C. Pantofaru, K. Hsiao, G. Bradski, P. Brook, E. Dreyfuss, A side of data with my robot: three datasets for mobile manipulation in human environments, in: Special Issue: Towards a WWW for Robots, vol. 18(2), IEEE Robotics & Automation Magazine, 2011, pp. 44–57.

[17] Y. Lu, D. Song, H. Li, J. Liu, Automatic recognition of spurious surface in building exterior survey, in: IEEE CASE, 2013, pp. 1047–1052.

[18] A.-A. Agha-mohammadi, D. Song, Robust recognition of planar mirrored walls using a single view, in: IEEE ICRA, 2011, pp. 1186–1191.

[19] A. Ramisa, D. Aldavert, S. Vasudevan, R. Toledo, R.L. de Mántaras, Evaluation of three vision based object perception methods for a mobile robot, J. Intell. Robot. Syst. 68 (2) (2012) 185–208.

[20] W. Garage, The PR2, April 26, 2011, http://www.willowgarage.com/pages/pr2/overview.

CHAPTER 4.2

A Side of Data with My Robot[1]

Matei Ciocarlie[1], Caroline Pantofaru[1], Kaijen Hsiao[1], Gary Bradski[1], Peter Brook[2], Ethan Dreyfuss[3]

[1]*Willow Garage Inc., Menlo Park, CA, USA;* [2]*University of Washington, Seattle, WA, USA;* [3]*Redwood Systems, Fremont, CA, USA*

Chapter Outline

---

[1]

Household Service Robotics. http://dx.doi.org/10.1016/B978-0-12-800881-2.00013-X

The consideration of data set design, collection, and distribution methodology is becoming increasingly important as robots move out of fully controlled settings, such as assembly lines, into unstructured environments. Extensive knowledge bases and data sets will potentially offer a means of coping with the variability inherent in the real world. In this study, we introduce three new data sets related to mobile manipulation in human environments. The first set contains a large corpus of robot sensor data collected in typical office environments. Using a crowd-sourcing approach, this set has been annotated with ground-truth information outlining people in camera images. The second data set consists of three-dimensional (3-D) models for a number of graspable objects commonly encountered in households and offices. Using a simulator, we have identified on each of these objects a large number of grasp points for a parallel jaw gripper. This information has been used to attempt a large number of grasping tasks using a real robot. The third data set contains extensive proprioceptive and ground truth information regarding the outcome of these tasks.

All three data sets presented in this chapter share a common framework, both in software [the robot operating system (ROS)] and in hardware [the personal robot 2 (PR2)]. This allows us to compare and contrast them from multiple points of view, including data collection tools, annotation methodology, and applications.

Unlike its counterpart from the factory floor, a robot operating in an unstructured environment can expect to be confronted by the unexpected. Generality is an important quality for robots intended to work in typical human settings. Such a robot must be able to navigate around and interact with people, objects, and obstacles in the environment, with a level of generality reflecting typical situations of daily living or working. In such cases, an extensive knowledge base, containing and possibly synthesizing information from multiple relevant scenarios, can be a valuable resource for robots aiming to cope with the variability of the human world.

Recent years have seen a growing consensus that one of the keys to robotic applications in unstructured environments lies in collaboration and reusable functionality [1,2]. A result has been the emergence of a number of platforms and frameworks for sharing operational building blocks, usually in the form of code modules, with functionality ranging from low-level hardware drivers to complex algorithms. By using a set of now well-established guidelines, such as stable, documented interfaces and standardized communication protocols, this type of collaboration has accelerated development toward complex applications. However, a similar set of methods for sharing and reusing data has been slower to emerge.

In this chapter, we present three data sets that use the same robot framework, comprising the ROS [1,3] and PR2 platform [4]. While sharing the underlying software and hardware architecture, they address different components of a mobile manipulation task: interacting with humans and grasping objects. They also highlight some of the different choices available for creating and using data sets for robots. As such, this comparison endeavors to begin a dialog on the format of data sets for robots. The three data sets, exemplified in Figure 1, are as follows:

- the Moving People, Moving Platform data set, containing robot perception data in office environments with an emphasis on person detection
- the Household Objects and Grasps data set, containing 3-D models of objects common in the household and office environments, as well as a large set of grasp points for each model precomputed in a simulated environment
- the Grasp Playpen data set, containing both proprioceptive data from the robot's sensors and ground-truth information from a human operator as the robot performs a large number of grasping tasks.

While new data sets can be independently made available through code or application releases, they can also provide stable interfaces for algorithm development. The intention is not to tie data sets to specific code instances. Rather, both the data set and code can follow rigorous (yet possibly independent) release cycles, while explicitly tagging compatibility between specific versions (e.g., Algorithm 1.0 has been trained on Data

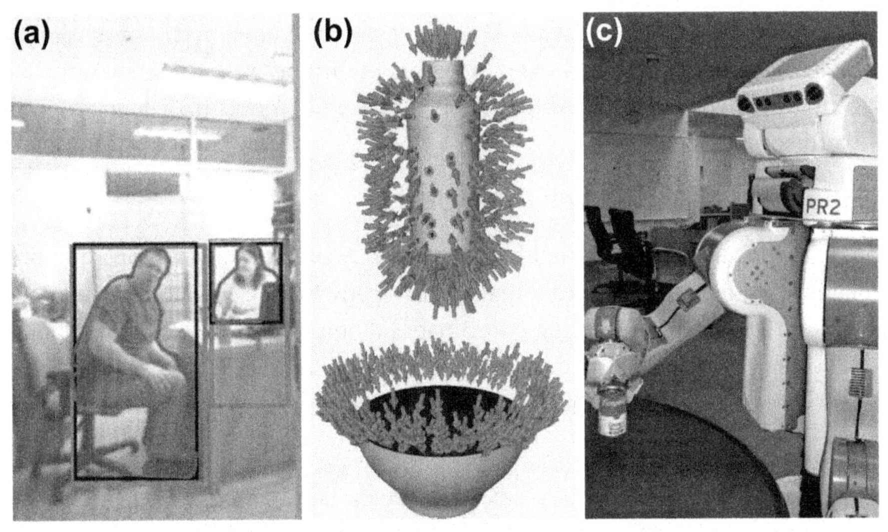

Figure 1

Examples from the data sets presented in this study. (a) Section of a camera image annotated with people's locations and outlines. (b) 3-D models of household objects with grasp point information (depicted by green arrows) generated in simulation. (c) The PR2 robot executing a grasp while recording visual and proprioceptive information. (For interpretation of the color in this figure legend, the reader is referred to the online version of this book.)

3.2). The potential benefits of using such a release model for data sets include the following:

- defining a stable interface to the data set component of a release will allow external researchers to provide their own modified and/or extended versions of the data to the community, knowing that it will be directly usable by anyone running the algorithmic component
- similarly, a common data set and interface can enable a direct comparison of multiple algorithms [5]
- a self-contained distribution, combining a compatible code release and the sensor data needed to test and use them, can increase the research and development community by including groups that do not have access to hardware platforms.

The number of mobile manipulation platforms capable of combining perception and action is constantly rising; as a result, the methods by which we choose to share and distribute data are becoming increasingly important. In an ideal situation, a robot confronted with an unknown scenario will be able to draw on similar experiences from a different robot and then finally contribute its own data back to the community. The context for this knowledge transfer can be online (with the robot itself polling and then sending data back to a repository) or offline (with centralized information from multiple robots used as training data for more general algorithms). Other choices include the format and contents of the data itself (which can be raw sensor data or the result of task-specific processing), the source of annotations and other metadata (expert or novice human users or automated processing algorithms), etc. These choices will become highly relevant as we move toward a network of publicly accessible knowledge repositories for robots and their programmers, which will be discussed later.

The Moving People, Moving Platform Data Set.

PRs operate in environments populated by people. They can interact with people on many levels by planning to navigate toward a person, navigating to avoid a specific person, navigating around a crowd, performing coordinated manipulation tasks such as object handoff, or avoiding contact with a person in a tabletop manipulation scenario. For all of these interactions to be successful, people must be perceived in an accurate and timely manner.

Training and evaluating perception strategies requires a large amount of data. This section presents the Moving People, Moving Platform data set [6], which contains robot sensor data of people in office environments. This data set is available at http://bags.willowgarage.com/downloads/people_dataset.html.

The data set is intended for use in offline training and testing of multisensor person detection and tracking algorithms that are part of larger planning, navigation, and manipulation systems. Typical distances between the people and the robot are in the range of 0.5–5 m. Thus, the data are more interesting for navigation scenarios such as locating people with whom to interact than in tabletop manipulation scenarios.

4.2.1 Related Work

The main motivation for creating this data set was to encourage research into indoor, mobile robot perception of people. There is a large amount of literature in the computer vision community on detecting people outdoors, from cars, in surveillance imagery, or in still images and movies on the Internet. Examples of such data sets are described below. In contrast, PRs often function indoors. There is currently a lack of multimodal data for creating and evaluating algorithms for detecting people indoors from a mobile platform. This is what the vacuum Moving People, Moving Platform data set aims to fill.

Two of the most widely used data sets for detecting and segmenting people in single images from the Internet are the Institut National de Recherche en Informatique et en Automatique (INRIA) person data set [7] and PASCAL visual object challenge data set [5]. Both data sets contain a large number of images, as well as bounding boxes annotating the extent of each person. The PASCAL data set also contains precise outlines of each person. Neither data set, however, contains video, stereo, or any other sensor information commonly available to robots. The people are contained in extremely varied environments (indoors, outdoors, in vehicles, etc.) People in the INRIA data set are in upright poses referred to as *pedestrians* (e.g., standing, walking, leaning, etc.) On the other hand, poses in the PASCAL data set are unrestricted. For the office scenarios considered in this chapter, people are often not pedestrians. However, their poses are also not random.

Data sets of surveillance data, including [8] and the Technische Universiät München (TUM) kitchen data set [9], are characterized by stationary cameras, often mounted above people's heads. Algorithms using these data sets make strong use of background priors and subtractions.

Articulated limb tracking is beyond the scope of this chapter but should be mentioned. Data sets such as the Carnegie Mellon University (*CMU*) MoCap [10] and HumanEva-II [11] are strongly constrained by a small environment, simple background, and in the case of the CMU data set, tight, uncomfortable clothing.

Detecting people from cars has been a focus of late in the research community. The Daimler pedestrian data set [12] and Caltech pedestrian data set [13] contain monocular video data taken from cameras attached to car windshields. Pedestrians are annotated with bounding boxes denoting the visible portions of their bodies, as well as bounding boxes denoting the predicted entire extent of their bodies, including occluded portions. In contrast to our scenario, the people in this data set are pedestrians outdoors, and the cameras in the cars are moving quickly. Similar to our scenario, the sensor is mounted in a moving platform.

In contrast to the above examples, the Moving People, Moving Platform data set contains a large amount of data of people in office environments, indoors, in a realistic variety of poses, wearing their own clothing, taken from multiple sensors onboard a moving robot platform.

4.2.2 Contents and Collection Methodology

4.2.2.1 Collection Methodology

Data sets can be collected in many ways, and the collection methodology has an impact on both the type of data available and its accuracy. For the Moving People, Moving Platform data set, data were collected by teleoperating the PR2 to drive through four different office environments, recording data from onboard sensors. The robot's physical presence in the environment affected the data collected.

Teleoperation generates a different data set than that by autonomous robot navigation; however, it was a compromise required to obtain entry into other companies' offices. Teleoperation also allowed online decisions about when to start and pause data collection, limiting data set size and avoiding repetitive data such as empty hallways. However, it also opened the door to operator bias.

During collection, the subjects were asked to go about their normal daily routine. The approaching robot could be clearly heard and so could not take people by surprise. Some people ignored the robot, while others were distracted by the novelty and stopped to take photographs or talk to the operator. The operator minimized tainting of the data, although some images of people with camera phones were included for realism (as this scenario often occurs at robot demos).

Capturing natural human behavior is difficult, as discussed in [14]. A novel robot causes unnatural behavior (such as taking a photograph) but is entertaining, and people are patient. On the other hand, as displayed toward the end of our data collection sessions, a robot cohabitating with humans for an extended time allows more natural behavior to emerge, but the constant monitoring presence leads to impatience and annoyance.

4.2.3 Contents: Robot Sensor Data

Given that this data set is intended for offline training and testing, data set size and random access speed are of minimal concern. In fact, providing as much raw data as possible is beneficial to algorithm development. The raw sensor data were therefore stored in ROS format bag files [3]. The images contain Bayer patterns and are not rectified, the laser scans are not filtered for shadow points or other errors, and the image de-Bayering and rectification information is stored with the data. ROS bags make it easy to visualize, process, and run data in simulated real-time within a ROS system. The following list summarizes the contents of the data set, with an example in Figure 2. Figure 3 shows the robot's sensors used for data set collection:

- a total of 2.5 h of data in four different indoor office environments
- 70 GB of compressed data (118 GB uncompressed)
- images from a wide field of view (FoV), color stereo cameras located approximately 1.4 m off the ground, at 25 Hz (640 × 480).
- images from narrow FoV, monochrome stereo cameras located approximately 1.4 m off the ground, at 25 Hz (640 × 480)
- Bayer pattern, rectification, and stereo calibration information for each stereo camera pair laser scans from a planar laser approximately 0.5 ft off the ground, with a frequency of 40 Hz
- laser scans from a planar laser on a tilting platform approximately 1.2 m off the ground, at 20 Hz
- the robot's odometry and transformations between robot coordinate frames.

4.2.4 Annotations and Annotation Methodology

4.2.4.1 Annotation

All annotations in the data set correspond to a de-Bayered, rectified version of the image from the left camera of the wide FoV stereo camera pair. Approximately one-third of the frames were annotated, providing approximately 38,000 annotated images. Table 1 presents the annotation statistics. Annotations take one of three forms: exact outlines of the visible parts of people, bounding boxes of the visible parts computed from the outlines, and bounding boxes of the predicted full extent of people, including occluded parts. Annotation examples can be found in Figure 4. These design decisions were driven by the desire for consistency with previous computer vision data sets, as well as the restrictions imposed by the use of Amazon's Mechanical Turk marketplace for labeling, which will be discussed in the following subsection.

Within the data set ROS bags, annotations are provided as ROS messages, time synchronized with their corresponding images. To align an annotation with an image, the users must de-Bayer and rectify the images. Since the annotations were created on the rectified images, the camera parameters may not be changed after annotation, but the algorithm used for de-Bayering may be improved. In addition, to complement the non-ROS data set distribution, XML-format annotations are provided with the single image files.

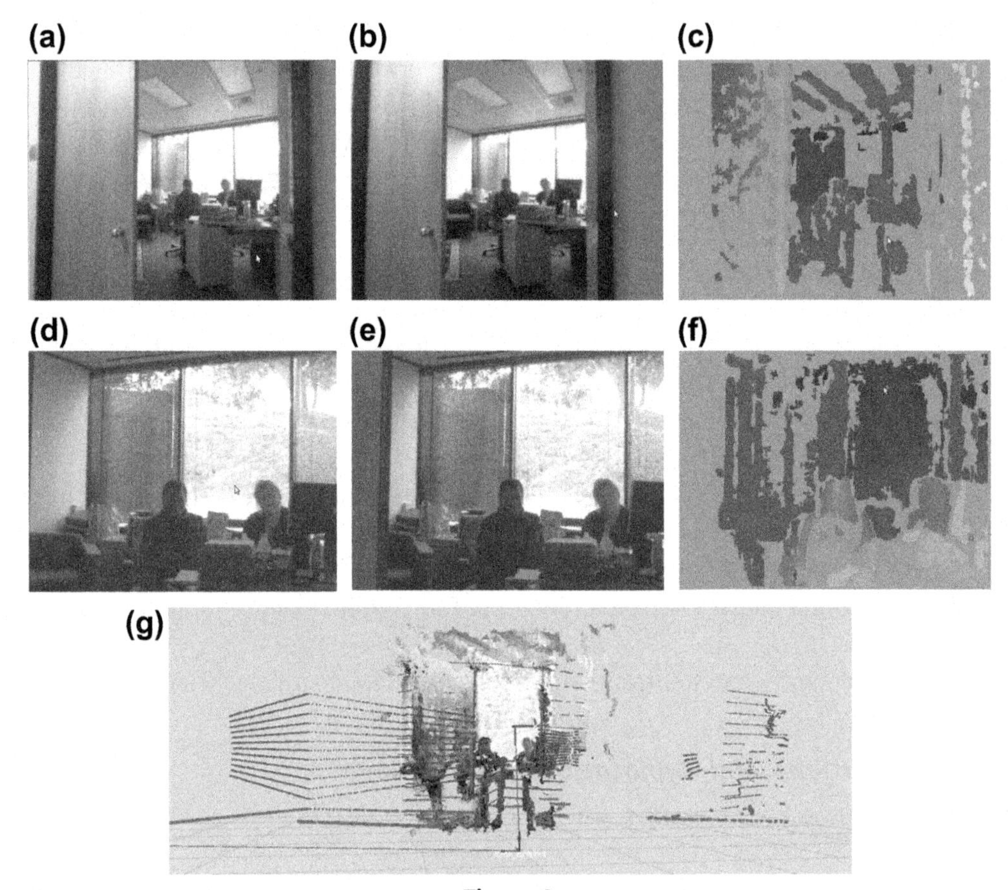

Figure 2

A snapshot of data in the Moving People, Moving Platform data set. While raw data in a robotics-specific format such as ROS bags is preferred by the robotics community, it is valuable to consider other research communities who may contribute solutions. For example, the computer vision community pursues research into person detection that is applicable to robotics scenarios. To encourage participation in solving this robotics challenge, the data set is also presented in a format familiar to the vision community: PNG-format images. In the current offering of the data set, the PNG images are de-Bayered and rectified to correspond to the annotations; however, they could also be offered in their raw form. (a) Wide FoV stereo left camera, (b) wide FoV stereo right camera, (c) wide FoV false-color depth, (d) narrow FoV stereo left camera, (e) narrow FoV stereo right camera, (f) narrow FoV stereo false-color depth, and (g) 3-D visualization. Red/green/blue axes: robot's base and camera frames. Red dots, data from the planar laser on the robot base; blue dots, 0.5 s of scans from the tilting laser; and the true-color point clouds are from the stereo cameras. Photo courtesy of Willow Garage, Inc. (For interpretation of the color in this figure legend, the reader is referred to the online version of this book.)

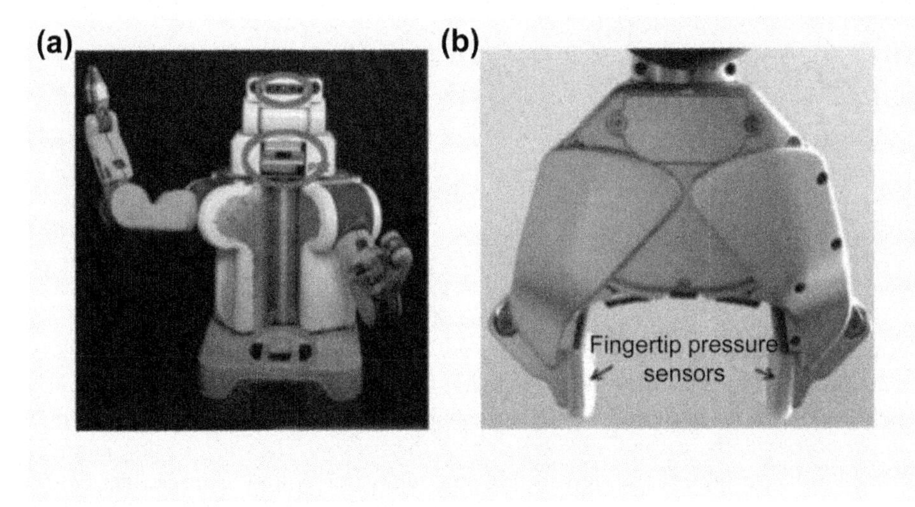

Figure 3
(a) The PR2 robot with sensors used for collecting the Moving People, Moving Platform data set circled in red. From top to bottom: the wide FoV stereo camera pair and the narrow FoV stereo camera pair interleaved on the head, tilting 2-D laser, and planar 2-D laser atop the robot's base. (b) The PR2 gripper and tactile sensors used for collecting data during grasp execution. Photo courtesy of Willow Garage, Inc. (For interpretation of the color in this figure legend, the reader is referred to the online version of this book.)

4.2.5 Annotation Methodology

Annotation of the data set was crowd-sourced using Amazon's Mechanical Turk marketplace [15]. The use of an Internet workforce allowed a large data set to be created relatively quickly but also had implications for the annotations. The workers were untrained and anonymous. Untrained workers are most familiar with rectified, de-Bayered images, and so the robot sensor data were presented as such. As discussed in the previous subsection, image-based annotations are generally incomplete for a robotics application.

Two separate tasks were presented to workers. In the first task, workers were presented with a single image and asked, for each person in the image, to draw a box around the entire person, even if parts of the person were occluded in the image. The visible parts

Table 1: Contents of the Moving People, Moving Platform data set

	Number of Images		
	Total	Labeled	With People
Training files	57,754	21,064	13,417
Testing files	50,370	16,646	–
Total	108,124	37,710	–

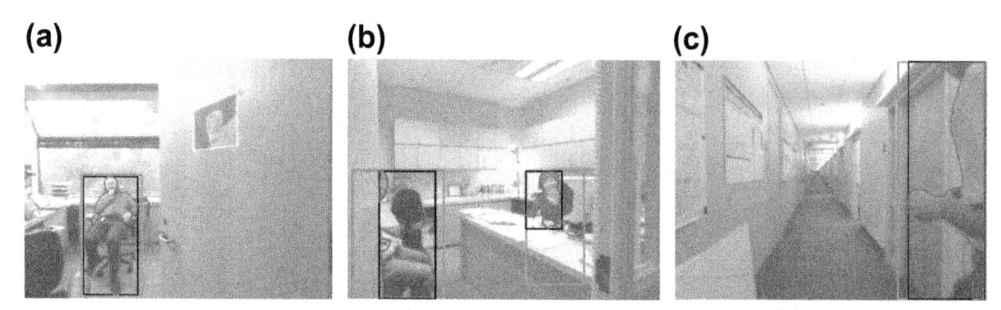

Figure 4
Examples of ground-truth labels in the Moving People, Moving Platform data set. The images have been manipulated to improve outline visibility; they are brighter and have less contrast than the originals. The green bounding box is the predicted full extent of the person. The black bounding box corresponds to the visible portion of the person. The red polygon is an accurate outline of the visible portion of the person. Photo courtesy of Willow Garage, Inc. (a) Office 1. (b) Office 2. (c) Corridor. (For interpretation of the color in this figure legend, the reader is referred to the online version of this book.)

of the person were reliably contained within the outline; however, variability occurred in the portion of the bounding box surrounding occluded parts of the person. This variability could be seen between consecutive frames in the video. In the vast majority of cases, however, workers agreed on the general direction and location of missing body parts. For example, if a person in an image sat at a desk with their legs occluded by the desk, all of the annotations predicted that there were legs behind the desk, below the visible upper body, but the annotations differed in the position of the feet at the bottom of the bounding box.

In the second task, workers were required to draw an accurate, polygonal outline of the visible parts of a single person in an enlarged image. The workers were presented with both the original image and an enlarged image of the predicted bounding box of a single person (as annotated by workers in the first task). An example of the interface is shown in Figure 5. As this task was more constrained, the resulting annotations had less variability.

Mechanical Turk is a large community of workers of varying skills and intent; hence, quality control of results is an important issue. Mechanical Turk allows an employer to refuse to pay or ban underperforming workers. These acts, however, are frowned upon by the worker community who communicates regularly through message boards, resulting in a decreased and angry workforce. Thus, it is important to avoid refusing payment or banning workers whenever possible. The following are lessons learned in our quest for accurate annotations.

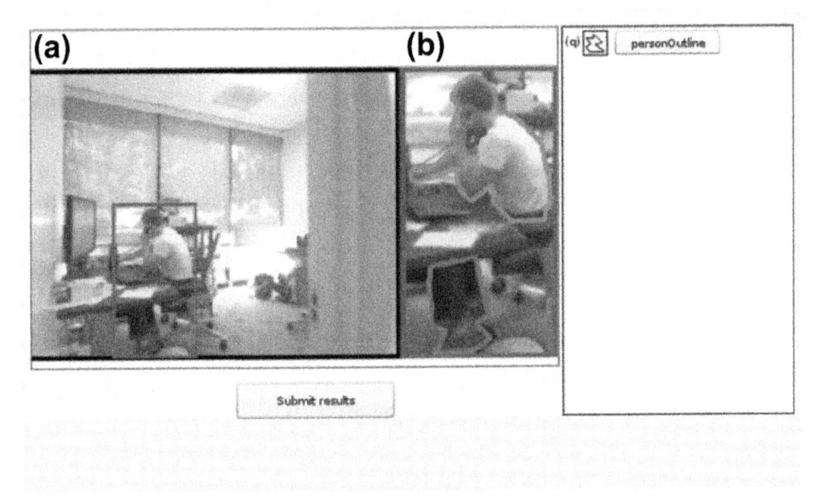

Figure 5
The Mechanical Turk interface for annotating outlines of people for the Moving People, Moving Platform data set. (a) Workers were presented with the original image with a bounding box annotation of one person (by another worker), and (b) an enlarged view of the bounding box on the right. The worker drew a polygonal outline of the person in (b). *Photo courtesy of Willow Garage, Inc.*

- Lesson 1: Interface design can directly improve annotation accuracy. For the outline annotations described in this chapter, the workers were presented with an enlarged view of the person's bounding box. Small errors at this enlarged scale were irrelevant at the original image size.
- Lesson 2: Clear, succinct instructions improve annotation quality. Workers often skim instructions, so pictures with examples of good and bad results are more effective than text.
- Lesson 3: Qualification tests are valuable. Requiring workers to take a multiple choice test to qualify to work on a task significantly improved annotation quality. The simple tests for these tasks verified full comprehension of the instructions and were effective tools for removing unmotivated workers.
- Lesson 4: The effective worker pool for a task is small. For each of the two labeling tasks, each image annotation could be performed by a different worker, implying that hundreds of workers would complete the thousands of jobs. This hypothesis was incorrect: approximately 20 workers completed more than 95% of the work. It appears that workers mitigate training time by performing many similar jobs. This also implies that a workforce can be loyal, so it is worthwhile to train and treat them well, which leads to the final lesson.
- Lesson 5: Personalized worker evaluation increases annotation quality. Initially, workers graded their peers' annotations. Unfortunately, since grading was an easier task than annotating, it attracted less motivated workers. In addition, loyal annotators were upset

by the lack of personal feedback. Grading the graders does not scale, and failing to notice a malicious grader leads to numerous misgraded annotations. These facts encouraged us to grade the annotations personally and write lengthy comments to workers making consistent mistakes. The workers were extremely receptive to this approach, quickly correcting their mistakes, thus significantly reducing duplication of work. Overall, personalized feedback for the small number of workers reduced our own workload.

There are other ways to identify incorrect annotations; however, they were not applicable in this situation. For example, the completely automated public Turing test to tell computers and humans apart (reCAPTCHA) style [16] of presenting two annotations and grading the second based on the first assumes that the errors are consistent. For the annotation task in the Moving People, Moving Platform data set, however, errors resulted from misunderstanding the instructions for a particular image scenario (e.g., a person truncated by the image border). Unless both of the images presented contain the same scenario(s), the redundancy of having two images cannot be exploited.

4.2.6 Applications

This data set is exclusively intended for offline training and testing of person detection and tracking algorithms from a robot perspective. The use of multiple sensor modalities, odometry, and situational information is encouraged. Some possible components that could be tested using this data set are face detection, person detection, human pose fitting, and human tracking. Examples of information beyond that offered by other data sets that could be extracted and used for algorithm training include the appearances of people in multiple robot sensors, typical human poses in office environments (e.g., sitting and standing), illumination conditions (e.g., heavily back-lit offices with windows), scene features (e.g., ceilings, desks, and walls), and how people move around the robot. This is just a small sample of the applications for this data set.

4.2.7 Future Work

It is important to take a moment to discuss the possible constraints on algorithm design imposed by the annotation format and methodology. Two-dimensional (2-D) outlines can only be accurate in the image orientation and resolution. Robots, however, operate in three dimensions. Given that stereo camera information is noisy, it is unclear how to effectively project information from a 2-D image into the 3-D world. The introduction of more reliable instantaneous depth sensors may ameliorate this problem. However, even a device such as the Microsoft Kinect sensor [17] is restricted to one viewpoint. Algorithms developed on such a data set can only provide incomplete information. A format for 3-D annotations that can be obtained from an untrained workforce is an open area for research.

Short-term work for this data set will be focused on obtaining additional types of annotations. It would be informative to have semantic labels for the data set such as whether the person is truncated, occluded, etc., and pose information such as whether the person is standing, sitting, etc. Future data sets may focus on perceiving people during interaction scenarios such as object handoff. Additional data from new sensors, such as the Microsoft Kinect, would also enhance the data set.

Finally, an additional interesting data set could be constructed containing relationships between people and objects, including spatial relationships and human grasps and manipulations of different objects. Object affordances could enhance the other data sets described in this chapter.

4.2.7.1 The Household Objects and Grasps Data Set

A PR's ability to navigate around and interact with people can be complemented by its ability to grasp and manipulate objects from the environment, aiming to enable complete applications in domestic settings. In this section, we describe a data set that is part of a complete architecture for performing pick-and-place tasks in unstructured (or semistructured) human environments. The algorithmic components of this architecture, developed using the ROS framework, provide abilities such as object segmentation and recognition, motion planning with collision avoidance, and grasp execution using tactile feedback. For more details, we refer the reader to our chapter describing the individual code modules as well as their integration [18]. The knowledge base, which is the main focus of this chapter, contains relevant information for object recognition and grasping for a large set of common household objects.

The objects and grasps data set is available in the form of a relational database, using the SQL standard. This provides optimized relational queries, both for using the data online and managing it offline, as well as low-level serialization functionality for most major languages. Unlike the data set described in the previous section, the Household Objects and Grasps set is intended for both offline use during training stages and online use at execution time; in fact, our current algorithms primarily use the second of these options.

An alternative for using this data set, albeit indirectly, is in the form of remote ROS services. A ROS application typically consists of a collection of individual nodes, communicating and exchanging information. The transmission control protocol/Internet protocol (TCP/IP) transport layer removes physical restrictions, allowing a robot to communicate with a ROS node situated in a remote physical location. All the data described in this section are used as the backend for publicly available ROS services running on a dedicated accessible server, using an application programming interface defined in terms of high-level application requirements (e.g., grasp planning). Complete information for using this option, as well as regular downloads for local use of the same data, is available at http://www.ros.org/wiki/household_objects_database.

4.2.8 Related Work

The database component of our architecture was directly inspired by the Columbia grasp database (CGDB) [19,20], released together with processing software integrated with the GraspIt! simulator [21]. The CGDB contains object shape and grasp information for a very large ($n = 7256$) set of general shapes from the Princeton Shape Benchmark [22]. The data set presented here is smaller in scope ($n = 7256$), referring only to actual graspable objects from the real world, and is integrated with a complete manipulation pipeline on the PR2 robot.

While the number of grasp-related data sets that have been released to the community is relatively small, previous research provides a rich set of data-driven algorithms for grasping and manipulation. The problems that are targeted range from grasp point identification [23] to dexterous grasp planning [24,25] and grasping animations [26,27], to name only a few. In this study, we are primarily concerned with the creation and distribution of the data set itself, and the possible directions for future similar data sets used as online or offline resources for multiple robots.

4.2.9 Contents and Collection Methodology

One of the guiding principles for building this database was to enable other researchers to replicate our physical experiments and to build on our results. The database was constructed using physical objects that are generally available from major retailers (while this current release is biased toward U.S. based retailers, we hope that a future release can include international ones as well). The objects were divided into three categories: for the first two categories, all objects were obtained from a single retailer (IKEA and Target, respectively), while the third category contained a set of household objects commonly available in most retail stores. Most objects were chosen to be naturally graspable using a single hand (e.g., glasses, bowls, and cans); a few were chosen as use cases for two-hand manipulation problems (e.g., power drills).

For each object, we acquired a 3-D model of its surface (as a triangular mesh). To the best of our knowledge, no off-the-shelf tool exists that can be used to acquire such models for a large set of objects in a cost and time effective way. To perform the task, we used two different methods, each with its own advantages and limitations:

- For those objects that are rotationally symmetric about an axis, we segmented a silhouette of the object against a known background and used rotational symmetry to generate a complete mesh. This method can generate high-resolution, very precise models but is only applicable to rotationally symmetrical objects.
- For all other objects, we used the commercially available tool 3DSOM (Creative Dimension Software Ltd., U.K.). 3DSOM builds a model from multiple object silhouettes and cannot resolve object concavities and indentations.

Overall, for each object, the database contains the following core information:

- the maker and model name (where available)
- the product barcode (where available)
- a category tag (e.g., glass, bowl, etc.)
- a 3-D model of the object surface, as a triangular mesh.

For each object in the database, we used the *GraspIt*! simulator to compute a large number of grasp points for the PR2 gripper (shown in Figure 3). We note that, in our current release, the definition of a good grasp is specific to this gripper, requiring both finger pads to be aligned with the surface of the object (finger pad surfaces contacting with parallel normal vectors) and further rewarding postures where the palm of the gripper is close to the object as well. In the next section, we will discuss a data-driven method for relating the value of this quality metric to real-world probability of success for a given grasp.

Our grasp planning tool used a simulated annealing optimization, performed in simulation, to search for grip-per poses relative to the object that satisfied this quality metric. For each object, this optimization was allowed to run over 4 h, and all the grasps satisfying our requirements were saved in the database; an example of this process is shown in Figure 6 (note that the stochastic nature of our planning method explains the lack of symmetry in the set of database grasps, even in the case of a symmetrical object). This process resulted in an average of 600 grasp points for each object. In the database, each grasp contains the following information:

- the pose of the gripper relative to the object
- the value of the gripper degree of freedom, determining the gripper opening
- the value of the quality metric used to distinguish good grasps.

The overall data set size, combining both model and grasp information, is 76 and 12 MB uncompressed and compressed, respectively.

4.2.10 Annotations and Annotation Methodology

Unlike the other two data sets presented in this chapter, the models and grasps set does not contain any human-generated information. However, grasp points derived using our autonomous algorithm have one important limitation: they do not take into account object-specific semantic information or intended use. This could mean a grasp that places one finger inside a cup or bowl or prevents a tool from being used. To alleviate this problem, an automated algorithm could take into account more recent methods for considering intended object use [28]. Alternatively, a human operator could be used to demonstrate usable grasps [29]. The scale of the data set, however, precludes the use of few expert operators, while a crowd-sourcing approach, similar to the one discussed in the previous section in the context of labeling persons, raises the difficulty of specifying 6-D grasp points with simple input methods such as a point-and-click interface.

4.2.11 Applications

The database described in this study was integrated in a complete architecture for performing pick-and-place tasks on the PR2 robot. A full description of all the components used for this task is beyond the scope of this chapter. Here, we present

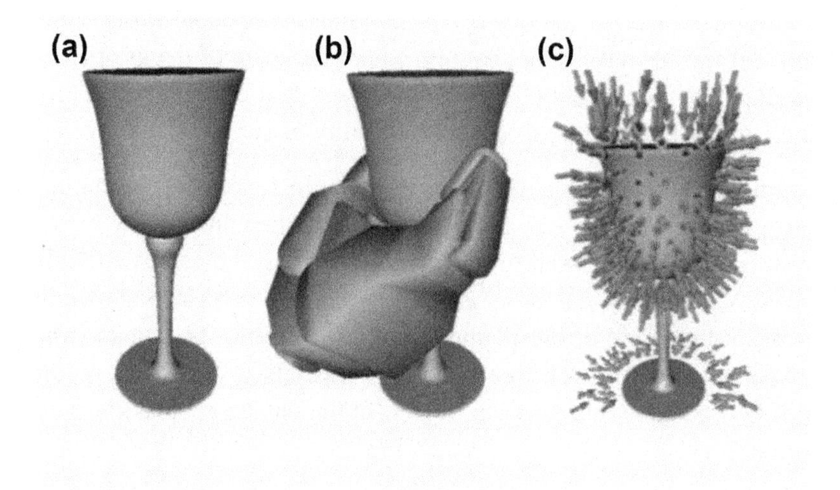

Figure 6
Grasp planning in simulation on a database model. (a) The object model; (b) grasp example using the PR2 gripper; and (c) the complete set of precomputed grasps for the PR2 gripper. Each arrow shows one grasp: the arrow location shows the position of the center of the leading face of the palm, while its orientation shows the gripper approach direction. Gripper roll around the approach direction is not shown.

a high-level overview with a focus on the interaction with the database; for more details on the other components, we refer the reader to [18].

In general, a pick-and-place task begins with a sensor image of the object(s) to be grasped in the form of a point cloud acquired using a pair of stereo cameras. Once an object is segmented, a recognition module attempts to find a match in the database, using an iterative matching technique similar to the iterative closest point (ICP) algorithm [30]. We note that this recognition method uses only the 3-D surface models of the objects stored in the database. Our data-driven analysis discussed in the next section has also been used to quantify the results of this method and relate the recognition quality metric to ground-truth results.

If a match is found between the target object and a data-base model, a grasp planning component will query the database for all precomputed grasp points of the recognized object. Since these grasp points were precomputed in the absence of other obstacles and with no arm kinematic constraints, an additional module checks each grasp for feasibility in the current environment. Once a grasp is deemed feasible, the motion planner generates an arm trajectory for achieving the grasp position, and the grasp is executed. An example of a grasp executed using the PR2 robot is shown in Figure 7. For additional quantitative analysis of the performance of this manipulation framework, we refer the reader to [18].

The manipulation pipeline can also operate on novel objects. In this case, the database-backed grasp planner is replaced by an online planner able to compute grasp points based only on the perceived point cloud from an object; grasps from this grasp planner are used

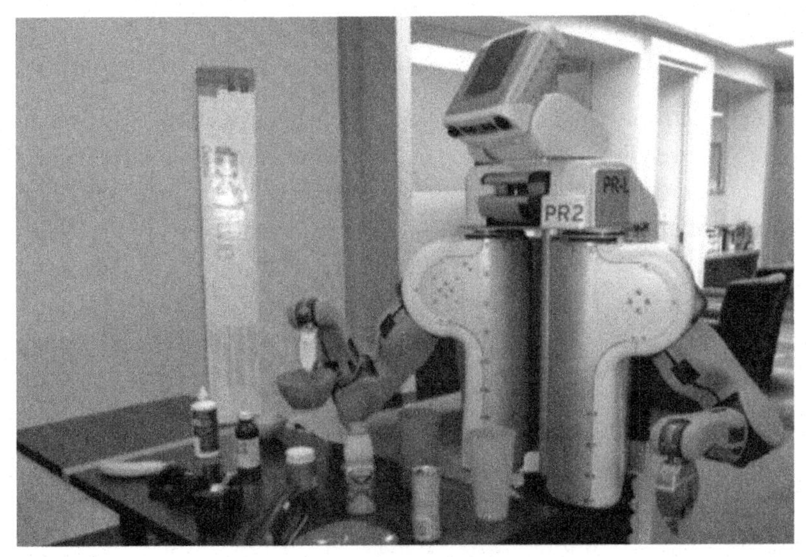

Figure 7
The PR2 robot performing a grasping task on an object recognized from the model database.
Photo courtesy of Willow Garage, Inc.

in addition to the precomputed grasps to generate the Grasp Playpen database described in the next section. Grasp execution for unknown objects is performed using tactile feedback to compensate for unexpected contacts. We believe that a robot operating in an unstructured environment should be able to handle unknown scenarios while still exploiting high-level perception results and prior knowledge when these are available. This dual ability also opens up a number of promising avenues for autonomous exploration and model acquisition that we will discuss later.

4.2.12 Future Work

We believe that the data set that we have introduced, while useful for achieving a baseline for reliable pick-and-place tasks, can also serve as a foundation for more complex applications. Efforts are currently underway to

- improve the quality of the data set itself, e.g., by using 3-D model capture methods that can correctly model concavities or model small and sharp object features at better resolution
- improve the data collection process, aiming to make it faster, less operator intensive, or both
- use the large computational budgets afforded by offline execution to extract more relevant features from the data, which can in turn be stored in the database
- extend the data set to include grasp information for some of the robotic hands most commonly used in the research community
- develop novel algorithms that can make use of this data at run time
- improve the accessibility and usability of the data set for the community at large.

One option for automatic acquisition of high-quality 3-D models for a wide range of objects is to use high-resolution stereo data, able to resolve concavities and indentations, in combination with a pan-tilt unit. Object appearance data can be extended to also contain 2-D images, from a wide range of viewpoints. This information can then be used to precompute relevant features, both 2-D and 3-D, such as speeded up robust features (SURF) [31], point feature histogram [32], or viewpoint feature histogram [33]. This will enable the use of more powerful and general object recognition methods.

The grasp planning process outlined here for the PR2 gripper can be extended to other robot hands as well. For more dexterous models, a different grasp quality metric can be used, taking into account multifingered grasps, such as metrics based on the Grasp Wrench Space [34]. The Columbia grasp database also shows how large scale offline grasp planning is feasible even for highly dexterous hands, with many degrees of freedom [19].

The grasp information contained in the database can be exploited to increase the reliability of object pickup tasks. An example of relevant offline analysis is the study of how each grasp in the set is affected by potential execution errors, stemming from imperfect robot calibration or incorrect object recognition or pose detection. Our preliminary results show that we can indeed rank grasps by their robustness to execution errors; an example is shown in Figure 8. In its current implementation, this analysis is computationally intensive, but it can be performed offline and the results stored in the database for online use.

4.2.12.1 The Grasp Playpen Data Set

Using the pick-and-place architecture described in the previous section, we have set up a framework that we call the Grasp Playpen for evaluating grasps of objects using the PR2 gripper and recording relevant data throughout the entire process. In this framework, the robot performed grasps of objects from the household objects data set placed at known locations in the environment, enabling us to collect ground-truth information for object shape, object pose, and grasps attempted. Furthermore, the robot attempted to not only grasp the object but also shake it and transport it around in an attempt to estimate how robust the grasp is. Such data are useful for offline training, testing, and parameter estimation for both object recognition and grasp planning and evaluation algorithms.

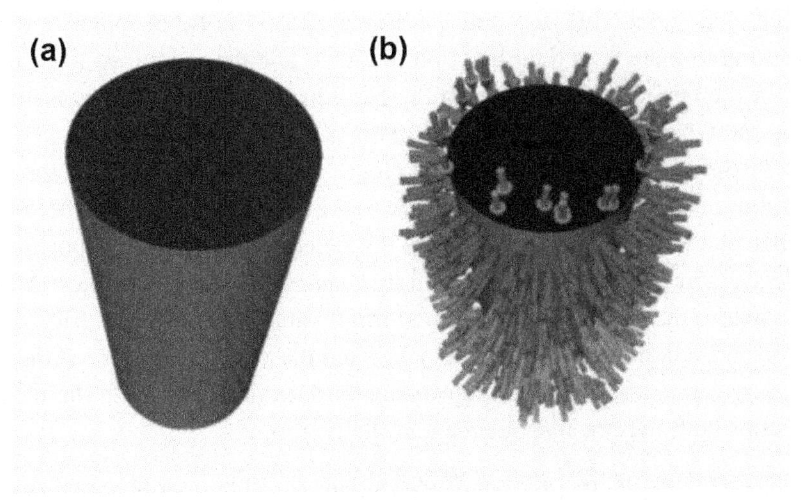

Figure 8
Quantifying grasp robustness to execution errors, from low (red markers) to high (green markers). Note that grasps in the narrow region of the cup are seen as more robust to errors, as the object fits more easily within the gripper. (a) The cup to be grasped and (b) grasp robustness analysis. (For interpretation of the color in this figure legend, the reader is referred to the online version of this book.)

The Grasp Playpen data set can be downloaded for use at http://bags.willowgarage.com/downloads/grasp_playpen_dataset/grasp_ playpen_dataset.html.

4.2.13 Related Work

Although there has been a significant amount of research that uses data from a large number of grasps to either learn how to grasp or evaluate grasp features, it has generally not been accompanied by releases of the data itself. For instance, Balasubramanian et al. [35] use a similar procedure of grasping and shaking objects to evaluate the importance of various features used in grasp evaluation such as orthogonality. Detry et al. [36] execute a large number of grasps with a robot to refine estimated grasp affordances for a small number of objects. However, none of the resulting data appears to be publicly available. Saxena et al. [23] have released a labeled training set of images of objects labeled with the 2-D location of the grasping point in each image; however, the applicability of such data is limited. The semantic database of 3-D objects from TU Muenchen [9] contains point cloud and stereo camera images from different views for a variety of objects placed on a rotating table, but the objects are not meshed and the data set contains no data related to grasping.

4.2.14 Contents and Collection Methodology

Each grasp recording documents one attempt to pickup a single object in a known location, placed alone on a table, as shown in Figure 1(c). The robot selects a random grasp by (1) trying to recognize the object on the table and using a grasp from the stored set of grasps for the best detected model in the Household Objects and Grasps database (planned using the *GraspIt*! simulator) or (2) using a grasp from a set generated by the novel-object grasp planner based on the point cloud. It then tries to execute the grasp. To estimate the robustness of the grasp chosen, the robot first attempts to lift the object off the table. If that succeeds, it will slowly rotate the object into a sideways position, then shake the object vigorously along two axes in turn, and then move the object off and away from the table and to the side of the robot, and finally attempt to place it back on the other side of the table. Visual and proprioceptive data from the robot are recorded during each phase of the grasp sequence; the robot automatically detects if and when the object is dropped and stops both the grasp sequence and the recording.

In total, the data set contains recordings of 490 grasps of 30 known objects from the Household Objects and Grasps data set, collected using three different PR2 robots over a three-week period. Most of these objects are shown in Figure 9. Each grasp recording includes both visual and proprioceptive data. The data set also contains 150 additional images and point clouds of a total of 44 known objects from the Household Objects and Grasps data set, including the 30 objects used for grasping. An example of the point cloud

Figure 9
(a) A subset of the objects used in the Grasp Playpen data set's grasp recordings. (b) The point cloud for a nondairy creamer bottle, with the appropriate model mesh overlaid in the recorded ground-truth pose. *Photo courtesy of Willow Garage, Inc.*

with its ground-truth model mesh overlaid is shown in Figure 9. Recorded data are stored as ROS-for-mat bag files, as in the Moving People, Moving Platform data set. Each grasp also has an associated text file summarizing the phase of the grasp reached without dropping the object, as well as any annotations added by the person.

Each grasp recording contains visual data of the object on the table prior to the grasp, from two different views obtained by moving the head:

- images and point clouds from the narrow FoV, monochrome stereo cameras (640 × 480)
- images from the wide FoV, color stereo cameras (640 × 480)
- images from the gigabit color camera (2448 × 2050)
- the robot's head angles and camera frames
- During the grasp sequence, the recorded data contain narrow and wide FoV stereo camera images (640 × 480, 1 Hz)
- grasping arm forearm camera images (640 × 480, 5 Hz)
- grasping arm fingertip pressure array data (25 Hz)
- grasping arm accelerometer data (33.3 kHz)
- the robot's joint angles (1.3 kHz)
- the robot's camera and link frames (100 Hz)
- the requested pose of the gripper for the grasp

The average size of all the recorded data for one grasp sequence (compressed or uncompressed) is approximately 500 MB; images and point clouds alone are approximately 80 MB.

4.2.15 Annotations and Annotation Methodology

The most important annotations for this data set contain the ground-truth model identification number (ID) and a pose for each object. Each object is placed in a randomly generated, known location on the table by carefully aligning the point cloud for the object (as seen through the robot's stereo cameras) with a visualization of the object mesh in the desired location. The location of the object is thus known to be within operator precision for placing the object and is recorded as ground truth.

Further annotations to the grasps are added to indicate whether the object hit the table while being moved to the side or being placed, whether the object rotated significantly in the grasp or was placed in an incorrect orientation, and whether the grasp was stopped due to a robot or software error.

4.2.16 Applications

The recorded data from the Grasp Playpen data set are useful for evaluating and modeling the performance of object detection, grasp planning, and grasp evaluation algorithms.

For the ICP-like object detection algorithm described in the "Applications" section of "The Household Objects and Grasps Data Set" section, we have used the recorded object point clouds along with their ground-truth model IDs (and the results of running object detection) to create a model for how often we get a correct detection (identify the correct object model ID) for different returned values of the detection algorithm's match error, which is the average distance between each stereo point cloud point and the proposed object's mesh. The resulting naive Bayes model is shown in Figure 10, along with a

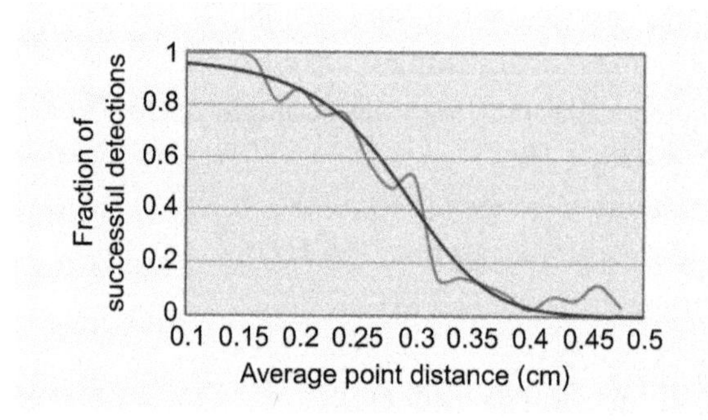

Figure 10
Correct object recognition rates versus the object detector's match error (average point distance) for our object recognizer. The blue line shows data from the Grasp Playpen data set, and the black line shows the Naive Bayes model chosen to approximate it. (For interpretation of the color in this figure legend, the reader is referred to the online version of this book.)

smoothed histogram of the actual proportion of correct detections seen in the Grasp Playpen data set.

For the GraspIt! quality metric described in Section 4.2.9, we have used the grasps that were actually executed, along with whether they were successful or not (and GraspIt!'s estimated grasp quality for those grasps, based on the ground-truth model and pose), to model how often grasps succeed or fail in real life for different quality values returned by GraspIt! Histogrammed data from the Grasp Playpen data set are shown in Figure 11, along with the piecewise-linear model for grasp quality chosen to represent it.

We have also used just the recorded object point clouds to estimate how well other grasp planners and grasp evaluation algorithms do on real (partial) sensor data. Because we have the ground-truth model ID and pose, we can use a geometric simulator such as GraspIt! to estimate how good an arbitrary grasp is on the true object geometry. Thus, we can ask a new grasp planner to generate grasps for a given object point cloud, and then evaluate in GraspIt! how likely that grasp is to succeed (with energy values translated into probabilities via the model described above). Or we can generate grasps using any grasp planner or at random and ask a new grasp evaluator to say how good it thinks each grasp is (based on just seeing the point cloud), and again use the ground-truth model pose/geometry to compare those values to GraspIt!'s success probability estimates. This allows us to generate data on arbitrarily large numbers of grasps, rather than just the 490 recorded grasps; we have used this technique ourselves to evaluate new grasp planners and evaluators, as well as to create models for them and perform feature-weight optimization.

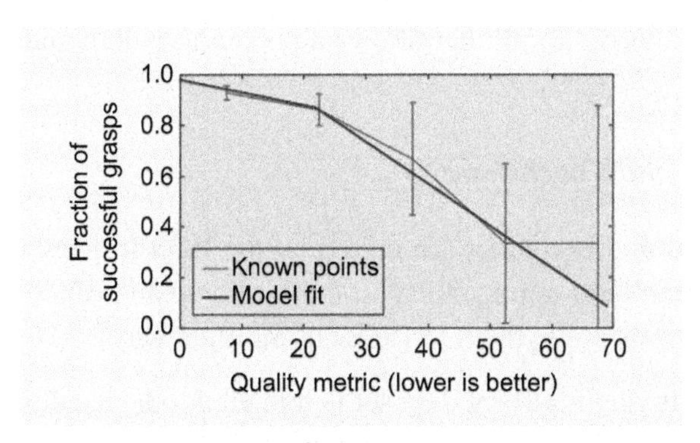

Figure 11
Experimental grasp success percentages versus GraspIt!'s grasp quality metric for the PR2 gripper. The blue line shows binned data from all 490 grasps in the Grasp Playpen data set, and the black line shows the piecewise-linear model chosen to approximate it. Blue error bars show 95% confidence on the mean, computed using bootstrap sampling. (For interpretation of the color in this figure legend, the reader is referred to the online version of this book.)

4.2.16.1 Future Work

Because we use random grasps planned using our available grasp planners to grasp the objects presented to the robot and because those grasps tend to be of high quality, approximately 90% of the grasps in the data set succeed in at least lifting the object. Thus, although the data are useful for differentiating very robust grasps from only marginal grasps, we would require more data on grasp failures to better elucidate the differences between marginal and bad grasps. In the future, we plan to obtain data for more random/less good grasps. We also plan to obtain data for more complex/cluttered scenes than just single objects on a table.

Other planned or possible uses of the data include

- testing object recognition and pose estimation algorithms
- trying to predict when a collision has occurred based on the recorded accelerometer data from grasps in which the object hit the table
- testing in-hand object tracking algorithms
- learning graspable features and weights for grasp features from image and point cloud data.

Obtaining grasp recordings by manually placing objects in the manner used for the Grasp Playpen data set is a fairly labor-intensive method. Killpack and Kemp have recently released code and the mechanical design for a PR2 playpen [37] that allows one to record grasps using the PR2 in a semiautomated manner. Currently, there is no mechanism for determining the ground-truth pose of the object being grasped, which is necessary for many of the proposed applications of the Grasp Playpen data set. However, automatically generated grasp recordings, if done with objects with known models, could be annotated using Mechanical Turk, using a tool that allows a person to match and pose the correct object model.

4.2.16.2 Discussion and Conclusions

The data sets discussed in this chapter are united by the ROS framework, their collection via the PR2 platform and their applicability to indoor home and office scenarios. The data sets' applications, however, force them to differ in multiple ways.

The Moving People, Moving Platform data set is intended to be used in an offline knowledge transfer context. In other words, robots are meant to utilize the data in batch format to train person detection algorithms, and then once again in batch format to evaluate these algorithms. This offline mechanism implies that access speed and data set size are not of primary importance when considering the data set format and contents. This allows the data to be presented in its raw, loss-less format. Off-line training is best performed with large amounts of data and annotation, and the nature of the annotations in this case required human input. These factors led to using humans in a crowd-sourced environment as a source of annotations. All these requirements were met within the ROS framework by using ROS bag files and providing the data on the Internet for batch download.

The Household Objects and Grasps data set are primarily used in an online knowledge transfer context. This implies that the format and contents need to support fast random access, both in retrieving the data from the Internet and accessing individual data elements within the data set. Thus, the data are stored in a relational database. The information is also compressed whenever possible to grasp points or object meshes instead of full object images or scans. Computing grasp points appropriate to a robot is performed automatically and offline using the GraspIt! simulator. No additional annotations from human sources are provided. The relational database containing this data set has an interface within the ROS framework, allowing a running robot system to access the data online.

The Grasp Playpen data set provides an additional venue for grasp information, but this time the knowledge transfer is intended to happen in an offline context. As in the Moving People, Moving Platform data set, the data does not need to be accessed quickly, and the size of the data set is less important. This allows for storage in raw format in ROS bags, and the contents are less restricted, including images, point clouds, and additional sensor data for later exploration. Finally, given the broader potential uses of this data set, the source of annotations is both automatic, generated by the robot as it successfully or unsuccessfully manipulates an object, and manual, with human annotations in text files. Once again, the data are available for batch download and can be viewed within the ROS framework.

The knowledge transfer context, the format and contents of the data, and the source of annotations are only some of the important characteristics of robotic data sets. We have expanded on them in this study as they are particularly relevant to the releases presented here; there are, however, a number of additional issues to consider when designing data sets. An incomplete list includes the following: are there other communities who could offer interesting input into the data, such as the computer vision community for the Moving People, Moving Platform data set? What is the correct accuracy level? Can the data set be easily expanded? Is it possible to add in additional sensor modalities or annotation modalities, perhaps in the way that the Grasp Playpen data set extends the Household Objects and Grasps data set? Does the data reflect the realistic conditions in which a scenario will be encountered? Can the objects in the household objects data set be recognized in clutter, or do people normally act as they do in the moving people data set? Finally, does there need to be a temporal component to the data, such as people or objects appearing differently at night versus during the day? This is only a small sample of the questions that should be asked.

Data set collection and annotation for mobile robots is typically a time and resource-intensive task, and the data sets presented here are no exception. Furthermore, obtaining such data sets requires access to a robot such as the PR2, which are not available to everyone. In light of the effort and resources required, we hope that by releasing these data sets, we can allow others to access useful data for their own research that they would not otherwise be able to obtain.

A particularly compelling direction of research considers the possibility of robots automatically augmenting and sharing data sets as they operate in their normal environments. People regularly draw on online information when faced with a new environment, getting data such as directions and product information from ubiquitous mobile communication devices. In a similar way, robots can share their experiences in an online manner, and some of the technology described in this chapter can enable this exchange. For example, a robot can regularly collect sensor data from its surroundings, use a crowd-sourcing method to annotate it, and contribute it back to the Moving People, Moving Platform data set.

The grasping pipeline presented here can serve as a foundation for fully automatic model acquisition: a robot can grasp a previously unseen object, inspect it from multiple viewpoints, and acquire a complete model, using techniques such as the ones presented in [38]. A robot could also learn from past pickup trials. Additional metadata, such as object classes, labels, or outlines in sensor data can be obtained online using a crowd-sourcing similar to the one used for the Moving People, Moving Platform data set. Visual and proprioceptive information from any attempted grasp can be added to the Grasp Playpen data set. Numerous other possibilities exist as we move toward a set of online resources for robots.

Data set design is a complex subject, but collecting and presenting data in an organized and cohesive manner is the key to progress in robotics. The data sets presented in this chapter are a small step toward useful mobile manipulation platforms operating in human environments. By continuing to collect and distribute data in open formats such as ROS, a diverse array of future algorithms and robots can learn from experience.

References

[1] M. Quigley, B. Gerkey, K. Conley, J. Faust, T. Foote, J. Leibs, et al., ROS: an open-source robot operating system, in: Proc. Int. Conf. Robotics and Automation Workshop on Open-Source Software, 2009.

[2] P. Fitzpatrick, G. Metta, L. Natale, Towards long-lived robot genes, Robot. Auton. Syst. 56 (1) (2008) 29–45.

[3] W. Garage, ROS Wiki, 2011, April 26. Available from: http://www.ros.org.

[4] W. Garage, The PR2, 2011, April 26. Available from: http://www.willowgarage.com/pages/pr2/overview.

[5] M. Everingham, L. Van Gool, C.K.I. Williams, J. Winn, A. Zisserman, The pascal visual object classes (VOC) challenge, Int. J. Comput. Vis. 88 (2) (June 2010) 303–338. Available from: http://pascallin.ecs.soton.ac.uk/challenges/VOC/.

[6] C. Pantofaru, The Moving People, Moving Platform Dataset, 2010. Available from: http://bags.willowgarage.com/downloads/people_dataset.html.
[7] N. Dalal, B. Triggs, Histograms of oriented gradients for human detection, in: Proc. IEEE Conf. Computer Vision and Pattern Recognition (CVPR), vol. 1, 2005, pp. 886–893.
[8] F. Fleuret, J. Berclaz, R. Lengagne, P. Fua, Multicamera people tracking with a probabilistic occupancy map, IEEE T. Pattern Anal. 30 (2) (February 2008) 267–282.
[9] M. Tenorth, J. Bandouch, M. Beetz, The TUM kitchen data set of everday manipulation activities for motion tracking and action recognition, in: Proc. IEEE Int. Workshop on Tracking Humans for the Evaluation of Their Motion in Image Sequences (THEMIS), 2009, pp. 1089–1096.
[10] CMU Graphics Lab, Motion Capture Database 26, (April 26, 2011). Available from: http://mocap.cs.cmu.edu/.
[11] L. Sigal, A. Balan, M. Black, HumanEva: synchronized video and motion capture dataset and baseline algorithm for evaluation of articulated human motion, Int. J. Comput. Vis. 87 (1) (2010) 4–27.
[12] M. Enzweiler, D.M. Gavrila, Monocular pedestrian detection: survey and experiments, IEEE T. Pattern Anal. 31 (12) (2009) 2179–2195. Available from: http://www.gavrila.net/Research/Pedestrian_Detection/Daimler_Pedestrian_Benchmark_D/Daimler_Pedestrian_Detection_B/daimler_pedestrian_detection_b_html.
[13] P. Dollár, C. Wojek, B. Schiele, P. Perona, Pedestrian detection: a benchmark presented at IEEE Intl., in: Intl. Conf. Computer Vision and Pattern Recognition, 2009, June. Available from: http://www.vision.caltech.edu/Image_Datasets/CaltechPedestrians/.
[14] C. Pantofaru, User observation & dataset collection for robot training, in: Proc. ACM/IEEE Conference on Human Robot Interaction (HRI), 2011, pp. 217–218.
[15] Amazon Mechanical Turk, April 26, 2011. Available from: https://www.mturk.com.
[16] L. von Ahn, B. Murer, C. McMillen, D. Abraham, M. Blum, reCAPTCHA: human-based character recognition via web security measures, Science 321 (September 2008) 1465–1468.
[17] MicrosoftCorp, Kinect for Xbox 360, Redmond, Washington, 2011.
[18] M. Ciocarlie, K. Hsiao, E. Jones, S. Chitta, R.B. Rusu, I.A. Sucan, Towards reliable grasping and manipulation in household environments, in: Proc. Int. Symp. Experimental Robotics, 2010.
[19] C. Goldfeder, M. Ciocarlie, H. Dang, P. Allen, The Columbia grasp database, in: Proc. Int. Conf. Robotics and Automation, 2009, pp. 1710–1716.
[20] C. Goldfeder, M. Ciocarlie, J. Peretzman, H. Dang, P. Allen, Data-driven grasping with partial sensor data, in: Proc. Int. Conf. Intelligent Robots and Systems, 2009, pp. 1278–1283.
[21] A. Miller, P.K. Allen, GraspIt! a versatile simulator for robotic grasping, IEEE Robot. Automat. Mag. 11 (4) (2004) 110–122.
[22] P. Shilane, P. Min, M. Kazhdan, T. Funkhouser, The princeton shape benchmark, in: Shape Model. Appl, 2004. Available from: http://dx.doi.org/10.1109/SMI.2004.1314504.
[23] A. Saxena, J. Driemeyer, A. Ng, Robotic grasping of novel objects using vision, Int. J. Robot. Res. 27 (2) (2008) 157–173.
[24] A. Morales, T. Asfour, P. Azad, S. Knoop, R. Dillmann, Integrated grasp planning and visual object localization for a humanoid robot with five-fingered hands, in: Proc. IEEE/RSJ Int. Conf. Intelligent Robots and Systems (IROS), 2006, pp. 5663–5668.
[25] Y. Li, J.L. Fu, N.S. Pollard, Data-driven grasp synthesis using shape matching and task-based pruning, IEEE T. Vis. Comput. Gr. 13 (4) (2007) 732–747.
[26] Y. Aydin, M. Nakajima, Database guided computer animation of human grasping using forward and inverse kinematics, Comput. Gr. 23 (1999) 145–154.
[27] K. Yamane, J. Kuffner, J. Hodgins, Synthesizing animations of human manipulation tasks, ACM T. Graphic. 23 (3) (2004) 532–539.
[28] D. Song, K. Huebner, V. Kyrki, D. Kragic, Learning task constraints for robot grasping using graphical models, in: Proc. IEEE/RSJ Intl. Conf. Intelligent Robots and Systems, 2010, pp. 1579–1585.
[29] C. de Granville, J. Southerland, A. Fagg, Learning grasp affordances through human demonstration, in: Proc. Intl. Conf. Development and Learning, 2006.

[30] P.J. Besl, M.I. Warren, A method for registration of 3-D shapes, IEEE T. Pattern Anal. 14 (2) (1992) 239–256.
[31] H. Bay, A. Ess, T. Tuytelaars, L.V. Gool, SURF:speeded up robust features, Comput. Vis. Image Und. 110 (3) (2008) 346–359.
[32] R.B. Rusu, N. Blodow, M. Beetz, Fast point feature histograms (FPFH) for 3D registration presented at Proc, in: Int. Conf. Robotics and Automation, 2009. Available from: http://files.rbrusu.com/publications/Rusu09ICRA.pdf.
[33] R.B. Rusu, G. Bradski, R. Thibaux, J. Hsu, Fast 3D recognition and pose using the viewpoint feature histogram, in: Proc. Int. Conf. Intelligent Robots and Systems, 2010, pp. 2155–2162.
[34] C. Ferrari, J. Canny, Planning optimal grasps, in: Proc. IEEE Int. Conf. Robotics and Automation, 1992, pp. 2290–2295.
[35] R. Balasubramanian, L. Xu, P. Brook, J. Smith, Y. Matsuoka, Human-guided grasp measures improve grasp robustness on a physical robot, in: Proc. ICRA, 2010, pp. 2294–2301.
[36] R. Detry, E. Baseski, M. Popovic, Y. Touati, N. Krueger, O. Kroemer, Learning object-specific grasp affordance densities, in: Proc. Int. Conf. Development and Learning, 2009, pp. 1–7.
[37] M. Killpack, C. Kemp, ROS Wiki Page for the Pr2_playpen Package, 2011, April 11. Available from: http://www.ros.org/wiki/pr2_playpen.
[38] M. Krainin, P. Henry, X. Ren, D. Fox, Manipulator and object tracking for in hand model acquisition, in: Proc. Int. Conf. Robotics and Automation, 2010 Workshop on Best Practice in 3D Perception and Modeling for Mobile Manipulation, 2010.

CHAPTER 4.3

Robust Recognition of Planar Mirrored Walls

Yan Lu, Dezhen Song

Department of Computer Science and Engineering, Texas A&M University, College Station, TX, USA

Chapter Outline

4.3.1 Introduction

The fast development of service robots has advanced robot work space from factory floors to our daily life. One important new task is to employ robots to perform building surveys to assist in building energy retrofits, because buildings account for around 40% of energy usage [1]. In the survey, robots need to recognize reflective surfaces to provide a proper estimation of a building's thermal load. Unfortunately, highly reflective surfaces, such as glassy building exteriors and mirrored walls, challenge almost every type of sensor including laser range finders, sonar arrays, and cameras, because light and sound signals simply bounce off the surfaces, which become invisible to the sensors. Detecting these surfaces is also necessary to avoid collisions in robot navigation.

We report a method for this new planar mirror detection problem (PMDP) using two views from an onboard camera. First, we derive geometric constraints for corresponding

Household Service Robotics. http://dx.doi.org/10.1016/B978-0-12-800881-2.00014-1

real–virtual features across two views. The constraints include (1) the mirror normal as a function of vanishing points of lines connecting the real–virtual feature point pairs and (2) the mirror depth in a closed-form format derived from a mirror plane-induced homography. We also address the issue that popular feature detectors, such as a scale-invariant feature transform (SIFT), are not reflection invariant by combining a secondary reflection with an affine scale-invariant feature transform (ASIFT). Based on the results, we employ a random sample consensus (RANSAC) framework to develop a robust mirror detection algorithm. We have implemented the algorithm and tested it in both in-lab and field settings. The algorithm achieved an overall accuracy rate of 91.0%.

4.3.2 Related Work

PMDP is not a simple plane reconstruction problem using three-dimensional (3D) vision. It relates to many areas, including intelligence level tests in the artificial intelligence (AI) community, planar catadioptric stereo (PCS) systems, construction of specular surfaces, and reflection-invariant feature extractions.

In AI and animal behavior communities, researchers often assess intelligence levels based on the subject's ability to detect a mirror, or its own reflection [2,3]. In the well-known mirror and mark test, a subject has a mark that cannot be directly seen but is visible in the mirror. If the subject increases the exploration and self-direction actions toward the mark, it means that the subject recognizes the mirror image as self. Existing results show that chimpanzees [3], gorillas [4], dolphins [5], and magpies [6] have evident self-recognition in front of mirrors, but monkeys do not [7]. We do not yet have mirror and mark tests for robots. It is clearly not a trivial problem. Initial related results focus on robot self-recognition [8,9], using motion and appearance, which is not as difficult as recognizing a mirror when a robot cannot see its own reflection. Such cases are not unusual because the robot cannot see itself when approaching a mirror from the side. Our approach addresses this problem by exploring symmetry in the scene.

Mirror detection is also related to PCS systems in computer vision. A PCS system usually consists of a static camera and one or more planar mirrors with the aim of achieving stereo or structure from motion [10–12]. Because detecting a mirror pose is just a calibration problem in PCS systems, in-lab settings and calibration patterns (e.g., checkerboard) can be used here. However, this is not a viable approach when robots need to detect mirror surfaces in situ.

In a way, planar mirror detection can be viewed as a special case of specular surface construction. Existing approaches rely on active sensing by changing lighting [13–15] and polarity [16–19] or assuming curvature of the mirror [20]. These approaches are difficult to adapt for robots because natural lighting can easily overwhelm the setup. To avoid this issue, we use features from images.

SIFT [21] is well known for its invariance to image scaling and translation and partial invariance to affine distortion. However, it is not reflection-invariant and thus cannot be applied to our problem. As extensions of SIFT, descriptors invariant to mirror reflection have been designed by modifying the SIFT descriptor structure at the expense of distinctiveness, such as MI-SIFT [22] and FIND [23]. They still cannot fit our need because our feature correspondence involves not only a reflection difference but also a significant projective distortion induced by perspective changes. On the other hand, descriptors invariant to affine transforms can handle large perspective changes (e.g., [24,25]). Among these affine invariant descriptors, ASIFT [26] shows promising performance and becomes our choice for a feature transformation. Later we show how to make an ASIFT reflection-invariant.

In a previous work [27], our group investigated the problem of estimating the orientation of a mirror plane using a single view. However, the depth information cannot be extracted from a single view and it limits the detection ability.

4.3.3 Problem Definition

To define our problem and focus on the most relevant issues, we make the following assumptions:

- Each view captures a real scene and its mirror reflection, and the scene is feature-rich.
- The intrinsic camera matrix is known to be K.
- The baseline distance $|t|$ between two views is known. The distance is usually short and can be measured by onboard sensors such as an inertial measurement unit. If $|t|$ is unknown, our method still applies, but the depth result is measured in a ratio instead of as an absolute value.

We also have the following conventions in notation. Let I and $\{I\}$ be the image and the image coordinate system (ICS) for the first view, respectively. I' and $\{I'\}$ are defined similarly for the second view. The camera coordinate system (CCS) is right-handed, with the origin C at the camera center, and the z-axis along the principal axis. With respect to the CCS of the first view, we define:

- $\pi_m = (\mathrm{n}_m^\mathrm{T}, d_m)^\mathrm{T}$ as the mirror plane, where n_m is a 3×1 unit vector indicating the normal of π_m, and d_m is the plane depth (i.e., the distance from C to π_m),
- $\mathrm{X}_{\mathrm{r}i}$ as the i-th real 3D point and $\mathrm{X}_{\mathrm{v}i}$ as its mirror reflection (a virtual point),
- $\mathrm{x}_{\mathrm{r}i}$ and $\mathrm{x}_{\mathrm{v}i}$ as the projections of $\mathrm{X}_{\mathrm{r}i}$ and $\mathrm{X}_{\mathrm{v}i}$ in $\{I\}$, respectively, and
- $\mathrm{X}_{\mathrm{r}i} \leftrightarrow \mathrm{X}_{\mathrm{v}i}$ as a 3D real—virtual (R-V) pair and $\mathrm{x}_{\mathrm{r}i} \leftrightarrow \mathrm{x}_{\mathrm{v}i}$ as a 2D R-V pair.

In the CCS of the second view, notations differ from their counterparts in the CCS of the first view by adding a superscript $'$, e.g., n'_m, $\mathrm{x}'_{\mathrm{r}i}$, and $\mathrm{x}'_{\mathrm{v}i}$. It is worth noting that there is a new type of correspondence between the 2D R-V pairs in both views, which is denoted in a quadruple format: $\mathrm{Q}_i = \{\mathrm{x}_{\mathrm{r}i}, \mathrm{x}_{\mathrm{v}i}, \mathrm{x}'_{\mathrm{r}i}, \mathrm{x}'_{\mathrm{v}i}\}$.

Also, all the above notations about points are represented in homogeneous coordinates, whereas their inhomogeneous counterparts are denoted by adding a tilde on their top, e.g., $\tilde{\mathrm{x}}_{ri}$.

With assumptions and notations defined, our PMDP is:

Definition 1. Given two views I and I', the camera calibration matrix K and the camera translation distance $|t|$, determine if there is a mirror. If so, estimate π_m.

4.3.4 Modeling

We begin with analyzing the geometric relationship between noise-free feature points. The geometric relationship will be used in a RANSAC framework later to filter noisy inputs. The noise-free feature inputs here are quadruples $\{\mathrm{Q}_i\}$. The geometric relationship is constraints on quadruples induced by 3D reflection and the imaging process. As a result, π_m will be derived as a function of the quadruples in two stages: orientation and depth. First, we solve the mirror orientation using quadruples.

Lemma 1. Given two quadruples, Q_i and Q_j, the mirror normal with respect to both CCSs can be obtained as follows,

$$\mathrm{n}_m = \mathrm{K}^{-1}(\mathrm{x}_{\mathrm{r}i} \times \mathrm{x}_{\mathrm{v}i}) \times (\mathrm{x}_{\mathrm{r}j} \times \mathrm{x}_{\mathrm{v}j}), \quad \mathrm{n}'_m = \mathrm{K}^{-1}(\mathrm{x}'_{\mathrm{r}i} \times \mathrm{x}'_{\mathrm{v}i}) \times (\mathrm{x}'_{\mathrm{r}j} \times \mathrm{x}'_{\mathrm{v}j}), \tag{1}$$

where the symbol '$\times$' represents the cross product.

Proof. Consider the geometric relationship in Figure 1. As a convention, we define $\overleftrightarrow{AB}$ as the line passing through points A and B. From the property of planar mirror reflection, we have $\overleftrightarrow{\mathrm{X}_{\mathrm{r}i}\mathrm{X}_{\mathrm{v}i}} \perp \pi_m$, $\overleftrightarrow{\mathrm{X}_{\mathrm{r}j}\mathrm{X}_{\mathrm{v}j}} \perp \pi_m$, and thus $\overleftrightarrow{\mathrm{X}_{\mathrm{r}i}\mathrm{X}_{\mathrm{v}i}} // \overleftrightarrow{\mathrm{X}_{\mathrm{r}j}\mathrm{X}_{\mathrm{v}j}}$. After a projective transformation, the projections of $\overleftrightarrow{\mathrm{X}_{\mathrm{r}i}\mathrm{X}_{\mathrm{v}i}}$ and $\overleftrightarrow{\mathrm{X}_{\mathrm{r}j}\mathrm{X}_{\mathrm{v}j}}$ in $\{I\}$ and $\{I'\}$ would intersect at a vanishing point v (or v$'$) in the corresponding ICS:

$$(\mathrm{x}_{\mathrm{r}i} \times \mathrm{x}_{\mathrm{v}i}) \times (\mathrm{x}_{\mathrm{r}j} \times \mathrm{x}_{\mathrm{v}j}) = \mathrm{v}, \quad (\mathrm{x}'_{\mathrm{r}i} \times \mathrm{x}'_{\mathrm{v}i}) \times (\mathrm{x}'_{\mathrm{r}j} \times \mathrm{x}'_{\mathrm{v}j}) = \mathrm{v}'. \tag{2}$$

On the other hand, v can be viewed as the projection of n_m in $\{I\}$

$$\mathrm{v} = \mathrm{K}n_m, \quad \text{and similarly,} \quad \mathrm{v}' = \mathrm{K}n'_m. \tag{3}$$

Combining Eqns (2) and (3), we obtain Eqn (1).

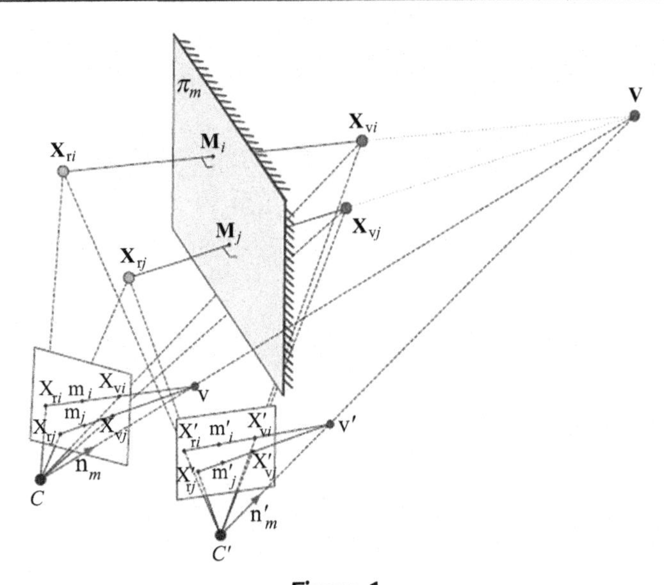

Figure 1
A perspective illustration of the geometry relationship between real–virtual pairs across two views.

The second step is to derive the mirror depth d_m. From epipolar geometry, we can obtain the camera rotation matrix R and translation vector t by decomposing the essential matrix [28]. A straightforward way of computing the equation of π_m is by reconstructing 3D points via triangulation. However, we will show a homography-based method that avoids the triangulation process.

Our method involves the homography between the corresponding middle points of R-V pairs in two views. Let $\overline{AB}$ denote the line segment defined by points A and B in the rest of the chapter. Denote the midpoint of $\overleftrightarrow{X_{ri}X_{vi}}$ by M_i, and its projection in $\{I\}$ by m_i (see Figure 1 for examples). m_i can be obtained using a cross ratio, which is detailed in the following lemma:

Lemma 2. Given quadruple Q_i, the projection m_i of the midpoint M_i of $\overleftrightarrow{X_{ri}X_{vi}}$ is determined as follows,

$$\tilde{m}_i = (1-a)\tilde{x}_{ri} + a\tilde{x}_{vi}, \quad \text{and } a = \frac{|\overline{x_{ri}v}|}{2|\overline{x_{ri}v}| - |\overline{x_{ri}x_{vi}}|}, \tag{4}$$

where $|\cdot|$ denotes the length of the line segment.

Proof. Consider the projection from $\overleftrightarrow{X_{ri}X_{vi}}$ to $\overleftrightarrow{x_{ri}x_{vi}}$. A basic invariant in this projection is the cross ratio of the four collinear points X_{ri}, M_i, X_{vi}, and V:

$$\frac{\left|\overline{x_{ri}m_i}\right|\left|\overline{x_{ri}v}\right|}{\left|\overline{x_{ri}x_{vi}}\right|\left|\overline{m_i v}\right|} = \frac{\left|\overline{X_{ri}M_i}\right|\left|\overline{X_{vi}V}\right|}{\left|\overline{X_{ri}X_{vi}}\right|\left|\overline{M_i V}\right|} = \frac{1}{2} \tag{5}$$

Representing m_i as $\tilde{m}_i = (1-a)\tilde{x}_{ri} + a\tilde{x}_{vi}, 0 \le a \le 1,$ in the inhomogeneous coordinate, we have

$$\begin{aligned} \left|\overline{x_{ri}m_i}\right| &= a\left|\overline{x_{ri}x_{vi}}\right|, \\ \left|\overline{m_i v}\right| &= \left|\overline{x_{ri}v}\right| - a\left|\overline{x_{ri}x_{vi}}\right|. \end{aligned} \tag{6}$$

Substituting Eqn (6) into Eqn (5) gives the final result in Eqn (4).

We now can derive the mirror depth with m_i.

Lemma 3. Given quadruple Q_i and mirror normal n_m, the mirror depth is

$$d_m = \left([m_i']_\times KRK^{-1}m_i\right)^\dagger [m_i']_\times Ktn_m^T K^{-1}m_i \tag{7}$$

where $(\cdot)^\dagger$ denotes the pseudo-inverse operation, and $[m_i']_\times$ is a skew-symmetric matrix,

$$\begin{bmatrix} 0 & -m_{i3}' & m_{i2}' \\ m_{i3}' & 0 & -m_{i1}' \\ -m_{i2}' & m_{i1}' & 0 \end{bmatrix} \tag{8}$$

Proof. Observe that M_i lies on the plane π_m. Then m_i and m_i' must obey a homography, $m_i' = Hm_i$, induced by π_m, where H can be expressed as [28]

$$H = K\left(R - \frac{1}{d_m}tn_m^T\right)K^{-1} \tag{9}$$

H has one degree of freedom (DOF) because only d_m is unknown.

m_i and m_i' can be computed from Q_i using Eqn (4). Because
$m_i' = Hm_i = K\left(R - \frac{1}{d_m}tn_m^T\right)K^{-1}m_i$, we have

$$m_i' \times K\left(R - \frac{1}{d_m}tn_m^T\right)K^{-1}m_i = [m_i']_\times KRK^{-1}m_i - [m_i']K\frac{1}{d_m}tn_m^T K^{-1}m_i = 0$$

and then we have

$$[m_i']_\times KRK^{-1}m_i d_m = [m_i']_\times Ktn_m^T K^{-1}m_i$$

The above system of equations is overdetermined because the rank of $[m_i']_\times$ is 2. Thus, the least-square solution of d_m is given by Eqn (7), which is also an exact solution when the system is noise-free.

4.3.5 Algorithm

Section 4.3.4 provides a geometric relationship for noise-free quadruples. To complete the algorithm, we need to select a correct feature transformation and verify the geometric relationship with respect to noisy features using the well-accepted RANSAC framework. First, let us detail the feature detection method selection in a quadruple extraction.

4.3.5.1 Quadruple Extraction

To form a quadruple, we need two kinds of point correspondences: cross-view correspondence, e.g., $x_{ri} \leftrightarrow x_{ri}'$, and R-V pair correspondence, e.g., $x_{ri} \leftrightarrow x_{vi}$. The former can be handled by standard feature extraction methods, such as SIFT, as long as the perspective change is not significant. However, the latter is nontrivial because $x_{ri} \leftrightarrow x_{vi}$ involves an improper transformation in 3D (between X_{ri} and X_{vi}).

Therefore, the key to this problem is how to find features and their correspondence under the improper transformation. We need to convert the reflection to a rigid body transformation such that existing feature extraction and matching algorithms can be employed. The intuition of our approach comes from a special scenario in which a secondary mirror π_s is placed in the same plane as π_m but in the opposite orientation. Letting X_{si} be the result of X_{ri} after a consecutive reflection about π_m and π_s, it is clear that X_{si} is exactly the same point as X_{ri}, which makes matching their projections in image a trivial problem.

In fact, it is proven that two consecutive reflections lead to a rigid body transformation regardless of the mirror configuration [11]. Therefore, the position of π_s can be arbitrarily chosen. The remaining problem is that introducing the secondary mirror in 3D is difficult to implement because it requires the 3D positions of points to perform the secondary reflection, which is not viable. Fortunately, a special group of π_s allows the 3D reflection to be reduced to 2D image flipping about an arbitrary axis in the ICS, which is independent of point positions.

Lemma 4. If π_s contains the camera principal axis, then for any X_{si}, its projection x_{si} can be obtained by flipping x_{vi} about an axis in the ICS.

Proof. Denote π_I as the image plane as illustrated in Figure 2.

Because it contains the principal axis (i.e., the z-axis), π_s can be expressed as $\left(n_s^T, 0\right)^T$, where $n_s = (n_x, n_y, 0)^T$. Because X_{vi} and X_{si} are symmetrical about π_s, we have

$$X_{si} = TX_{vi} \tag{10}$$

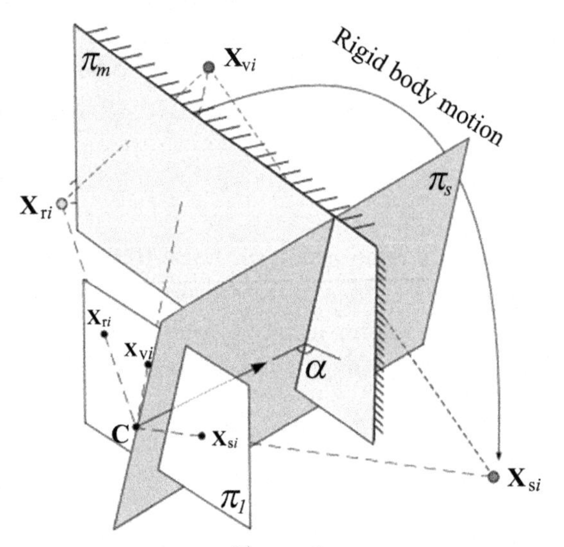

Figure 2
The configuration of π_I, π_s, and π_m. π_s is placed to contain the camera principal axis.

where $T = \begin{bmatrix} I_3 - 2n_s n_s^T & 0 \\ 0 & 1 \end{bmatrix}$.

Moreover, we have the projection relationship

$$x_{vi} = PX_{vi}, \quad x_{si} = PX_{si}, \tag{11}$$

where $P = [K|0]$ is the projection matrix.

Combining Eqns (10) and (11) gives

$$x_{si} = PTP^{\dagger} x_{vi} = \begin{bmatrix} I_2 - 2n_{s1:2}n_{s1:2}^T & 0 \\ 0 & 1 \end{bmatrix} x_{vi} \tag{12}$$

where $P^{\dagger}$ is the pseudo-inverse of P and $n_{s1:2} = (n_x, n_y)^T$.

Equation (12) implies that x_{vi} and x_{si} are symmetrical about an axis with a normal $n_{s1:2}$, which is actually the intersection of π_I and π_s. This completes the proof.

Lemma 4 allows us to find the correspondence between $x_{ri} \leftrightarrow x_{si}$ instead of that of $x_{ri} \leftrightarrow x_{vi}$. Furthermore, the axis of flipping can be arbitrarily chosen because there is only a planar rotation difference between the resulting images with different flipping axes.

Although the image flipping process solves the improper transformation issue, it also introduces a new challenge for feature matching. The problem is that the rotation angle θ of the resulting rigid body motion (between X_{ri} and X_{si}) is as large as two times the angle α between the principal axis and π_m [11] (see Figure 2). Therefore, as the value of θ varies with α in different cases, it can easily lead to a significant perspective change, which often fails the standard SIFT algorithm.

To handle this problem, we employ an affine invariant feature extraction algorithm ASIFT, which has advantages over SIFT when dealing with large perspective changes. Once a correspondence $\mathrm{x}_{ri} \leftrightarrow \mathrm{x}_{si}$ is identified, the R-V pair $\mathrm{x}_{ri} \leftrightarrow \mathrm{x}_{vi}$ is readily established based on the known mapping between x_{ri}, x_{vi}, and x_{si}. Algorithm 1 summarizes how quadruples are constructed.

4.3.5.2 Maximum Likelihood Estimation (MLE)

To apply the RANSAC framework, we need to estimate n_m, n'_m, and d_m using the quadruples $\{\mathrm{Q}_i\}$ from the inlier set by minimizing a cost function. Assuming measurement errors are Gaussian, then the estimation is MLE, if the reprojection error is employed as the cost function. Let us derive this metric.

For Q_i, let $\chi_i = (\tilde{\mathrm{x}}_{ri}, \tilde{\mathrm{y}}_{ri}, \tilde{\mathrm{x}}_{vi}, \tilde{\mathrm{y}}_{vi}, \tilde{\mathrm{x}}'_{ri}, \tilde{\mathrm{y}}'_{ri}, \tilde{\mathrm{x}}'_{vi}, \tilde{\mathrm{y}}'_{vi})^{\mathrm{T}}$ be an 8-vector formed by concatenating the inhomogeneous coordinates of x_{ri}, x_{vi}, x'_{ri}, and x'_{vi}. Given points χ_i in the measurement space ¡^{8}, the task of estimating n_m, n'_m, and d_m becomes finding a variety that passes through the points χ_i in ¡^{8}. Because of noise, it is impossible to fit a variety exactly. In this case, let v be the variety corresponding to n_m, n'_m, and d_m, and let $\hat{\chi}_i$ be the closest point to χ_i lying on v.

Algorithm 1: ASIFT-based Quadruple Extraction

Input : Two images I and I'
Output: A set of quadruples $\{\mathbf{Q}_k\}$

1. flip I left-right (or up-down) to get I_f;
2. find matches $\{\mathbf{x}_{ri} \leftrightarrow \mathbf{x}_{si}\}$ between I and I_f using ASIFT;
3. map $\mathbf{x}_{si}$ in I_f back to $\mathbf{x}_{vi}$ in I to establish R-V correspondences $\{\mathbf{x}_{ri} \leftrightarrow \mathbf{x}_{vi}\}$;
4. apply steps 1-3 to I' to obtain $\{\mathbf{x}'_{rj} \leftrightarrow \mathbf{x}'_{vj}\}$;
5. find cross-view matches $\{\mathbf{x}_{rk} \leftrightarrow \mathbf{x}'_{rk}\}$ from between $\{\mathbf{x}_{ri} \leftrightarrow \mathbf{x}_{vi}\}$ and $\{\mathbf{x}'_{rj} \leftrightarrow \mathbf{x}'_{vj}\}$ using putative matching of ASIFT;
6. construct quadruples $\{\mathbf{Q}_k\}$ from $\{\mathbf{x}_{ri} \leftrightarrow \mathbf{x}_{vi}\}$ and $\{\mathbf{x}'_{rj} \leftrightarrow \mathbf{x}'_{vj}\}$ according to $\{\mathbf{x}_{rk} \leftrightarrow \mathbf{x}'_{rk}\}$;
7. **return** $\{\mathbf{Q}_k\}$;

Given n_m, n'_m, and d_m, define

$$C_v(\hat{\chi}_i) := \begin{bmatrix} (\hat{\mathrm{x}}_{ri} \times \hat{\mathrm{x}}_{vi})^{\mathrm{T}} \mathrm{K}\mathrm{n}_m \\ (\hat{\mathrm{x}}'_{ri} \times \hat{\mathrm{x}}'_{vi})^{\mathrm{T}} \mathrm{K}\mathrm{n}'_m \\ \hat{\mathrm{m}}'_i \times \mathrm{H}\hat{\mathrm{m}}_i \end{bmatrix},$$

where H, $\hat{\mathrm{m}}_i$, and $\hat{\mathrm{m}}'_i$ are intermediate variables computed using Eqns (9) and (4), respectively, and $\hat{\mathrm{x}}_{ri} = (\hat{\tilde{\mathrm{x}}}_{ri}, \hat{\tilde{\mathrm{y}}}_{ri}, 1)^{\mathrm{T}}$, and similarly for $\hat{\mathrm{x}}_{vi}$, $\hat{\mathrm{x}}'_{ri}$, and $\hat{\mathrm{x}}'_{vi}$. Then the MLE method is to find n_m, n'_m, d_m, and $\hat{\chi}_i$ that minimize the error function,

$$\sum_i \|\chi_i - \hat{\chi}_i\|^2_{\Sigma_i}, \tag{13}$$

subject to $C_v(\chi_i) = 0$, $\forall i$, where Σ_i is the covariance of χ_i, and $||\cdot||_\Sigma$ represents the Mahalanobis distance.

Although minimizing the reprojection error is the MLE, it involves solving a high-dimensional nonlinear optimization problem, which is quite complex and time-consuming. To speed up the algorithm, we derive the Sampson error approximation. Instead of finding the closest point $\hat{\chi}_i$ on the variety v to the measurement χ_i, the Sampson error function estimates a first-order approximation to $\hat{\chi}_i$. For given n_m, n'_m, and d_m, any point χ_i lying on v will satisfy $C_v(\chi_i) = 0$. Then the Sampson approximation to Eqn (13) is $\sum_i \varepsilon_i^{\mathrm{T}} \left(\mathrm{J}_i \Sigma_i \mathrm{J}_i^{\mathrm{T}}\right)^{-1} \varepsilon_i$, where $\varepsilon_i = C_v(\chi_i)$ and $\mathrm{J}_i = \frac{\partial C_v}{\partial \chi_i}$.

4.3.5.3 Applying the RANSAC Framework

We are now ready to apply the RANSAC to the set S of quadruples to estimate π_m. The whole algorithm is summarized in Algorithm 2. There are two thresholds used: inlier—outlier threshold τ_{d} and mirror detection threshold τ_{n}. Threshold τ_{d} in step 9 is used to determine whether the quadruple belongs to the current inlier set. τ_{d} is chosen based on 8-DOF of the decision variables. With a preset probability threshold of 0.95, $\tau_{\mathrm{d}} = \sqrt{15.51\sigma^2}$ according to [28], where σ is the standard deviation of the measurement error for the feature points.

Algorithm 2: Robust Mirror Estimation using RANSAC

Input : Two images I and I'
Output: Mirror plane π_m or ***no mirror***

1 obtain a set S of quadruples using Algorithm 1;
2 $N = \infty$;
3 **for** $k \leftarrow 1$ **to** N **do**
4   randomly sample 2 quadruples from S;
5   compute $\mathrm{n}_m^{(k)}$, $\mathrm{n}'^{(k)}_m$ and $d_m^{(k)}$ using (1) and (7);
6     $\mathcal{I}_k = \emptyset$;   // initialize inlier set
7   **for** $\mathrm{Q}_i \in S$ **do**
8         $D_i = \sqrt{\epsilon_i^{\mathrm{T}} (\mathrm{J}_i \Sigma_i \mathrm{J}_i^{\mathrm{T}})^{-1} \epsilon_i}$;
9   **if** $|D_i| < \tau_d$ **then**
10            $\mathcal{I}_k = \mathcal{I}_k \cup \mathrm{Q}_i$;
11   update N using (4.18) from [28] (Page 119);
12 $k^\star = \arg\max_k |\mathcal{I}_k|$;
13 $\mathcal{I}^\star = \mathcal{I}_{k^\star}$;
14 **if** $|\mathcal{I}^\star| < \tau_n$ **then**
15   **return** ***no mirror***;
16 **else**
17   re-estimate n_m, n'_m and d_m with $\mathcal{I}^\star$ by minimizing Sampson error using the Levenberg-Marquardt algorithm;
18   (guided matching): find correspondence inliers consistent with the optimal estimation;
19   **return** π_m;

The algorithm returns "no mirror" when the size of the maximum inlier set is smaller than τ_n (Step 15). τ_n will be determined experimentally through in-lab tests in Section 4.3.6-A. In step 11, the maximum sample iteration N is chosen adaptively (page 119 of [28]). Steps 17 and 18 of Algorithm 2 can be iterated until the number of correspondence inliers is stable.

4.3.6 Experiments

We have implemented the proposed algorithm using MatLab under a Windows 7 operating system. For the ASIFT algorithm, we used the open source implementation in [29]. Images were taken by a precalibrated Vivicam 7020 camera with a resolution of 640 × 480 pixels. We first tested the algorithms in our lab to determine the algorithm accuracy under the controlled settings and to determine the proper threshold before the extensive field tests.

4.3.6.1 In-Lab Tests

Figure 3(a) illustrates the setup of the in-lab tests. Define α as the angle between the camera optical axis and the mirror plane π_m. This is usually the robot approach angle toward the mirror plane. It is important to know how α affects the estimation accuracy of π_m for collision avoidance purposes. Data are collected in six different α values ranging from 5° to 60°. The scene structure is kept the same during the test (see Figure 3(b)), with abundant features. The baseline distance between the first and the second views is 25.4 cm while maintaining the same optical axis. Ground truth data are obtained using physical measurements.

Figure 4 illustrates that both angular errors of the mirror plane normal and relative depth errors are reasonably small under different α values. Note that 100 trials were carried out for each α setting. The results are desirable because errors are not sensitive to α values. Note that we have not performed experiments for cases with large angle values (i.e., $\alpha > 60°$). At large angles, the camera/robot almost faces the mirror directly. Because a regular camera has a horizontal field of view larger than 55°, the robot can see itself in the mirror. For such cases, the problem becomes trivial because it is reduced to a self-appearance-based mirror detection, which is less challenging.

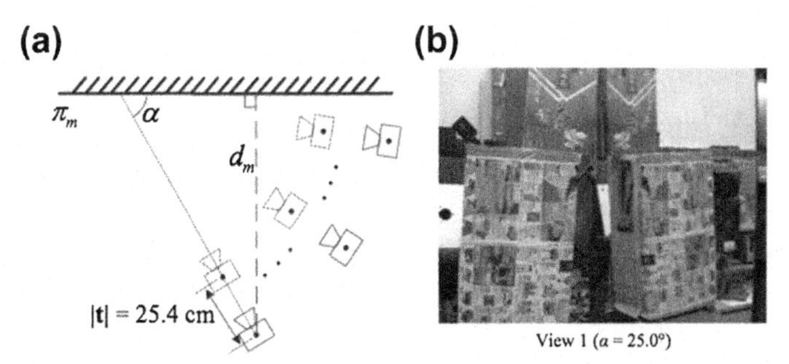

Figure 3
In-lab experiment setup. (a) Experiment configurations. (b) A sample view.

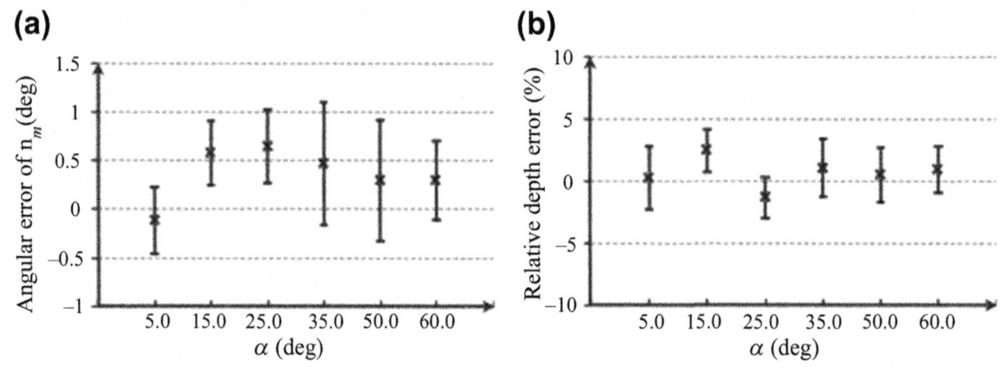

Figure 4

The accuracy of the estimated mirror plane with respect to α values. (a) Angular error of the mirror normal. (b) Relative depth error for the mirror plane. The vertical bar and the middle cross represent the one standard deviation range and sample mean, respectively.

The second experiment is to explore the relationship between the quadruple inlier number and the estimation accuracy and hence determine the threshold τ_n in Algorithm 2. We use the same data set from the first experiment. For every pair of images, the mirror parameters are computed each time as the number of quadruple inliers is changed by incrementally adjusting the ASIFT feature detection threshold. Then we group the estimation results according to their corresponding quadruple inlier numbers and compare the estimation error across groups. The results are shown in Figure 5. As expected, the standard deviation of estimation generally decreases as the quadruple inlier number increases. When the quadruple inlier number drops below 6, the estimation accuracy becomes untrustable owing to its large standard deviation. Hence we set $\tau_n = 6$ for our field tests.

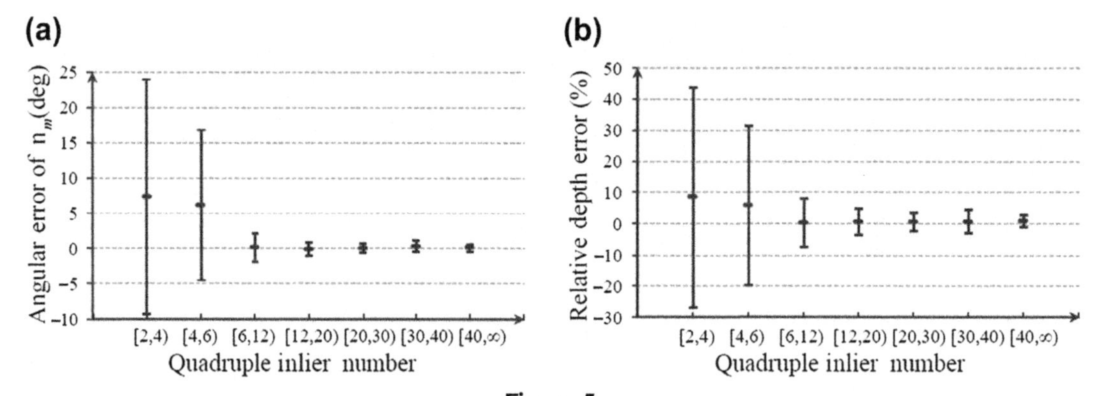

Figure 5

The mean and standard deviation plot of mirror plane parameters versus number of quadruple inliers. (a) Angular error of the mirror normal. (b) Relative depth error for the mirror plane. The vertical bar represents one standard deviation range.

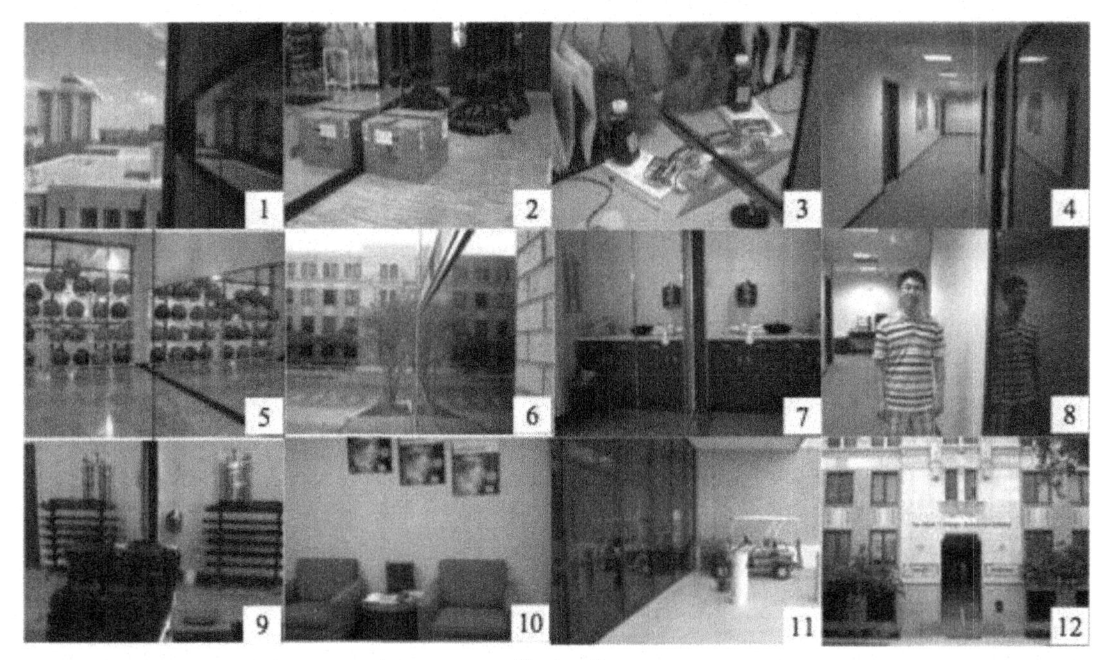

Figure 6
Sample images from the data set.

4.3.6.2 Field Tests

We have tested our algorithm in the field. A data set of 100 pairs of images was collected from real world scenes with or without mirrored walls, such as gymnasiums, corridors, campuses, and shopping malls (see Figure 6). In the data set, 50% of the image pairs contain mirrored walls such as wall mirrors, window glasses, and water surfaces.

The detection result is presented in a confusion matrix in Table 1, in which "Positive" indicates the existence of mirrored walls. In the confusion matrix, the true positive rate and true negative rate are both high, indicating a desirable recognition ability. The false positive cases are typically caused by objects with strong symmetrical appearances, e.g., sample image 12 in Figure 6. The false negative cases are mainly due to a lack of features in the scene. The overall detection accuracy is 91.0%.

Table 1: Field test results

		Predicted	
		Positive	**Negative**
Actual	Positive	45	5
	Negative	4	46

4.3.7 Conclusion and Future Work

We addressed PMDP using two views from an onboard camera. First, we derived geometric constraints for corresponding real–virtual features across two views. Based on the geometric constraints, we employed a RANSAC framework and ASIFT to develop a robust mirror detection algorithm. We implemented the algorithm and tested it in both in-lab and field settings. The algorithm achieved an overall accuracy of 91.0%. In the future, we will study how to segment the mirror region out of the background image. This would be important for recognizing objects such as glassy doors or windows.

Acknowledgments

The authors thank D. Shell and N. Amato for their insightful discussions and C. Kim, W. Li, H. Ge, M. Hielsberg, S. Guo, Z. Ma, and X. Liu for their input and contributions to the Networked Robots Laboratory in Texas A&M University.

References

[1] NSTC, National Science and Technology Council, Federal Research and Development Agenda for Net-zero Energy, High-performance Green Buildings, October 2008. http://www.bfrl.nist.gov/buildingtechnology/documents/FederalRDAgendaforNetZeroEnergyHighPerformanceGreenBuildings.pdf.
[2] W.G. Walter, An imitation of life, Sci. Am. 182 (2) (1950) 42–45.
[3] G. Gallup, Chimpanzees: self-recognition, Science 167 (3914) (January 1970) 86–87.
[4] F. Patterson, W. Gordon, The case for the personhood of gorillas, in: The Great Ape Project: Equality Beyond Humanity, Fourth Estate, London, 1993, pp. 58–77.
[5] D. Reiss, L. Marino, Mirror self-recognition in the bottlenose dolphin: a case of cognitive convergence, PNAS 98 (10) (May 2001) 5937–5942.
[6] H. Prior, A. Schwarz, O. Gntrkn, Mirror-induced behavior in the magpie (pica pica): evidence of self-recognition, PLoS Biol. 6 (8) (2008) e202, 208.
[7] M. Hauser, C. Miller, K. Liu, R. Gupta, Cotton-top tamarins (saguinus oedipus) fail to show mirror-guided self-exploration, Am. J. Primatol. 53 (3) (2001) 131–137.
[8] P. Michel, K. Gold, B. Scassellati, Motion-based robotic self-recognition, in: Intelligent Robots and Systems, 2004. (IROS 2004), in: Proceedings. 2004 IEEE/RSJ International Conference on, vol. 3, IEEE, September 2004. Sendai, Japan, pp. 2763–2768.
[9] P. Haikonen, Reflections of consciousness: the mirror test, in: Proceedings of the 2007 AAAI Fall Symposium on Consciousness, November 2007. Arlington, VA, pp. 67–71.
[10] J. Gluckman, S. Nayar, Planar catadioptric stereo: geometry and calibration, in: Computer Vision and Pattern Recognition, 1999. IEEE Computer Society Conference on, vol. 1, IEEE, June 1999. Ft. Collins, CO.
[11] J. Gluckman, S. Nayar, Catadioptric stereo using planar mirrors, Int. J. Comput. Vis. 44 (1) (2001) 65–79.
[12] G. Mariottini, S. Scheggi, F. Morbidi, D. Prattichizzo, Planar catadioptric stereo: single and multi-view geometry for calibration and localization, in: Robotics and Automation, 2009. ICRA'09. IEEE International Conference on, IEEE, May 2009. Kobe, Japan, pp. 1510–1515.
[13] K. Reiner, K. Donner, Stereo vision on specular surfaces, in: Proceedings of 4th IASTED International Conference on Visualization, Imaging and Image Processing, September 2004, Marbella, Spain.

[14] K.N. Kutulakos, E. Steger, A theory of refractive and specular 3D shape by light-path triangulation, Int. J. Comput. Vis. 76 (1) (January 2008) 13–29.
[15] S. Rozenfeld, I. Shimshoni, M. Lindenbaum, Dense mirroring surface recovery from 1d homographies and sparse correspondences, IEEE Trans. Pattern Anal. Mach. Intell. 33 (2) (February 2011) 325–337.
[16] M. Ferraton, C. Stolz, F. Meriaudeau, Surface reconstruction of transparent objects by polarization imaging, in: IEEE International Conference on Signal Image Technology and Internet Based Systems, November 2008. Bali, Indonesia, pp. 474–479.
[17] D. Miyazaki, M. Kagesawa, K. Ikeuchi, Transparent surface modeling from a pair of polarization images, IEEE Trans. Pattern Anal. Mach. Intell. 26 (1) (January 2004) 73–82.
[18] S. Rahmann, N. Canterakis, Reconstruction of specular surfaces using polarization imaging, in: IEEE Computer Society Conference on Computer Vision and Pattern Recognition (CVPR'01), December 2001, Kauai, HI.
[19] O. Morel, C. Stolz, F. Meriaudeau, P. Gorria, Active lighting applied to 3D reconstruction of specular metallic surfaces by polarization imaging, Appl. Opt. 45 (17) (January 2006) 4062–4068.
[20] M. Oren, S.K. Nayar, A theory of specular surface geometry, Int. J. Comput. Vis. 24 (2) (September 1997) 105–124.
[21] D. Lowe, Distinctive image features from scale-invariant keypoints, Int. J. Comput. Vis. 60 (4) (November 2004) 91–110.
[22] R. Ma, J. Chen, Z. Su, MI-SIFT: mirror and inversion invariant generalization for sift descriptor, in: Proceedings of the ACM International Conference on Image and Video Retrieval, ACM, July 2010. Xi'an, China, pp. 228–235.
[23] X. Guo, X. Cao, FIND: a neat flip invariant descriptor, in: 2010 International Conference on Pattern Recognition, IEEE, August 2010. Istanbul, Turkey, pp. 515–518.
[24] J. Matas, O. Chum, M. Urban, T. Pajdla, Robust wide-baseline stereo from maximally stable extremal regions, Image Vis. Comput. 22 (10) (September 2004) 761–767.
[25] K. Mikolajczyk, C. Schmid, Scale & affine invariant interest point detectors, Int. J. Comput. Vis. 60 (1) (2004) 63–86.
[26] J. Morel, G. Yu, ASIFT: a new framework for fully affine invariant image comparison, SIAM J. Imaging Sci. 2 (2) (April 2009) 438–469.
[27] A. Agha-mohammadi, D. Song, Robust recognition of planar mirrored walls using a single view, in: Robotics and Automation, Proceedings 2011 IEEE International Conference on, IEEE, Shanghai, China, May 2011, pp. 1186–1191.
[28] R. Hartley, A. Zisserman, Multiple View Geometry in Computer Vision, Cambridge University Press, Cambridge, 2003.
[29] G. Yuand, J.-M. Morel, ASIFT: An Algorithm for Fully Affine Invariant Comparison, 2011. http://www.ipol.im/pub/algo/my affine sift/.

CHAPTER 4.4

Evaluation of Three Vision Based Object Perception Methods for a Mobile Robot[1]

Arnau Ramisa[1], David Aldavert[2], Shrihari Vasudevan[3], Ricardo Toledo[2], Ramon Lopez de Mantaras[4]

[1]Institute of Robotics and Industrial Informatics, IRI UPC-CSIC, Barcelona, Spain; [2]Departament Ciències de la Computació, Universitat Autònoma de Barcelona, Bellaterra, Spain; [3]Center for Field Robotics, The University of Sydney, Sydney, Australia; [4]Artificial Intelligence Research Institute (IIIA) of the Spanish National Research Council (CSIC), Bellaterra, Spain

Chapter Outline

4.4.1 Introduction

Currently there is a big push towards complex cognitive capabilities in robotics research. One central requirement towards these capabilities is to be able to identify higher level features like objects, doors, etc.

[1] With kind permission from Springer Science + Business Media: Journal of Intelligent and Robotic Systems, Evaluation of Three Vision Based Object Perception Methods for a Mobile Robot, Vol. 68, 2012, pp. 185–208, Arnau Ramisa.

Household Service Robotics. http://dx.doi.org/10.1016/B978-0-12-800881-2.00015-3

Although impressive results are obtained by modern object recognition and classification methods, a lightweight object perception method, suitable for mobile robots and able to learn new objects in an easy and autonomous way is still lacking.

Works such as those of Vasudevan et al. [1], Martinez et al. [2], Galindo et al. [3] or Jensfelt et al. [4] investigate underlying representations of spatial cognition for autonomous mobile robots based on (or enhanced with) day-to-day objects. Visual object perception applied to mobile robotics is a common defining characteristic of all of these works. The success of such works is primarily decided by the strength of the perception system in place. With this in mind, this work evaluates several object perception methods for autonomous mobile robotics.

Although different modalities of perception (e.g., laser range-finder, color camera, time-of-flight camera, haptics) can be used, in this chapter we focus on passive vision, as it is interesting for several reasons, such as affordable cost, autonomy, compatibility with human environments or richness of perceived information.

Recently several methods have been quite successful in particular instances of addressing this problem, such as detecting frontal faces or cars [5], or in datasets that concentrate on a particular issue (e.g., classification in the Caltech-101 [6] dataset). However in more challenging datasets, like the detection competition of the Pascal Visual Object Challenge (VOC) [7], the methods presented typically achieve a low average precision. This low performance is not surprising, since object recognition in real scenes is one of the most challenging problems in computer vision [8]. The visual appearance of objects can change enormously due to different viewpoints, occlusions, illumination variations or sensor noise. Furthermore, objects are not presented alone to the vision system, but they are immersed in an environment with other elements, which clutter the scene and make recognition more complicated.

In a mobile robotics scenario a new challenge is added to the list: computational complexity. In a dynamic world, information about the objects in the scene can become obsolete even before it is ready to be used if the recognition algorithm is not fast enough.

In the present chapter, our intent is to survey some well established object recognition systems, comment on their applicability to robotics and evaluate them in a mobile robotics scenario. The selected methods are the SIFT object recognition algorithm [9], the Bag of Features [10], and the Viola and Jones boosted cascade of classifiers [5], and they were chosen taking into consideration issues relevant to our objective, for example its ability to detect at the same time they recognize, its speed or scalability and the difficulty of training the system. From the obtained results we extract our conclusions and propose several modifications to improve the performance of these methods. Namely, we propose improvements to increase the precision of the SIFT object recognition method, and a segmentation approach to make the Bag of Features method suitable for detection in interactive time. We also benchmark the proposed methods against the typically used Viola and Jones classifier. Finally, we perform extensive tests with the selected methods in our publicly available dataset[2] to assess their performance in a mobile robotics setting.

The three methods are fundamentally different in that they address recognition, classification and detection (three core problems of visual perception), but still can also be tailored to the other objectives. We compare and benchmark these three successful vision approaches for use in real mobile robotics experiments, providing an useful guide for roboticists who need to enable their robots with object recognition capabilities. The selected algorithms are evaluated under different issues, namely:

Detection: Having the ability to detect where the image is located in the object. In most situations, large portions of the image are occupied by background objects that introduce unwanted information which may confuse the object recognition method.
Classification: A highly desirable capability for an object detection method is to be able to generalize and recognize previously unseen instances of a particular class.
Occlusions: Usually a clear shot of the object to recognize will not be available to the robot. An object recognition method must be able to deal with only partial information of the object.
Texture: Objects with a rich texture are typically easier to recognize than those only defined by their shape and color. We want to evaluate the behavior of each method with both types of objects.
Repetitive patterns: Some objects, such as a chessboard, present repetitive patterns that cause problems in methods that have a data association stage.
Training set resolution: Large images generate more features at different scales (especially for smaller ones) that are undoubtedly useful for object recognition. However, if training images have a resolution much higher than test images, descriptors may become too different.
Training set size: Most methods can benefit from a larger and better annotated training set. However, building such a dataset is time-consuming. We want to assess which is the least amount of training information that each method requires to obtain its best results.
Trainable online: It is very desirable to have a method that can be easily trained for new objects, especially if it can be done by a normal user while the robot is running.
Run-Time: One of the most important limitations of the scenario we are considering is the computation time. We want to measure the frame-rate at which comparable implementations of each method can work.

[2] Available for download at http://www.iiia.csic.es/~aramisa/iiia30.html.

Detection accuracy: Computing accurately the location of the object can significantly benefit other tasks such as grasping or navigation. We are interested in quantifying the precision of the object detection in the object recognition algorithm according to the ground truth.

Although different parts of object recognition methods (e.g., feature detectors and descriptors, machine learning methods) have been extensively compared in literature, to our knowledge there is no work that compares the performance of complete object recognition methods in a practically hard application like mobile robotics.

Probably the work most related with ours is the one of [11], where four methods (SIFT and KPCA + SVM with texture and color features) were combined in an object recognition/classification task for human–robot interaction. The appropriate method for each class of object was chosen automatically from the nine combinations of task/method/features available, and models of the learned objects were improved during interaction with the user (pictured as a handicapped person in the paper). This work was, however, more focused on building a working object classification method suitable for the particular task of human–robot interaction with feedback from the human user, and not in evaluating each particular method in a standardized way. Furthermore, no quantitative results were reported for the experiments with the robot.

Mikolajczyk et al. [12,13] do a comprehensive comparison of interest region detectors and descriptors in the context of keypoint matching. Although this works are undoubtedly related with the one presented here, the objectives of the comparison are notably different: while Mikolajczyk et al. measured the repeatability of the region detectors and the matching precision of the region descriptors, here we focus on the performance of three well-known object recognition methods in the very specific setting of mobile robotics.

Recently a competition named "Solutions in Perception" has been put forward with the support of Willow Garage, with similar motives that validate our work: There is no reliable "gold standard" method for object recognition in robotics, even with limited capabilities, that allows you to build robotic applications based on its results.

The aim of this competition is to analyze what is *actually* doable with current vision machinery in a real robotics scenario. The objective of this chapter is similar, but as opposed to the competition, we focus on providing an in-detail analysis of the methods performance, and restrict ourselves to passive color cameras, while the competition allows you to use any type of sensor.

The rest of the chapter is divided as follows: First, Table 1 shows the conclusions reached in this work regarding the applicability of the evaluated methods in the mobile robot vision domain. Next, in Section 4.4.2 comes an overview of the datasets used in our

Table 1: Qualitative summary of results found in our experiments

	SIFT	Vocabulary Tree	Cascade of Simple Classifiers
Detection	Can detect objects under in-plane rotation, scale changes and small out-of-plane rotations	Must be complemented with a sliding windows approach, a segmentation algorithm or an interest operator	Is able to determine the most probable bounding box of the object
Pose estimation	Up to an affine transformation	Presence/absence only	Presence/absence only
Classification (intra-class variation and generalization)	No	Yes	Yes
Occlusions	Tolerates it as long as at least 3 points can be reliably matched (depends on amount of texture)	Showed good tolerance to occlusions	Low tolerance to occlusions
Repetitive patterns	No	Yes	Yes
Minimum training set size	One image	Tens of images	Hundreds or thousands of images
Training set resolution	VGA resolution is sufficient	Benefits from higher resolution of training data	VGA resolution is sufficient
Trainable online	Easy, requires clear picture of the object, and takes a few seconds to re-train the matching tree	Easy, requires a few good views of the object, and a few seconds to re-train the inverted files	Hard, requires a large collection of training images and up to a few hours of training for certain objects
Run-time	Less than a second per image	Two seconds per image with a segmentation algorithm included	Less than a second per image

experimentation. In Sections 4.4.3—4.4.5 the different object recognition algorithms are briefly described and the experiments done to arrive at the conclusions for each are presented. Finally, in Section 4.4.6 the results obtained with the three methods are discussed, and in Section 4.4.7, the conclusions of the chapter are presented and continuation lines proposed.

4.4.2 Datasets and Performance Metrics

In order to evaluate the methods in a realistic mobile robots setting, we have created the IIIA30 dataset, which consists of three sequences of different lengths acquired by our mobile robot while navigating at approximately 50 cm/s in a laboratory type environment and approximately twenty good quality images for training taken with a standard digital

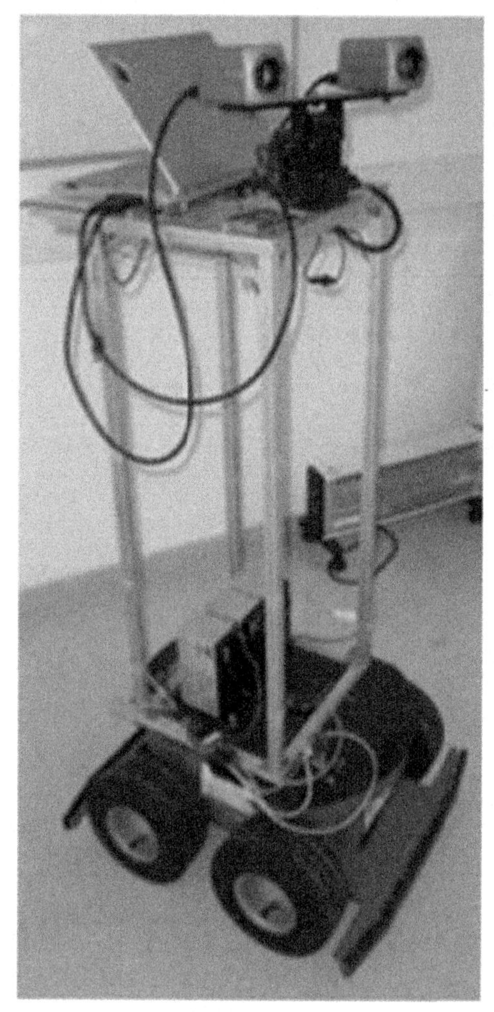

Figure 1
Robotic platform used in the experiments.

camera. The camera mounted in the robot is a Sony DFW-VL500 and the image size is 640×480 pixels. In Figure 1 the robotic platform used can be seen. The environment has not been modified in any way and the object instances in the test images are affected by lightning changes, blur caused by the motion of the robot, occlusion and large viewpoint, and scale changes.

We have considered a total of 30 categories (29 objects and a background) that appear in the sequences. The objects have been selected to cover a wide range of characteristics: some are textured and flat, like the posters, while others are textureless and only defined by their shape. Figure 2(a) shows the training images for all the object categories, and

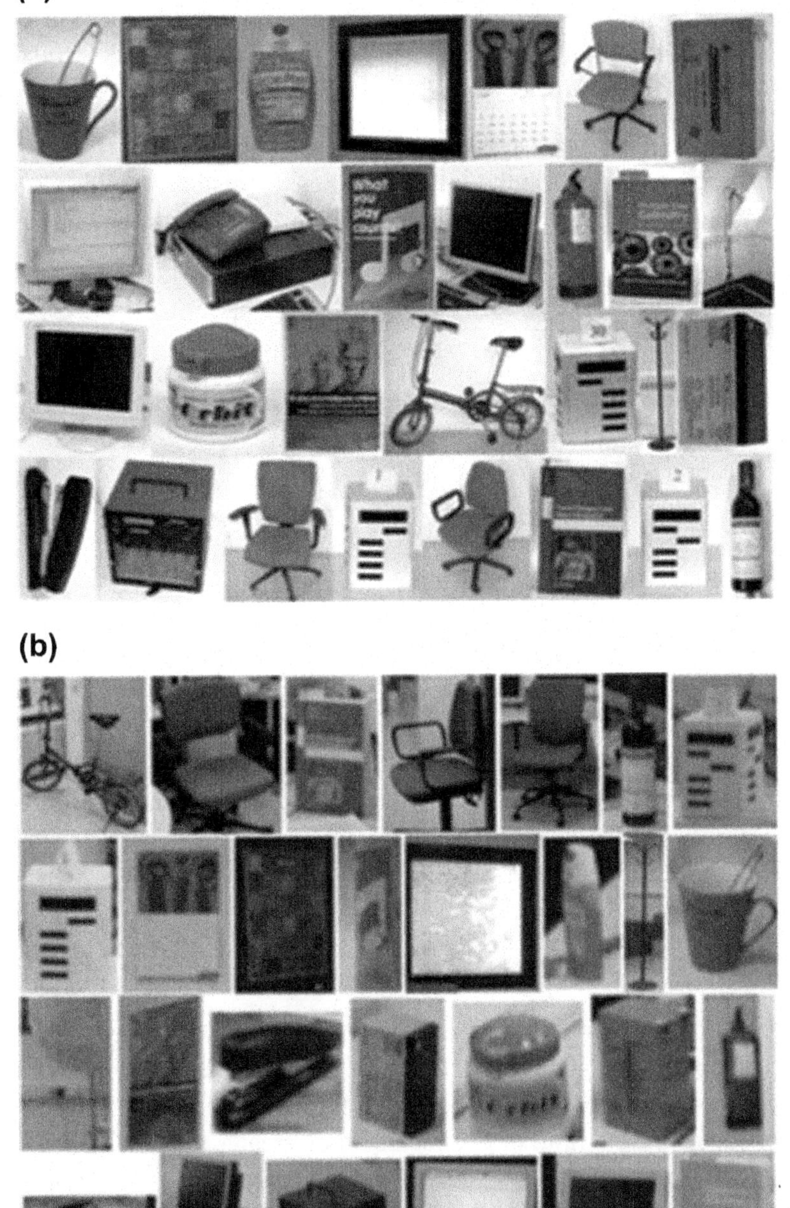

Figure 2
(a) Training images for the IIIA30 dataset. (b) Cropped instances of objects from the test images (for visualization). The actual testing is performed in the full images unless otherwise stated.

Figure 2(b) shows some cropped object instances from the test images (please keep in mind that testing of the methods is done in the full images). Each occurrence of an object in the video sequences has been manually annotated in each frame to construct the ground truth, along with its particular image characteristics (e.g., blurred, occluded, etc.).

In order to evaluate the performance of the different methods we used several standard metrics that are briefly explained in the following lines. Precision is defined as the ratio of true positives among all the positively labeled examples, and reflects how accurate our classifier is.

$$\text{Pre} = \frac{\text{True Positives}}{\text{False Positives} + \text{True Positives}} \tag{1}$$

Recall measures the percentage of true positives that our classifier has been able to label as such. Namely,

$$\text{Rec} = \frac{\text{True Positives}}{\text{False Negatives} + \text{True Positives}} \tag{2}$$

Since it is equally important to perform well in both metrics, we also considered the *f*-measure metric:

$$f\text{-measure} = \frac{2 \cdot \text{Precision} \cdot \text{Recall}}{\text{Precision} + \text{Recall}} \tag{3}$$

This measure assigns a single score to an operating point of our classifier weighting equally precision and recall, and is also known as f_1-measure or balanced *f*-score. If the costs of a false positive and a false negative are asymmetric, the general *f*-measure can be used by adjusting the β parameter:

$$f_g\text{-measure} = \frac{(1+\beta^2) \cdot \text{Precision} \cdot \text{Recall}}{\beta^2 \cdot \text{Precision} + \text{Recall}} \tag{4}$$

In the object detection experiments, we have used the Pascal VOC object detection criterion [7] to determine if a given detection is a false or a true positive. In brief, to consider an object as a true positive, the bounding boxes of the ground truth and the detected instance must have a ratio of overlap equal or greater than 50% according to the following equation:

$$\frac{BB_{\text{gt}} \cap BB_{\text{detected}}}{BB_{\text{gt}} \cup BB_{\text{detected}}} \geq 0.5 \tag{5}$$

where BB_{gt} and BB_{detected} stand for the ground truth and detected object bounding box respectively. For objects marked as occluded only the visible part has been annotated in the ground truth, but the SIFT object recognition method will still try to adjust the detection bounding box for the whole object based only in the visible part. Since the type of annotation is not compatible with the output of the SIFT algorithm, for the case of objects marked as occluded, we have modified the above formula in the following way:

$$\frac{BB_{\text{gt}} \cap BB_{\text{detected}}}{BB_{\text{gt}}} \geq 0.5 \tag{6}$$

As can be seen in the previous equation, it is only required that the detected object bounding box overlaps 50% of the ground truth bounding box.

Apart from the IIIA30 dataset, in order to test and adjust the parameters of the Vocabulary Tree object recognition method, we have used two pre-segmented image databases:

ASL: The ASL recognition dataset[3] consists of nine household objects from the Autonomous Systems Lab of the ETHZ [14]. It consists of around 20 training images per object from several viewpoints and 36 unsegmented test images with several instances of the objects, some of them with illumination changes or partial occlusions. The training images have been taken with a standard digital camera at a resolution of 2 megapixels, while the test images have been acquired with a STHMDCS2VAR/C stereo head by Videre design at the maximum possible resolution (1.2 megapixels). A segmented version of the training object instances has also been used in some experiments, and is referred to as *segmented* ASL. Some images of the segmented version can be seen in Figure 3.

Caltech10: This is a subset of the Caltech 101 dataset [6], widely used in computer vision literature. We have taken 100 random images of the 10 most populated object categories, namely: planes (lateral), bonsais, chandeliers, faces (frontal), pianos, tortoises, sails, leopards, motorbikes, and clocks as seen in Figure 4. Training and testing subsets are determined randomly in each test. Experiments with this dataset have been done following the setup of [15]: 30 random training images and the rest for testing.

4.4.3 Lowe's SIFT

The Lowe's SIFT object recognition approach is a view-centered object detection and recognition system with some interesting characteristics for mobile robots, most significant of which is the ability to detect and recognize objects in an unsegmented image. Another interesting feature is the Best-Bin-First algorithm used for approximated fast matching, which reduces the search time by two orders of magnitude for a database of 100,000 keypoints for a 5% loss in the number of correct matches [9]. Figure 5 shows an overview of our implementation of the SIFT object recognition algorithm steps.

[3] http://www.iiia.csic.es/~aramisa/datasets/asl.html.

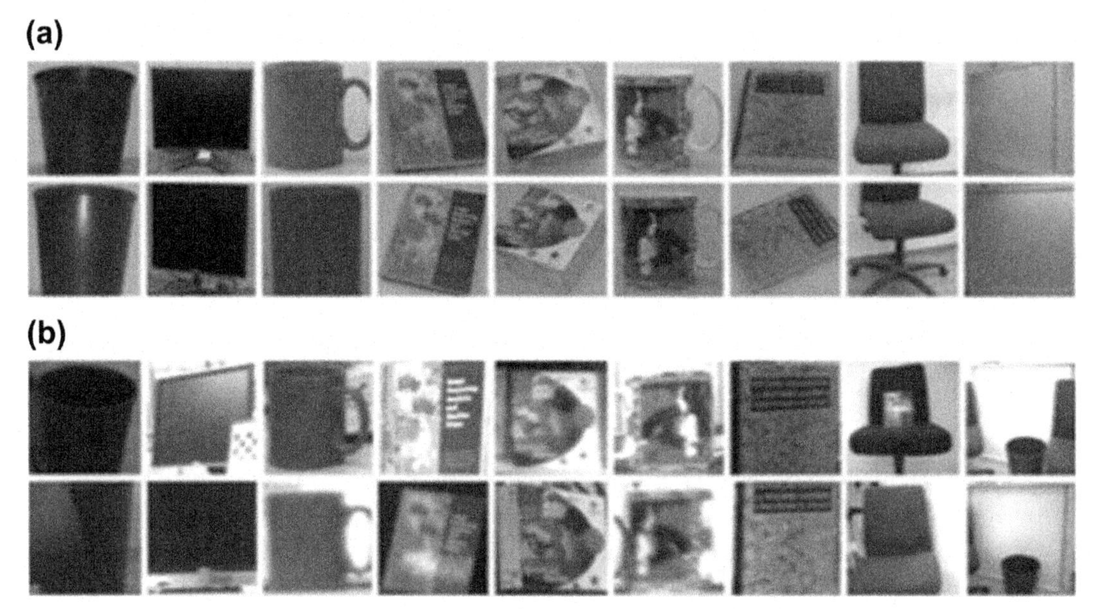

Figure 3
Segmented ASL dataset images. (a) Training. (b) Testing.

Figure 4
Images from Caltech10 dataset.

The first stage of the approach consists on matching individually the SIFT descriptors of the features detected in a test image to the ones stored in the object database using the Euclidean distance. As a way to reject false correspondences, only those query descriptors for which the best match is isolated from the second best and the rest of database descriptors are retained. In Figure 6, the matching features between a test and model images can be seen. The presence of some outliers (incorrect pairings of query and database features) can also be observed.

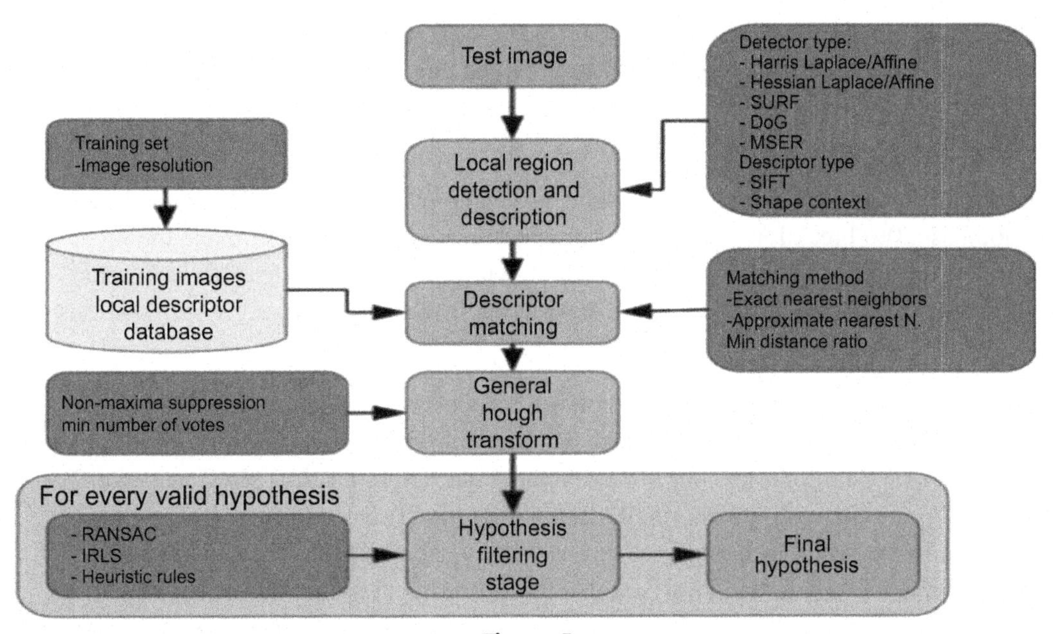

Figure 5
Diagram of the Lowe's SIFT method with all the tests performed shown as *purple boxes*, *Orange ones* refer to steps of the method and *green* to input/output of the algorithm. (For interpretation of the references to color in this figure legend, the reader is referred to the online version of this book.)

Figure 6
Matching stage in the SIFT object recognition method.

Once a set of matches is found, the Generalized Hough Transform is used to cluster each match of every database image depending on its particular transformation (translation, rotation and scale change). Although imprecise, this step generates a number of initial coherent hypotheses and removes a notable portion of the outliers that could potentially confuse more precise but also more sensitive methods. All clusters with at least three matches for a particular training object are accepted, and fed to the next stage: the Least Squares method, used to improve the estimation of the affine transformation between the model and the test images.

This approach has been modified in several ways in our experiments: The least squares method has a 0% breakdown point (i.e., any false correspondence will make the model fitting method fail or give sub-optimal results), which is a rather unfeasible restriction since we have found it is normal to still have some false matches in a given hypothesis after the Hough Transform.

To alleviate this limitation, instead of the least squares, we have used the Iteratively Reweighted Least Squares (IRLS), which we have found to perform well in practice at a reasonable speed. Furthermore we have evaluated the RANdom SAmple Consensus (RANSAC), another well-known model fitting algorithm, to substitute or complement the IRLS. The RANSAC algorithm iteratively tests the support of models estimated using minimal subsets of points randomly sampled from the input data. Finally, we have incorporated some domain knowledge by defining several heuristic rules on the parameters of the estimated affine transformation to reject those clearly beyond plausibility. Namely:

Hypotheses with object centers that are too close.
Hypotheses that have a ratio between the x and y scales below a threshold.

For evaluating the method, one image per category from the training image set is used. As there are several parameters to adjust in this method, we used the first sequence of the IIIA30 dataset (IIIA30-1) as test data to perform an extensive cross-validation over detector and descriptor type, training image size, matching method, distance ratio to the second nearest neighbor for rejecting matches, non-maxima suppression and minimum number of votes in the Hough Transform and hypothesis verification and refinement methods. Since this chapter is too extensive to be included here, details are provided online for the interested reader.[4] The following is a brief summary of the most relevant results obtained with the cross-validation.

Taking into account all combinations, the best recall obtained has been 0.45 with the Hessian Laplace detector and the less restrictive settings possible. However this configuration suffered from a really low precision, just 0.03.

The best precision score has been 0.94, and has been obtained also with the Hessian Laplace detector, with a restrictive distance ratio to accept matches: 0.5. The recall of this combination was 0.14. The same precision value but with a lower recall has been obtained with the SURF and Hessian Affine detectors.

[4] http://www.iiia.csic.es/~aramisa/datasets/iiia30_results/results.html.

Looking at the combinations that had a best balance between recall and precision (best *f*-measure), the top performing ones obtained 0.39 also with the Hessian Laplace detector (0.29 recall and 0.63 precision). However, even though the approximate nearest neighbors are used, each image takes around 2 s to be processed.

Given the objectives of this work, the most relevant way to analyze the results consists in prioritizing the time component and selecting the fastest parameter settings.

As a runtime greater than 1 s is not acceptable for our purposes, the combinations that improved the *f*-measure with respect to faster combinations for those close to 1 s for images have been selected as interesting. Table 2 shows the parameters of the chosen combinations.

Once the parameter combinations that best suited our purposes were found, we evaluated them in all the test sequences.

4.4.3.1 Evaluation of Selected Configurations

This section presents the results obtained applying the parameter combinations previously selected to all the sequences in the dataset. In general all possible combinations of parameters performed better in well textured and flat objects, like books or posters. For example the *Hartley book* or the *calendar* had an average recall across the six configurations (see Table 2 for the configuration parameters) of 0.78 and 0.54 respectively. This is not surprising as the SIFT descriptor assumes local planarity, and depth discontinuities can severely degrade descriptor similarities. On average, textured objects achieved a recall of 0.53 and a precision of 0.79 across all sequences. Objects only defined by shape and color were in general harder or even impossible to detect, as can be seen[5] in Table 3. Recall for these types of objects was only 0.05 on average. Configuration 6, which used the Hessian Laplace detector, exhibited a notably better performance for some objects of this type thanks to its higher number of detected regions. For example, the chair that obtained a recall of 0.54, or the *rack* that obtained a 0.77 recall using this feature detector. Finally, and somewhat surprisingly, objects with a repetitive texture such as the *landmark cubes* (see Figure 2) had a quite good recall of 0.46 on average. Furthermore, the result becomes even better if we take into consideration, that besides the self-similarity, all three *landmark cubes* were also similar to one another.

Regarding the image quality parameters (see Table 4), all combinations behaved in a similar manner: the best recall, as expected, was obtained by images not affected by blur, occlusions or strong illumination changes. From the different disturbances, what was

[5] For space reasons, only part of the table was included. The full Table can be found in http://www.iiia.csic.es/~aramisa/datasets/iiia30_results/results.html.

Table 2: Detailed configuration parameters and results for the six representative configurations in increasing time order

Method	Distance Ratio	Detector	Min. Matches	HT Method	RANSAC	Approx-NN	IRLS	Heuristics	Time (s)	Recall	Precision	*f*-Measure
Config 1	0.8	SURF	5	NMS	No	Yes	Yes	No	0.37	0.15	0.51	0.23
Config 2	0.8	SURF	3	NMS	Yes	Yes	Yes	Yes	0.42	0.14	0.87	0.24
Config 3	0.8	DoG	10	NMS	No	Yes	Yes	No	0.52	0.17	0.47	0.25
Config 4	0.8	DoG	10	NMS	Yes	Yes	Yes	Yes	0.55	0.19	0.9	0.28
Config 5	0.8	DoG	5	NMS	Yes	Yes	Yes	Yes	0.60	0.19	0.87	0.31
Config 6	0.8	HesLap	10	NMS	Yes	Yes	Yes	Yes	2.03	0.28	0.64	0.39

They have been chosen for providing the best results in a sufficiently short time.

Table 3: Object-wise recall and precision for all combinations

Object	Config 1		Config 2		Config 3		Config 4		Config 5		Config 6	
	Rec	Pre	Rec	Pre	Rec	Pre	Rec	Pre	Rec	Pre	Rec	Pre
Grey battery	0	0	0	0	0	0	0	0	0	0	0	0
Bicycle	**0.54**	0.52	0.52	**1.00**	0.33	0.52	0.36	0.89	0.38	0.90	0.33	0.62
Hartley book	0.58	**0.93**	0.58	**0.93**	0.86	0.77	0.88	0.88	**0.95**	0.85	0.81	0.73
Calendar	0.44	0.65	0.35	**0.86**	0.56	0.66	0.56	0.79	0.56	0.79	**0.79**	0.71
Chair 1	0.03	0.08	0.02	0.33	0	0	0	0	0.01	1.00	**0.54**	**1.00**
Charger	0.03	0.20	0.03	**0.50**	0	0	0	0	0	0	**0.18**	0.14
Cube 2	0.62	0.28	0.67	**0.67**	0.71	0.11	**0.76**	0.59	0.76	0.55	0.52	0.38
Monitor 3	0	0	0	0	0	0	0	0	0	0	**0.02**	**0.33**
Poster spices	0.38	0.77	0.42	**0.94**	0.54	0.79	0.53	0.87	**0.58**	0.87	0.56	0.92
Rack	0.26	0.59	0.26	**1.00**	0.10	0.80	0.10	1.00	0.23	1.00	**0.77**	0.79

Bold face marks best performance in precision and recall for each object.

Table 4: Recall depending on image characteristics

Object	Config 1	Config 2	Config 3	Config 4	Config 5	Config 6
Normal	0.26	0.25	0.26	0.28	0.3	0.33
Blur	0.1	0.1	0.16	0.15	0.18	0.25
Occluded	0.16	0.14	0.14	0.12	0.14	0.34
Illumination	0	0	0.06	0.06	0.06	0.06
Blur + Occl	0.06	0.04	0.08	0.06	0.09	0.14
Occl + Illum	0.08	0.08	0.08	0.08	0.08	0.06
Blur + Illum	0	0	0	0	0	0

Normal stands for object instances with good image quality and blur for blurred images due to motion, illumination indicates that the object instance is in a highlight or shadow and therefore has low contrast. Finally the last three rows indicate that the object instance suffers from two different problems at the same time.

tolerated best was occlusion, followed by blur and then by illumination. Combinations of problems also had a demolishing effect in the method performance as seen in the last three rows of Table 4, being the worst cases for the combinations of *blur* and *illumination* which had 0 recall. Object instance size (for objects with a bounding box defining an area bigger than 5000 pixels) did not seem to have such an impact in performance as image quality has. The performance with objects of smaller area has not yet been rigorously analyzed and is left for future work. As can be seen in the results, RANSAC and the heuristics significantly improved precision without affecting recall.

Finally, we have validated the detection accuracy by the ratio of overlap between the ground truth bounding box and the detected object instance as calculated in Eqn (5). As can be seen in Figure 7, on average 70% of true positives have a ratio of overlap greater than 80%, regardless of the parameter combination. Furthermore, we found no appreciable

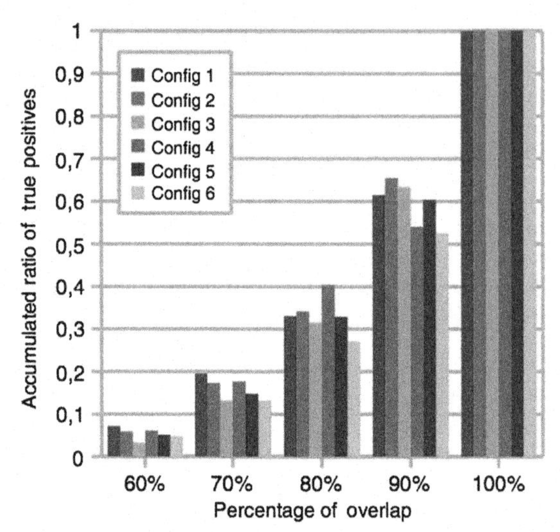

Figure 7

Accumulated frequencies for ratio of overlap between the ground truth bounding box and the detected bounding box for correctly found objects (true positives). An object is considered correctly detected if the ratio of overlap between the bounding boxes computed with Eqn (5) is 50% or more.

advantage on detection accuracy for any object type or viewing conditions; although a more in-depth analysis of this should be addressed in future work.

As a means to provide a context to the results obtained with the six selected configurations (i.e., how good are they with respect to what can be obtained without taking into account the execution time), we compare them to the best overall recall and precision values obtained with the SIFT object recognition method. Table 5 displays the averaged precision and recall values of the four configurations that obtained the overall best recall and the four that obtained the overall best precision, as well as the six selected configurations. As can be seen in the table, the attained recall in the selected configurations was 20% lower than the maximum possible, independently of the type of objects. Precision is more affected by the amount of texture, and differences with respect to the top performing configurations ranged from 17% to 38%.

4.4.3.2 Discussion

Experiments show that, by using the SIFT object recognition approach with the proposed modifications, it is possible to precisely detect, considering all image degradations, around 60% of well-textured object instances with a precision close to 0.9 in our challenging dataset at approximately one frame per second in 640 $\times$ 480 pixel images with our not fully optimized implementation. Even detectors known to sacrifice repeatability (probability of finding the same feature region in slightly different viewing conditions) for

Table 5: Average recall and precision of the configurations that where selected for having the best values according to these two measures in the last section

	Best Recall		Best Precision		Selected Config.	
	Mean	Std	Mean	Std	Mean	Std
			Repetitively Textured Objects			
Recall	0.65	0.09	0.16	0.01	0.46	0.05
Precision	0.02	0.01	0.75	0.15	0.43	0.24
			Textured Objects			
Recall	0.70	0.03	0.28	0.03	0.53	0.10
Precision	0.05	0.02	0.96	0.02	0.79	0.09
			Not Textured Objects			
Recall	0.21	0.01	0.01	0.01	0.05	0.04
Precision	0.03	0.01	0.62	0.32	0.24	0.21

Also average results among the six selected configurations are shown for comparison. Standard deviation is provided to illustrate scatter between the selected configurations. Objects are grouped in the three "level of texture" categories in the following way: the three cubes form the repetitively textured category, the two books, the calendar and the three posters form the textured category, and the rest fall into the non textured category.

speed such as the SURF obtain reasonable results. Performance degrades for objects with repetitive textures or no texture at all. Regarding image disturbances, the approach resisted occlusions well, since the SIFT object recognition method is able to estimate a reliable transformation (as long as a minimum number of correct matches is found, three by default), but not so well for blur due to motion or deficient illumination.

The step of the algorithm that takes most of the processing time is the descriptor matching, as it has a complexity of $O(N \cdot M \cdot D)$ comparisons, where N is the number of features in the new test image, M is the number of features in the training dataset and D is the dimension of the descriptor vector. Approximate matching strategies, such as [16] used in this work, make the SIFT object recognition method suitable for robotic applications by largely reducing its computational cost. In our experiments we experienced only a 0.01 loss in the f-measure for a speed-up up to 35 times. The training time of the approximate nearest neighbor algorithm was typically of a few seconds. Furthermore, an implementation tailored to performance should be able to achieve even faster rates. A drawback of the SIFT object recognition method is that it is not robust to viewpoint change. It would be interesting to evaluate how enhancing the method with 3D view clustering as described in [17] affects the results, as it should introduce robustness to this type of transformation.

4.4.4 Vocabulary Tree Method

The Vocabulary Tree approach [10] to object classification is based on the bag of words document retrieval method, that represents the subject of a document by the frequency in which certain words appear in the text. This technique has been adapted to visual object classification substituting the words with local descriptors such as SIFT computed on image features [18,19].

Although recently many approaches have been proposed following the *bag of words* model, we have selected this particular one because scalability to large numbers of objects in a computationally efficient way is addressed, which is a key feature in mobile robotics. Figure 8 shows the main steps of the [10] algorithm. First the local feature descriptors are extracted from a test image, and a visual vocabulary is used to quantize those features into *visual words*.

A hierarchical vocabulary tree is used instead of a linear dictionary, as it allows us to code a larger number of visual features and simultaneously reduce the look-up time to

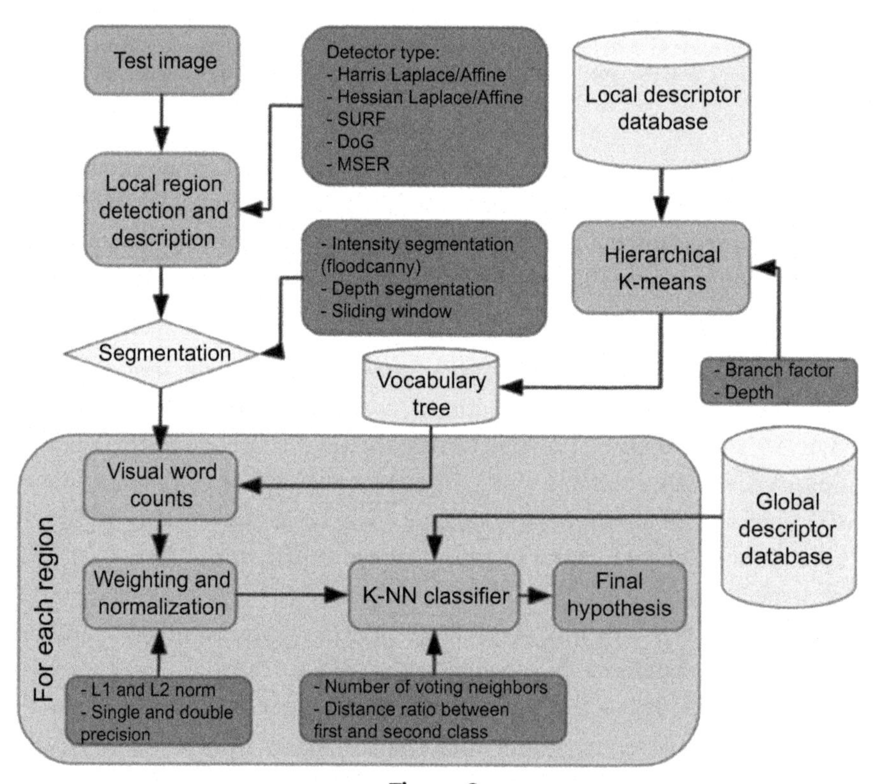

Figure 8

Diagram of the vocabulary tree method. Modifications to the original algorithm have *yellow background* and tests performed are shown as *purple boxes*. As before, *orange boxes* refer to steps of the method and green to input/output of the algorithm. (For interpretation of the references to color in this figure legend, the reader is referred to the online version of this book.)

logarithmic in the number of leaves. The vocabulary tree is built using hierarchical k-means clustering, where the parameter k defines the branch factor of the tree instead of the final number of clusters like in the flat (standard) k-means. On the negative side, using such hierarchical dictionaries causes aliasing in feature space that can reduce the performance of the approach.

Then, the visual words are weighted in accordance to its discriminative power with the term frequency-inverse document frequency (TF-IDF) scheme to improve retrieval performance. Let n_i be the number of descriptors corresponding to the code word i found in the query image and m_i the number of descriptors corresponding to the same code word for a given training image, and let q and d be the histogram signatures of the query and database images, then the histogram bins q_i and d_i can be defined as:

$$q_i = n_i \omega_i, \quad d_i = m_i \omega_i \tag{7}$$

where ω_i is the weight assigned to node i. A measure based in entropy is used to define the weights:

$$\omega_i = \ln\left(\frac{N}{N_i}\right) \tag{8}$$

where N is the number of images in the database, and N_i is the number of images in the database with at least one descriptor vector path through node i. Since signatures will be normalized before comparison, the resulting schema is the TF-IDF.

To compare a new query image with a database image, the following score function is used:

$$s(q,d) = \left\| \frac{q}{\|q\|} - \frac{d}{\|d\|} \right\| \tag{9}$$

The normalization can be in any desired norm, but the L1-norm (also known as the "Manhattan" distance) was found to perform better both by [10] and in our experiments. The class of the object in the query image is determined as the dominant one in the k nearest neighbors from the database images.

The second speed-up proposed by Nister and Stewenius consists of using inverted files to organize the database of training images. In an *inverted files* structure each leaf node contains the ID number of the images whose signature value for this particular leaf is not zero. To take advantage of this representation, and assuming that the signatures have been previously normalized, the previous equation can be simplified making the distance computation only dependent on the non-zero elements both in the query and database vectors. With this distance formulation one can use the inverted files and, for each node, accumulate to the sum only for the training signatures that have non-zero value.

If signatures are normalized using the L2 norm (i.e., the Euclidean distance), the distance computation can be simplified further to:

$$\|q - d\|_2^2 = 2 - 2 \sum_{i|q_i \neq 0, d_i \neq 0} q_i d_i \tag{10}$$

and since we are primarily interested in the ranking of the distances, we can simply accumulate the products and sort the results of the different images in descending order.

The main drawback of the Vocabulary Tree method is that it needs at least a rough segmentation of the object to be recognized. The most straightforward solution to overcome this limitation is to divide the input image using a grid of fixed overlapping regions and process each region independently. Alternatively, we propose a fast segmentation algorithm to generate a set of meaningful regions that can later be recognized with the vocabulary tree method.

The first option has the advantage of simplicity and universality: Results do not depend on a particular method or set of segmentation parameters, but just on the positions and shapes of the windows evaluated. However a square or rectangular window usually does not fit correctly the shape of the object we want to detect and, in consequence, background information is introduced. Furthermore, if we want to exhaustively search the image, in the order of $O(n^4)$ overlapping windows will have to be defined, where n is the number of pixels of the image. This will be extremely time-consuming, and also fusing the classification output of the different windows into meaningful hypotheses is a non-trivial task. One way that could theoretically speed-up the sliding window process is using integral images [20]. This strategy consists on first computing an integral image (i.e., accumulated frequencies of visual word occurrences starting from an image corner, usually top-left) for every visual word in the vocabulary tree. Having the integral image pre-computed for all visual words, the histogram of visual word counts for an arbitrary sub-window can be computed with four operations instead of having to test if every detected feature falls inside the boundaries of the sub-window. Let I_i be the integral image of a query image for node i of the vocabulary tree, then the histogram H of visual words counts for a given sub-window W can be computed in the following way:

$$H_i = I_i(W_{br}) + I_i(W_{tl}) - I_i(W_{tr}) - I_i(W_{bl}) \tag{11}$$

for all i, where W_{br}, W_{tl}, W_{tr} and W_{bl} are respectively the bottom right, top left, top right and bottom left coordinates of W.

The computational complexity of determining the visual word counts for an arbitrary sub-window is therefore $O(4 \cdot \varphi)$ operations, where φ is the size of the vocabulary. Doing the same without integral images has a complexity of $O(5 \cdot \eta)$, where η is the number of visual words found in the test image. From this, it is clear that integral images are a speed-up as long as φ is significantly smaller than η (e.g., in case of dense feature extraction from the image with a small vocabulary).

The second alternative is using a segmentation method to divide the image into a set of regions that must be recognized. Various options exist for this task which can be broadly classified as intensity based and, if stereo pairs of images are available, depth based. In this work we have evaluated one method of each type. Namely, an intensity based method similar to the watershed algorithm, and a depth based one.

4.4.4.1 Intensity-Based Segmentation

The **intensity based** method we propose, that we called *floodcanny*, consists on first applying the Canny edge detector [21] to the image, and using the resulting edges as hard boundaries in a *flood filling* segmentation process. In contrast with conventional watershed methods, in our method seed points are not local minima of the image, but are arbitrarily chosen from the set of unlabeled points; and a limit in brightness difference is imposed both for lower as well as for higher intensity values with respect to the seed point. For each candidate region of an acceptable size (in our experiments, having an area bigger than 900 pixels), a set of five sub-windows of different size centered in the segmented area are defined and evaluated. In general, it is intuitive to think that, the more accurate the segmentation of the image passed to the classifier is, the better will be the results of the object recognition method. More specifically, methods that can overcome highlights, shadows or weak reflections as the one proposed by [22] have a potential to provide more meaningful regions for the classifier, and the combination of such type of methods with appearance-based classifiers is an area of great interest, that we would address in a future work. For the present work however, we have used only our proposed *floodcanny* method, which, despite its simplicity, achieved good segmentation results as can be seen in Figure 9. Furthermore, it is fast to apply (less than 30 ms for a 640×480 image), which is very convenient given our objectives.

4.4.4.2 Depth-Based Segmentation

The second segmentation alternative proposed consisted of directly matching features between the left and right image to detect areas of constant depth. Since the geometry of the stereo cameras is known a *priori*, epipolar geometry constraints can be used together with the scale and orientation of a given feature to reduce the set of possible matches. To determine the possible location of the objects in the environment, a grid of 3D cells of different sizes is used. Reprojected features cast a vote for a cell of a grid if it lies within the 3D cell coordinates. Cells that have a minimum number of votes are reprojected to the image and added as a candidate window. It seems tempting to directly use the matched features to construct the histogram of feature word counts, as it would reduce the amount

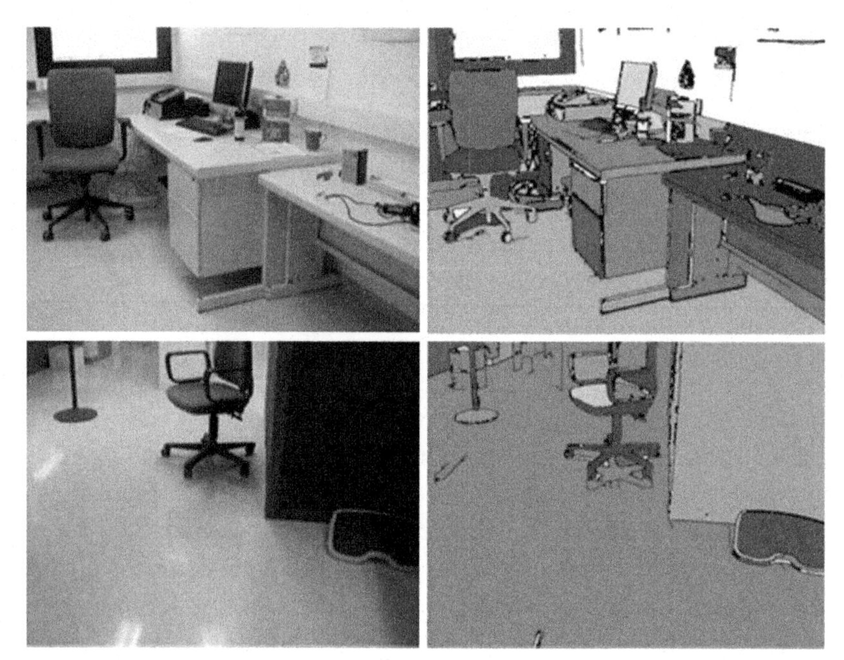

Figure 9
Results of the segmentation process using the *floodcanny* method. The first column shows the original images and the second column the segmented regions. Each *color* represents a different region, and Canny edges are superimposed for clarity. (For interpretation of the references to color in this figure legend, the reader is referred to the online version of this book.)

of background introduced in the visual word counts histogram. However, there is no guarantee that all features of the object have been detected in both images and matched, and the effects of missing important object features are potentially worse than introducing a small amount of background. Therefore we considered it more adequate to accept all visual words close to a set of valid matches.

4.4.4.3 Experimental Results

As in Section 4.4.3, an extensive cross-validation study has been conducted to evaluate the range of parameters of the method. For brevity here we only include the most relevant results and refer the interested reader to [23], which contains all the experimental details of experiments that address:

1. Floating point precision (single/double)
2. Histogram normalization method
3. Effect in computational time of inverted files
4. Quality and number of training images

5. Different segmentation methods (i.e., sliding windows, intensity-based and depth-based segmentation)
6. The effect of different widths and depths of the vocabulary tree
7. Number of nearest neighbors in the kNN classifier
8. Different types of feature detectors
9. Additional tests with manually pre-segmented image datasets.

Detection with Segmentation We have evaluated the proposed *floodcanny* intensity based segmentation algorithm and the depth based segmentation approach described earlier.

We applied the *floodcanny* to the first sequence of the IIIA30 dataset with good results. For each region sufficiently large, a set of five windows of different sizes, centered at the detected region is defined. As can be seen in Figure 10, the number of false positives has decreased from thousands to only tens.

Despite this result, the proposed segmentation scheme is not optimal, as it usually works better for large and textureless objects, which can be segmented as a big single region. Contrarily, small and textured objects pose a problem to the *floodcanny* method, as no single large enough region can be found.

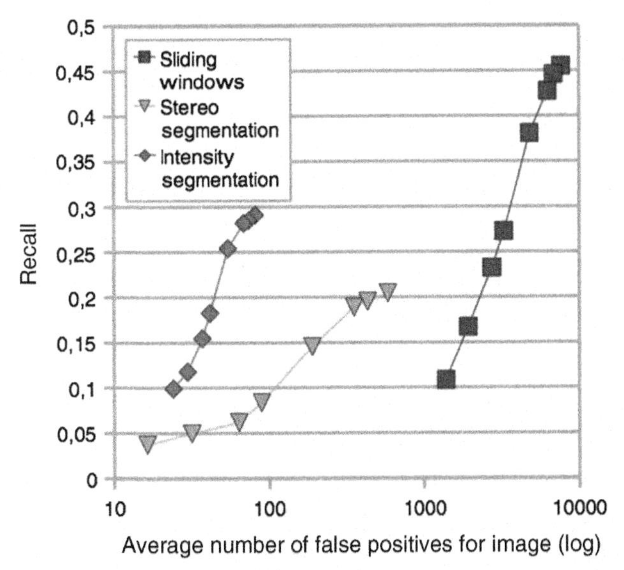

Figure 10

Results of applying intensity segmentation (the *floodcanny algorithm*), stereo segmentation and sliding windows to generate the sub-windows to evaluate at the first sequence of the IIIA30 dataset. For the three experiments the DoG detector and a tree with branch factor 10 and depth 4 have been used.

Table 6: Precision and recall for some interesting objects of the IIIA30 dataset in the final vocabulary tree experiment (i.e., tree with branch factor 9 and depth 4, and features found with the Hessian Affine detector)

	10 nn		10 nn with Filtering $\delta = 0.8$		5 nn		1 nn		10 nn with Relaxed Overlap	
Objects	**Rec**	**Prec**	**Rec**	**Prec**	**Rec**	**Prec**	**Rec**	**Prec**	**Rec**	**Prec**
Grey battery	0.36	0.01	0.32	0.02	0.32	0.01	0.36	0.01	0.60	0.02
Bicycle	0.67	0	0.59	0	0.58	0.01	0.49	0.01	0.70	0
Hartley book	0.21	0	0.21	0	0.19	0	0.21	0	0.81	0.01
Calendar	0.18	0	0.09	0	0.15	0	0.12	0	0.53	0.01
Chair 1	0.70	0.05	0.69	0.06	0.72	0.05	0.78	0.06	0.71	0.06
Charger	0.11	0	0	0	0	0	0	0	0.11	0
Cube 2	0.11	0	0.11	0	0.11	0	0.17	0	0.28	0.01
Monitor 3	0.77	0.16	0.77	0.17	0.66	0.14	0.71	0.09	0.93	0.21
Poster spices	0.46	0.02	0.46	0.02	0.35	0.02	0.46	0.03	0.59	0.03
Rack	0.60	0.06	0.58	0.07	0.60	0.07	0.58	0.06	0.82	0.09

Different choices of parameters for the classifier are displayed. Also, the last column, shows the results obtained using Eqn (6) instead of Eqn (5) to measure overlap.

Regarding the depth segmentation, Figure 10 also shows the results for this experiment. Although the maximum attained recall is lower than that of sliding windows, it must be noted that, at a similar level of recall, false positives are much lower.

4.4.4.4 Evaluation of Selected Configuration

In this section we summarize the results obtained with the parameter configurations selected in the cross-validation study on all the test sequences.

Except for recall, which is better for the Vocabulary Tree method, the SIFT object recognition has better results in all other aspects related to robotics.

As can be seen in Table 6, with the segmentation schema adopted in this final experiment, we have obtained a recall better than with the SIFT method for untextured objects.[6] Unfortunately small and textured objects are harder to detect with the current segmentation, as they usually do not generate a large enough uniform region. However this is not a weakness of the Vocabulary Tree method but of the segmentation approach.

[6] For space reasons, only part of the table was included. The full Table can be found in http://www.iiia.csic.es/~aramisa/datasets/iiia30_results/results.html.

Table 7: Precision and recall depending on texture level of the objects in the final experiment with the Vocabulary Tree

	10 nn	10 nn–0.8	5 nn	1 nn	10 nn–Relaxed
		Repetitively Textured Objects			
Recall	0.18	0.18	0.21	0.23	0.29
Free	0	0	0	0	0.01
		Textured Objects			
Recall	0.29	0.27	0.26	0.28	0.53
Prec	0.02	0.02	0.02	0.02	0.02
		Not Textured Objects			
Recall	0.29	0.26	0.27	0.29	0.39
Prec	0.03	0.03	0.03	0.03	0.04

The objects are grouped in the same way as in Table 5. The title 10 nn–0.8 stands for 10 nearest neighbors with filtering $\delta = 0.8$, and 10 nn-relaxed for 10 nearest neighbors with relaxed overlap.
From Ref. [10]

Table 8: Recall depending on image characteristics

	10 nn	10 nn–0.8	5 nn	1 nn	10 nn–Relaxed
Normal	0.24	0.23	0.24	0.25	0.45
Blur	0.29	0.28	0.28	0.3	0.46
Occluded	0.64	0.61	0.62	0.62	0.64
Illumination	0.06	0.06	0.06	0.11	0.11
Blur + Occl	0.43	0.41	0.43	0.46	0.43
Occl + Illum	0.11	0.11	0.08	0.08	0.11
Blur + Illum	0.14	0	0	0	0.14

Normal stands for object instances with good image quality and blur for blurred images due to motion, illumination indicates that the object instance is in a highlight or shadow and therefore has low contrast. Finally the last three rows indicate that the object instance suffers from two different problems at the same time.

Objects like computer monitors, chairs or an umbrella had a recall comparable to that of textured objects. As can be seen in Table 7, a similar recall was obtained for the objects of types textured and not textured. A slightly worse recall was obtained for the repetitively textured objects, but we believe it is mostly because of the segmentation method.

Regarding the image quality parameters (see Table 8), the occluded objects obtained a higher recall level, but this was because, as mentioned in the previous discussion, the sliding windows approach taken in this experiment does not enforce a precise detection and, therefore, Eqn (5) discards hypotheses correctly detecting object instances. When Eqn (6) was used for all objects, instead of restricting it only to the occluded ones, recall for objects with *normal* and *blurred* viewing conditions is increased. The percentage of detected objects with a degree of overlap from 90% to 100% between the found and the ground truth bounding box was increased by 14%, showing that, although not precisely, the considered windows did overlap almost the whole object region.

4.4.4.5 Discussion

With the selected configurations we obtained an average recall of 30%. More importantly, this approach has been able to detect objects that the SIFT could not find because of its restrictive matching stage. However, also 60 false positives per image on average were detected with the selected configuration, which represents a precision of 2% on average.

In the light of the performed experiments, it seems clear that the Vocabulary Tree method cannot be directly applied to a mobile robotics scenario, but some strategy to reduce the number of false positives is necessary. In addition to reducing false positives to acceptable levels, it is necessary to accelerate the detection step in order to process images coming from the robot's cameras at an acceptable rate. In terms of training time, constructing the vocabulary tree and the inverted files can cost up to several minutes, depending on the vocabulary size. Improving the segmentation strategy, or using a technique such as the one presented in [24] could help improve the accuracy.

Nevertheless, we found that the Vocabulary Tree method was able to detect objects that were inevitably missed by the SIFT Object Recognition method. Furthermore, new and promising *bag of features* type approaches are currently being proposed, such as the aforementioned [25] approach, the one by [26] and especially the one by [27]. In a future work we plan to evaluate some of these methods.

4.4.5 Viola–Jones Boosting

A third commonly used object recognition method is the cascade of weak classifiers proposed by Viola and Jones [5]. This method constructs a cascade of simple classifiers (i.e., simple Haar-like features in a certain position inside a bounding box) using a learning algorithm based on AdaBoost. Speed was of primary importance to the authors of [5], and therefore every step of the algorithm was designed with efficiency in mind. The method uses rectangular Haar-like features as input from the image, computed using Integral Images, which makes it a constant time operation regardless of the scale or type of feature. Then, a learning process that selects the most discriminative features constructs a cascade where each node is a filter that evaluates the presence of a single Haar-like feature with a given scale at a certain position in the selected region. The most discriminative filters are selected to be in the first stages of the cascade to discard windows not having the object of interest as soon as possible. At classification time, the image is explored using sliding windows. However, thanks to the cascade structure of the

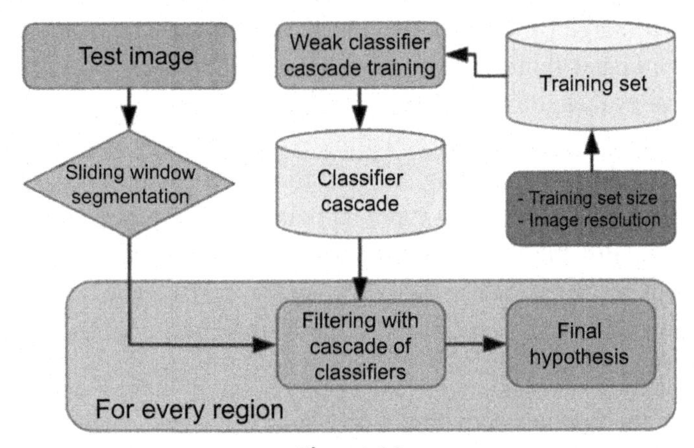

Figure 11
Diagram of the Viola and Jones cascade of weak classifiers method, with tests shown as *purple boxes*. *Orange boxes* refer to steps of the method and *green* to input/output of the algorithm. (For interpretation of the references to color in this figure legend, the reader is referred to the online version of this book.)

classifier it's only at interesting areas where processor time is really spent. Figure 11 shows the main steps of the method.

Notwithstanding its well known advantages, this approach suffers from significant limitations. The most important one being the amount of data required to train a competent classifier for a given class. Usually hundreds of positive and negative examples are required (e.g., in Ref. [28] 5000 positive examples, derived using random transformations from 1000 original training images, and 3000 negative examples where used for the task of frontal face recognition). Another known drawback is that a fixed aspect ratio of the objects is assumed with this method, which may not be constant for certain classes of objects (e.g., cars). Another drawback is the difficulty of generalizing the approach above 10 objects at a time [29]. Finally, the tolerance of the method to changes in the point of view is limited to about 20°. In spite of these limitations, the Viola and Jones object detector has had remarkably successful and is widely used, especially for the tasks of car and frontal face detection.

Since the publication of the original work by Viola and Jones, many improvements to the method have appeared, for example to address the case of multi-view object recognition [30,31].

4.4.5.1 Experimental Results

In this work the original method has been evaluated using a publicly available implementation.[5]

Training Set Size and Image Quality As previously mentioned, one of the most important limitations of the Viola and Jones object recognition method is the size of the training set. In this work we have evaluated three different training sets. The first one consists of images extracted from the ground truth bounding boxes from test sequence IIIA30-3. The second one consists of the same training set used for the Vocabulary Tree experiments (20 good quality training images per object type) and additional synthetic views generated from these images. Finally, the third training set is a mix between good quality images extracted from videos recorded with a digital camera (for 21 objects, between 700 and 1200 manually segmented images per object), and a single training image plus 1000 new synthetic views (for eight objects).

The dataset used for the first test only had a few images for each type of object: 50 to 70 images per class. In Table 9 the results obtained for sequences IIIA30-1 and IIIA30-2 are shown. With so few training data, the Viola and Jones classifier is able to find only some instances for objects of 11 out of the 29 categories. This performance is expected due to the limited amount of training data.

Table 10 shows the results obtained with the twenty training images used in the Vocabulary Tree experiments, but further enhancing the set by synthetically generating a 100 extra images for each training sample. As it can be seen, the usage of high-quality images and the synthetic views significantly improved the results.

Finally, Table 11 shows the results obtained using the third training set, which consisted of hundreds of good quality images extracted from video recordings done with a conventional

Table 9: Recall and precision values obtained training the Viola and Jones object detector using images extracted from the IIIA30-3 sequence and evaluating in sequences IIIA30-1 and IIIA30-2

Object	Recall	Prec	Object	Recall	Prec
Grey battery	0.0	0.0	Monitor 2	0.14	0.14
Red battery	0.28	0.02	Monitor 3	0.03	0.01
Bicycle	0.46	0.07	Orbit box	0.03	0.01
Ponce book	0.0	0.0	Dentifrice	0.0	0.0
Hartley book	0.03	0.01	Poster CMPI	0.17	0.15
Calendar	0.19	0.01	Phone	0.0	0.0
Chair 1	0.11	0.22	Poster mystrands	0.36	0.27
Chair 2	0.71	0.05	Poster spices	0.46	0.06
Chair 3	0.0	0.0	Rack	0.0	0.0
Charger	0.0	0.0	Red cup	0.0	0.0
Cube 1	0.0	0.0	Stapler	0.03	0.01
Cube 2	0.0	0.0	Umbrella	0.03	0.02
Cube 3	0.0	0.0	Window	0.36	0.2
Extinguisher	0.0	0.0	Wine bottle	0.0	0.0
Monitor 1	0.0	0.0			

Table 10: Recall and precision values for each object category for the Viola and Jones object detector when using the same training set as with the bag of features with synthetically generated images

Object	Recall	Prec	Object	Recall	Prec
Grey battery	0.01	0.02	Monitor 2	0.41	0.20
Red battery	0.08	0.04	Monitor 3	0.40	0.18
Bicycle	0.01	0.10	Orbit box	0.10	0.16
Ponce book	0.08	0.31	Dentifrice	0.01	0.03
Hartley book	0.04	0.08	Poster CMPI	0.10	0.05
Calendar	0.11	0.27	Phone	0.07	0.08
Chair 1	0.02	0.30	Poster mystrands	0.71	0.12
Chair 2	0.01	0.34	Poster spices	0.05	0.05
Chair 3	0.02	0.05	Rack	0.06	0.55
Charger	0.0	0.08	Red cup	0.01	0.05
Cube 1	0.06	0.21	Stapler	0.02	0.20
Cube 2	0.0	0.56	Umbrella	0.05	0.58
Cube 3	0.03	0.24	Window	0.10	0.08
Extinguisher	0.09	0.13	Wine bottle	0.03	0.32
Monitor 1	0.02	0.01			

camera. A conclusion that can be quickly inferred from the table is the decrease in performance caused by occlusions. Even objects that achieve a good recall and precision with good viewing conditions, fail in the case of occlusions. In contrast, blurring and illumination variations did not affect performance significantly. Regarding the object types, (textured, untextured and repetitively textured) textured objects obtained an overall recall of 26% and precision of 33%, similar to that of repetitively textured objects (24% recall and 36% precision). Finally, untextured objects obtained 14% of recall and 19% precision. With this dataset, the average *f*-measure obtained is higher than the one obtained with the bag of features object detection method.

The performance on the posters is surprisingly low in comparison to the other two methods. The explanation could be the large changes in point of view that the posters suffer through the video sequences. The time necessary to apply the classifiers for all the classes to one test image is 728 ms on average.

4.4.5.2 Discussion

Despite the use of very simple image features, the Viola and Jones cascade of classifiers attains a good level of precision and recall for most of the objects in a very low runtime. Its main drawbacks are the large, in comparison with the other evaluated techniques, training dataset required to obtain a good level of performance, and the limited robustness to changes in the point of view and occlusions of the method. Training time also increases

Table 11: Recall and precision values for each object category using the Viola and Jones object detector

	All		Non-Occluded		Occluded	
Object	Recall	Prec	Recall	Prec	Recall	Prec
Grey battery	0.36	0.24	0.41	0.24	0.0	0.0
Red battery	0.37	0.82	0.44	0.82	0.0	0.0
Bicycle	0.0	0.0	0.0	0.0	0.0	0.0
Ponce book	0.81	0.88	0.86	0.86	0.25	0.02
Hartley book	0.66	0.94	0.70	0.94	0.0	0.0
Calendar*	0.33	0.08	0.38	0.08	0.0	0.0
Chair 1	0.0	0.0	0.0	0.0	0.0	0.0
Chair 2*	0.0	0.0	0.0	0.0	0.0	0.0
Chair 3	0.0	0.0	0.0	0.0	0.0	0.0
Charger	0.12	0.08	0.12	0.08	0.0	0.0
Cube 1	0.22	0.43	0.23	0.29	0.2	0.15
Cube 2	0.23	0.11	0.20	0.09	0.34	0.03
Cube 3	0.28	0.53	0.37	0.48	0.09	0.06
Extinguisher	0.0	0.0	0.0	0.0	0.0	0.0
Monitor 1*	0.0	0.0	0.0	0.0	0.0	0.0
Monitor 2*	0.23	0.57	0.39	0.57	0.0	0.0
Monitor 3*	0.04	0.13	0.05	0.13	0.0	0.0
Orbit box*	0.15	0.03	0.17	0.03	0.0	0.0
Dentifrice	0.0	0.0	0.0	0.0	0.0	0.0
Poster CMPI	0.11	0.34	0.19	0.34	0.0	0.0
Phone	0.05	0.09	0.0	0.0	0.3	0.09
Poster Mystrands	0.0	0.0	0.0	0.0	0.0	0.0
Poster spices	0.04	0.38	0.12	0.38	0.0	0.0
Rack	0.0	0.0	0.0	0.0	0.0	0.0
Red cup	0.89	0.89	0.89	0.89	0.0	0.0
Stapler	0.24	0.21	0.24	0.21	0.0	0.0
Umbrella	0.0	0.0	0.0	0.0	0.0	0.0
Window	0.03	0.40	0.10	0.40	0.0	0.0
Wine bottle*	0.10	0.06	0.10	0.06	0.0	0.0

with the size of the dataset, and can be several hours. Furthermore, some theoretically "easy" objects, such as posters, proved to be troublesome to the Viola and Jones method. This is probably due to overfitting to some particular view, or to too much variability of the Haar feature distribution when changing the point of view, where the method was unable to find any recognizable regular pattern.

Nevertheless, the idea of a boosted cascade of weak classifiers is not limited to the very fast but simple Haar features, but any kind of classifier can be used for that matter. A very interesting alternative is using linear SVMs as weak classifiers, since it allows adding a non-linear layer to an already efficient linear classifier. Such an idea has been already successfully applied in a few cases [32,33], and we believe it is a very interesting line to investigate.

4.4.6 Discussion

The first evaluated method is the SIFT object recognition method, proposed by [9]. Many issues including training image quality, approximate local descriptor matching or false hypotheses filtering methods are evaluated in a subset of the proposed dataset. Furthermore, we propose and evaluate several modifications to the original schema to increase the detected objects and reduce the computational time.

The parameter settings that attained best overall results are subsequently tested in the rest of the dataset and are carefully evaluated to have a clear picture of the responses that can be expected from this method with respect to untextured objects or image degradations.

Next, a similar evaluation is carried out on for the second method, the Vocabulary Tree proposed by [10]. In the case of the Viola and Jones cascade of weak classifiers, the used implementation directly offers a thoroughly evaluated selection of parameters, and the main variable we have evaluated is the training set size.

From the results obtained, it can be seen that with the present implementation of the methods, the SIFT object recognition method adapts better to the performance requirements of a robotics application. Furthermore, it is easy to train, since a single good quality image sufficed to attain good recall and precision levels. However, although this method is resistant to occlusion and reasonable levels of motion blur, its usage is mostly restricted to flat well textured objects. Also, classification (generalizing to unseen object instances of the same class) is not possible with this approach.

On the other hand, the Vocabulary Tree method has obtained good recognition rates both for textured and untextured objects, but too many false positives per image were found. Finally, the Viola and Jones method offers both a good recall (specially for low-textured objects) and execution speed, but is very sensitive to occlusions and the simple features used seem to be unable to cope with the most richly textured objects in case of strong changes in point of view.

Although we have evaluated the proposed object recognition methods in a wide range of dimensions, one that is lacking is a more in-depth study of how the composition and size of the training set affects the overall results. For example, having similar objects, as the different monitors or chairs in the IIIA30 dataset, can cause confusion to the methods. Therefore future work will address the evaluation of different sub-sets of target objects.

The main limitation of the SIFT object recognition method is that only the first nearest neighbor of each test image feature is considered in the subsequent stages. This restriction makes the SIFT method very fast, but at the same time makes it unable to detect objects with repetitive textures. Other approaches with direct matching, like that of [34], overcome this by allowing every feature to vote for all feasible object hypotheses given the feature position and orientation. Evaluating these types of methods, or modifying the SIFT to accept several hypotheses for each test image feature, would be an interesting line of continuation of this study.

The sliding windows approach could be improved by allowing windows with a good probability of a correct detection to inhibit neighboring and/or overlapping windows, or simply keeping the best window for a given object would clearly reduce the number of false positives.

Regarding the segmentation schema, we believe that results can be improved by adopting more reliable techniques, being able to resist highlights and shadows. Besides, textured areas pose a problem to the segmentation algorithm as, with the current technique, no windows will be cast in scattered areas. It would be interesting to test if a Monte Carlo approach to fuse neighboring regions can help alleviate the problem without significantly affecting the computational time. Also a voting mechanism to detect areas with a high number of small regions can be attempted.

The Viola and Jones approach was the fastest of the three in execution time and, as mentioned earlier, it obtained a reasonable level of precision and recall—especially for the low-textured objects, but at the cost of a significantly larger training effort, both in computational cost and labeled data, than the other two methods. In addition, object instances with occlusions had a performance notably lower in comparison.

More powerful features, like the ones used for the other two methods, or the popular HOGs [35], could also be used in the Viola and Jones cascade of classifiers. However that would increase the computational cost of the method. In order to handle the viewpoint, change extensions have been proposed to the method [31,36], especially using error–correcting output codes [37]. It would be interesting to evaluate the impact on the performance of these extensions.

4.4.7 Conclusions

Object perception capabilities are a key element in building robots able to develop useful tasks in generic, unprepared, human environments. Unfortunately, state of the art papers in computer vision do not evaluate the algorithms with the problems faced in mobile robotics. In this chapter we have contributed an evaluation of three object recognition algorithms in the difficult problem of object recognition in a mobile robot: the SIFT object recognition method, the Vocabulary Tree and a boosted cascade of weak classifiers. In contrast with the case of high-quality static Flickr photos, images acquired by a moving robot are likely to be of low resolution, unfocused and affected by problems like bad framing, motion blur or inadequate illumination, due to the short dynamic range of the camera. The three methods have been thoroughly evaluated in a dataset obtained by our mobile robot while navigating in an unprepared indoor environment. Finally, in order to improve the performance of the methods, we have also proposed several improvements.

This work aims to be a practical help for roboticists that want to enable their mobile robots with visual object recognition capabilities, highlighting the advantages and drawbacks of each method and commenting on its applicability in practical scenarios. Furthermore, relevant enhancements for the methods existent in literature (e.g., support for 3D models in the SIFT object recognition method) are reported.

We have created a challenging dataset of video sequences with our mobile robot while moving in an office type environment. These sequences have been acquired at a resolution of 640×480 pixels with the robot cameras, and are full of blurred images due to motion, large viewpoint and scale changes and object occlusions.

In summary: Three fundamentally different methods, each one a representative of a successful established paradigm for visual object perception, have been evaluated for the particular task of object detection in a mobile robot platform. Furthermore, a number of variations or improvements to the selected methods are being actively produced and evaluated.

Future work includes evaluating more state-of-the-art methods for object recognition, such as those of Philbin et al. [38], Collet et al. [39] or Felzenszwalb et al. [40]. We intend to continue working on this problem, and publish the results in a "Part 2" article.

Acknowledgments

This work was supported by the following grants: JAE Doc of the CSIC, FEDER European Social funds, AGAUR grant 2009-SGR-1434, the Government of Spain under research programme Consolider Ingenio 2010: MIPRCV (CSD2007-00018) and MICINN project TIN2011-25606 (SiMeVé), Rio Tinto Centre for Mine Automation, and the ARC Centre of Excellence programme, funded by the Australian Research Council and the New South Wales State Government.

References

[1] S. Vasudevan, S. Gachter, V. Nguyen, R. Siegwart, Cognitive maps for mobile robots an object based approach, Robot. Auton. Syst. 55 (5) (2007) 359–371 (From Sensors to Human Spatial Concepts).

[2] O. Martinez Mozos, R. Triebel, P. Jensfelt, A. Rottmann, W. Burgard, Supervised semantic labeling of places using information extracted from sensor data, Robot. Auton. Syst. 55 (5) (2007) 391–402.

[3] C. Galindo, A. Saffiotti, S. Coradeschi, P. Buschka, J. Fernandez-Madrigal, J. González, Multihierarchical semantic maps for mobile robotics, in: IEEE/RSJ International Conference on Intelligent Robots and Systems, IROS, 2005, pp. 2278–2283.

[4] P. Jensfelt, S. Ekvall, D. Kragic, D. Aarno, Augmenting slam with object detection in a service robot framework, in: The 15th IEEE International Symposium on Robot and Human Interactive Communication, 2006, Roman, 2006, pp. 741–746.

[5] P. Viola, M. Jones, Rapid object detection using a boosted cascade of simple features, in: IEEE Conference on Computer Vision and Pattern Recognition, vol. 1, 2001, p. 511.
[6] L. Fei-Fei, R. Fergus, P. Perona, Learning generative visual models from few training examples: an incremental Bayesian approach tested on 101 object categories, in: Workshop on Generative-Model Based Vision. IEEE Computer Society, 2004.
[7] M. Everingham, L. Van Gool, C.K.I. Williams, J. Winn, A. Zisserman, The PASCAL Visual Object Classes Challenge 2007 (VOC2007) Results, 2007. http://www.pascalnetwork.org/challenges/VOC/voc2007/workshop/index.html.
[8] N. Pinto, D.D. Cox, J.J. Dicarlo, Why is real-world visual object recognition hard? PLoS Comput. Biol. 4 (1) (2008) e27+. http://dx.doi.org/10.1371/journal.pcbi.0040027.
[9] D.G. Lowe, Distinctive image features from scaleinvariant keypoints, Int. J. Comput. Vis. 60 (2) (2004) 91–110.
[10] D. Nister, H. Stewenius, Scalable recognition with a vocabulary tree, in: IEEE Conference on Computer Vision and Pattern Recognition, vol. 2, 2006, pp. 2161–2168.
[11] A. Mansur, Y. Kuno, Specific and class object recognition for service robots through autonomous and interactive methods, IEICE Trans. Inf. Syst. E91-D (6) (2008) 1793–1803. http://dx.doi.org/10.1093/ietisy/e91-d.6.1793.
[12] K. Mikolajczyk, C. Schmid, A performance evaluation of local descriptors, IEEE Trans. Pattern Anal. Mach. Intell. 27 (10) (2005) 1615–1630.
[13] K. Mikolajczyk, T. Tuytelaars, C. Schmid, A. Zisserman, J. Matas, F. Schaffalitzky, T. Kadir, L.V. Gool, A comparison of affine region detectors, Int. J. Comput. Vis. 65 (1/2) (2005) 43–72.
[14] A. Ramisa, S. Vasudevan, D. Scaramuzza, R.L. de Mántaras, R. Siegwart, A tale of two object recognition methods for mobile robots, in: A. Gasteratos, M. Vincze, J.K. Tsotsos (Eds.), ICVS Lecture Notes in Computer Science, May 12–15, 2008, vol. 5008, pp. 353–362. http://dblp.uni-trier.de/db/conf/icvs/icvs2008.html#RamisaVSMS08.
[15] K. Grauman, T. Darrell, The pyramid match kernel: discriminative classification with sets of image features, in: International Conference on Computer Vision, 2005, pp. 1458–1465.
[16] M. Muja, D. Lowe, Fast approximate nearest neighbors with automatic algorithm configuration, in: International Conference on Computer Vision Theory and Applications (VISAPP'09), 2009.
[17] D.G. Lowe, Object recognition from local scaleinvariant features, in: International Conference on Computer Vision, vol. 2, 1999, p. 1150.
[18] G. Csurka, C. Bray, C. Dance, L. Fan, Visual categorization with bags of keypoints, in: Workshop on Statistical Learning in Computer Vision, ECCV, 2004, pp. 1–22.
[19] J. Sivic, A. Zisserman, Video google: a text retrieval approach to object matching in videos, Int. Conf. Comput. Vis. 2 (2003) 1470–1477.
[20] F. Porikli, Integral histogram: a fast way to extract histograms in cartesian spaces, in: IEEE Conference on Computer Vision and Pattern Recognition, vol. 1, 2005, pp. 829–836.
[21] J. Canny, A computational approach to edge detection, IEEE Trans. Pattern Anal. Mach. Intell. 6 (1986) 679–698.
[22] E. Vazquez, J. van de Weijer, R. Baldrich, Image segmentation in the presence of shadows and highlights, in: European Conference on Computer Vision, vol. 4, 2008, pp. 1–14.
[23] A. Ramisa, Localization and Object Recognition for Mobile Robots (Ph.D. Thesis), Autonomous University of Barcelona, 2009.
[24] R. Bianchi, A. Ramisa, R. Mantaras, Automatic selection of object recognition methods using reinforcement learning, in: Recent Advances in Machine Learning, vol. 262, Springer Studies in Computational Inteligence, 2010, pp. 421–439 (dedicated to the memory of Prof. Ryszard S. Michalski).
[25] B. Fulkerson, A. Vedaldi, S. Soatto, Localizing objects with smart dictionaries, in: European Conference on Computer Vision, 2008, pp. 179–192.
[26] F. Moosmann, E. Nowak, F. Jurie, Randomized clustering forests for image classification, IEEE Trans. Pattern Anal. Mach. Intell. 30 (9) (2008) 1632–1646.

[27] C.H. Lampert, M.B. Blaschko, T. Hofmann, Beyond sliding windows: object localization by efficient subwindow search, in: IEEE Conference on Computer Vision and Pattern Recognition, 2008.
[28] R. Lienhart, E. Kuranov, V. Pisarevsky, Empirical analysis of detection cascades of boosted classifiers for rapid object detection, in: DAGM 25th Pattern Recognition Symposium, 2003, pp. 297−304.
[29] A. Torralba, K. Murphy, W. Freeman, Sharing visual features for multiclass and multiview object detection, IEEE. Trans. Pattern Anal. Mach. Intell. 29 (2007) 854−869.
[30] C. Huang, H. Ai, B. Wu, S. Lao, Boosting nested cascade detector for multi-view face detection, in: International Conference on Pattern Recognition, 2004, pp. 415−418.
[31] M. Jones, P. Viola, Fast multi-view face detection, in: IEEE Conference on Computer Vision and Pattern Recognition, 2003.
[32] D. Aldavert, A. Ramisa, R. Toledo, R. Mantaras, Fast and robust object segmentation with the integral linear classifier, in: IEEE Conference on Computer Vision and Pattern Recognition, 2010, pp. 1046−1053. http://dx.doi.org/10.1109/CVPR.2010.5540098.
[33] P. Viola, M. Jones, D. Snow, Detecting pedestrians using patterns of motion and appearance, Int. J. Comput. Vis. 63 (2005) 153161.
[34] B. Leibe, A. Leonardis, B. Schiele, Robust object detection with interleaved categorization and segmentation, Int. J. Comput. Vis. 77 (1−3) (2008) 259−289.
[35] N. Dalal, B. Triggs, Histograms of oriented gradients for human detection, in: IEEE Conference on Computer Vision and Pattern Recognition, 2005, pp. 886−893.
[36] Z. Zhang, M. Li, S.Z. Li, H. Zhang, Multi-view face detection with floatboost, in: IEEE Workshop on Applications of Computer Vision, 2002, p. 184. http://dx.doi.org/10.1109/ACV.2002.1182179.
[37] T.G. Dietterich, G. Bakiri, Solving multiclass learning problems via error-correcting output codes, J. Artif. Intell. Res. 2 (1995) 263−286.
[38] J. Philbin, O. Chum, M. Isard, J. Sivic, A. Zisserman, Object retrieval with large vocabularies and fast spatial matching, in: IEEE Conference on Computer Vision and Pattern Recognition, 2007, pp. 1−8.
[39] A. Collet, D. Berenson, S. Srinivasa, D. Ferguson, Object recognition and full pose registration from a single image for robotic manipulation, in: IEEE International Conference on Robotics and Automation, 2009, pp. 48−55.
[40] P. Felzenszwalb, D. McAllester, D. Ramanan, A discriminatively trained, multiscale, deformable part model, in: IEEE Conference on Computer Vision and Pattern Recognition, 2008, pp. 1−8.

PART 5

Grasping and Manipulation

CHAPTER 5.1

The State of the Art in Grasping and Manipulation for Household Service

Yuandong Sun[1], **Huihuan Qian**[1,2], **Yangsheng Xu**[1,2]

[1]Department of Mechanical and Automation Engineering, The Chinese University of Hong Kong, Hong Kong SAR, China; [2]Shenzhen Institutes of Advanced Technology, Chinese Academy of Sciences, Shenzhen, China

Chapter Outline

Imagine a scenario as follows. A robot is now outside our apartment building and is going to serve us in our apartment. The robot first slides a magnetic card through a card reader to open the main entrance of the apartment. Then it moves to the elevator and presses the button "going up." It takes the elevator to the floor where we live. Then it moves out of the elevator and finds our front door. Finally, it opens the door and enters our apartment. Inside our apartment, the robot provides several services, such as making coffee, washing dishes, and folding laundry. Moreover, it charges itself when the battery is low.

Such a scenario will commonly happen in our future life when robotic grasping and manipulation start to play a more important role. Robotic grasping and manipulation comprise three subtasks, i.e., target detection, planning, and control. The robot first needs to detect the target. Here "detect" means to find (or recognize) the target and locate it. The target could be the object to be manipulated (e.g., electrical plugs) or the destination

Household Service Robotics. http://dx.doi.org/10.1016/B978-0-12-800881-2.00016-5

of the object (electrical sockets). Then the robot approaches the target in an optimized trajectory. In the process of approaching and grasping, the robot should be controlled to perform the task correctly and smoothly.

In this chapter, we review the approaches to accomplish these three subtasks, i.e., target detection, planning, and control.

5.1.1 Target Detection

Target detection is the first step in robotic grasping and manipulation. In this section, we provide a sufficient review of target detection, although there will be some overlap between target detection and object recognition, which was discussed in the previous chapter.

The approaches to target detection are classified into four categories: (1) laser-based approaches, (2) vision-based approaches, (3) Kinect-based approaches, and (4) other approaches. The categorization depends on the sensors that are used.

5.1.1.1 Laser-Based Approaches

A. Jain et al. [1] simply used a laser pointer to illuminate a door handle, after which the robot was able to locate the door handle autonomously. They proposed another method in [2], which used a three-dimensional (3D) point cloud. The door detection and handle detection algorithms analyzed only the points that fell within a predefined volume of interest (Figure 1).

R. B. Rusu et al. [3] used a tilting laser sensor to acquire a 3D point cloud P. P is then down-sampled to P_d using a fast octree structure. P_d is continued to be down-sampled to

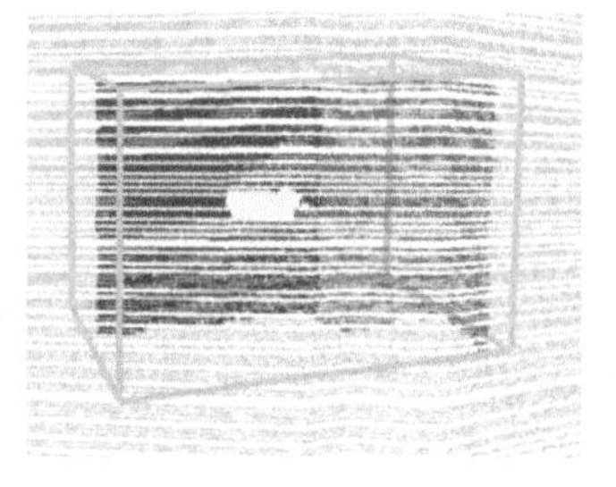

Figure 1
Left: Coordinate axes parallel to the axes of the frame with the origin at the robot that is used by the door and door handle detection algorithms. Middle: Image of a door and handle. Right: The output of the door handle segmentation algorithm. The volume of interest (VOI) is the yellow box, the door is blue, the door handle is yellow, and the points inside the VOI are dark gray. The point cloud outside the VOI is light gray [2]. (For interpretation of the references to color in this figure legend, the reader is referred to the online version of this book.)

Figure 2
Door and handle identification example in a 3D point cloud acquired using the tilting Hokuyo laser [3].

P_Z. The surface normal of all the points in P_Z is approximately perpendicular to the world z-axis. P_Z is then split into several clusters, which are the potential door candidates. By fitting planes using the randomized M-estimator sample consensus robust estimator, doors are located. By combining a 3D point cloud with the geometric information of the door, the handle on the door could be located (Figure 2). This method was also used in [4] for laser-based door detection.

M. Bansal et al. [5] proposed an algorithm for real-time door and stair detection. The 3D point cloud is acquired by the laser in a streaming manner. They proposed efficient data structures and algorithms to minimize memory copying and access. M.B. Ansal et al. [6] proposed a similar strategy to handle a 3D point cloud stream. A 3D point cloud is first stored in the memory in a dense 3D voxel grid with a relatively large voxel size. Then a data-chunking mechanism is employed to process the data in the immediate vicinity of the robot. This architecture requires a very small memory and minimal data copying (Figure 3).

5.1.1.2 Vision-Based Approaches

Vision-based approaches use either a stereo camera pair or a single camera.

J. Bohren et al. [7] used a narrow stereo camera pair on the robot PR2 to obtain a dense stereo point cloud (Figure 4). The point cloud is filtered and clustered using the method proposed in [3]. A fridge door and handle can be detected thereafter. J. Sturm et al. [8] used an active stereo camera instead of a passive stereo camera as used in [7]. A compact, high-power LED device projects a fixed pattern. The pattern is seen by the stereo camera and the projection matrices can then be found. The projection matrices are used to establish a 3D point cloud. A RANSAC (random sample consensus) algorithm is then used

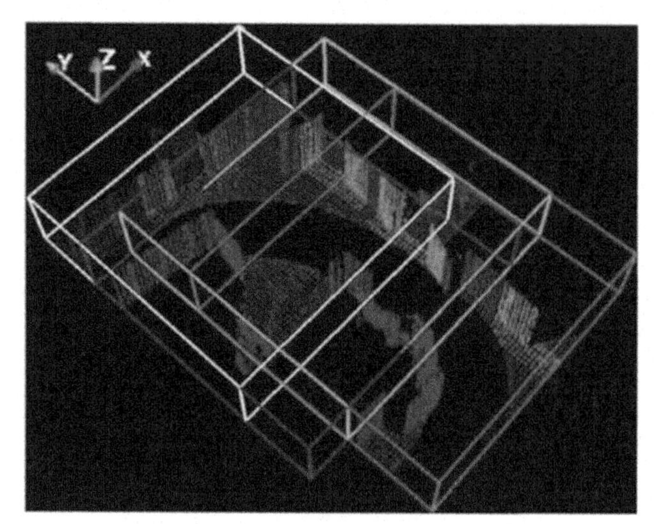

Figure 3
Chunking example. The four positions of the chunk as the robot traverses a curved path around an office corridor are shown. The bounding boxes are color coded in order of time: red, green, blue, and yellow. The offsets show a roll in X from green to blue and rolls in Y in all other cases. Also shown are the composite point clouds with red points marking detected doors and blue points marking unlabeled structures [6]. (For interpretation of the references to color in this figure legend, the reader is referred to the online version of this book.)

Figure 4
The output of the fridge handle detector from the narrow stereo, showing the handle points (red) and the handle orientation (top right). The left side presents the point cloud obtained using the tilting laser sensor, with a transparent model of the PR2 robot in the foreground [7]. (For interpretation of the references to color in this figure legend, the reader is referred to the online version of this book.)

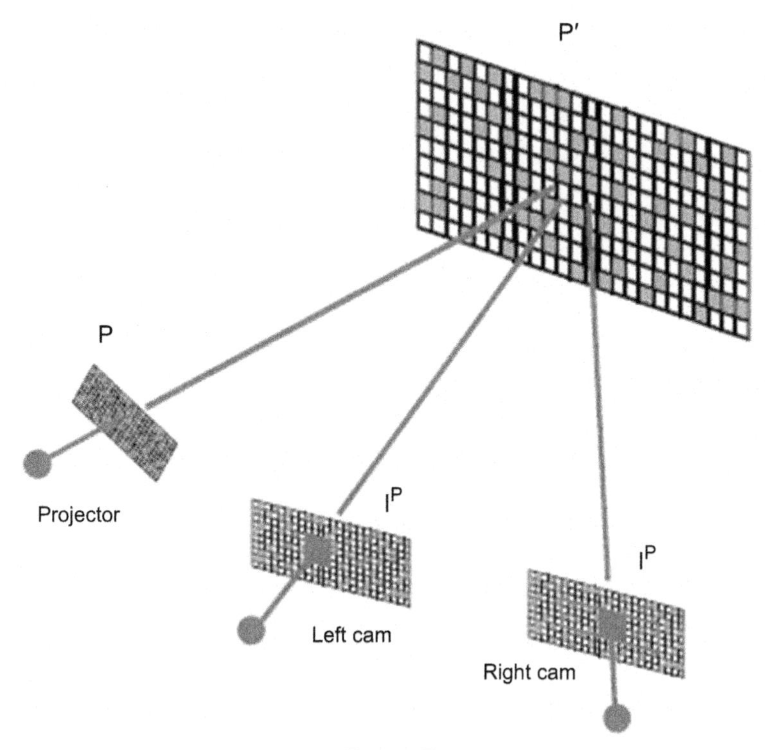

Figure 5
The projector and stereo camera system. A pattern P is projected onto a surface to produce P′, which is imaged by a left and right camera. To compute depth, the small red block in the left camera image is matched against a range of blocks in the right image at the same vertical offset, indicated by the outlined rectangle [8]. (For interpretation of the references to color in this figure legend, the reader is referred to the online version of this book.)

to locate different planes for further actions (Figure 5). N. Kwak et al. [9] first located the knob in 2D images using a stereo camera. Then they constructed a 3D position of the knob by stereo geometry.

D. Ignakov et al. [10] also extracted a 3D point cloud for target detection. But their point cloud is extracted from images taken with a single camera. Optical flow is used to calculate the point cloud. A RANSAC algorithm is applied on the point cloud to locate the door and the handle. Instead of building a full 3D model of the object, A. Saxena et al. [11] solved 3D positions of grasping points. A single camera takes images from different angles to solve the 3D positions. These 3D positions are sufficient to accomplish the grasping task (Figure 6).

5.1.1.3 Kinect-Based Approaches

Kinect is a motion-sensing device developed by Microsoft. It integrates a range sensor with a camera. It solves the depth problem in monocular vision and is much cheaper

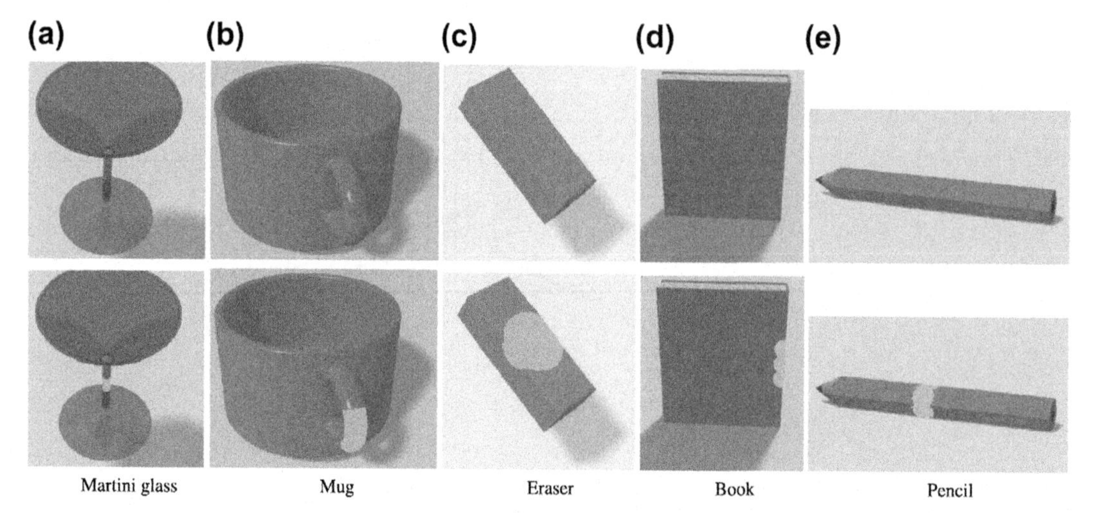

Figure 6
The images (top row) with the corresponding labels (highlighted in the bottom row) of the five object classes used for training. The classes of objects used for training were (a) martini glasses, (b) mugs, (c) whiteboard erasers, (d) books, and (e) pencils [11].

than a laser sensor. Therefore, Kinect has been widely used in research related to computer vision.

J. Stueckler et al. [12] used Kinect to develop real-time 3D information about the environment, such as a table (Figure 7). They applied RANSAC on a 3D point cloud to determine the surface of the table. Then the objects above the table could be detected easily. T. Rühr et al. [13] proposed two approaches using Kinect to detect handles. To detect handles without specularity, they applied a refined version of the approach proposed in [3]. If the handles are too thin or specular, the range sensor fails to measure the

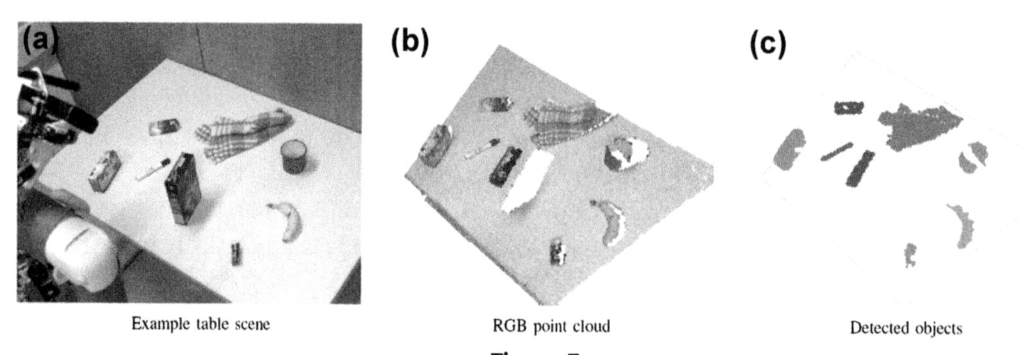

Figure 7
(a) Example table-top setting. (b) Raw point cloud from the Kinect with RGB information. (c) Each detected object is marked with a random color [12].

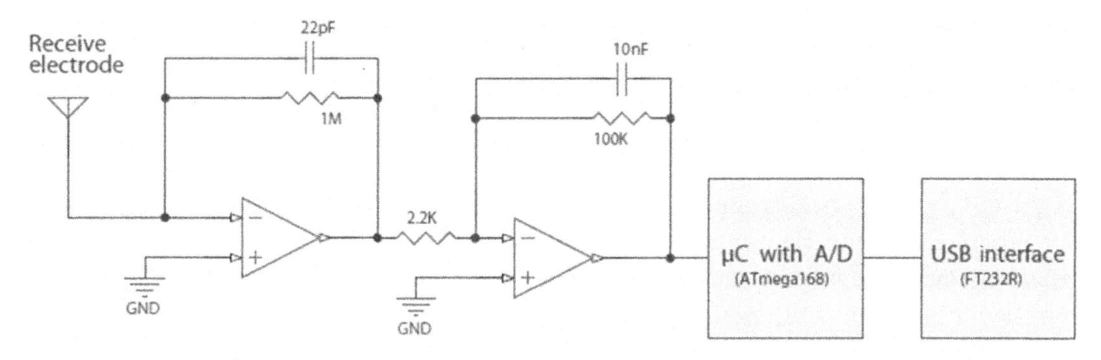

Figure 8
Schematic of circuit used to sense 60-Hz electric fields emitted from a standard electrical socket [15].

distance, so that leaves "holes" in the depth images. To fill the "holes," they performed a series of dilation and erosion operations. A. Herzog et al. [14] extracted the grasp height map from a 3D point cloud captured by Kinect. The height map was used to determine the properties of regions surrounding the object, i.e., surface, void, background, and occlusion.

5.1.1.4 Other Approaches

For autonomous charging, B. Mayton et al. [15] detected the sockets by sensing the electric field emitted from the sockets. They developed their own sensor electronics to identify electric signals (Figure 8).

There are some approaches that combine various sensors to accomplish target detection. E. Klingbeil et al. [16] first located the visual key points of the door and handle in 2D images. Then their 3D locations were obtained through a 2D laser scan (in the horizontal plane) along with a vertical-wall assumption. To achieve robust handle detection, W. Meeussen et al. [4] used two different methods for handle detection in parallel: laser-based and stereo-vision-based approaches. If both methods agree on the handle location, then it is a successful detection. M. J. Schuster et al. [17] proposed a method to distinguish surfaces from clutter so that the robot could successfully place objects on the cluttered surfaces. They collected data from a laser and a camera. After extracting the range and color features, they used an Ada Boost classifier to classify surfaces and clutter. Thereafter, they used a RANSAC plane-fitting algorithm to eliminate outlier points, which are classified into surfaces.

5.1.2 Planning

To accomplish grasping and manipulation, the planning requires two aspects: motion planning and grasping planning. Motion planning is used to execute a collision-free trajectory to reach the object or move the object. Grasping planning is used to grasp the object firmly without damage.

5.1.2.1 Motion Planning

For the task of door opening, in [16], the 3D features of the handle are used to geometrically compute 3D waypoints. These waypoints can be used to plan a path for the end effector. The waypoints are then transformed into a configuration space, and a probabilistic road map [18] motion planning algorithm is used to generate a smooth path. J. Sturm et al. [19] proposed an approach to predict the trajectory of the end effector based on previous experience (Figure 9). The kinematic model of an articulated object is estimated based on the trajectory of the end effector. Then this model is used to predict the future trajectory of the end effector. S. Chitta et al. [20] proposed a coordinated arm-base motion planning for door opening. This approach uses a graph-based representation of the task. Then a graph search algorithm, such as A*, can be used for motion planning.

For object placement, K. Harada et al. [21] first constructed a polygon model of the surrounding environment and then clustered the model. Then they determined the position and orientation of the object placed in the environment through several tests, such as the convexity test, the cluster inclusion test, and the stability test. When there is no continuous space large enough for direct placement, A. Cosgun et al. [22] presented a planner that finds a sequence of linear pushes that clears the necessary space (Figure 10).

K. Lakshmanan et al. [23] presented an algorithm for laundry folding. Their motion planning approach plans are directly at the level of robotic primitives. If a sequence of desired folds is given, the approach outputs a sequence of robotic primitives for laundry folding.

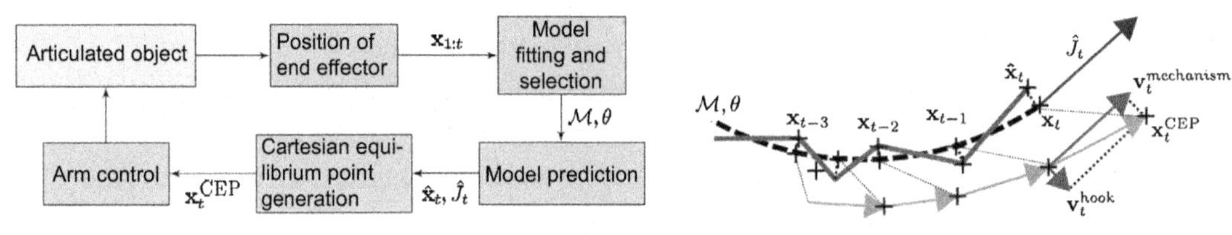

Figure 9

Left: Overall control structure. The robot observes the trajectory $x_{1:t}$ of its end effector. From that, it estimates the model M, θ and projects the current pose $\hat{x}_t$ onto the model and estimates the Jacobian $\hat{J}_t$ at this point. Using these estimates, it generates the next Cartesian equilibrium point x_t^{CEP}. Right: Example trajectory illustrated for the right cabinet door (top view). The generated CEP trajectory smoothly pulls the door open, while keeping the hook robustly on the handle [19].

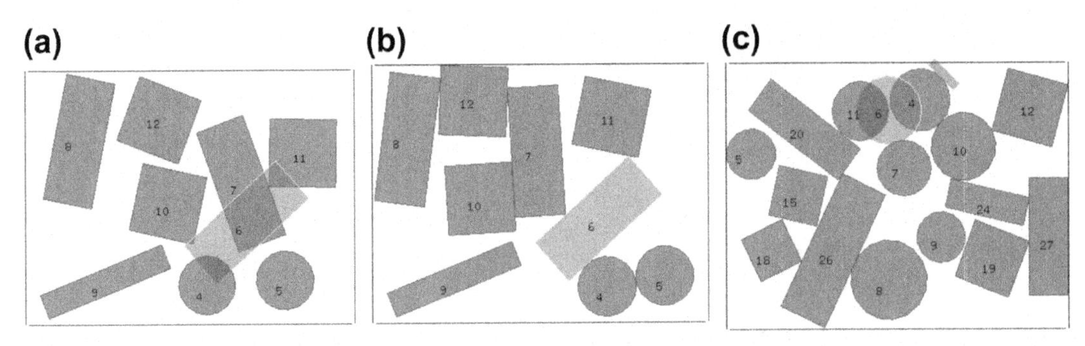

Figure 10

(a) A successful run from the all-indirectly pushable scenario. o_6 is the candidate goal footprint, table clutter is 39.5%. (b) Table configuration after plan execution on (a); 24 total pushes searched in 4.8 s resulting in a plan of four pushes. (c) A failed run from the half-indirectly pushable scenario. Clutter was 51.7%, total number of pushes searched was 2505, requiring 43.6 s of execution [22].

R. B. Rusuy et al. [24] proposed a sampling-based motion planner, which was also used in [7]. The principle of sampling-based motion planning is to sample the state space of the robot and maintain a data structure of the samples. This data structure is then used to guide the robot to the goal. They first established the dynamic obstacle map. Then they used a sampling-based motion planner called kinematic motion planning by interior–exterior cell exploration [25]. I. A. Sucan et al. [26] proposed a data structure called task motion multigraphs for a sampling-based motion planner. V. Sukhoy et al. [27] proposed an approach to slide a magnetic card through a card reader. This approach can detect motion constraints in real-time to avoid bending the card too much. H. Dang et al. [28] proposed learning from a human demonstration approach to manipulate everyday objects. The robot first decomposes a demonstrated task into sequential manipulation primitives. Then it constructs an ask descriptor for further manipulation.

5.1.2.2 Grasping Planning

N. Hudson et al. [29] separated grasping planning into free-space grasping and table grasping (Figure 11). The free-space grasping planner is similar to "GraspIt!" [30,31]. The table grasping planner uses a reactive controller to exploit the geometry of the table by placing the fingers in a special configuration. Then the robot grasps the object. F. Stulp et al. [32] presented an approach that enables the robot to acquire movement primitives, which are implemented as dynamic movement primitives [33]. The probabilistic model-free reinforcement learning algorithm (policy improvement with path integrals—PI^2 [34]) is then used to optimize the chance of grasping an object.

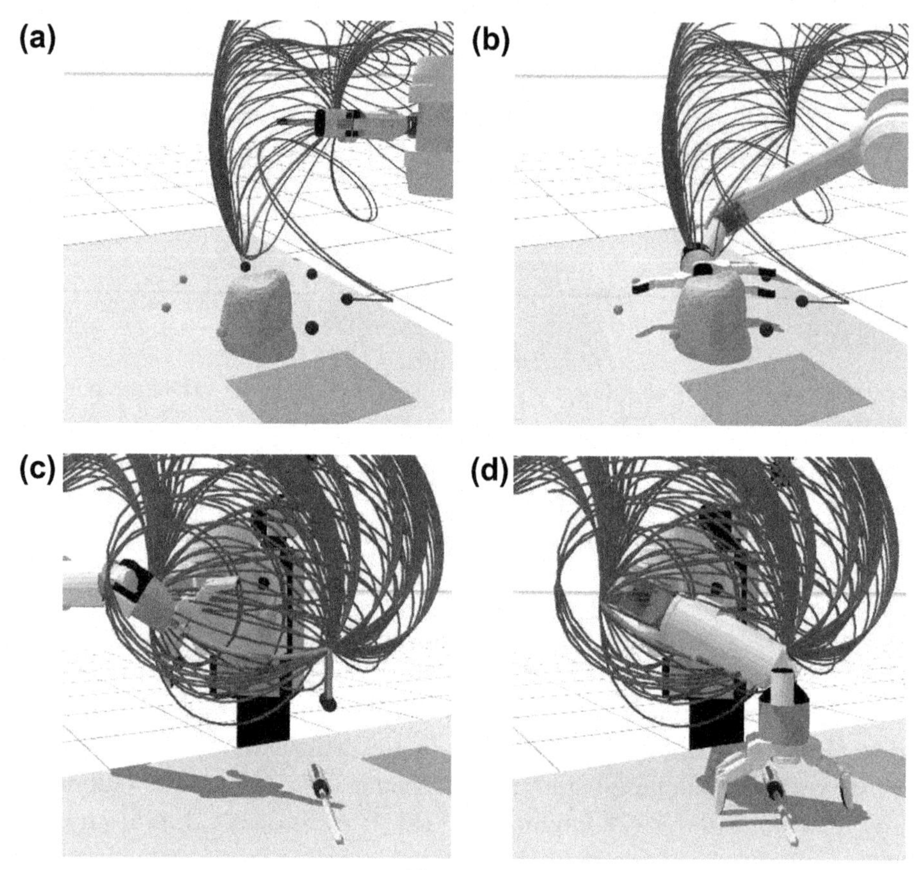

Figure 11

Examples of grasp sets for two different objects. Free-space grasp sets use the object geometry to identify candidate manipulator configurations. Table grasp sets alternatively use the table and the object geometry to determine possible wrist configurations. Collision-free wrist poses with valid inverse kinematic solutions for the arm are shown as blue spheres. Invalid wrist poses are shown as orange spheres [29]. (a) Free-space grasp set generation for a rock (premotion). (b) Free-space grasp set generation for a rock (post-motion). (c) Table grasp set generation for a screwdriver (pre-motion). (d) Table grasp for a screwdriver (post-motion).

H. Dang et al. [28] proposed a task-oriented grasp planning. They used a 6 degrees of freedom pose sensor to track the positions of a human wrist during the learning phase. Then an *eigengrasp* grasp planner [31] is used to solve the kinematics of the robotic hand. J. Stueckler et al. [12] developed flexible grasping from either the side of or above the object. The planned grasps are executed using parameterized motion primitives.

5.1.3 Control

Control strategies can be classified into three categories: (1) position control, (2) force control, and (3) compliant control, such as impedance control or hybrid position/force control.

5.1.3.1 Position Control

H. Arisumi et al. [35] used position control to manipulate a knob and also a swing door. A. A. Moughlbay et al. [36] used model-based tracking techniques to apply 3D visual serving on both localization and manipulation tasks.

5.1.3.2 Force Control

R. Leontie et al. [37] used a proprioceptive force feedback control to balance the load on a two-armed robot. H. Dang et al. [38] proposed a method to deal with stable grasping under pose uncertainty using tactile sensing data. A. Jain et al. [39] proposed a model of predictive control only depending on tactile sensing. No prior knowledge of an explicit model of the environment is required.

5.1.3.3 Compliant Control

A. Jain et al. [2,40,41] used an implementation of impedance control, equilibrium point control, to manipulate everyday objects, such as to open doors (Figure 12). T. Winiarski et al. [42] used a redundant robot arm with an impedance control to open a cabinet door.

W. Chung et al. [43] proposed a hybrid position and force control strategy for a multifingered robot hand to open doors (Figure 13). J. M. Romano et al. [44] proposed

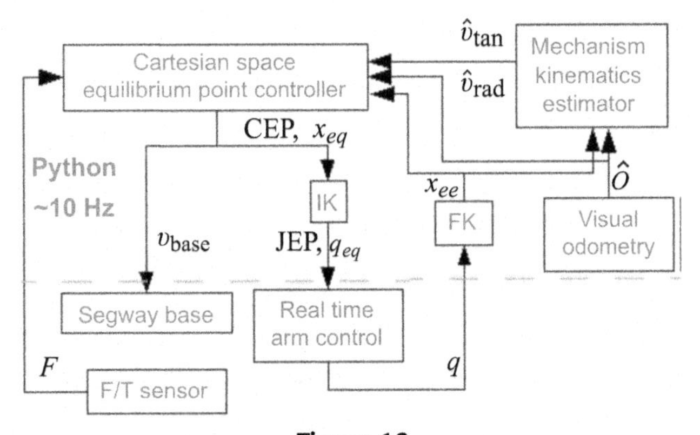

Figure 12
Figure showing the overall control structure [40].

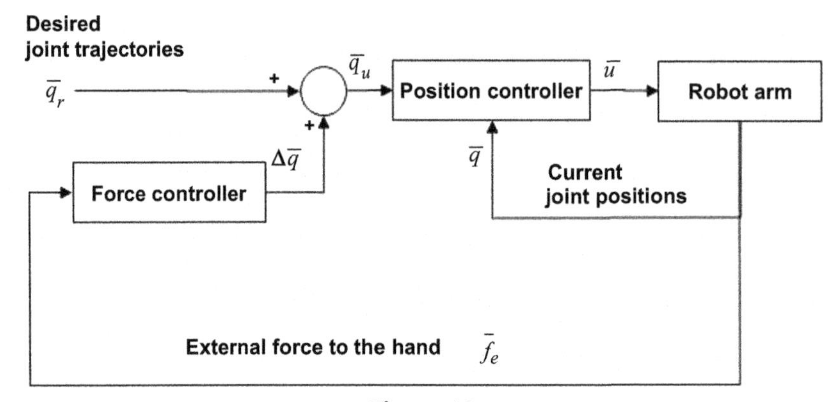

Figure 13
Block diagram of hybrid force/position control strategy [43].

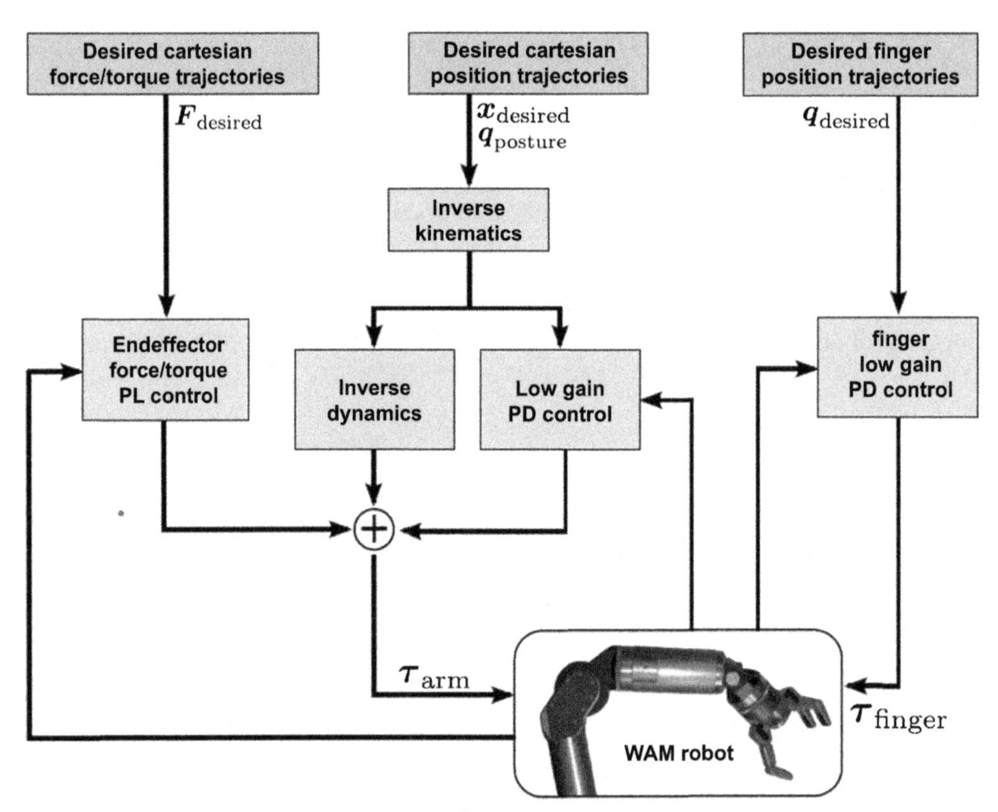

Figure 14
Overview of the controllers used in the experiments [45].

a position and force control strategy for a grasping task. M. Kalakrishnan et al. [45] used the PI^2 algorithm to learn force/torque profiles. These force/torque profiles are controlled in combination with the controlled position trajectories (Figure 14).

5.1.4 Summary of Case Studies

The remainder of this chapter elaborates three case studies of grasping and manipulation for household service.

- Laundry folding
  S. Miller et al. [46] presented an algorithm for laundry folding. The core of the algorithm is a quasi-static cloth model. The model helps to avoid the complex dynamics of cloth under significant parts of the state space. If a 2D cloth polygon and a desired sequence of folds are given, the algorithm outputs a motion plan for a deemed g-fold at the level of the gripper motion. They implemented their algorithm on a Willow Garage PR2. The robot first identifies a clothing article spread out on a table, using a model-based optimization approach. Then it executes the computed folding sequence, which is defined for four clothing categories: towels, pants, short-sleeved shirts, and long-sleeved shirts. Meanwhile, the robot can visually track its progress over successive folds.
- Visual servoing
  D. Kragic et al. [47] used computational vision to manipulate objects in everyday settings. The process includes recognition/detection, transportation, and grasping. The robot first recognizes the object, e.g., a cup. Then the authors used coarse visual servoing to control the robot to the vicinity of the cup. The robot needs to control the motion of both the platform and the arm to keep the cup in the field of view, while approaching the object. Finally, the authors used model-based visual servoing to estimate the pose of the object and servo to the object to allow grasping.
- Cognitive control
  K. Kawamura et al. [48] used a multi-agent-based cognitive control to develop the next generation of robots with robust sensorimotor intelligence. The authors elaborated the progress made on cognitive architecture and control, working memory training, and a self-motivated, internal state-based action selection mechanism.

References

[1] A. Jain, C.C. Kemp, Behaviors for robust door opening and doorway traversal with a force-sensing mobile manipulator, in: RSS Workshop on Robot Manipulation: Intelligence in Human Environments, 2008.

[2] A. Jain, C.C. Kemp, Behavior-based door opening with equilibrium point control, in: RSS 2009 Workshop: Mobile Manipulation in Human Environments, 2009.

[3] R.B. Rusu, W. Meeussen, S. Chitta, M. Beetz, Laser-based perception for door and handle identification, in: Proceedings of the International Conference on Advanced Robotics, June 2009.

[4] W. Meeussen, M. Wise, S. Glaser, S. Chitta, C. McGann, P. Mihelich, E.M. Eppstein, M. Muja, V. Eruhimov, T. Foote, J. Hsu, R.B. Rusu, B. Marthi, G. Bradski, K. Konolige, B. Gerkey, E. Berger, Autonomous door opening and plugging in with a personal robot, in: Proceedings of IEEE International Conference on Robotics and Automation, 729–736, Anchorage, Alaska, USA, May 3–8, 2010.

[5] M. Bansal, B. Southall, B. Matei, J. Eledath, H. Sawhney, Lidar-based door and stair detection from a mobile robot, in: Proceedings of SPIE 7692, May 7, 2010. Unmanned Systems Technology XII.

[6] M.B. Ansal, B. Matei, B. Southall, J. Eledath, H. Sawhney, A lidar streaming architecture for Mobile robotics with application to 3D structure characterization, in: Proceedings of IEEE International Conference on Robotics and Automation, 1803–1810, Shanghai, China, May 9–13, 2011.

[7] J. Bohren, R.B. Rusu, E.G. Jones, E.M. Eppstein, C. Pantofaru, M. Wise, L. Mösenlechner, W. Meeussen, S. Holzer, Towards autonomous robotic butlers: lessons learned with the PR2, in: Proceedings of IEEE International Conference on Robotics and Automation, 5568–5575, Shanghai, China, May 9–13, 2011.

[8] J. Sturm, K. Konolige, C. Stachniss, W. Burgard, Vision-based detection for learning articulation models of cabinet doors and drawers in household environments, in: Proceedings of IEEE International Conference on Robotics and Automation, 362–368, Anchorage, Alaska, USA, May 3–8, 2010.

[9] N. Kwak, H. Arisumi, K. Yokoi, Visual recognition of a door and its knob for a humanoid robot, in: Proceedings of IEEE International Conference on Robotics and Automation, 2079–2084, Shanghai, China, May 9–13, 2011.

[10] D. Ignakov, G. Okouneva, G. Liu, Localization of a door handle of unknown geometry using a single camera for door-opening with a mobile manipulator, Auton. Robot. 33 (4) (November 2012) 415–426.

[11] A. Saxena, J. Driemeyer, A.Y. Ng, Robotic grasping of novel objects using vision, Int. J. Robot. Res. 27 (2) (February 2008) 157–173.

[12] J. Stueckler, R. Steffens, D. Holz, S. Behnke, Real-time 3D perception and efficient grasp planning for everyday manipulation tasks, in: Proceedings of the European Conference on Mobile Robots (ECMR), 177–182, Örebro, Sweden, September 2011.

[13] T. Rühr, J. Sturm, D. Pangercic, M. Beetz, D. Cremers, A generalized framework for opening doors and drawers in kitchen environments, in: Proceedings of IEEE International Conference on Robotics and Automation, 3852–3858, Saint Paul, Minnesota, USA, May 14–18, 2012.

[14] A. Herzog, P. Pastor, M. Kalakrishnan, L. Righetti, J. Bohg, T. Asfour, S. Schaal, Learning of grasp selection based on shape-templates, Auton. Robot 36 (1–2) (January 2014) 51–65.

[15] B. Mayton, L. LeGrand, J.R. Smith, Robot, feed thyself: plugging in to unmodified electrical outlets by sensing emitted AC electric fields, in: Proceedings of IEEE International Conference on Robotics and Automation, 715–722, Anchorage, Alaska, USA, May 3–8, 2010.

[16] E. Klingbeil, A. Saxena, A.Y. Ng, Learning to open new doors, in: Proceedings of IEEE/RSJ International Conference on Intelligent Robots and Systems, 2751–2757, Taipei, Taiwan, October 18–22, 2010.

[17] M.J. Schuster, J. Okerman, H. Nguyen, J.M. Rehg, C.C. Kemp, Perceiving clutter and surfaces for object placement in indoor environments, in: Proceedings of IEEE-ras International Conference on Humanoid Robots, 152–159, Nashville, TN, USA, December 6–8, 2010.

[18] F. Schwarzer, M. Saha, J.-C. Latombe, Adaptive dynamic collision checking for single and multiple articulated robots in complex environments, IEEE T. Robot. 21 (3) (2005) 338–353.

[19] J. Sturm, A. Jain, C. Stachniss, C.C. Kemp, W. Burgard, Operating articulated objects based on experience, in: Proceedings of IEEE/RSJ International Conference on Intelligent Robots and Systems, 2739–2744, Taipei, Taiwan, October 18–22, 2010.

[20] S. Chitta, B. Coheny, M. Likhachev, Planning for autonomous door opening with a mobile manipulator, in: Proceedings of International Conference on Robotics and Automation, 1799–1806, Anchorage, Alaska, USA, May 3–8, 2010.

[21] K. Harada, T. Tsuji, K. Nagata, N. Yamanobe, H. Onda, T. Yoshimi, Y. Kawai, Object placement planner for robotic pick and place tasks, in: Proceedings of IEEE/RSJ International Conference on Intelligent Robots and Systems, 980–985, Vilamoura, Algarve, Portugal, October 7–12, 2012.

[22] A. Cosgun, T. Hermans, V. Emeli, M. Stilman, Push planning for object placement on cluttered table surfaces, in: Proceedings of IEEE/RSJ International Conference on Intelligent Robots and Systems, 4627–4632, San Francisco, CA, USA, September 25–30, 2011.

[23] K. Lakshmanan, A. Sachdev, Z. Xie, D. Berenson, K. Goldberg, P. Abbeel, A constraint-aware motion planning algorithm for robotic folding of clothes, in: Experimental Robotics, Springer Tracts in Advanced Robotics, vol. 88, 2013, pp. 547–562.

[24] R.B. Rusuy, I.A. Sucan, B. Gerkeyz, S. Chittaz, M. Beetzy, L.E. Kavraki, Real-time perception-guided motion planning for a personal robot, in: Proceedings of IEEE/RSJ International Conference on Intelligent Robots and Systems, 4245–4252, St. Louis, USA, October 11–15, 2009.

[25] I.A. Sucan, L.E. Kavraki, Kinodynamic motion planning by interior-exterior cell exploration, in: Proceedings of International Workshop on the Algorithmic Foundations of Robotics, Guanajuato, Mexico, December 2008.

[26] I.A. Sucan, L.E. Kavraki, Mobile manipulation: encoding motion planning Options using task motion multigraphs, in: Proceedings of IEEE International Conference on Robotics and Automation, 5492–5498, Shanghai, China, May 9–13, 2011.

[27] V. Sukhoy, V. Georgiev, T. Wegter, R. Sweidan, A. Stoytchev, Learning to slide a magnetic card through a card reader, in: Proceedings of IEEE International Conference on Robotics and Automation, 2398–2404, Saint Paul, Minnesota, USA, May 14–18, 2012.

[28] H. Dang, P.K. Allen, Robot learning of everyday object manipulations via human demonstration, in: Proceedings of IEEE/RSJ International Conference on Intelligent Robots and Systems, 1284–1289, Taipei, Taiwan, October 18–22, 2010.

[29] N. Hudson, T. Howard, J. Ma, A. Jain, M. Bajracharya, S. Myint, C. Kuo, L. Matthies, P. Backes, P. Hebert, T. Fuchs, J. Burdick, End-to-end dexterous manipulation with deliberate interactive estimation, in: Proceedings of IEEE International Conference on Robotics and Automation, 2371–2378, Saint Paul, Minnesota, USA, May 14–18, 2012.

[30] A. Miller, S. Knoop, H. Christensen, P. Allen, Automatic grasp planning using shape primitives, in: Proceedings of International Conference on Robotics and Automation, 1824–1829, Taipei, Taiwan, September 14–19, 2003.

[31] A. Miller, P. Allen, Graspit! A versatile simulator for robotic grasping, IEEE Robot. Autom. Mag. 11 (4) (2004) 110–122.

[32] F. Stulp, E. Theodorou, J. Buchli, S. Schaal, Learning to grasp under uncertainty, in: Proceedings of IEEE International Conference on Robotics and Automation, 5703–5708, Shanghai, China, May 9–13, 2011.

[33] A. Ijspeert, J. Nakanishi, S. Schaal, Movement imitation with nonlinear dynamical systems in humanoid robots, in: Proceedings of IEEE International Conference on Robotics and Automation, 1398–1403,Washington, DC, May 11–15, 2002.

[34] E. Theodorou, J. Buchli, S. Schaal, A generalized path integral approach to reinforcement learning, J. Mach. Learn. Res. 11 (November 2010) 3137–3181.

[35] H. Arisumi, N. Kwak, K. Yokoi, Systematic touch scheme for a humanoid robot to grasp a door knob, in: Proceedings of IEEE International Conference on Robotics and Automation, 3324–3331, Shanghai, China, May 9–13, 2011.

[36] A. A. Moughlbay, E. Cervera, P. Martinet, Real-time model based visual servoing tasks on a humanoid robot, in: Intelligent Autonomous Systems 12, Advances in Intelligent Systems and Computing, vol. 193, 2013, pp. 321–333.

[37] R. Leontie, E. Drumwright, D.A. Shell, R. Simha, Load equalization on a two-armed robot via proprioceptive sensing, in: Experimental Robotics, Springer Tracts in Advanced Robotics, vol. 88, 2013, pp. 499–513.

[38] H. Dang, P.K. Allen, Stable grasping under pose uncertainty using tactile feedback, Auton. Robot. 32 (4) (April 2013) 458–482.

[39] A. Jain, M.D. Killpack, A. Edsinger, C. C Kemp, Reaching in clutter with whole-arm tactile sensing, Int. J. Robot. Res. 32 (4) (2013) 458–482.

[40] A. Jain, C.C. Kemp, Pulling open doors and drawers: coordinating an omni-directional base and a compliant arm with equilibrium point control, in: Proceedings of International Conference on Robotics and Automation, 1807–1814, Anchorage, Alaska, USA, May 3–8, 2010.
[41] A. Jain, C.C. Kemp, Improving robot manipulation with data-driven object-centric models of everyday forces, Auton. Robot. 35 (2–3) (2013) 143–159.
[42] T. Winiarski, K. Banachowicz, Opening a door with a redundant impedance controlled robot, in: Proceedings of International Workshop on Robot Motion and Control, 221–226,Wasowo, Poland, July 3–5, 2013.
[43] W. Chung, C. Rhee, Y. Shim, H. Lee, S. Park, Door-opening control of a service robot using the multifingered robot hand, IEEE Trans. Ind. Electron. 56 (10) (2009) 3975–3984.
[44] J.M. Romano, K. Hsiao, G. Niemeyer, S. Chitta, K.J. Kuchenbecker, Human-inspired robotic grasp control with tactile sensing, IEEE Trans. Robot. 27 (6) (2011) 1067–1079.
[45] M. Kalakrishnan, L. Righetti, P. Pastor, S. Schaal, Learning force control policies for compliant manipulation, in: Proceedings of IEEE/RSJ International Conference on Intelligent Robots and Systems, 4639–4644, San Francisco, CA, USA, September 25–30, 2011.
[46] S. Miller, J. van den Berg, M. Fritz, T. Darrell, K. Goldberg, P. Abbeel, A geometric approach to robotic laundry folding, Int. J. Robot. Res. 31 (2) (2011) 249–267.
[47] D. Kragic, H.I. Christensen, Robust visual servoing, Int. J. Robot. Res. 22 (10) (2003) 923–939.
[48] K. Kawamura, S. Gordon, P. Ratanaswasd, C. Garber, E. Erdemir, Implementation of cognitive control for robots, in: Proceedings of the 4th COE Workshop on Human Adaptive Mechatronics (HAM), 2007.

CHAPTER 5.2

A Geometric Approach to Robotic Laundry Folding[1]

Stephen Miller[1], Jur van den Berg[2], Mario Fritz[3], Trevor Darrell[1], Ken Goldberg[1], Pieter Abbeel[1]

[1]*Electrical Engineering and Computer Science, College of Engineering, University of California, Berkeley, CA, USA;* [2]*School of Computing, University of Utah, Salt Lake City, UT, USA;* [3]*Max-Planck Institute for Informatics, Saarbrücken, Germany*

Chapter Outline

---

[1] Stephen Miller, Jur van den Berg, Mario Fritz, Trevor Darrell, Ken Goldberg and Pieter Abbeel, International Journal of Robotics Research, Vol. 31, No. 2, pp. 249–267, copyright © 2013 by SAGE Publications, reprinted by Permission of SAGE.

Household Service Robotics. http://dx.doi.org/10.1016/B978-0-12-800881-2.00017-7

5.2.1 Introduction

With the twentieth century advent of personal computers, the dream of the future was one of convenience and autonomy: of intelligent machines automating the monotonous rotes of daily life.

Few tedious tasks are as universal to the human experience as household chores. No utopian future would be complete, then, without household robots relieving humans of these tasks: doing the dishes, sweeping the floors, setting the table, and doing the laundry. In this chapter, we explore the latter challenge. Washing machines and dryers have automated much of the process, but one clear bottleneck remains: autonomous laundry folding.

While advances continue to be made in the field of household robotics, this vision has proven particularly difficult to realize. This is largely due to the vast state space in which such a robot is expected to operate. In addition to the well-established problems of perception and manipulation in unconstrained environments, the task of laundry folding poses a unique challenge as not only the environment, but the very object which must be manipulated, is itself highly complex. Cloth is non-rigid, flexible, and deformable, with an infinite-dimensional configuration space. It may be found in an innumerable variety of poses, rendering the perceptual tasks of classification and pose-estimation extremely difficult. Furthermore, the dynamics of cloth are difficult to capture in even the most sophisticated simulators, posing great challenges to the manipulation planning task.

We do not, in this work, intend to provide a brute-force mechanism for planning under such complexity. Rather, we aim to simplify the problem by carving out a particular subset of cloth configurations which, under a number of governing assumptions, may be represented by a few parameters while retaining predictable behavior during robotic interaction. In so doing, we build on the results of fellow researchers, Balkcom and Mason [1] and Bell [2].

In particular, we exploit the use of gravity by considering clothing articles which may be separated into at most two parts: one part which lies flat on a horizontal surface and one (possibly empty) which hangs vertically from the robot's grippers parallel to the gravity vector, separated by a single line which we deem the *baseline*. In so doing, we replace an infinitely dense mesh with a finite-sided polygon, and reduce the recognition and planning tasks to shape-fitting and two-dimensional geometry, respectively.

The work that follows considers three questions:

1. What is this simplified model, and under what assumptions is it a valid approximation for cloth behavior?
2. Given a cloth polygon, how can a robot be made to execute folds on it?
3. Given a single image of a clothing article, how can such a polygonal representation be extracted?

The remainder of this chapter is organized as follows. In Section 5.2.2 we discuss work related to both the manipulation and detection of deformable objects. In Section 5.2.3 we define the folding problem, and present a subset of cloth configurations under which the folding task becomes purely geometric. We additionally discuss the assumptions under which our predictions may reasonably hold. In Section 5.2.4 we describe an algorithm to compute the manipulation necessary to allow a robot to execute a folding sequence, using a set of primitives deemed *g-folds*. In Section 5.2.5 we devise a perceptual scheme for extracting, from a single image of a clothing article, a category-level polygonal representation on which the above algorithm may act; one which both classifies the presented article, and projects its structure into a simpler polygon which may be more easily folded. In Section 5.2.6 we show the experimental results of our approach, both in folding clothing articles when the category and cloth polygon are known, and in inferring this knowledge visually when unknown. We combine these tools to implement a complete folding system on a Willow Garage PR2 robot. We conclude in Section 5.2.7.

This chapter brings together two previously separate bodies of work. The g-fold formalism and subsequent motion planning strategies were first presented by Berg et al. [3]. The task of classifying and recognizing the pose of clothing was done by Miller et al. [4]. We now present, for the first time, our complete framework for robotic laundry folding (Figure 1).

5.2.2 Related Work

5.2.2.1 Manipulation

In the work of Bell and Balkcom [5], grasp points necessary to immobilize a polygonal non-stretchable piece of cloth are computed. Gravity is used by Bell [2] to reduce the number of grasp points required to hold cloth in a predictable configuration, potentially with a single fold, to two grippers. We extend this work and include a folding surface. We

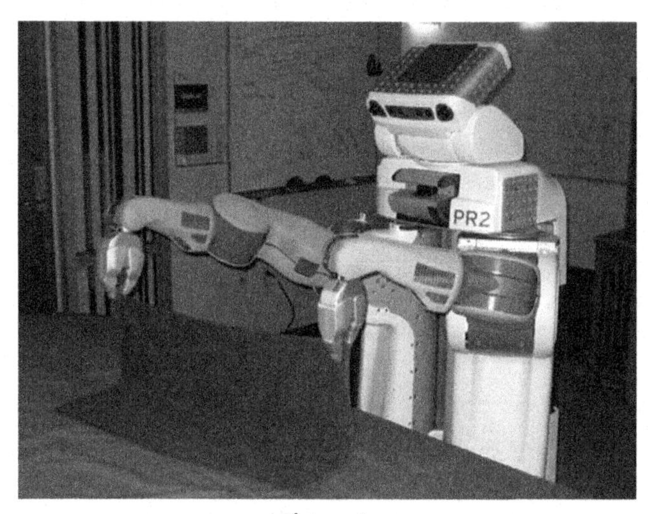

Figure 1
The PR2 robotic platform (developed by Willow Garage) performing a g-fold on a towel.

assume that points that are lying on a table are fixed by friction and gravity, and need not be grasped. Bell's work also demonstrates how to fold a T-shirt using the Japanese method[2]; this fold can be achieved by grasping the cloth at three points without re-grasping.

Fahantidis et al. [6] discuss robotic handling of cloth material with application to a number of specific folds. The work of Osawa et al. [7] also discusses a specific folding manipulation. The work of Maitin-Shepard et al. [8] deals specifically with folding towels. This work focuses on visual detection of the vertices of the towel, and uses a scripted motion to achieve folds using a PR-2 robot. We build on the results of this work in our experiments.

Some prior work also describes robots using tools and the design of special purpose end-effectors as a step toward laundry folding. For example, Osawa et al. [7] developed a robot capable of using a "flip-fold" for folding and a plate for straightening out wrinkles. Salleh et al. [9] present an inchworm gripper for tracing the edge of a piece of clothing. A range of gripper designs is presented by Monkman [10].

There is also quite a large body of work on *cloth simulation*, which simulates the behavior of cloth under manipulation forces using the laws of physics, including those of Baraff and Witkin [11], Bridson et al. [12], and Choi and Ko [13]. In our work, we manipulate cloth such that it behaves quasi-statically, allowing us to reason about the geometry of the cloth, while avoiding complex physics or dynamics.

[2] As popularized by the video "Japanese way of folding T-shirts!" at http://www.youtube.com/watch?v=b5AWq5aBjgE. Original footage 2006, uploaded to YouTube 2010.

Folding has been extensively studied in the context of *origami*. Balkcom and Mason [1] consider a model of paper where unfolded regions are considered to be rigid facets connected by creases which form "hinges," and detail a folding procedure which respects the assumptions of this model. Our approach is similar, in that we also consider a subset of the configuration space where the dynamics are simpler. However, unlike the hinge model, our cloth model assumes full flexibility and requires no bending energy. While this assumption yields a notably different fold manipulation, the formalism of the fold lines themselves is similar, and we draw from results in paper folding in our own work. Applications of paper folding outside origami include box folding as presented by Liu and Dai [14] and metal bending as presented by Gupta et al. [15], where the material model is essentially the same as that of paper.

5.2.2.2 Perception

Estimating the configuration of a clothing article can be seen as an instance of an articulated pose estimation task. Classic articulated pose estimation methods iteratively fit or track an articulated model, updating the pose of individual part segments subject to the overall body constraints. Early methods such as those of Bregler and Malik [16] were based on optic flow and linearized exponential map-based constraints; subsequent approaches include the efficient sampling methods of Sidenbladh et al. [17], exemplar methods of Demirdjian et al. [18], regression strategies of Urtasun and Darrell [19], and subspace optimization methods of Salzmann and Urtasun [20]. Borgefors [21] used an energy-optimization strategy to match edge points between two images.

Related models for fully non-rigid shape modeling and estimation are typically based on a learned manifold, e.g., active appearance models as proposed by Cootes and Taylor [22]. Few methods investigate clothing explicitly. Notable exceptions to this are the recent work of Guan et al. [23], which expands the SCAPE manifold-based model of Anguelov et al. [24] to include a model of three-dimensional clothing forms; the work in person tracking systems by Rosenhahn et al. [25], which attempts to account for clothing variation, and methods for estimating folds in deformable surfaces proposed by Salzmann and Fua [26]. In this work we adopt a much simpler model, and propose schemes for direct optimization of specific shape forms that can be directly related to ensuing manipulation.

Fahantidis et al. [6] describe the isolated executions of grasping a spread-out material, folding a spread-out material, laying out a piece of material that was already being held, and flattening wrinkles. Their perception system relies on a library of exact, polygonal shape models of the instances considered and then matches the sequence of extracted edges.

There is a body of work on recognizing categories of clothing. For example, Osawa et al. [27], and Hamajima and Kakikura [28] present approaches to spreading out a piece of clothing using two robot arms and then classifying its category.

Yamakazi and Inaba [29] present an algorithm that recognizes wrinkles in images, which in turn enables them to detect clothes in a scene. Kobori et al. [30] have extended this attening and spreading clothing. Kiflattening and spreading clothing. Kita et al. [31] fit the geometry of the silhouette of a hanging piece of clothing to the geometry of a mass spring model of the same piece of clothing and are able to infer some three-dimensional information about the piece of clothing merely from its silhouette.

5.2.3 Problem Description

Let us begin with a description of the folding task. We assume gravity is acting in the downward vertical ($-z$) direction and a sufficiently large planar table in the horizontal (xy) plane. We assume the article of clothing can be fully described by a simple *polygon* (convex or non-convex) initially lying on the horizontal surface. We are given the initial n vertices of the polygonal cloth in a counterclockwise order.

We make the following assumptions on the cloth material:

1. The cloth has infinite flexibility. There is no energy contribution from bending.
2. The cloth is non-stretchable. No geodesic path lengths can be increased.
3. The cloth has no slip between either the surface on which it lies or itself.
4. The cloth has zero thickness.
5. The cloth is subject to gravity.
6. The behavior of the cloth is quasi-static: the effects of inertia are negligible.
   At the core of our approach is the following additional assumption, which we call the *downward tendency assumption*:
7. If the cloth is "released" from any gripped state, no point of the cloth will ever move upwards as a result of only gravity and internal forces within the cloth.[3]

The above assumptions do not directly follow from physics, rather they are an approximation which seems to match the behavior of reasonably shaped cloth, such as everyday clothing articles, surprisingly well, allowing us to reason purely about the geometry of the cloth: the state space consists of just configurations, and cloth motion is readily determined from hand motion.

The downward-tendency assumption allows the cloth to be held by the grippers such that one section lies horizontally on the surface and another section hangs vertically. The line that separates the horizontal and the vertical parts is called the *baseline*. To ensure deterministic behavior of the cloth, the grippers must be arranged such that the vertical section does not deform, i.e., such that it does not change its shape with respect to the

[3] While this assumption tends to hold for typical shapes, it is not always true. An example of where the assumption is not accurate is for an exotic family of shapes called pinwheels, as is proven by Bell [2].

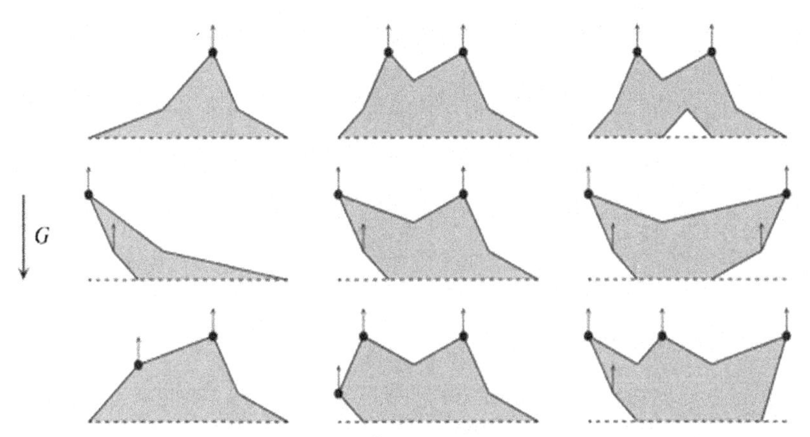

Figure 2

Examples of vertical parts of cloths in various configurations. In order for the cloth not to deform, all convex vertices not at the baseline at which the negative gravity vector (small arrows) does not point into the cloth must be grasped. These vertices are indicated by the dots.

original (potentially stacked) geometry. The points that are lying on the surface (including those on the baseline) are immobilized, as they cannot move in the plane due to friction and will not move upward per the downward-tendency assumption, so they need not be grasped. Figure 2 shows an example, where points of the cloth are held by grippers.

To ensure that the vertical part of the cloth does not deform, we employ the following theorem:

Theorem 1. A vertically hanging cloth polygon is immobilized when every convex vertex of the cloth at which the negative gravity vector does not point into the cloth polygon is fixed (i.e., be held by a gripper or be part of the baseline).

Proof. See Appendix A.

A *g-fold* (*g* refers to gravity) is specified by a directed line segment in the plane that partitions the polygon into two parts, one to be folded over another. A g-fold is successfully achieved when the part of the polygon to the *left* of the directed line segment is folded across the line segment and placed horizontally on top of the other part, while maintaining the following property:

- At all times during a folding procedure, every part of the cloth is either horizontal or vertical, and the grippers hold points on the vertical part such that it does not deform (see Figure 3).[4]

[4] Not all folds can be achieved using a g-fold. In terms of origami, it is only possible to execute a *valley fold* on an upright portion of cloth, or a *mountain fold* on a flipped portion. While this renders many folds possible, more complex ones, such as *reverse folds*, cannot be done in this way. See Balkcom and Mason [1].

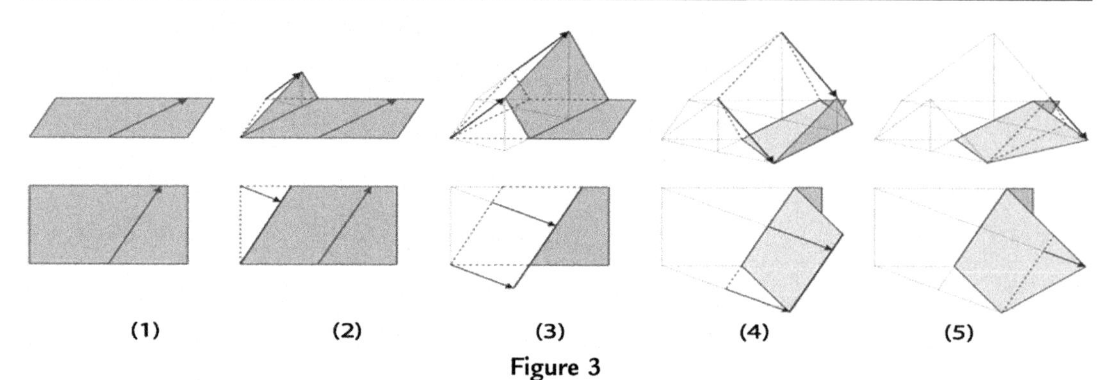

Figure 3

The motion of two grippers (arrows) successfully performing the first g-fold specified in Figure 5(a) shown both in a three-dimensional view and top view. At all times, all parts of the cloth are either vertical or horizontal and the cloth does not deform during the manipulation. The boundary between the vertical part and the horizontal part of the cloth is called the *baseline*.

This ensures that the cloth is in a fully predictable configuration according to our material model at all times during the folding procedure.

A *g-fold sequence* is a sequence of g-folds as illustrated in Figure 4. After the initial g-fold, the *stacked* geometry of cloth allows us to specify two types of g-fold: a "red" g-fold and a "blue" g-fold. A blue g-fold is specified by a line segment partitioning the polygon formed by the *silhouette* of the stacked geometry into two parts, and is successfully achieved by folding the (entire) geometry left of the line segment. A red g-fold is similarly specified, but only applies to the geometry that was folded in the previous g-fold (see Figure 5).

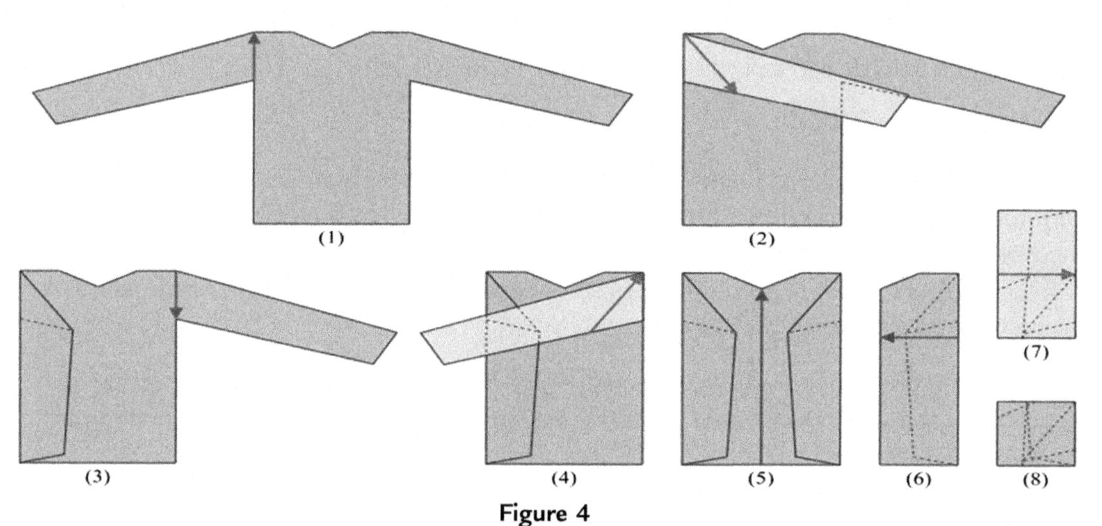

Figure 4

Folding a long sleeve into a square using a sequence of seven g-folds. "Red" g-folds apply to the geometry that was folded in the preceding g-fold. "Blue" g-folds apply to the entire stacked geometry. (For interpretation of the references to color in this figure legend, the reader is referred to the online version of this book.)

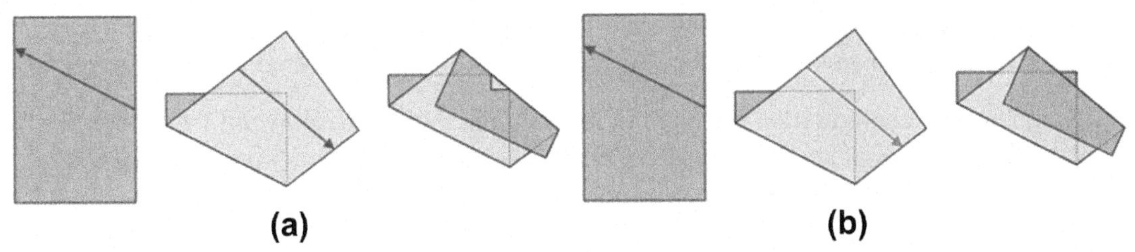

Figure 5

(a) A g-fold is specified by a directed line segment partitioning the (stacked) geometry into two parts. The g-fold is successfully achieved when the part of the geometry left of the line segment is folded around the line segment. A sequence of two g-folds is shown here. (b) A g-fold sequence similar to (a), but the second g-fold (a red g-fold) is specified such that it only applies to the part of the cloth that was folded in the previous g-fold. (For interpretation of the references to color in this figure legend, the reader is referred to the online version of this book.)

We are given a robot with k point grippers that can grasp the cloth at any point on the *boundary* of the polygon formed by the silhouette of the stacked geometry. At each such point, the gripper will grasp all layers of the stack at that point (i.e., it is not capable of distinguishing between layers). Each of the grippers is able to move independently above the xy-plane and we assume that gripper motion is exact.

The problem we discuss in this chapter is then defined as follows. Given a specification of a sequence of g-folds, determine whether each of the folds are feasible given the constraints and, if so, compute the number of grippers needed and the manipulation motion for each of the grippers to achieve the g-folds.

5.2.4 Fold Execution

In this section, we describe the algorithm that addresses the problem as formulated above. We first discuss single g-folds on unstacked geometry (Section 5.2.4.1) and then sequences of g-folds and stacked geometry (Section 5.2.4.2).

5.2.4.1 Single g-folds on Unstacked Geometry

Here we discuss the case of performing a single g-fold of the original (unstacked) polygon. During the manipulation, the cloth must be separated in a vertical part and a horizontal part at all times. The line separating the vertical part and the horizontal part is called the *baseline*.

Given a polygonal cloth and a specification of a g-fold by a directed line segment (e.g., the first g-fold of Figure 5(a)), we plan the manipulation as follows. The manipulation consists of two phases. In the first phase, the part of the cloth that needs to be folded is brought vertical above the line segment specifying the g-fold (see Figure 3(a)–(c)).

In the second phase, the g-fold is completed by manipulating the cloth such that the vertical part is laid down on the surface with its original normal reversed (Figure 3(c)–(e)).

Let us look at the configuration the cloth is in when the part of the polygon left of the line segment is fully vertical above the line segment (see Figure 3(c)). Each convex vertex at which the negative gravity vector does not point into the cloth must be grasped by a gripper. This set of vertices can be determined in $O(n)$ time, if n is the number of vertices of the cloth. We show here that the entire g-fold can be performed by grasping only vertices in this set.

The first phase of the g-fold is bringing the part of polygon that is folded vertically above the line segment specifying the g-fold. We do this as shown in Figure 3(a)–(c), manipulating the cloth such that the baseline of the vertical part is parallel to the line segment at all times. Initially, the "baseline" is outside the cloth polygon (meaning that there is no vertical part) and is moved linearly toward the line segment specifying the g-fold.

In the second phase, the g-fold is completed by laying down the vertical part of the cloth using a mirrored manipulation in which the baseline is again parallel to the line segment at all times. Initially the baseline is at the line segment specifying the g-fold and is moved linearly outward until the baseline is outside the folded part of the polygon (see Figure 3(c)–(e)).

The corresponding motions of the grippers holding the vertices can be computed as follows. Let us assume without loss of generality that the line segment specifying the g-fold coincides with the x-axis and points in the positive x-direction. Hence, the part of the polygon above the x-axis needs to be folded. Each convex vertex of this part in which the positive y-vector points outside of the cloth in its initial configuration needs to be held by a gripper at some point during the manipulation. We denote this set of vertices by V. Let y^* be the maximum of the y-coordinates of the vertices in V. Now, we let the baseline, which is parallel to the x-axis at all times, move "down" with speed 1, starting at $y_b = y^*$, where y_b denotes the y-coordinate of the baseline. Let the initial planar coordinates of a vertex $v \in V$ be(x_v,y_v). As soon as the baseline passesy_v, vertex v starts to be manipulated. When the baseline passes $-y_v$, vertex v stops being manipulated. During the manipulation, the vertex is held precisely above the baseline. In general, the three-dimensional coordinate $(x(y_b),y(y_b),z(y_b))$ of the gripper holding vertex v as a function of the y-coordinate of the baseline is given by:

$$x(y_b) = x_v \tag{1}$$

$$y(y_b) = y_b \tag{2}$$

$$z(y_b) = y_v - |y_b| \tag{3}$$

For $y_b \in [y_v, -y_v]$. Outside of this interval, the vertex is part of the horizontal part of the cloth and does not need to be grasped by a gripper. This reasoning applies to all vertices $v \in V$. When the baseline has reached $-y^*$, all vertices have been laid down and the g-fold is completed.

As a result, we do not need to grasp any vertex outside of V at any point during the manipulation, where V is the set of vertices that need to be grasped in the configuration where the part of the cloth that is folded is vertical above the line segment specifying the g-fold (in this case the x-axis). At all other points in time the vertical part is a subset that has exactly the same orientation with respect to gravity, so the same amount of vertices, or fewer, needs to be grasped. Hence, the set of vertices that need to be grasped and the motions of them can be computed in $O(n)$ time.

5.2.4.2 Sequences of g-folds and Stacked Geometry

Here we discuss the case of folding an already folded geometry. First, we discuss how to represent a folded, stacked geometry. Let us look at the example of the long-sleeve T-shirt of Figure 4, and in particular at the geometry of the cloth after five g-folds. The creases of the folds have subdivided the original polygon into facets (see Figure 6(a)). With each such facet, we maintain two values: an integer indicating the height of the facet in the stacked geometry (1 is the lowest) and a transformation matrix indicating how the facet is transformed from the original geometry to the folded geometry. Each transformation matrix is a product of a subset of the matrices F_i that each correspond to the mirroring in the line segment specifying the ith g-fold. In Figure 6(b), we show the lines of each of the g-folds with the associated matrix F_i.

Given the representation of the current stacked geometry and a line segment specifying a new g-fold, we show how we manipulate the cloth to successfully perform the g-fold or

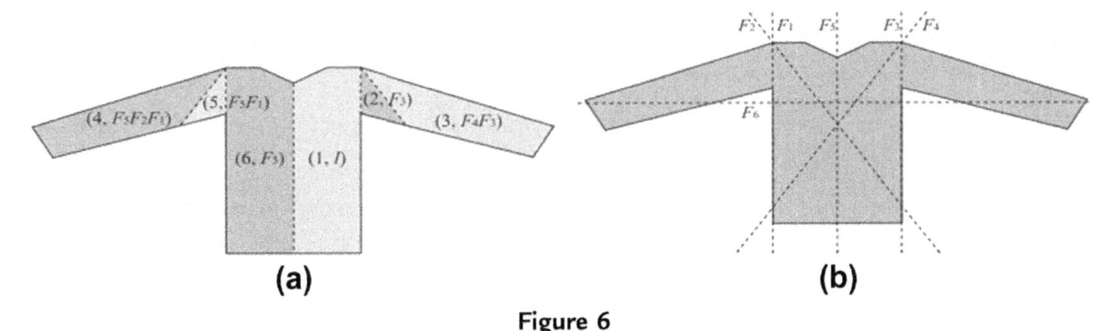

Figure 6

(a) The representation of a folded stacked geometry. The example shown here is the long-sleeved shirt of Figure 4 after five g-folds. With each facet, the stack height (integer) and a transformation matrix is stored. (b) Each transformation matrix F_i corresponds to mirroring the geometry in the line segment specifying the ith g-fold.

report that the g-fold is infeasible. We assume that the line segment specifying the g-fold partitions the silhouette of the stacked geometry into two parts (i.e., a blue g-fold). Let us look at the sixth specified g-fold in the long-sleeve T-shirt example, which folds the geometry of Figure 6.

Each facet of the geometry (in its folded configuration) is either fully to the left of the line segment, fully to the right, or intersected by the line segment specifying the g-fold. The facets intersected by the line segment are subdivided into two new facets, both initially borrowing the data (the stack height and the transformation matrix) of the original facet. Now, each facet will either be folded, or will not be folded. Figure 7 shows the new geometry in the long-sleeve T-shirt example after subdividing the facets by the line segment specifying the g-fold. The gray facets need to be folded.

As in the case of folding planar geometry, for each facet each convex vertex at which the gravity vector points outside of the facet at the time it is above the line segment specifying the g-fold should be held by a gripper, and each non-convex vertex or convex vertices where the negative gravity vector points inside the facet need not be held by a gripper. If a vertex is part of multiple facets, and according to at least one facet it needs not be held by a gripper, it does not need to be held by a gripper.

For the T-shirt example, the vertices that need to be grasped are shown using dots in Figure 7 and labeled $v_1,\ldots,v_7$. Applying the transformation matrices stored with the incident facet to each of the vertices shows that v_1, v_3, v_5, and v_7 will coincide in the plane. As a gripper will grasp all layers the geometry, only one gripper is necessary to

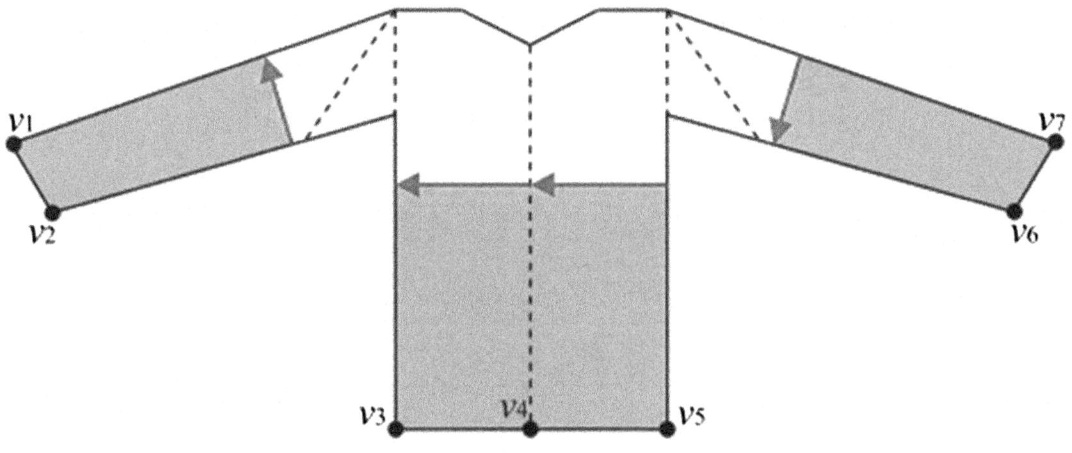

Figure 7

The geometry in the long-sleeve T-shirt example after subdividing the facets by the line segment specifying the sixth g-fold. The gray facets need to be folded. The convex vertices for which the negative gravity vector points outside of the facet are shown using dots.

hold these vertices. Vertex v_4 also needs to be held by a gripper. Vertices v_2 and v_6 remain, but they need *not* be grasped. We draw this conclusion for the following reason. As can be seen in Figure 4, these vertices are fully *covered*. That is, the vertex is "hidden" behind other facets of the cloth both below and above it in the stacked geometry. As we assume that the friction between two pieces of the cloth is infinite, this vertex will not be able to deform as a result of gravity, and need not be grasped. Using the heights stored at each facet, we can compute for each vertex whether it is covered or not.

This defines fully what vertices need to be grasped to achieve a g-fold of stacked geometry. If any such vertex is not on the boundary of the silhouette of the stacked geometry, the g-fold is infeasible (for example, the second g-fold of Figure 5(a) is infeasible for this reason). The 3-D motion of the grippers can be computed in the same way as for planar geometry, as discussed in Section 5.2.4.1. The running time for computing the vertices that need to be grasped is in principle exponential in the number of g-folds that preceded, as in the worst case i g-folds create 2^i facets. If we consider the number of g-folds a constant, the set of vertices that need to be grasped can be identified in $O(n)$ time.

After the g-fold is executed, we need to update the data fields of the facets that were folded in the geometry: each of their associated transformation matrices is *pre-multiplied* by the matrix F_i corresponding to a mirroring in the line segment specifying the g-fold (F_6 in Figure 6(b) for the T-shirt example). The stack height of these facets is updated as follows: the order of the heights of all facets that are folded is *reversed*, and these facets are put on top of the stack. In the example of Figure 7, the facets that are folded have heights 4, 6, 1, and three before the g-fold, and heights 8, 7, 10, and 9 after the g-fold, respectively.

The above procedure can be executed in series for a sequence of g-folds. Initially, the geometry has one facet (the original polygon) with height one and transformation matrix I (the identity matrix). If a g-fold is specified to only apply to the folded part of the geometry of the last g-fold (a "red" g-fold), the procedure is the same, but only applies to those facets that were folded in the last g-fold. We allow these kinds of g-folds as a special primitive if they need the same set of vertices to be grasped as the previous g-fold. Even if the vertices that are grasped are not on the boundary of the silhouette of the geometry, the g-fold can be achieved by not releasing the vertices after the previous g-fold. This enriches the set of feasible fold primitives.

5.2.5 Determining the Cloth Polygon

The above sections establish a robust framework for manipulating a clothing article given an approximate polygonal representation of its configuration. In this section, we examine the problem of visually inferring this representation, by both classifying which type of clothing article (e.g., towel, pants, or shirt) is present in a single image, and identifying an annotated polygon by which it can be approximated when folding. We additionally consider the problem of visually tracking folds, to gauge progress and accuracy throughout the folding procedure.

Spread crudely on a table, real-world clothing items do not perfectly resemble simple polygons. They contain curves rather than straight lines, corners which are rounded rather than sharp, and small intricacies which no two articles of a given class will necessarily share. Rather than reason explicitly about these complex shapes, we wish to transpose them as best we can into a shape which we know how to fold: in our experiments, this will be one of the four polygons detailed in Figure 8, either spread out or partially folded.

To do so, we employ a top-down approach to pose estimation: if a particular class of polygon is desired, let the article's shape be approximated by the best-fitting instance of that class, governed by some choice of distance metric. This draws upon the template-matching approach of Borgefors [21], in which a polygonal template is iteratively fit to an observed contour. As will be explored in Section 5.2.6.2, however, a priori knowledge of clothing structure (for instance, the symmetry between left and right sleeves) may be exploited to greatly improve the resulting fit. We therefore consider an augmented representation, deemed the parametrized shape model, which preserves this constrained internal structure, as well as a scheme for optimizing fit while incrementally relaxing these constraints.

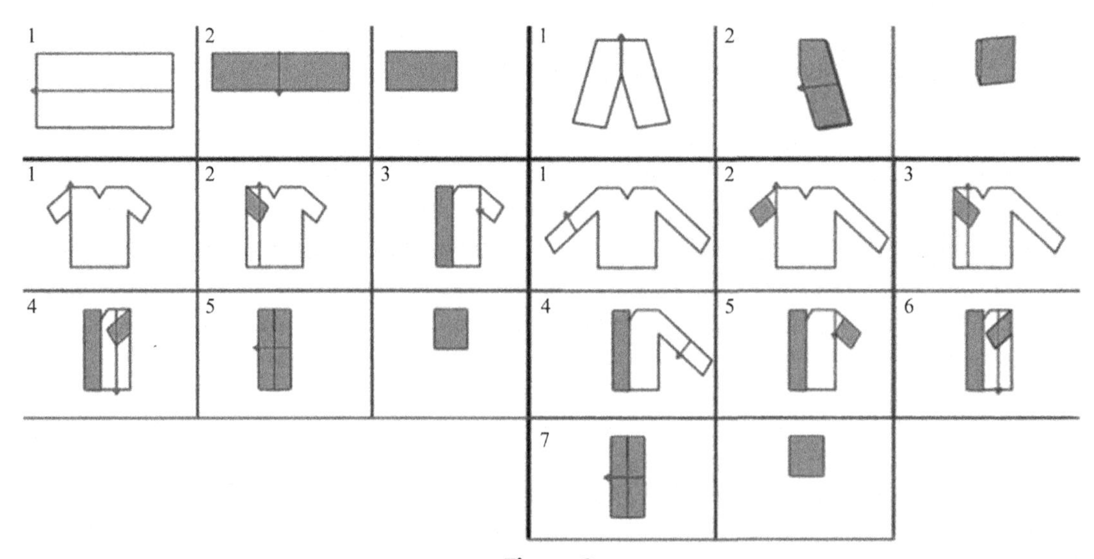

Figure 8

The sequences of folds used in our experiments. Note that the long-sleeved fold is identical to the short-sleeved fold, with added folds for tucking in the sleeves. This may well be seen as a single primitive, parametrized about the sleeve length.

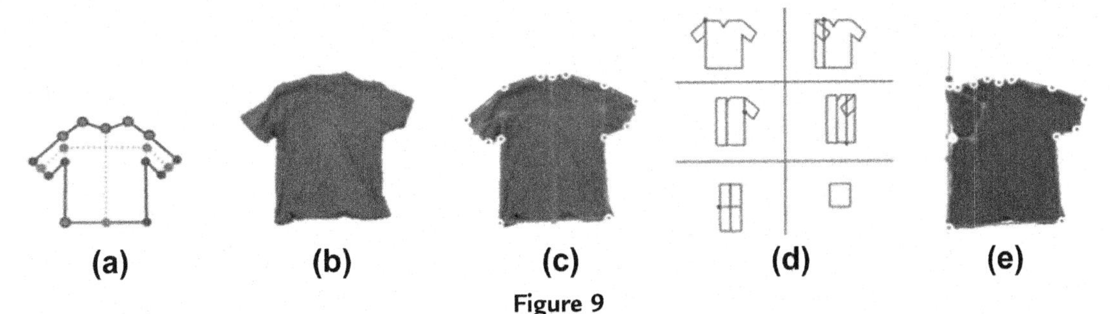

Figure 9
(a) A parametrized shape model for a T-shirt. Red indicates a skeletal parameter, blue indicates a landmark point. (b) An example instance. (c) The result from running our approach: our model (pink) is overlayed onto the image of the clothing it is trying to fit. The determined landmark points are shown as white dots. (d) An example of a parametrized specification of a folding sequence. All fold lines shown are defined relative to the parametrized shape model. (e) Result from running our approach when a fold line is present. (For interpretation of the references to color in this figure legend, the reader is referred to the online version of this book.)

In Section 5.2.5.1 we formalize the notion of a parameterized shape model. For every clothing category, we attempt to find a minimal set of parameters which can describe the range of shapes it may take on. Every legal setting of these parameters defines a polygon. We call the vertices of this polygon the landmark points, which may be used as input to the folding algorithm in Section 5.2.4. Figure 9(a) shows a parameterized shape model for T-shirts as well as the polygon associated with this particular model instantiation. We further augment this representation to include folded versions of these articles.

In Sections 5.2.5.2 and 5.2.5.3 we propose an optimization strategy for determining the parameters which best fit an observed contour. In Section 5.2.5.4, we demonstrate how this fit may also be used to classify the clothing article. What results is a class-level description of the observed clothing article, and a polygon to represent it.

5.2.5.1 Parametrized Shape Models

We define a model M by the following components:

A landmark generator.

$$M_{LG} : \{P \in \mathbb{R}^{p}\} \rightarrow \{L \in \mathbb{R}^{2\times\ell}\}$$

which takes a parameter vector P as input, and returns the associated collection of landmark points, L.[5]

[5] In this work, the only information used is the resulting polygon, and hence all articles have landmark points which lie on the contour. In general, however, there is no reason that this must be the case: any point, whether virtual or on the contour, may be considered a landmark.

A contour generator.

$$M_{CG} : \{P \in ¡^{p}\} \rightarrow \{C \in ¡^{2 \times c}\}$$

which takes a set of scalar parameters P as input, and returns the contour of the polygon which would arise from the given parameters, with a fixed number of samples per side.

A legal input set.

$$M_L \subseteq ¡^{p}$$

which defines the set of parameters in which M may reasonably be found.

A transformation operator.

$$M_T : \{P \in ¡^{p}, T \in ¡^{2}, \theta \in ¡, s \in ¡\} \rightarrow \{P' \in ¡^{p}\}$$

which transforms a set of parameters in such a way that the resultant contour M_{CG} will be translated, rotated, and scaled by the given values of T, θ, and s.

5.2.5.1.1 Skeletal models

To capture the structure of the clothing, we parametrize a model about a set of interior (or *skeletal*) points, as well as features which detail the distance from the interior points to the contour. These may include landmark vertices, displacements between a skeletal vertex and its nearest edge, or scalars such as height and width.

Figure 9(a) shows an example of a skeletal model for a T-shirt; a more detailed list of the parameters, as well as all other skeletal models, may be found in Appendix B. The parameters are highlighted in red, and the landmark points are highlighted in blue. A red point with a blue outline indicates a landmark point which is itself a parameter. The generated contour is outlined in black. The legal input set is detailed in Appendix B.3.

5.2.5.1.2 Folded models

Once the pose of the spread-out article has been determined, we wish to visually track the progress and accuracy of our folding procedure. To any model M^0, we may wish to add a single fold line. We thus define a *folded model*, such that.

$$P^{\text{folded}} = \left[\Theta \middle| P^0\right]$$
$$M_L^{\text{folded}} = ¡^{4} \times M_L^0,$$

where all parameters of the original model P^0 are allowed to vary and, in addition, the parameters Θ specify a directed line segment about which the model is to be folded. The resulting landmark points are computed by folding the polygon specified by $M_{LG}^0(P^0)$ about this line. Note that there is no restriction on what sort of model M^0 is. This allows us to specify folds recursively. If M^0 is unfolded, M^{folded} will contain a single fold. If M^0 contains a single fold, M^{folded} will contain two, and so on.[6]

If we are certain the clothing article did not move during a folding operation, we may reduce this task to finding a single fold line on a known polygon, rather than determining both simultaneously. We therefore define a *static folded model*, such that.

$$P^{\text{folded}} = [\Theta]$$
$$M_L^{\text{folded}} \equiv \text{¡}^4.$$

5.2.5.2 Energy Function

We now aim to find the parameters which optimally fit a given image. Our approach extracts the contour of the clothing article in the image and uses an energy function which favors a contour fit. We define the energy E as follows:

$E(P) = (\alpha) \times \overline{d}(M_{CG}(P) \rightarrow C) + (1 + \alpha) \times \overline{d}(C \rightarrow M_{CG}(P))$, where $\overline{d}(A \rightarrow B)$ is the average nearest-neighbor distance[7] from A to B:

$$\overline{d}(A \rightarrow B) \equiv \frac{1}{|A|} \sum_{a \in A} \underset{b \in B}{\operatorname{argmin}} \|b - a\|$$

The parameter α is used to adjust the way in which the model fits to the contour. If α is too low, the model will attempt to fit every point of the contour, often overfitting to deviations such as wrinkles. If α is too high, the model may cease to cover the contour at all, fixating instead on a single portion. We have found that setting $\alpha = 0.5$ is sufficient to counter both negative tendencies.

5.2.5.3 Energy Optimization

Our energy-optimization follows a coarse-to-fine strategy, in which the parameter space begins small and increases as the procedure continues. It first only considers translation, rotation and scale, then considers all parameters but enforces certain symmetry constraints amongst them, and finally optimizes over all parameters without the symmetry constraints.

---

[6] One may wonder why Θ has four dimensions rather than the expected two. This is because a fold line is not a line per se. Rather, it is a directed line segment, which may intersect one portion of cloth without intersecting another, colinear portion.

[7] We additionally considered the use of dynamic time warping [36,37] in our distance metric. The results, however, showed little improvement, so for the sake of simplicity and computational efficiency, we restrict our approach to nearest-neighbor.

5.2.5.3.1 Initialization

PCA approach To infer the necessary translation, rotation, and scale, we rely on a principal component analysis (PCA) of the observed contour, and contour defined by the model.

We first compute the initial model contour as.

$$M_c = M_{CG}(P_0).$$

We then calculate the centers of mass of the observed contour and the model contour: c_0 and c_m, respectively. We then compute the relative translation between the two contours.

$$T = c_0 - c_m.$$

We then perform PCA to estimate the principal axes of each contour, denoted by a_0 and a_m. We compute the relative angle between the two axes.

$$\theta = \arccos(a_0 \cdot a_m).$$

Finally, for each contour we find the point of intersection between the top of the contour and its principal axis, denoted t_0 and t_m. We compute the relative scale between the two contours as.

$$s = \frac{\|t_0 - c_0\|}{\|t_m - c_m\|}$$

which is approximately the ratio of the heights of the two contours. The resultant contour $M_c(P)$ will be centered about c_0, and scaled and rotated such that $t_0 = t_m$.[8]

Having computed these three values, we then update our model estimate such that.

$$\mathrm{P}' \leftarrow M_T(P, T, \theta, s).$$

Multi-angle approach We additionally consider a second approach, in which the optimization is run with multiple initializations, attempting all possible rotations within a granularity of $\delta\theta$. Upon completion, the fitted model which yields the lowest energy function is chosen, and all others are discarded. The method for choosing translation and scale is the same as in the PCA approach.

5.2.5.3.2 Optimization

To ensure the best possible fit, our standard approach performs the optimization in three phases: orientation, symmetric, and asymmetric.

---

[8] Thus described, the PCA approach leaves an ambiguity in terms of which direction is assumed to be "up" on the principal axis. To resolve this, we attempt both upright and upside-down initializations, and choose the minimum-cost result after the optimization is complete.

In the *orientation phase*, all parameters are held relatively fixed, with only one external degree of freedom: θ, which defines the net rotation of the contour points about the center of gravity of the model. This phase is only run when using the PCA-based initialization, and it tends to improve the orientation estimate as it considers the entire contour, rather than just its principal component. When using the multi-angle initialization we found it better to skip the orientation phase as it reduced the variety of orientations explored.

In the *symmetric phase*, the model is free to translate, rotate, scale, or deform within the limits determined by its legal input set, as long as left–right symmetry is maintained. In terms of implementation, this is done by optimizing over a subset of the model parameters, those which describe the left and center portions of the model, and computing the implied values for the remaining right parameters such that symmetry is enforced.

In the *asymmetric phase*, all parameters are optimized over, and the model is free to translate, rotate, scale, or deform within the limits determined by its legal input set.

For the numerical optimization, we use coordinate-wise descent over the parameters: evaluating the gradients numerically (rather than analytically) and maintaining an adaptive step size for each parameter. This algorithm is presented in detail in Appendix C.

To enforce legality constraints on the parameters, we augment the energy function with a penalty for constraint violation. We first normalize the fit such that.

$$\forall P : 0 \leq E_{norm}(P) < 1.$$

To do so, we set.

$$E_{norm} = \frac{E}{E_{\max}}.$$

As a simple upper bound, E_{max} is set to $\sqrt{h^2 + w^2}$, where h and w denote the height and width of the image, respectively. This corresponds to the case in which the two contours are maximally distant given the size of the image.

We then define the structural penalty S as.

$$S(P) = \begin{cases} 0 & \textit{if } P \in M_L, \\ 1 & \textit{otherwise.} \end{cases}$$

The resulting energy function is then given by.

$$C(P) = E_{norm}(P) + S(P).$$

As the normalized energy E_{norm} lies between zero and one, the optimum of the cost function will never violate a constraint if a legal alternative exists.

5.2.5.4 Classification

For any image and specified model, the above procedure is able to return a set of fit parameters and an associated energy. By considering the value of the energy function as a measure of overall model fit, this provides a convenient means of category classification. When presented with an image and a set of possible categories, we run the above procedure multiple times, with one model associated with each category. The fitted model which results in the lowest final energy is selected, and the image is classified accordingly.

5.2.6 Experimental Results

In this section we describe the experimental verification of our approach. We begin by validating the reliability of our g-fold mechanism for folding in Section 5.2.6.1, using human annotated cloth polygons as input. We then evaluate the success of our perceptual tools for inferring this polygon in Section 5.2.6.2, run on a hand-compiled dataset of 400 clothing images. Finally, we combine the two components into an end-to-end robotic system in Section 5.2.6.3.

5.2.6.1 Clothing Manipulation

We validate the power of our g-fold framework by first implementing an open-loop laundry folding mechanism on a household robot. We first describe the setup, and then the corresponding results.

5.2.6.1.1 Experimental setup

We used the Willow Garage PR2 robotic platform developed by Wyrobek et al. [32]. The PR2 has two articulated seven-axis arms with parallel jaw grippers. We used a soft working surface, so the relatively thick grippers can easily get underneath the cloth. Our approach completely specifies end-effector position trajectories. It also specifies the orientation of the parallel jaw grippers' planes. We used a combination of native IK tools and a simple linear controller to plan the joint trajectories.

We experimented with the clothing articles shown in Figure 10. Whenever presented with a new, spread-out clothing article, a human user clicks on the vertices of the article in an image. This specifies the location of the cloth polygon.

To allow for arbitrary fold sequences, we give a human user the option of a manual fold specification. The user is presented with a graphical representation of the article, and the ability to draw arbitrary folds. Once a valid g-fold has been specified, the robot executes the fold, allowing the user to specify another. Figure 11 illustrates the fold sequence specification process through an example.

Figure 10
Three of each clothing category were used in conducting our experiments.

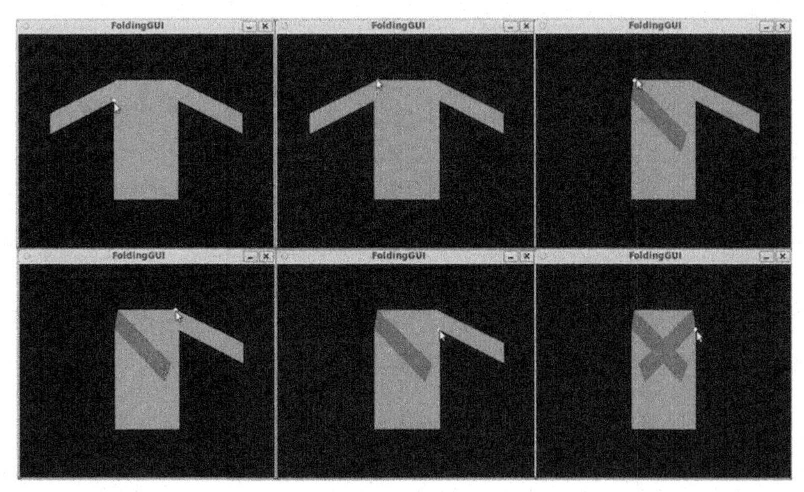

Figure 11
An example sequence of user-specified folds. The user first clicks on the left arm pit, then on the left shoulder to specify the first fold. The program then verifies that this is a valid g-fold for the chosen number of grippers. In this case it is, and it then shows the result after executing the g-fold (third image in the top row). Then the user specifies the next fold by two clicks, the program verifies whether it is a valid g-fold, and then shows the result after executing the g-fold.

To autonomously execute folds on known clothing categories, the program is also seeded with a set of folding primitives. When presented with a particular article of clothing, the user is given the option of calling one of these primitives. Once called, a sequence of folds is computed, parametrized on a number of features such as scaling, rotation, and side lengths. Figure 12 shows an example of a primitive being executed on a user-defined

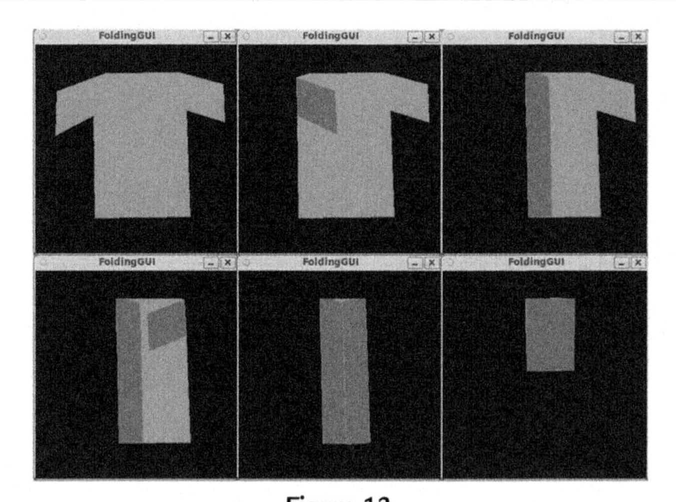

Figure 12
An example folding primitive, automatically executed on a T-shirt polygon. Note the clean fold, despite the imperfect symmetry of the original polygon.

polygon in the shape of a shirt. To ensure consistency across multiple trials, such primitives were used to execute the folds detailed in the following experimental results section.

While our approach assumes the cloth has zero resistance against bending, real cloth does indeed resist against bending. As a consequence, our approach outlined so far overestimates the number of grippers required to hold a piece of cloth in a predictable, spread-out configuration. Similarly, our robot grippers have non-zero size, also resulting in an overestimation of the number of grippers required. To account for both of these factors, our implementation offers the option to allocate a radius to each of our grippers, and we consider a point being gripped whenever it falls inside this radius. To compute the grasp points, we first compute the grasp points required for point grippers and infinitely flexible cloth. We then cluster these points using a simple greedy approach. We begin by attempting to position a circle of fixed radius in the coordinate frame such that it covers the maximum number of grasp points, while subsequently minimizing the average distance from each covered point to its center. This process is iterated until no point remains uncovered. For the duration of the fold, our grippers now follow the trajectory of the center of each cluster, rather than individual grasp points.

5.2.6.1.2 Experimental results

We tested our approach on four categories: towels, pants, short-sleeved shirts, and sweaters. Figure 8 shows the fold sequences used for each category. To verify the

Table 1: Experimental results of autonomous laundry folding

Category	Success Rate	Average Time (s)	Category	Success Rate	Average Time (s)
Towels	**9/9**	**200.0**	**Short-sleeved shirts**	**7/9**	**337.6**
Purple	3/3	215.6	Pink T-Shirt	2/3	332.8
Leopard	3/3	210.9	Blue T-Shirt	2/3	343.2
Yellow	3/3	173.5	White collared	3/3	337.6
Pants	**7/9**	**186.6**	**Long-sleeved tops**	**5/9**	**439.0**
Long Khaki	3/3	184.9	Long-sleeved shirt	2/3	400.7
Brown	1/3	185.9	Gray sweater	1/3	458.4
Short Khaki	3/3	189.1	Blue sweater	2/3	457.8

robustness of our approach, we tested on three instances of each category of clothing. These instances varied in size, proportion, thickness, and texture. At the beginning of each experimental trial, we provided the PR2 with the silhouette of the polygon through clicking on the vertices in two stereo images.

Table 1 shows success rates and timing on all clothing articles. Figure 13 shows the robot going through a sequence of folds.

As illustrated by the reported success rates, our method demonstrates a consistent level of reliability on real cloth, even when the manipulated fabric notably strays from the assumptions of our model. For instance, the g-fold method worked reasonably well on pants, despite the material's clear violation of the assumption of non-zero thickness, and a three-dimensional shape which was not quite polygonal. It was also able to fold a collared shirt quite neatly, despite that its rigid collar and buttons are not expressible in the language of our model. While these elements would likely be problematic if they intersected a g-fold, they can otherwise be ignored without issue.

Despite the simplifications inherent to our model, we have found it to match the behavior of real cloth quite closely in this setup. While human manipulation of cloth exploits a number of features which our model neglects, these features generally arise in states which our model considers unreachable. That is, handling true fabric often requires less caution than our g-fold model predicts, but rarely does it require more. Furthermore, even when unpredicted effects did arise, the final result was often not compromised.

Figure 13
The robot folding a T-shirt using our approach.

Although factors such as thickness may cause the cloth to deviate slightly from its predicted trajectory, most often in the form of "clumping" for thick fabrics, the resulting fold generally agrees with the model, particularly after smoothing. Much of our success can be attributed to a number of assumptions which were very closely met: namely, the lack of slip between the cloth and the table, and the lack of slip between the cloth and itself. The former allowed us to execute g-folds even when the modeled polygon did not perfectly match the silhouette of the cloth. As actual articles of clothing are not comprised solely of well-defined corners, this imprecision often resulted in a non-zero horizontal tension in the cloth during the folding procedure. However, as the friction between the cloth and the table far outweighs this tension, the cloth remained static. This allowed us to stabilize loose vertices by "sandwiching" them between two gripped portions of cloth. This technique, in combination with the robust gripping approach detailed above, allowed us to execute a number of folds (such as the shirt folds in Figure 8) which more closely resembled their standard human counterpart. With the exception of long-sleeved shirts, all sequences could theoretically be executed by a pair of point grippers. However, some relied on the ability to create perfect 90° angles, or precisely align two edges which (in actuality) were not entirely straight. Exact precision was impossible in both of these cases; but where there was danger of gravity influencing a slightly unsupported vertex, the friction of the cloth, in conjunction with its stiffness, often kept it in a stable configuration.

The trials were not, however, without error. Most often, failure was due to the limitations of our physical control, rather than a flaw in our model. For instance, 2/2 short-sleeved failures and 3/4 long-sleeved failures occurred at steps where the robot was required to grasp a portion of previously folded sleeve (short-sleeve steps 2 and 4, long-sleeve steps 3 and 6 in Figure 8). In each of these cases, the failure could be easily predicted from the location of the initial grasp. Either the robot did not reach far enough and grasped nothing, or reached too far and heavily bunched the cloth. These failures suggest a clear issue with our original implementation: namely, the reliance on open-loop control. While the initial position of each vertex is given, the location of a folded vertex must be derived geometrically. For this location to be correct, we must make two assumptions: that the cloth at hand is perfectly represented by the given polygon, and that the trajectory, once computed, can be exactly followed. Clearly, both are idealizations: the former disregards the multi-layered nature of all but towels (which saw a 100% success rate) and the latter is hindered by the inherent imprecision of any robotic mechanism. These errors greatly entail the need for a perceptual component which can track folds over time, as detailed in Section 5.2.5.1.2 and implemented in Section 5.2.6.3.

While overall folds were often executed correctly, the resulting article often contained minor imperfections, such as wrinkles. The robot was able to remove many of these via an open-loop smoothing motion. However, in order to make the fold truly neat, more advanced manipulations, such as ironing or precisely targeted smoothing motions, would most likely be necessary.

5.2.6.2 Clothing Detection

Using the methods detailed in Section 5.2.5, we designed a system able to infer the class and pose of a spread-out article of clothing. To do so, we defined a set of parametrized shape models which corresponded to each clothing class (Section 5.2.6.2.1). We then collected a dataset of clothing images (Section 5.2.6.2.2) and verified our detection algorithm on this dataset (Section 5.2.6.2.4).

5.2.6.2.1 Models used

For each of the four categories of clothing detailed above, we define an associated parametrized model: thus, to each article of clothing we attempt to fit a towel, pants, short-sleeved, and long-sleeved model. Each model defines a polygon which may be folded using one of the primitives detailed in Figure 8. The parameters and constraints of these models are discussed in detail in Appendix B.

As a baseline for performance comparison, we also define a polygonal model which is parametrized about.

$$p_{poly} = \left[l_1(x) l_1(y) \ldots l_\ell(x) l_\ell(y) \right]$$
$$L_{poly} = ¡^{2\ell}.$$

This model has no interior structure, and no legality constraints beyond self-intersection. For every clothing category, we construct a polygonal model whose initial landmark points are identical to those of the skeletal model for that category. This model provides a useful baseline for the performance of pure contour fitting, beginning with the same initialization and optimization techniques, but without taking any prior knowledge about clothing into consideration.

5.2.6.2.2 Data collection

To quantitatively gauge the accuracy of our approach, our shape-fitting code was run on a dataset of roughly 400 images, divided into four categories: towels, pants, short-sleeved shirts, and long-sleeved shirts. For each category, 10 representative articles of clothing were considered. These 40 articles varied greatly in size, proportion, and style (see Figure 14). Each article was then further placed in 10 or more poses, encompassing a variety of common spread-out configurations (see Figure 15).

Each object was initially photographed on a green table. To ensure rotational invariance, each image was transformed to a birdseye perspective, using OpenCV's checkerboard

Figure 14
The 40 articles of clothing in our dataset.

Figure 15
The article of clothing is put in various poses.

detector to locate the top-down frame. The background was then subtracted from each image. For most of these images, hue thresholding against the green background was sufficient: however, in cases where the complex texture of the clothing precluded hue thresholding, the Grabcut algorithm [33] was used to perform the subtraction, with foreground and background pixels manually selected by a user. Finally, the location of each landmark point was hand-annotated, to provide ground truth data for the model fitting task. The pipeline is illustrated in Figure 16.

5.2.6.2.3 Implementation details

We ran our experiments on a Lenovo Thinkpad, running an Intel Core 2 Extreme Processor. A typical model fit took roughly 30 s; for more complex procedures such as the four-phase multi-model approach for T-shirts, convergence would occasionally take up to 2.5 min. To rapidly compute the nearest-neighbor distances for the cost function, the Flann library [34] was used. The bulk of the image processing, including transformations, thresholding, and contour detection, was done with OpenCV [35].

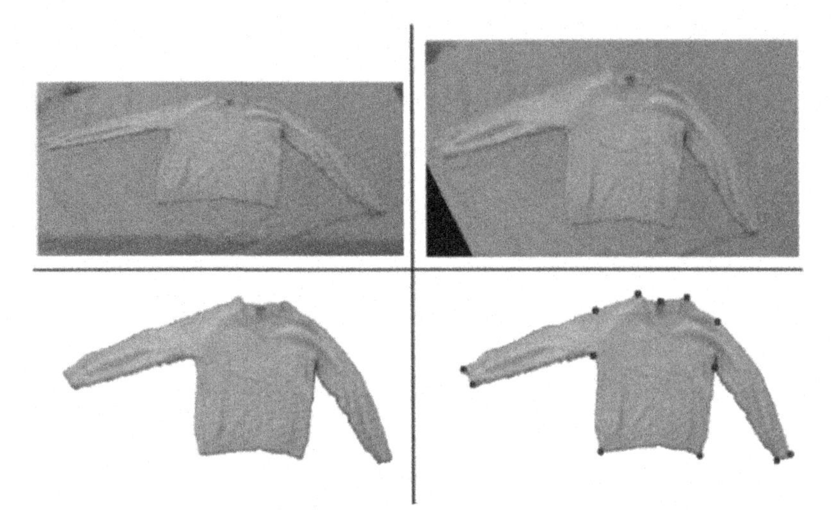

Figure 16
The dataset pipeline. Top left: Initially, the clothing is spread out on a green table. Top right: A birdseye transformation is then performed. Bottom left: The image is cropped, and the background is segmented out. Bottom right: To provide ground truth for the fitting procedure, the resulting image is hand-annotated.

5.2.6.2.4 Experimental results

Each image was first fit to the proper model according to its known category. Table 2 shows the accuracy of our approach on the 400 image dataset using both the PCA and multi-angle initializations, and the performance of the associated polygon model on the same set. These results are represented pictorially in Figure 17.

Our approach performs very well, obtaining typical accuracies of within eight pixels per landmark point and significantly outperforming the polygonal approach, the shortcomings of which are detailed in Figures 18 and 19.

Table 2: Results of fitting our skeletal models to the dataset. Model accuracy is measured as the average pixel distance from the predicted landmark point to the annotated landmark point

	Polygon Model		Skeletal Model (PCA)		Skeletal Model (Multi-angle) $\delta\,\theta = 10^\circ$	
Category	(pixels)	(cm)	(pixels)	(cm)	(pixels)	(cm)
Towels	2.89 ± 1.78	0.75 ± 0.46	2.89 ± 1.78	0.75 ± 0.46	2.86 ± 1.75	0.74 ± 0.45
Pants	14.91 ± 35.97	3.88 ± 9.35	4.23 ± 1.64	1.10 ± 0.43	4.13 ± 1.54	1.07 ± 0.40
Short sleeved	89.63 ± 44.88	23.30 ± 11.67	6.58 ± 3.14	1.71 ± 0.82	6.41 ± 3.05	1.67 ± 0.79
Long sleeved	14.77 ± 8.27	3.84 ± 2.15	7.09 ± 3.68	1.84 ± 0.96	8.06 ± 4.52	2.09 ± 1.17

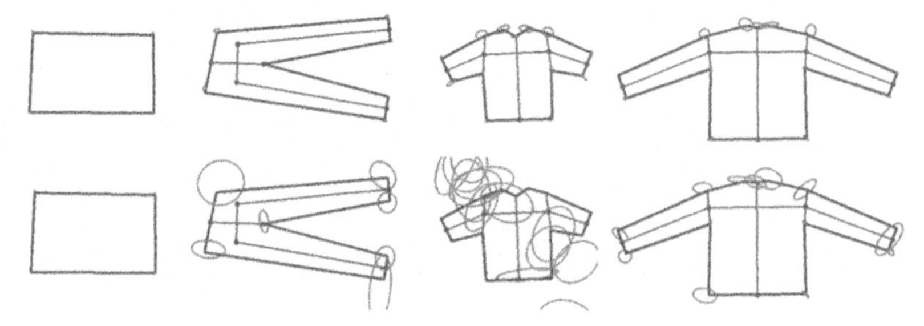

Figure 17

Comparison of individual landmark point errors. The center of the ellipses denotes mean error, and the size and skew their covariance, projected onto a canonical version of the article. Top: Pointwise error for skeletal models using the PCA approach. Bottom: Pointwise error for polygon models.

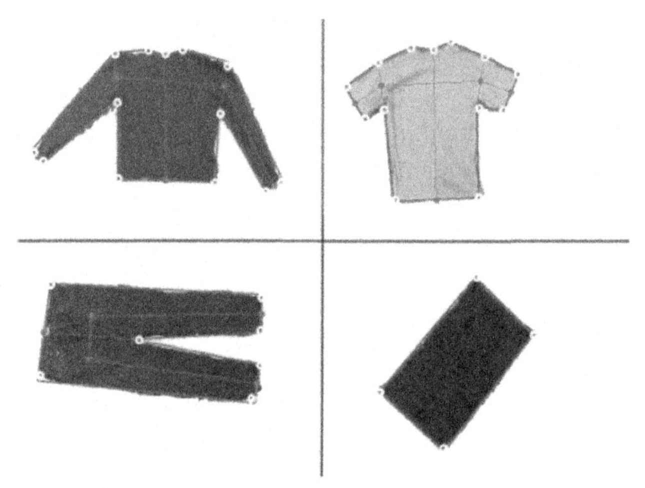

Figure 18

Example results of our approach of clothing.

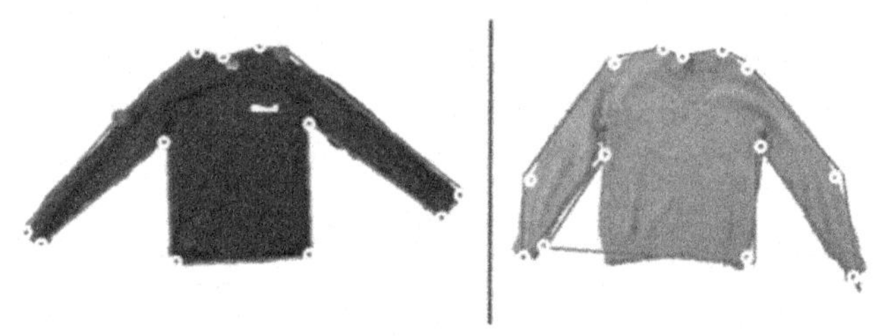

Figure 19

Failures of the polygon model. Left: Without more detailed structural information, the model is unable to detect more subtly defined points, such as the shoulder (detected shoulder points in red). Right: The unconstrained polygon approach will generally be attracted to the nearest edge; a poor initialization can easily ruin it.

Moreover, the relative gain of the skeletal approach on each category is quite telling. As the towel model is effectively structureless, there is no distinction between the two models, and hence no improvement. In the case of pants, the proximity between the two legs frequently caused the polygonal approach to attract to poor local minima; whereas the skeletal approach, with its implicit knowledge of structure, performed quite well. Short-sleeved shirts, being fairly homogeneous in shape, proved extremely difficult for the polygonal approach to fit, as can be readily seen in Figure 17. Despite the subtlety of shoulder and collar point locations, the longer sleeves of sweaters tend to sketch out a very clear polygonal shape; thus the polygon model performed somewhat reasonably, with most errors centered about shoulders, collars, and sleeve edges.

The results of the multi-angle approach were extremely consistent with that of PCA initialization, suggesting that the latter approach is sufficient for most purposes. Indeed, given the inherent ambiguity in landmark location and small number of examples on which the two differed, any perceived performance advantage would best be attributed to noise.

We then examined the case of the unknown clothing category. On 100% of test images, our method was able to accurately classify the clothing category. The classification scheme in Section 5.2.5.4 was used to distinguish shirts, pants, and towels. Thresholding the sleeve length at 35% of the shirt widths further distinguished all long-sleeved shirts from short-sleeved shirts. Therefore, the correct model is always chosen, and the performance is identical to the known, tabulated case.

Our approach, however, was not perfect. The location of collar points proved to be quite ambiguous, and was often incorrectly identified. Shoulders, while significantly localized by structural constraints, still proved a source of difficulty. Finally, the initialization was poor on a small number of instances, and in very rare cases was unable to be recovered.

5.2.6.3 A Combined End-to-End System

We then combined the perception system introduced in Section 5.2.6.2 with the folding system of Section 5.2.6.1 to provide the Willow Garage PR2 with a closed-loop folding system. The system runs as follows:

- A towel, pair of pants, short-sleeved or long-sleeved shirt begins spread out on a table in front of the PR2.
- Using the approach detailed in Section 5.2.6.2, the PR2 fits a skeletal model to the contour of the observed article. To avoid grasping virtual points, the landmark points are then relocated to their nearest neighbor on the observed contour.
- The PR2 then computes the parametrized fold primitive corresponding to the newly fit polygon.

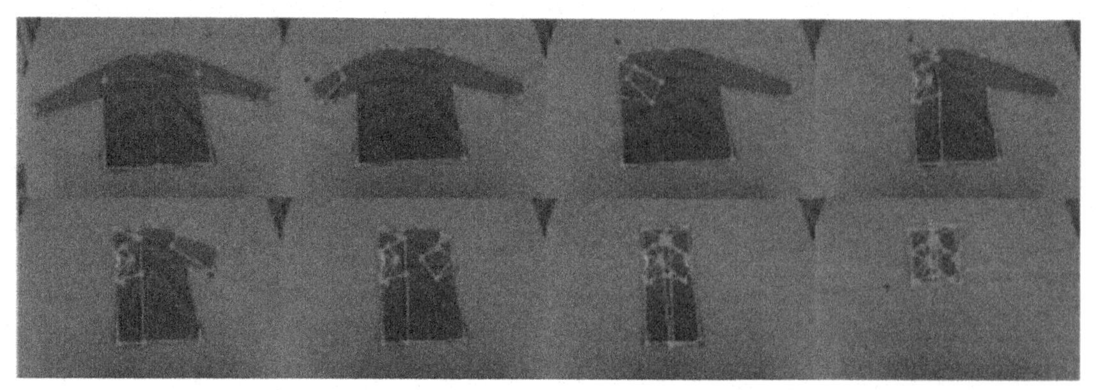

Figure 20
The robot fits a model to the initial configuration, then tracks each fold in the procedure.

- After each fold, the robot then re-examines the article. It then fits a static folded model to the newly observed contour, with initial model M^0 given by the previously determined landmark points, and parameters Θ seeded by the robot's intended fold.
- This is repeated until the article is folded.

Preliminary results have shown this approach to work consistently well, enabling fully automatic folding by the PR2. As suggested by the perception results in Section 5.2.6.2, the initial polygon detection phase has been able to eradicate human input with no notable deterioration in performance. Furthermore, the problem of failing to grasp previously folded portions of cloth, a frequent issue in the open-loop setup detailed in Section 5.2.6.1, is virtually eradicated; as errors are tracked the moment they occur, they are no longer compounded over time. (See Figure 20, for an example, of the tracking process.)

A number of factors, however, continue to hinder complete robustness. Most notably, the fold-tracking system works best when the view of the camera is stationary. Therefore, the robot remains stationary during the procedure, limiting the size of folded articles to the arm span of the robot. In addition, while grasp imprecision no longer compounds over successive folds, it remains a substantial issue: particularly on smaller articles, where the error relative to the size is often fairly high.

Videos of representative successful runs and continued progress, as well as software implementations of all aforementioned algorithms are available at http://rll.berkeley.edu/IJRR2011.

5.2.7 Conclusion and Future Work

We proposed a novel take on robotic laundry folding which averts the high dimensionality inherent to cloth by determining a set of necessary conditions under which its behavior is repeatable and known. In so doing, we greatly simplified the complexity of the system, and showed that even in this limited subspace, many folding procedures can be executed.

We further described the steps necessary to execute folds on polygonal cloth under these conditions, relying on intuitive geometric reasoning rather than computationally costly planning and simulation. Our experiments show that (1) this suffices to capture a number of interesting folds and (2) real cloth behaves benignly, even when moderately violating our assumptions.

We also provide an approach that equips a robotic system with the necessary perception capabilities; namely, the ability to visually infer a reasonable polygonal representation of a clothing article present in an image. We show that, via a model-based optimization approach, the pose and corresponding polygon of many common clothing articles can be reliably detected.

We experimentally demonstrated that this framework is capable of enabling a household robot to fold laundry. We tested the manipulation task in an open-loop setting and found it to be quite reliable. We tested the visual components on a large dataset of images, and found it to be highly accurate both in its ability to classify and infer the configuration of spread-out clothing. Finally, we combined the perception and manipulation tools on a robotic platform, and give a first look at an end-to-end system for robotic laundry folding.

Careful inspection of the gripper paths shows that a single very large parallel jaw gripper would suffice to execute a g-fold requiring an arbitrary number of point grippers. We plan to investigate a practical implementation of this idea for the PR2. However, a large gripper of this kind would significantly reduce the collision-free workspace volume.

In this study, the category-level folding primitives are specified by human users. Owing to the inherently aesthetic nature of the choice of primitive, such input may well be necessary. Yet it is interesting to consider the problem of automating this decision process, to allow for the folding of previously unseen article types in a reasonable way.

Our study assumes that, via some mechanism, an arbitrarily crumpled article of clothing may be spread out on a table. We are currently working on, and will continue to explore, a set of primitive actions which accomplish this task. We also look to expand our approach beyond folding to the entire laundry system.

Finally, while this study dealt explicitly with the task of laundry folding, we believe the tools put forth may generalize well beyond this to many deformable object manipulation challenges, such as ironing or bed-making. We hope that these results will provide another step forward on the long road toward the utopian future, where humans are never again nagged to "do their chores."

Funding

This work was supported in part by the NSF (award number IIS0904672), Willow Garage under the PR2 Beta Program, and a Feodor Lynen Fellowship granted by the Alexander von Humboldt Foundation.

Appendix A: Proof of Theorem 1

Let us assume for the purpose of the proof that the polygon lies/hangs in the xz – plane, and that gravity points in the $-z$ direction. The term above refers to having a higher z – coordinate.

From the work of Bell [2], we know that a non-stretchable planar tree is fully immobilized if each node of the tree of which its incident edges do not positively span ¡^{2} is fixed. Now, let us define an *upper string* of a polygon as a maximal sequence of edges of which the extreme vertices are convex vertices of the polygon, and no part of the polygon lies above the edges (see Figure 21(a)). A given polygon P can have multiple upper strings, but has at least one.

For each upper string that it holds at its convex vertices, the negative gravity vector points outside the polygon. As these convex vertices are fixed (by a gripper), the entire set of edges that the string consists of is immobilized. This can be seen by adding virtual vertical edges fixed in gravity pointing downward from the non-convex vertices, which make sure that the non-convex vertices cannot move upward (per the downward-tendency assumption). The incident edges of the non-convex vertices now positively span ¡^{2}, hence the entire string is immobilized.

Now, every point of the polygon P that can be connected to an upper string by a vertical line segment that is fully contained within P is immobilized. This is because this point

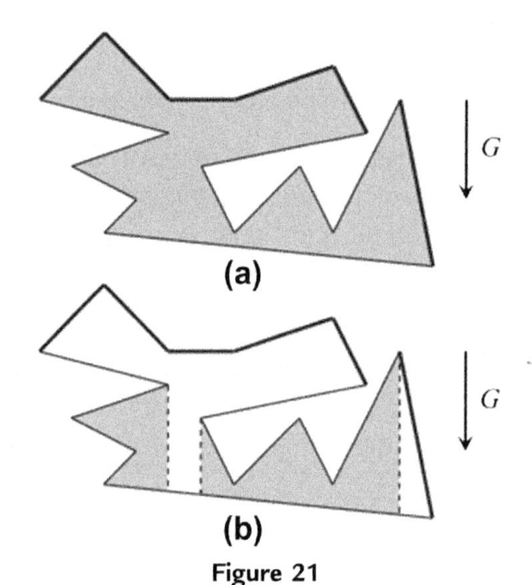

Figure 21

(a) A polygon with two upper strings is shown thick. (b) The white part of the polygon (including the vertical dashed edges) has proven immobilized. The gray part remains.

cannot move downward per the non-stretchability assumption (note that the upper string is immobilized), and it cannot move upward per the downward-tendency assumption. Hence, all such points can be "removed" from P: they have been proven immobilized. What remains is a smaller polygon P' (potentially consisting of multiple pieces) for which immobilization has not been proven (see Figure 21(b)). The smaller polygon P' has vertical edges that did not belong to the original polygon P. The points on these vertical edges are immobilized, including both incident vertices (of which the upper one may be a non-convex vertex of P that is convex in P'), as they vertically connect to the upper string.

Then, the proof recurses on the new polygon P', of which the convex vertices of the upper string(s) need to be fixed. Note that P' may have convex vertices that were non-convex in P. These need not be fixed, as they were already proven immobilized since they are part of the vertical edge of P'.

This proves the theorem. Note that the convex vertices where the negative gravity vector points into the polygon will never be part of an upper string at any phase of the proof, so they need not be fixed. Also, the recursion "terminates." This can be seen by considering the vertical trapezoidal decomposition of the original polygon P, which contains a finite number of trapezoids. In each recursion step, at least one trapezoid is removed from P, until the entire polygon has proven immobilized.

Appendix B: Shape Models Used

B.1 Towels

As there is little inherent structure to a towel, its skeletal model is simply parametrized about the location of its four vertices. Only one constraint was imposed, which is common to all of our models:

- The model contour cannot have any self-intersections.

See Figure 22 for details.

B.2 Pants

A skeletal model of pants was devised, whose parameters are shown in Figure 23.[9]

[9] In all of these models, the preferred representation of parameters was in Cartesian coordinates. We additionally explored optimizing directly over angles and lengths. In practice, however, the optimization worked best when all parameters were aperiodic and similarly scaled. Hence, whenever possible, a length/angle combination was represented by a two-dimensional point.

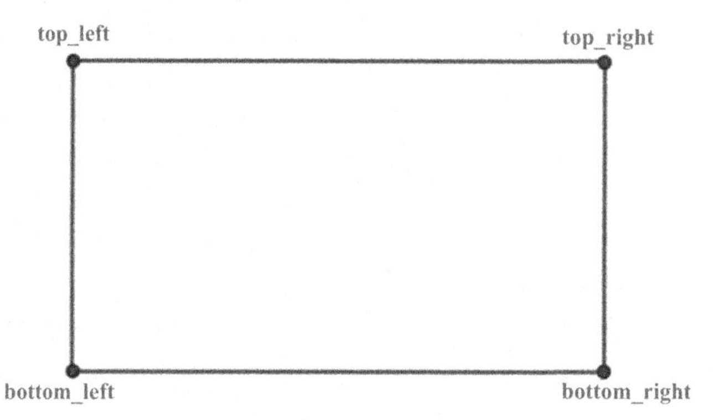

Figure 22
A towel model has eight total parameters, corresponding to four skeletal points. These are simply the four corners of the towel.

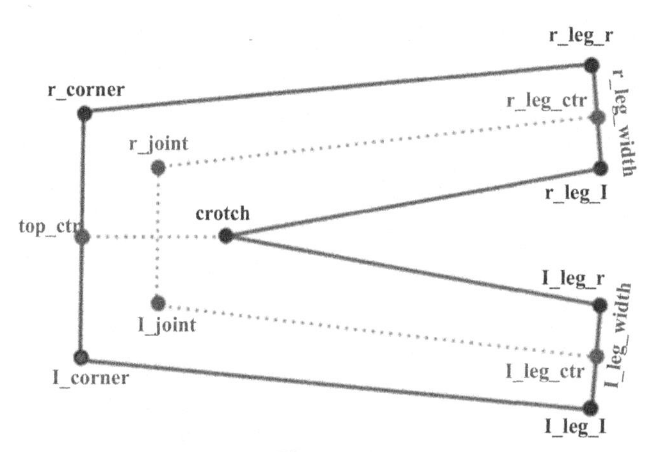

Figure 23
The pants skeleton is defined by 14 scalar parameters, corresponding to six skeletal points, and two scalar values, denoting the width of each pant leg. The remaining landmark points are generated as follows: the right corner is an extrapolation of the distance from the left corner to the top center; the crotch is the top center mirrored about the axis spanning the left and right joints; the leg corners are determined by the line perpendicular to the leg axis, at a distance specified by the leg width.

We found it was best to give the pants model as much freedom as possible. Therefore, only a small number of constraints were imposed, penalizing extreme deviations from the norm of[10]:

[10] For the precise numerical constraints of all of our models, see the attached code at http://rll.berkeley.edu/IJRR2011.

- the length of the legs relative to the height of the pants;
- the width of the legs relative to the width of the pants;
- the width of the pants relative to the height.

For the fitting of pants, two different initializations were attempted: the first with the legs virtually straight, and the second with the legs widely spaced. Both models were fit, and the one with the lowest final cost function was chosen.

B.3 Short-Sleeved Shirts

A skeletal model of short-sleeved shirts was also used, detailed in Figure 24.

In order to guide the optimization, a number of constraints were imposed, restricting:

- the location of the collar points with respect to the neck and shoulders;
- the location of the shoulders with respect to the armpits;
- the angle between the spine and horizontal axis;
- the relative size and angle of the sleeves;
- the width–height ratios of the sleeves and torso.

Two different initializations were attempted: the first with medium-length sleeves, and the second with extremely short sleeves. Both models were run, and the one with the lowest final cost function was chosen.

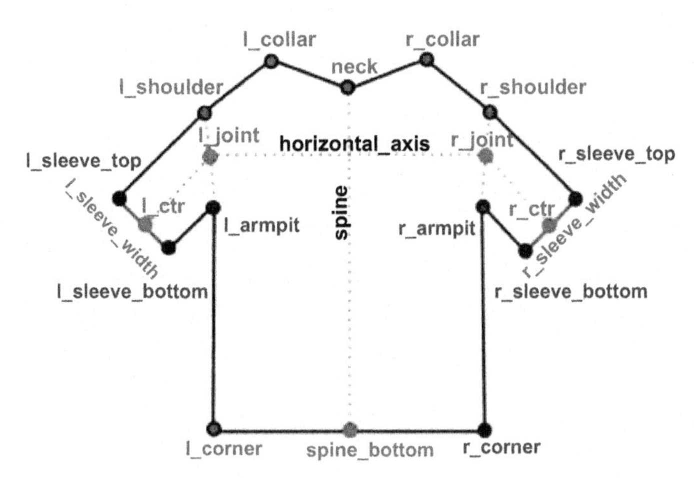

Figure 24

A short-sleeved shirt skeleton is defined by 24 parameters, corresponding to 11 skeletal points and two scalar parameters for sleeve width. The remaining landmark points are generated as follows: the right corner is found by extrapolating the line from the left corner to the spine bottom; the armpit is determined by extrapolating the line from the shoulder to the shoulder joint; the sleeve corners are determined by the line perpendicular to the sleeve axis, at a distance specified by the sleeve width.

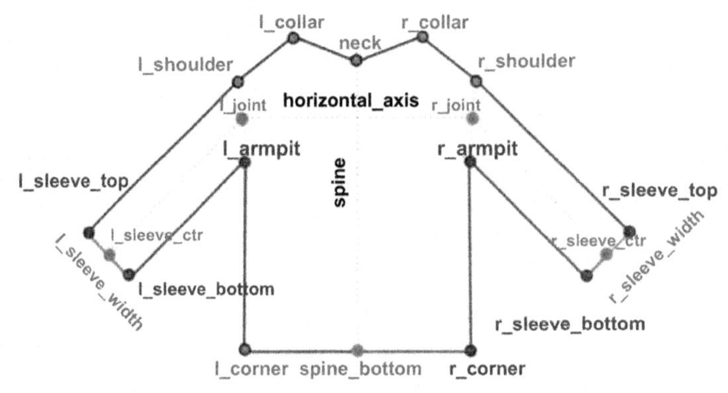

Figure 25
A long-sleeved shirt skeleton is defined by the same parameters as the short-sleeved skeleton.

In addition to the orientation, symmetric, and asymmetric phases of optimization, a fourth fine-tuning phase was run. In this phase, the location of all sleeve vertices were free to move, while the rest remained fixed. This was meant to account for the irregular shape of many T-shirt sleeves.

B.4 Long-Sleeved Shirts and Sweaters

The skeletal model for long-sleeved shirts is detailed in Figure 25.

This model is virtually identical to the short-sleeved model, with a single constraint added:

- Each sleeve must be at least twice as long as it is wide.

Only one initialization was used, with the arms at a downward angle.

As long-sleeved shirts have the potential for drastic asymmetry, both the orientation and symmetric phases of optimization proved to be non-useful, and occasionally damaging, the former settling on erroneous angles, and the latter on vastly incorrect poses. In some cases, the error was so great that the asymmetric phase could not correct for it. For this reason, only the asymmetric phase of optimization was used on this model.

Appendix C: Black Box Numerical Optimization

We employ a simple coordinate descent approach to optimization. For initial input parameters $P \in ¡^K$, score function $C(P)$, and initial step size δ:

```
δ1,...,δk ← δ
S ← C(P)
for iter← 1: 100 do
```

```
for i ∈ 1:K do

    P′ ← P
    P′_i ← P′_i + δ_i
    S′ ← C(P′)

    if S′ > S then
        P ← P′
        S ← S′
          δ_i ← − δ_i * 1.5
        else
          δ_i ← −δ_i
          P′ ← P
          P′_i ← P′_i + δ_i
          S′ ← C(P′)

      If S′ > S then

            P ← P′
            S ← S′
            δ_i ← δ_i * 1.5
          else
            δ_i ← −δ_i*0.5

if max|δ|<0.001 then
      break
    P_out ← P
with P_out the fit parameters.
```

References

[1] D. Balkcom, M. Mason, Robotic origami folding, Int. J. Robotics Res. 27 (2008) 613–627.

[2] M. Bell, Flexible Object Manipulation (Ph.D. thesis), Dartmouth College, 2010.

[3] J. van den Berg, S. Miller, K. Goldberg, P. Abbeel, Gravity-based robotic cloth folding, in: Proceedings 9th International Workshop on Algorithmic Foundations of Robotics (WAFR), 2010.

[4] S. Miller, M. Fritz, T. Darrell, P. Abbeel, Parametrized shape models for clothing, in: Proceedings of ICRA, Berkeley, CA, 2011.

[5] M. Bell, D. Balkcom, Grasping non-stretchable cloth polygons, Int. J. Robotics Res. 29 (2010) 775–784.

[6] N. Fahantidis, K. Paraschidis, V. Petridis, Z. Doulgeri, L. Petrou, G. Hasapis, Robot handling of flat textile materials, IEEE Robotics Automation Mag 4 (1997) 34–41.

[7] F. Osawa, H. Seki, Y. Kamiya, Clothes folding task by tool-using robot, J. Robotics Mechatron 18 (5) (2006) 618–625.

[8] J. Maitin-Shepard, M. Cusumano-Towner, J. Lei, P. Abbeel, Cloth grasp point detection based on multiple-view geometric cues with application to robotic towel folding, in: Proceedings IEEE International Conference on Robotics and Automation, 2010, 2010.

[9] K. Salleh, H. Seki, Y. Kamiya, M. Hikizu, Inchworm robot grippers in clothes manipulation optimizing the tracing algorithm, in: International Conference on Intelligent and Advanced Systems, 2007. ICIAS 2007, 2007, pp. 1051–1055.
[10] G.J. Monkman, Robot grippers for use with fibrous materials, Int. J. Robotics Res. 14 (1995) 144–151.
[11] D. Baraff, A. Witkin, Large steps in cloth simulation, in: Proceedings of SIGGRAPH 1998, 1998.
[12] R. Bridson, R. Fedkiw, J. Anderson, Robust treatment of collisions, contact, and friction for cloth animation, in: Proceedings of SIGGRAPH 2002, 2002.
[13] K. Choi, H. Ko, Stable but responsive cloth, in: Proceedings of SIGGRAPH 2002, 2002.
[14] J. Liu, J. Dai, An approach to carton-folding trajectory planning using dual robotic fingers, Robotics Autonomous Syst. 42 (2003) 47–63.
[15] S.K. Gupta, D. Bourne, K. Kim, S. Krishnan, Automated process planning for robotic sheet metal bending operations, J. Manufacturing Syst. 17 (1998) 338–360.
[16] C. Bregler, J. Malik, Tracking people with twists and exponential maps, in: Proceedings 1998 IEEE Computer Society Conference on Computer Vision and Pattern Recognition, 1998, pp. 8–15.
[17] H. Sidenbladh, M. Black, L. Sigal, Implicit probabilistic models of human motion for synthesis and tracking, in: Computer Vision – ECCV 2002, 2002, pp. 784–800.
[18] D. Demirdjian, T. Ko, T. Darrell, Constraining human body tracking, in: IEEE International Conference on Computer Vision, 2003.
[19] R. Urtasun, T. Darrell, Sparse probabilistic regression for activity-independent human pose inference, in: IEEE Conference on Computer Vision and Pattern Recognition, 2008. CVPR 2008, 2008, pp. 1–8.
[20] M. Salzmann, R. Urtasun, Combining discriminative and generative methods for 3D deformable surface and articulated pose reconstruction, in: Conference on Computer Vision and Pattern Recognition (CVPR), 2010.
[21] G. Borgefors, Hierarchical chamfer matching: a parametric edge matching algorithm, IEEE Trans. Pattern Anal. Machine Intell. 10 (1988) 849–865.
[22] T. Cootes, C. Taylor, Statistical Models of Appearance for Medical Image Analysis and Computer Vision, 2001. http://www.isbe.man.ac.uk/~bim/Papers/asm_aam_overview.pdf.
[23] P. Guan, O. Freifeld, M.J. Black, A 2D human body model dressed in eigen clothing (Lecture Notes in Computer Science, vol. 6311). European Conference on Computer Vision, ECCV, Part I, Springer, Berlin, 2010, pp. 285–298.
[24] D. Anguelov, P. Srinivasan, D. Koller, S. Thrun, J. Rodgers, J. Davis, SCAPE: shape completion and animation of people, ACM Trans. Graph 24 (3) (2005) 408–416.
[25] B. Rosenhahn, U. Kersting, K. Powell, R. Klette, G. Klette, H. Seidel, A system for articulated tracking incorporating a clothing model, Machine Vis. Appl. 18 (2007) 25–40.
[26] M. Salzmann, P. Fua, Reconstructing sharply folding surfaces: a convex formulation, in: Conference on Computer Vision and Pattern Recognition (CVPR), 2009.
[27] F. Osawa, H. Seki, Y. Kamiya, Unfolding of massive laundry and classification types by dual manipulator, JACIII 11 (2007) 457–463.
[28] K. Hamajima, M. Kakikura, Planning strategy for task of unfolding clothes, in: Proceedings of ICRA, vol. 32, 2000, pp. 145–152.
[29] K. Yamakazi, M. Inaba, A cloth detection method based on image wrinkle feature for daily assistive robots, in: IAPR Conference on Machine Vision Applications, 2009, pp. 366–369.
[30] H. Kobori, Y. Kakiuchi, K. Okada, M. Inaba, Recognition and motion primitives for autonomous clothes unfolding of humanoid robot, in: Proceedings of IAS 2010, 2010.
[31] Y. Kita, F. Saito, N. Kita, A deformable model driven visual method for handling clothes, in: Proceedings of ICRA, 2004.
[32] K. Wyrobek, E. Berger, H.F.M. Van der Loos, K. Salisbury, Towards a personal robotics development platform: rationale and design of an intrinsically safe personal robot, in: Proceedings of ICRA, 2008.
[33] C. Rother, V. Kolmogorov, A. Blake, "GrabCut": interactive foreground extraction using iterated graph cuts, ACM Trans. Graph 23 (2004) 309–314.

[34] M. Muja, D.G. Lowe, Fast approximate nearest neighbors with automatic algorithm configuration, in: International Conference on Computer Vision Theory and Application (VISSAPP), 2009.
[35] G. Bradski, The OpenCV Library, vol. 25, Dr Dobb's J Software Tools, 2000, pp. 120–123.
[36] S.B. Needleman, C.D. Wunsch, A general method applicable to the search for similarities in the amino acid sequence of two proteins, J. Mol. Biol. 48 (1970) 443–453.
[37] H. Sakoe, S. Chiba, Dynamic programming algorithm optimization for spoken word recognition, IEEE Trans. Acoust. Speech Signal Process. 26 (1978) 43–49.

CHAPTER 5.3

Robust Visual Servoing[1]

D. Kragic, H.I. Christensen

Center for Autonomous Systems, Royal Institute of Technology, Stockholm, Sweden

Chapter Outline

[1] D. Kragic and H.I. Christensen, International Journal of Robotics Research, Vol. 22, No. 10–11, pp. 923–939, copyright © 2013 by SAGE Publications, reprinted by Permission of SAGE.

Household Service Robotics. http://dx.doi.org/10.1016/B978-0-12-800881-2.00018-9

5.3.1 Introduction

Robotics is gradually expanding its application domain beyond manufacturing. In manufacturing settings, it is possible to engineer the environment so as to simplify detection and handling of objects. In an industrial context, this is often achieved through careful selection of background color and lighting. As the application domain is expanded, the use of engineering to simplify the perception problem is becoming more difficult. It is no longer possible to assume a given setup of lighting and a homogeneous background. Recent progress in *service robotics* shows a need to equip robots with facilities for operation in everyday settings where the design of computer vision methods to facilitate robust operation and to enable interaction with objects has to be reconsidered.

In this chapter we consider the use of computational vision for manipulation of objects in everyday settings. The process of manipulation of objects involves all aspects of recognition/detection, servoing to the object, alignment and grasping. Each of these processes has typically been considered independently or in relatively simple environments [1]. Through careful consideration of the task constraints and combination of multiple methods it is, however, possible to provide a system that exhibits robustness in realistic settings.

The chapter starts with a motivation that argues for integration of visual processes in Section 5.3.2. A key competence for robotic grasping is the recognition/detection of objects as outlined in Section 5.3.3. Once the object has been detected, the reaching phase requires use of a coarse servoing strategy which can be based on "simple" visual features. Unfortunately, "simple" visual features suffer from a limited generality requiring therefore an integration of multiple cues to achieve robustness as described in Section 5.3.4. Once the object has been recognized and an approximate alignment has been achieved, it is possible to use model-based methods for accurate interaction with the object as outlined in Section 5.3.5. Finally, all of these components are integrated into a complete system and used for a series of experiments. The results from these experiments are presented and discussed in Section 5.3.6. The overall approach and the associated results are discussed in Section 5.3.7.

5.3.2 Motivation

Robotic manipulation in an everyday setting is typically carried out in the context of a task such as "please, fetch the cup from the dinner table in the living room" or "please, fetch the rice package from the shelf" (see Figure 1). To execute such a manipulation task, we use a mobile platform with a manipulator on the top. The robot is equipped with

Figure 1
An example of a robot task: fetching a rice package from a shelf.

facilities for automatic localization and navigation [2], allowing it to arrive at a position in the vicinity of the dinner table. From there the robot is required to carry out the following.

1. Recognize a cup on the table (discussed in Section 5.3.3).
2. Transportation—servo to the vicinity of the cup, potentially involving both platform and manipulator motion (discussed in Section 5.3.4).
3. Estimate the pose (position and orientation) of the object (Section 5.3.5).
4. Alignment—servo to the object to allow grasping (Section 5.3.5).
5. Pick up the object and drive away.

In a typical scenario, the distance between the object and the on-board camera may vary significantly. This implies a significant uncertainty in terms of scale (i.e., the size of the object in the image). Based on the actual scale of the object in the image, a hierarchical strategy for visual servoing may be used. For distant objects ($\geq$1 m), there is no point in attempting to perform a full pose estimation, as the size of the object in the image typically will not provide the necessary information. A coarse position estimate is therefore adequate for an initial alignment.

A visual servoing task in general includes some form of (1) *positioning*, such as aligning the robot/gripper with the target, and (2) *tracking*, updating the position/pose of the image features/object. Typically, image information is used to measure the error between some current and reference/desired location. Image information used to perform the task is either (1) two-dimensional, using image plane coordinates, or (2) three-dimensional, retrieving the pose parameters of objects with respect to the camera/world/robot coordinate system. So, the robot is controlled using image information as either two- or three-dimensional, which classifies the visual servoing approaches as (1) image based, (2) position based or (3) hybrid visual servoing (2.5D servoing) [3].

If the object is far from the camera, due to the resolution/size of the object in the image, it is advantageous to use an image-based strategy to perform the initial alignment. Image-based servoing using crude image features such as center of mass is well suited for coarse alignment. A problem here is that visual cues such as color, edges, corners or fiducial marks in general are highly sensitive to variation in illumination, etc. and no single one of them will provide robust frame by frame estimates that allow continuous tracking. Through their integration it is however possible to achieve an increased robustness, as described in Section 5.3.4.

For the accurate alignment of the gripper with the object, we can utilize (1) a binocular camera pair, (2) a monocular image in combination with a geometric model, or (3) one camera in combination with a "structure-from-motion" approach. As the object has been recognized and its approximate image position is known, it is possible to utilize this information for efficient pose estimation using a wire-frame model. Model-based fitting is highly sensitive to surface texture and background noise. The availability of a good initial estimate does, however, allow a robust estimation of the pose, even in the presence of significant noise. Once a pose estimate is available, it is possible to use 2.5D or position-based servoing for the alignment of the gripper with the objects. This is even possible in the presence of significant perspective effects, which typically is the case when operating close to objects. Model-based tracking is in general highly sensitive to surface texture and initialization, but through its integration into a system context these problems are reduced to a minimum as outlined in Section 5.3.5.

Finally, as the gripper is approaching the object it is essential to use the earlier mentioned geometric model for grasp planning, an issue not discussed in detail here. For an introduction, consult [4] and [5].

5.3.3 Detection and Pose Estimation

Given a view of the scene, the robot should be able to find/recognize the object to be manipulated or give us an answer such as "I-did-not-find-the-cup". This task is very complicated and hard to generalize with respect to cluttered environments and types of objects. Depending on the implementation, the recognition process may provide the partial/full pose of the object which can in return be used for generating the robot control sequence. Object recognition is a long standing research issue in the field of computational vision [6]. We consider limited aspects of it, but demonstrate recognition in real environments, a problem that has received limited attention so far.

5.3.3.1 Appearance-Based Method

The object to be manipulated is recognized using the view-based Support Vector Machine system presented in [7]. The objective here is to detect and recognize everyday household objects in a scene.

The recognition step delivers the image position and approximate size of the image region occupied by the object (see Figure 2). This information is in our case used (1) to track the part of the image, the *window of attention*, occupied by the object while the robot approaches it, or (2) as the input to the pose estimation algorithm.

5.3.3.2 Feature-Based Method

In the case of a moving object, its appearance may change significantly between frames. For this reason, we have also exploited a feature or a cue-based method. Here, color and motion were used in an integrated framework where voting was used as an underlying integration strategy. This is discussed in more detail in the next section.

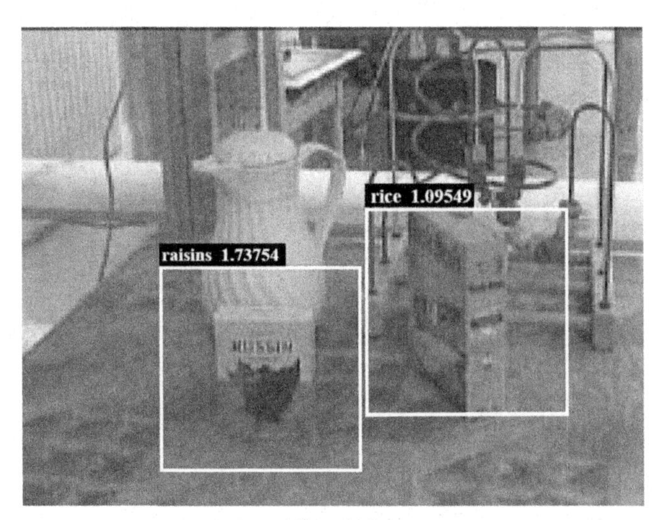

Figure 2
An example where the recognition system successfully recognizes two of the objects. Above each rectangle there is a value representing the recognition confidence which is proportional to the distance of the object to the hyperplane of the classifier. In order to make confidence measures comparable across all the classifiers, the distances are normalized with the margin of the hyperplane of the classifier. If the confidence is greater than 1, a positive detection/ recognition is assumed. If the confidence is less than 1, the classification is considered uncertain.

5.3.4 Transportation: Coarse Visual Servoing

While approaching the object, we want to keep in the field of view, or even centered, of the image. This implies that we have to estimate the position/velocity of the object in the image and use this information to control the mobile platform.

Our tracking algorithm employs the four-step *detect–match–update–predict loop* (Figure 3). The objective here is to track a part of an image (a region) between frames. The image position of its center is denoted by $\mathrm{p} = [x,y]^{\mathrm{T}}$. Hence, the state is $\mathrm{x} = [x \quad y \quad \dot{x} \quad \dot{y}]^{\mathrm{T}}$ where a piecewise constant white acceleration model is used [8]:

$$\begin{aligned} \mathrm{x}_{k+1} &= \mathrm{F}\mathrm{x}_k + \mathrm{G}v_k \\ z_k &= \mathrm{H}\mathrm{x}_k + w_k \end{aligned} \tag{1}$$

Here, v_k is a zero-mean white acceleration sequence, w_k is measurement noise and

$$\mathrm{F} = \begin{bmatrix} 1 & 0 & \Delta T & 0 \\ 0 & 1 & 0 & \Delta T \\ 0 & 0 & 1 & 0 \\ 0 & 0 & 0 & 1 \end{bmatrix}, \mathrm{G} = \begin{bmatrix} \frac{\Delta T^2}{2} & 0 \\ 0 & \frac{\Delta T^2}{2} \\ \Delta T & 0 \\ 0 & \Delta T \end{bmatrix}, \tag{2}$$

$$\mathrm{H} = \begin{bmatrix} 1 & 0 & 0 & 0 \\ 0 & 1 & 0 & 0 \end{bmatrix}$$

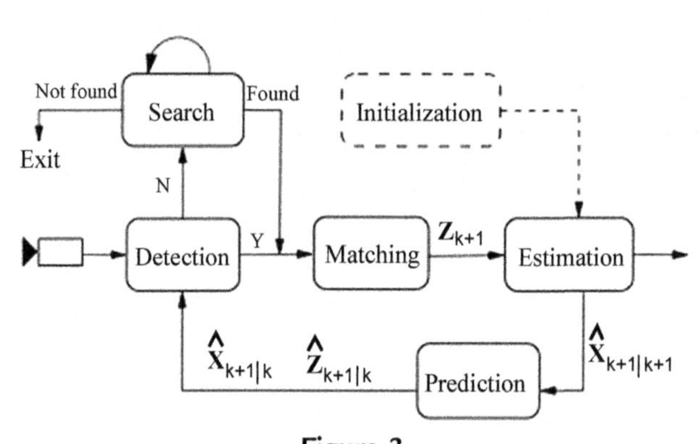

Figure 3
A schematic overview of the tracking system.

For prediction and estimation, the $\alpha{-}\beta$ filter is used [8]

$$\begin{aligned} \widehat{x}_{k+1|k} &= \mathrm{F}_k \widehat{x}_k \\ \widehat{z}_{k+1|k} &= \mathrm{H}\widehat{x}_{k+1|k} \\ \widehat{x}_{k+1|k+1} &= \widehat{x}_{k+1|k} + \mathrm{W}[z_{K+1} - \widehat{z}_{k+1|k}] \end{aligned} \tag{3}$$

with

$$\mathrm{W} = \begin{bmatrix} \alpha & 0 & \dfrac{\beta}{\Delta T} & 0 \\ 0 & \alpha & 0 & \dfrac{\beta}{\Delta T} \end{bmatrix}$$

where α and β are determined using steady-state analysis.

5.3.4.1 Voting

Voting, in general, may be viewed as a method to deal with n input data objects, c_i, having associated votes/weights w_i (n input data–vote pairs (c_i, w_i)) and producing the output data–vote pair (y, v) where y may be one of c_i or some mixed item. Hence, voting combines information from a number of sources and produces outputs which reflect the consensus of the information.

The reliability of the results depends on the information carried by the inputs and, as we will see, their number. Although there are many voting schemes proposed in literature, mean, majority and plurality voting are the most common ones. In terms of voting, a visual cue may be motion, color or disparity. Mathematically, a cue is formalized as a mapping from an action space, **A**, to the interval [0,1]:

$$\mathrm{c} : \mathbf{A} \rightarrow [0, 1]. \tag{4}$$

This mapping assigns a *vote* or a preference to each action $a \in \mathbf{A}$, which may, in the context of tracking, be considered as the position of the target. These votes are used by a *voter* or a *fusion center*, δ (**A**). Based on the ideas proposed in [9] and [10]; we define the following voting scheme:

DEFINITION 1. Weighted Plurality Approval Voting. For a group of homogeneous cues, $\mathbf{C} = \{c_1, \ldots, c_n\}$, where n is the number of cues and O_{c_i} is the output of a cue i, a weighted plurality approval scheme is defined as

$$\delta(a) = \sum_{i}^{n} w_i O_{c_i}(a) \tag{5}$$

where the most appropriate action is selected according to

$$a' = \arg\max\{\delta(a)|a \in \mathrm{A}\}. \tag{6}$$

5.3.4.2 Visual Cues

The cues considered in the integration process are as follows.

Correlation. The standard sum of squared differences (SSD) similarity metric is used and the position of the target is found as that giving the lowest dissimilarity score.

$$SSD(u, v) = \sum_{n} \sum_{m} [I(u+m, v+n) - T(m, n)]^2, \tag{7}$$

where $I(u, v)$ and $T(u, v)$ represent the gray-level values of the image and the template, respectively. To compensate for changes in the appearance of the tracked region, the template is updated on a 25 frames cycle.

Color. It has been shown in [11] that efficient and robust results can be achieved using the chromatic color space. Chromatic colors, known as "pure" colors without brightness, are obtained by normalizing each of the components by the total sum. Color is represented by r and g components, since the blue component is both the noisiest channel and it is redundant after the normalization.

Motion. Motion detection is based on computation of the temporal derivative and the image is segmented into regions of motions and regions of inactivity. This is estimated using image differencing

$$M[(u, v), k] = H[|I[(u, v), k] - I[(u, v), k-1]| - \Gamma] \tag{8}$$

where Γ is a fixed threshold and H is defined as

$$H(x) = \begin{cases} 0 : x \leq 0 \\ x : x > 0 \end{cases} \tag{9}$$

Intensity variation. In each frame, the following is estimated for all $m \times m$ (details about m are given in Section 5.3.4.4.2) regions inside the tracked window.

$$\sigma^2 = \frac{1}{m^2} \sum_{u} \sum_{v} \left[I(u, v) - \bar{I}(u, v)\right]^2 \tag{10}$$

where $\bar{I}(u, v)$ is the mean intensity value estimated for the window. For example, for a mainly uniform region, low variation is expected during tracking. On the other hand, if the region is rich in texture, large variation is expected. The level of texture is evaluated as proposed in [12].

5.3.4.3 Weighting

In Eqn (5) it is defined that the outputs from individual cues should be weighted by w_i. Consequently, the reliability of a cue should be estimated and its weight determined based on its ability to track the target. The reliability can be either (1) determined a priori and kept constant during tracking, or (2) estimated during tracking based on the cue's success in estimating the final result or based on how much it is in agreement with other cues. In our previous work, the following methods were evaluated [13]:

1. **Uniform weights**. Outputs of all cues are weighted equally, $w_i = 1/n$, where n is the number of cues.
2. **Texture-based weighting**. Weights are preset and depend on the spatial content of the region. For a highly textured region, we use: color (0.25), image differencing (0.3), correlation (0.25), intensity variation (0.2). For uniform regions, the weights are: color (0.45), image differencing (0.2), correlation (0.15), intensity variation (0.2). The weights were determined experimentally.
3. **One-step distance weighting**. The weighting factor, w_i, of a cue, c_i, at time step k depends on the distance from the predicted image position, $\hat{z}_{k|k-1}$. Initially, the distance is estimated as

$$d_i = \left\| z_k^i - \hat{z}_{k|k-1} \right\| \tag{11}$$

and errors are estimated as

$$e_i = \frac{d_i}{\sum_{i=1}^{n} d_i}. \tag{12}$$

Weights are then inversely proportional to the error with $\sum_{i=1}^{n} w_i = 1$.

4. **History-based distance weighting**. The weighting factor of a cue depends on its overall performance during the tracking sequence. The performance is evaluated by observing how many times the cue was in agreement with the rest of the cues. The following strategy is used.
   a. For each cue, c_i, examine if $\|z_k^i - z_k^j\| \langle d_T$ where $i, j = 1, \ldots, n$ and $i \neq j$. If this is true, $a_{ij} = 1$, otherwise $a_{ij} = 0$. Here, $a_{ij} = 1$ means that there is an agreement between the outputs of cues i and j at that voting cycle and d_T represents a distance threshold which is set in advance.
   b. Build the $(n-1)$ value set for each cue, $c_i : \{a_{ij} | j = 1, \ldots, n \text{ and } i \neq j\}$ and estimate sum $s_i = \sum_{j=1}^{n} a_{ij}$.

c. The accumulated values during N tracking cycles, $S_i = \sum_{k=1}^{N} s_i^k$, indicate how many times a cue, c_i, was in agreement with other cues. Weights are then simply proportional to this value:

$$w_i = \frac{S_i}{\sum_{i=1}^{n} S_i} \text{ with } \sum_{i}^{n} w_i = 1. \tag{13}$$

5.3.4.4 Implementation

We have investigated two approaches where voting is used for (1) *response fusion*, and (2) *action fusion*. The first approach makes use of "raw" responses from the employed visual cues in the image, which also represents the action space, **A**. Here, the response is represented either by a binary function (yes/no) answer, or in the interval [0,1] (these values are scaled between [0,255] to allow visual monitoring). The second approach uses a different action space represented by a *direction* and a *speed* (see Figure 4). Compared to the first approach, where the position of the tracked region is estimated, this approach can be viewed as estimating its velocity. Again, each cue votes for different actions from the action space, **A**, which is now the velocity space.

5.3.4.4.1 Initialization

According to Figure 3, a tracking sequence should be initiated by *detecting* the target object. If a recognition module is not available, another strategy can be used. In [14] it is proposed that *selectors* should be employed which are defined as heuristics that select regions possibly occupied by the target. When the system does not have definite state information about the target, it should actively search the state space to find it. Based on this idea, color and image differences (or foreground motion) may be used to detect the target in the first image. Again, if a recognition module is not available, these two cues may also be used in cases where the target either (1) has left the field of view, or (2) was occluded for a few frames. Our system searches the whole image for the target and once the target enters the image, tracking is regained.

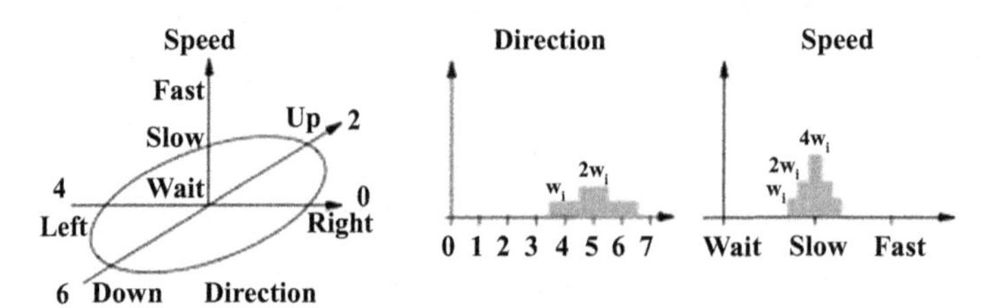

Figure 4
A schematic overview of the *action fusion* approach; the desired direction is (*down and left*) with a (*slow*) speed. The use of bins represents a *neighborhood voting scheme* which ensures that slight differences between different cues do not result in an unstable classification.

5.3.4.4.2 Response fusion approach

After the target is located, a template is initialized which is used by the correlation cue. In each frame, a color image of the scene is acquired. Inside the window of attention the response of each cue, denoted O_i, is evaluated (see Figure 5). Here, x represents a position:

Color. During tracking, all pixels whose color falls in the pre-trained color cluster are given a value between [0,255] depending on the distance from the center of the cluster $0 \leq O_{color}(\mathrm{x},k) \leq 255$ with

$$x \in \left[\widehat{z}_{k|k-1} - 0.5\mathrm{x}_w, \widehat{z}_{k|k-1} + 0.5\mathrm{x}_w\right] \tag{14}$$

where x_w is the size of the window of attention.

Motion. Using Eqns (8) and (9) with $\Gamma = 10$, the image is segmented into regions of motion and inactivity:

$$\begin{gathered} 0 \leq O_{motion}(\mathrm{x}, k) \leq 255 - \Gamma \quad \text{with} \\ \mathrm{x} \in \left[\widehat{z}_{k|k-1} - 0.5\mathrm{x}_w, \widehat{z}_{k|k-1} + 0.5\mathrm{x}_w\right]. \end{gathered} \tag{15}$$

Correlation. Since the correlation cue produces a single position estimate, the output is given by

$$\begin{gathered} O_{SSD}(x, k) = 255e^{\left(-\frac{\overline{x}^2}{2\sigma^2}\right)} \quad \text{with} \\ \sigma = 5, \quad x \in \left[z_{SSD} - 0.5x_w, z_{SSD} + 0.5x_w\right], \\ \overline{\mathrm{x}} \in \left[-0.5\mathrm{x}_w, 0.5\mathrm{x}_w\right] \end{gathered} \tag{16}$$

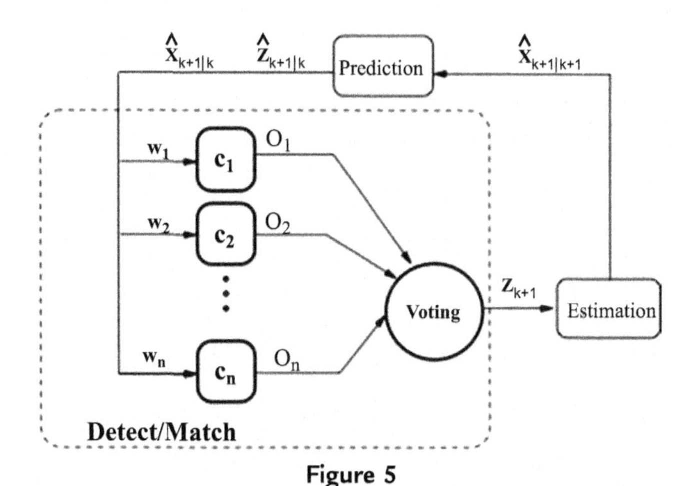

Figure 5
A schematic overview of the *response fusion* approach.

where the maximum of the Gaussian is centered at the peak of the SSD surface. The size of the search area depends on the estimated velocity of the region.

Intensity variation. The response of this cue is estimated according to Eqn (10). If a low variation is expected, all pixels inside an $m \times m$ region are given values (255-σ). If a large variation is expected, pixels are assigned a σ value directly. The size $m \times m$ of the subregions which are assigned the same value depends on the size of the window of attention with $n = 0.2\mathrm{x}_w$. Hence, for a 30×30 pixel window of attention, $m = 6$. The result is presented as follows:

$$0 \leq O_{\mathrm{var}}(\mathrm{x}, k) \leq 255 \quad \text{with}$$
$$\mathrm{x} \in \left[\widehat{z}_{k|k-1} - 0.5\mathrm{x}_w, \widehat{z}_{k|k-1} + 0.5\mathrm{x}_w\right] \tag{17}$$

Response fusion. The estimated responses are integrated using Eqn (5):

$$\delta(\mathrm{x}, k) = \sum_{i}^{n} w_i O_i(\mathrm{x}, k). \tag{18}$$

However, Eqn (6) cannot be directly used for selection, as there might be several pixels with same number of votes. Therefore, this equation is slightly modified:

$$\delta'(\mathrm{x}, k) = \begin{cases} 1: & \begin{array}{l} \mathit{if}\ \delta(\mathrm{x}, k) \text{is arg max}\{\delta(\mathrm{x}', k)|\mathrm{x}' \\ \in \left[\widehat{z}_{k|k-1} - 0.5\mathrm{x}_w, \widehat{z}_{k|k-1} + 0.5\mathrm{x}_w\right]\} \end{array} \\ 0: & \mathit{otherwise} \end{cases} \tag{19}$$

Finally, the new measurement z_k is given by the mean value (first moment) of $\delta'(\mathrm{x}, k)$, i.e., $\mathrm{z}_k = \overline{\delta'}(\mathrm{x}, k)$.

5.3.4.4.3 Action fusion approach

Here, the action space is defined by a direction d and speed s (see Figure 4). Both the direction and the speed are represented by histograms of discrete values where the direction is represented by eight values (see Figure 6):

$$\begin{array}{l} \mathrm{LD}\begin{bmatrix} -1 \\ 1 \end{bmatrix}, \mathrm{L}\begin{bmatrix} -1 \\ 0 \end{bmatrix}, \mathrm{LU}\begin{bmatrix} -1 \\ -1 \end{bmatrix}, \mathrm{U}\begin{bmatrix} 0 \\ -1 \end{bmatrix}, \\ \mathrm{RU}\begin{bmatrix} 1 \\ -1 \end{bmatrix}, \mathrm{R}\begin{bmatrix} 1 \\ 0 \end{bmatrix}, \mathrm{RD}\begin{bmatrix} 1 \\ 1 \end{bmatrix}, \mathrm{D}\begin{bmatrix} 0 \\ 1 \end{bmatrix}, \end{array} \tag{20}$$

with L, R, D, and U denoting left, right, down, and up, respectively. Speed is represented by 20 values with 0.5 pixel interval, which means that the maximum allowed displacement between successive frames is 10 pixels (this is easily made adaptive based on the estimated

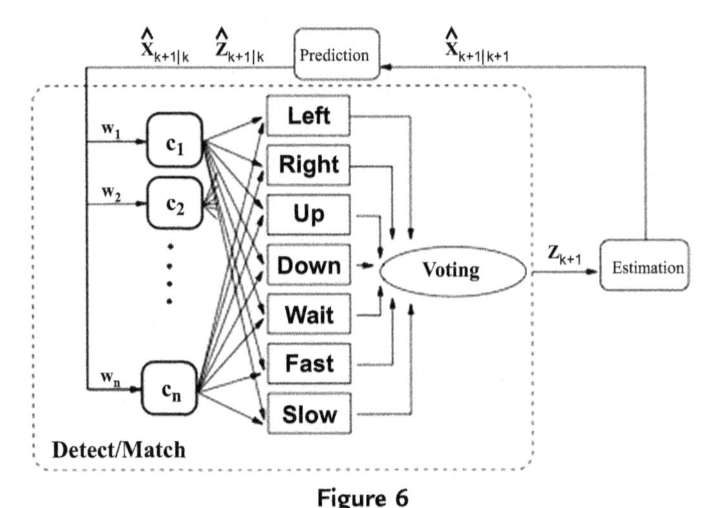

Figure 6
A schematic overview of the *action fusion* approach.

velocity). There are two reasons for choosing just eight values for the direction: (1) if the update rate is high or the inter-frame motion is slow, this approach will still give a reasonable accuracy and hence, a smooth performance; (2) by keeping the voting space rather small there is a higher chance that the cues will vote for the same action. Accordingly, each cue will vote for a desired direction and a desired speed. As presented in Figure 4, a *neighborhood voting scheme* is used to ensure that slight differences between different cues do not result in an unstable classification. Equation (3) is modified so that

$$\mathrm{H} = \begin{bmatrix} 0 & 0 & 1 & 0 \\ 0 & 0 & 0 & 1 \end{bmatrix} \quad and \quad \mathrm{W} = \begin{bmatrix} \alpha\Delta T & 0 & \beta & 0 \\ 0 & \alpha\Delta T & 0 & \beta \end{bmatrix}. \tag{21}$$

In each frame, the following is estimated for each cue:

Color. The response of the color cue is first estimated according to Eqn (14) followed by

$$\mathrm{a}_{color}(k) = \frac{\sum_{x} O_{color}(\mathrm{x}, k)\mathrm{x}(k)}{\sum_{\mathrm{x}} \mathrm{x}(k)} - \widehat{\mathrm{p}}_{k|k-1}$$
$$with \;\; \mathrm{x} \in \left[\widehat{\mathrm{p}}_{k|k-1} - 0.5\mathrm{x}_w, \widehat{\mathrm{p}}_{k|k-1} + 0.5\mathrm{x}_w\right] \tag{22}$$

where $\mathrm{a}_{color}(k)$ represents the desired action, and $\widehat{\mathrm{p}}_{k|k-1}$ is the predicted position of the tracked region. The same approach is used to obtain $\mathrm{a}_{motion}(k)$ and $\mathrm{a}_{var}(k)$.

Correlation. The minimum of the SSD surface is used as

$$\mathrm{a}_{SSD}(k) = \arg\min_{x} \left(SSD(\mathrm{x}, k)\right) - \widehat{\mathrm{p}}_{k|k-1} \tag{23}$$

where the size of the search area depends on the estimated velocity of the tracked region.

Action fusion. After the desired action, $a_i(k)$, for a cue is estimated, the cue produces the votes as follows:

$$\begin{aligned} &direction\ d_i = P(\text{sgn}(a_i)),\\ &speed\ s_i = \|a_i\|. \end{aligned} \tag{24}$$

Here, $P : \mathbf{x} \rightarrow \{0,1, ...,7\}$ is a scalar function that maps the two-dimensional direction vectors (see Eqn (20)) to one-dimensional values representing the bins of the direction histogram. Now, the estimated direction, d_i, and the speed, s_i, of a cue, c_i, with weight, w_i, are used to update the direction and speed of the histograms according to Figure 4 and Eqn (5). The new measurement is then estimated by multiplying the actions from each histogram which received the maximum number of votes according to Eqn (6)

$$z_K = S(\arg\max_d HD(d)) \arg\max_s HS(s) \tag{25}$$

where $S : x \rightarrow \left\{ \begin{bmatrix} -1 \\ 0 \end{bmatrix}, \ldots, \begin{bmatrix} -1 \\ 1 \end{bmatrix} \right\}$. The update and prediction steps are then performed using Eqns (21) and (3). The reason for choosing this particular representation instead of simply using a weighted sum of first moments of the responses of all cues is, as has been pointed out in [10]; that arbitration via vector addition can result in commands which are not satisfactory to any of the contributing cues.

5.3.4.5 Examples

The proposed methods have been investigated in detail in [13]. Here, we present one of the experiments where we evaluated the performance of the voting approaches as well as the performance of individual cues with respect to three sensor–object configurations typically used in visual servoing systems: (1) static sensor/moving object ("stand-alone camera system"); (2) moving sensor/static object ("eye-in-hand camera" servoing toward a static object); (3) moving sensor/moving object (camera system on a mobile platform or eye-in-hand camera servoing toward a moving object). The results are discussed through *accuracy* and *reliability* measures. The accuracy is expressed using an error measure which is a distance between the ground truth (chosen manually using a reference point on the object) and the currently estimated position of the reference point. The results are summarized through the mean square error and standard deviation in pixels. The measure of the reliability is on a yes/no basis depending on whether a cue (or the fused system) successfully tracks the target during a single experiment. The tracking is successful if the object is kept inside the window of attention during the entire test sequence.

Table 1: Qualitative results for various sensor–object configurations (in pixels)

	Static Sensor/Moving Object		Moving Sensor/Static Object		Moving Sensor/Moving Object	
	MSE	STD	MSE	STD	MSE	STD
RF	7	7	4	3	9	10
AF	7	9	4	10	13	25
Color	15	16	10	6	10	14
Diff.	23	26	Failed	Failed	Failed	Failed
SSD	25	27	12	13	17	21

AF, action fusion; RF, response fusion.

The two fusion approaches, as well as the individual cues, have been tested with respect to the ability to cope with occlusions of the target and to regain tracking after the target has left the field of view for a number of frames. The results are presented for correlation, color and image differences since the intensity variation cue cannot be used alone for tracking.

Accuracy (Table 1). As can be seen from the table, the best accuracy is achieved using the *response fusion* approach. Although the mean squared error (MSE) is similar for the *action fusion* approach in cases of *static sensor/moving object* and *moving sensor/static object* configurations, standard deviation (STD) is higher. The reason for this is the choice of the underlying voting space. For example, if the color cue shows a stable performance for a number of frames, its weight will be high compared to the other cues (or it might have been set to a high value from the beginning). In some cases, as in the case of a box of raisins presented in Figure 7, two colors are used at the same time. When an occlusion occurs, the position of the center of the mass of the color blob will change fast (and sometimes in different directions) which results in abrupt changes in both direction and speed. The other method, *response fusion*, on the other hand, does not suffer from this which results in a lower standard deviation value.

The comparison of the performance of the fusion approaches and the performance of the individual cues shows the necessity for fusion. Image differences alone cannot be used in cases of *moving sensor/static object* and *moving sensor/moving object* configurations since there is no ability to differ between the object and the background. As mentioned earlier,

Figure 7
Example images from a raisin package tracking.

during most of the sequences the target undergoes 3D motion which results in scale changes and rotations not modeled by SSD. It is obvious that these factors will affect this cue significantly, resulting in a large error as demonstrated in the table. This problem may be solved by using a better model (affine, quadratic; see [15]). It can also be seen that the color cue performed best of the individual cues. In the case of *moving sensor/static object*, after the tracking is initialized the color cue "sticks" to the object during the sequence and, since the background varies a little, the best accuracy is achieved compared to other configurations. During the other two configurations, the background will change containing also the same color as the target. This distracts the color tracker, resulting in increased error. The error is larger in the case of *static sensor/moving object* compared to *moving sensor/moving object* since in the test sequences the background included the target's color more often.

Reliability (Table 2). The reliability is expressed through a number of successful runs where the accuracy is obtained using the *texture-based weighting*. In Table 2, the obtained reliability results are ranked showing that color performs most reliably compared to other individual cues. In certain cases, especially when the influence of the background is not significant, this cue will perform satisfactorily. However, it will easily become distracted if the background takes a large portion of the window of attention and includes the target's color. Image differencing will depend on the size of the moving target with respect to the size of the window of attention and variations in lighting. In structured environments, however, this cue may perform well and may be considered in cases of a single moving target where the size of the target is small compared to the size of the image (or window of attention).

Figure 8 shows tracking accuracy for the proposed fusion approaches and for each of the cues individually. The plots and the table show the deviation from the ground truth value (in pixels). The target is a package of raisins. During this sequence, a number of occlusions occur (as demonstrated in the images), but the plots demonstrate a stable performance of the fusion approaches during the whole sequence. The color cue is, however, "fooled" by the box which is the same color as the target. The plots demonstrate

Table 2: Success rate for individual cues and fusion approaches

	Number of Successes	Number of Failures	%
RF voting	27	3	90
AF voting	22	8	73.3
Color	18	12	60
SSD	12	18	40
Diff.	7	23	23.3

AF, action fusion; RF, response fusion.

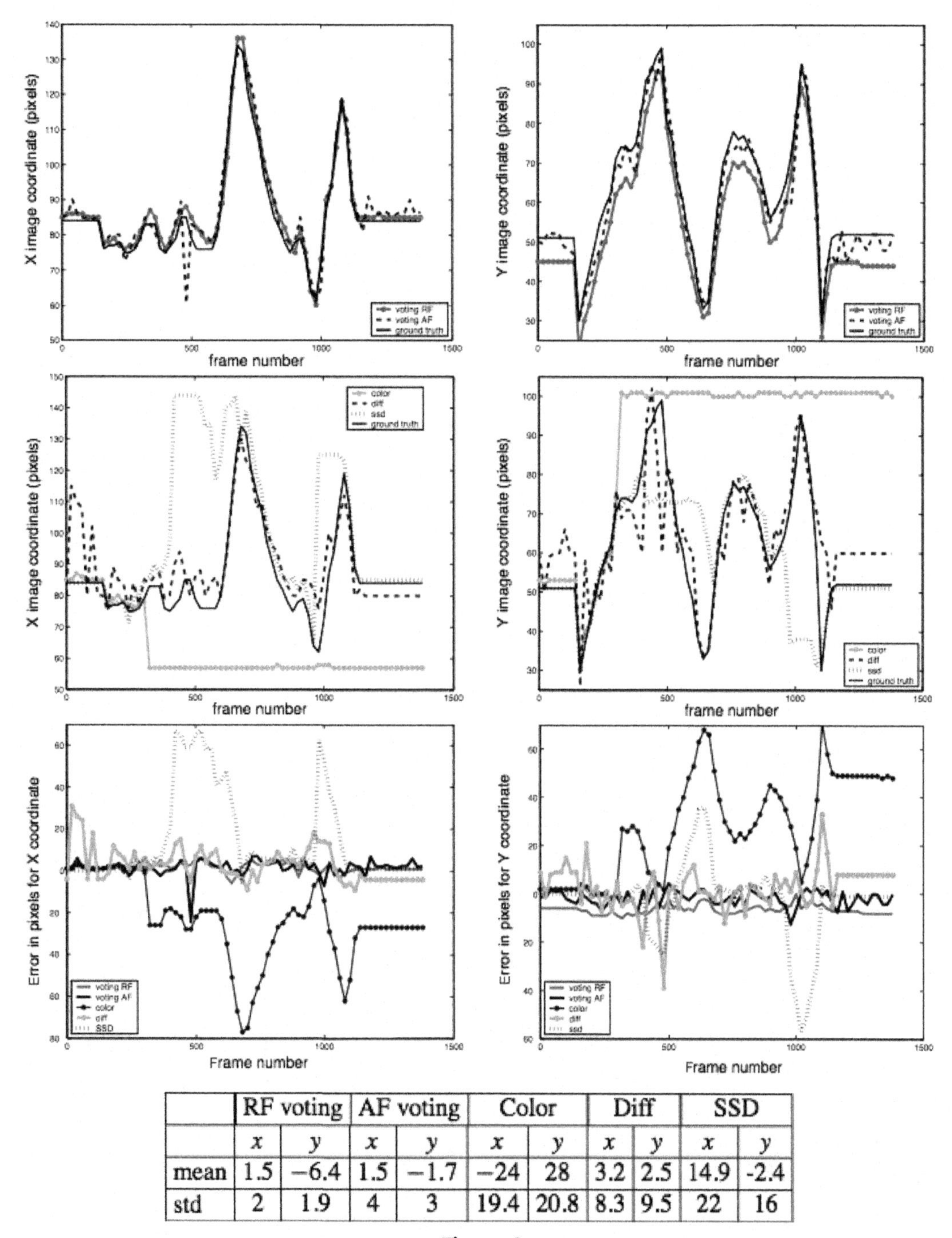

	RF voting		AF voting		Color		Diff		SSD	
	x	*y*	*x*	*y*	*x*	*y*	*x*	*y*	*x*	*y*
mean	1.5	−6.4	1.5	−1.7	−24	28	3.2	2.5	14.9	-2.4
std	2	1.9	4	3	19.4	20.8	8.3	9.5	22	16

Figure 8

The comparison between the ground truth, voting approaches and individual cues in case of occlusions (first two rows). Third row shows error plots for all approaches. The table represents the mean and standard deviation from the ground truth (in pixels). Some of the images from the tracking sequence are shown in Figure 7.

how this cue fails around frame 300 and never regains tracking after that. These two examples clearly demonstrate that tracking by fusion is far superior than any of the individual cues.

5.3.5 Model-Based Visual Servoing

Although suitable for a number of tasks, the previous approach lacks the ability to estimate position and orientation (pose) of the object. In terms of manipulation, it is usually required to accurately estimate the pose of the object, for example, to allow the alignment of the robot arm with the object or to generate a feasible pose for grasping. Using prior knowledge about the object, a special representation can further increase the robustness of the tracking system. Along with commonly used CAD models (wire-frame models), view-based and appearance-based representations may be employed [16].

A recent study of human visually guided grasps in situations similar to that typically used in visual servoing control has shown that the human visuomotor system takes into account the three-dimensional geometric features rather than the two-dimensional projected image of the target objects to plan and control the required movements [17]. These computations are more complex than those typically carried out in visual servoing systems and permit humans to operate in a large range of environments.

We have decided to integrate appearance-based and geometrical models in our model-based tracking system. Many similar systems use manual pose initialization where the correspondence between the model and object features is given by the user [18,19]. Although there are systems where this step is performed automatically, the approaches are time-consuming and not appealing for real-time applications [20,21]. One additional problem, in our case, is that the objects to be manipulated by the robot are highly textured (see Figure 9) and therefore not suited for matching approaches based on, for example, line features [22–24].

After the object has been recognized and its position in the image is known, an appearance-based method is employed to estimate its initial pose. The method we have implemented has been initially proposed in [25] where just three pose parameters have been estimated and used to move a robotic arm to a pre-defined pose with respect to the object. Compared to our approach, where the pose is expressed relative to the camera coordinate system, they express the pose relative to the current arm configuration, making the approach unsuitable for robots with a different number of degrees of freedom.

Compared to the system proposed in [26]; where the network has been entirely trained on simulated images, we use real images for training where no particular background was

Figure 9
Some of the objects we want the robot to manipulate.

considered. As pointed out in [26]; the illumination conditions (as well as the background) strongly affect the performance of their system and these cannot be easily obtained with simulated images. In addition, the idea of projecting just the wire-frame model to obtain training images cannot be employed in our case due to the texture of the objects. The system proposed in [23] also employs a feature-based approach where lines, corners and circles are used to provide the initial pose estimate. However, this initialization approach is not applicable in our case since, due to the geometry and textural properties, these features are not easy to extract with high certainty.

Our model-based tracking system is presented in Figure 10. During the *initialization* step, the initial pose of the object relative to the camera coordinate system is estimated. The main loop starts with a *prediction* step where the state of the object is predicted using the current pose estimate and a motion model. The visible parts of the object are then projected into the image (*projection and rendering* step). After the *detection* step, where a number of features are extracted in the vicinity of the projected ones, these new features are *matched* to the projected ones and used to estimate the new pose of the object. Finally, the calculated pose is input to the *update* step. The system has the ability to cope with partial occlusions of the object, and to successfully track the object even in the case of significant rotational motion.

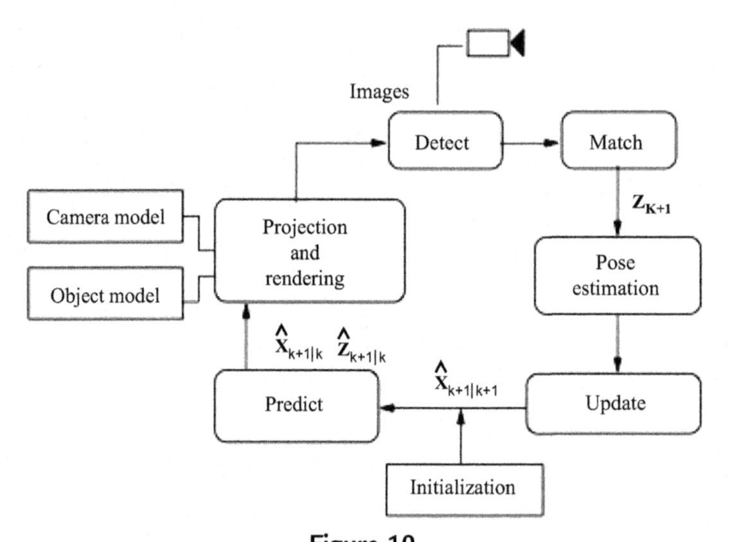

Figure 10
Block diagram of the model-based tracking system.

5.3.5.1 Prediction and Update

The system state vector consists of three parameters describing translation of the target, another three for orientation and an additional six for the velocities:

$$\mathrm{x} = \left[X, Y, Z, \phi, \varphi, \gamma, \dot{X}, \dot{Y}, \dot{Z}, \dot{\phi}, \dot{\varphi}, \dot{\gamma}\right]. \tag{26}$$

Here, φ, ψ, and γ represent roll, pitch, and yaw angles, respectively [27]. The following piecewise constant white acceleration model is considered [8]

$$\begin{aligned} \mathrm{x}_{k+1} &= \mathrm{F}\mathrm{x}_k + \mathrm{G}\mathrm{v}_k \\ \mathrm{z}_k &= \mathrm{H}\mathrm{x}_k + \mathrm{w}_k \end{aligned} \tag{27}$$

where v_k is a zero-mean white acceleration sequence, w_k is the measurement noise and

$$\mathrm{F} = \begin{bmatrix} \mathrm{I}_{6\times 6} & \Delta T \mathrm{I}_{6\times 6} \\ 0 & \mathrm{I}_{6\times 6} \end{bmatrix}, \quad \mathrm{G} = \begin{bmatrix} \dfrac{\Delta T^2}{2} \mathrm{I}_{6\times 6} \\ \Delta T \ \mathrm{I}_{6\times 6} \end{bmatrix}, \quad \mathrm{H} = \left[\mathrm{I}_{6\times 6} \quad | \quad 0\right]. \tag{28}$$

For prediction and update, the $\alpha-\beta$ filter is used:

$$\begin{aligned} &\widehat{\mathrm{x}}_{k+1|k} = \mathrm{F}_k \widehat{\mathrm{x}}_k \\ &\widehat{\mathrm{z}}_{k+1|k} = \mathrm{H}\widehat{\mathrm{x}}_{k+1|k} \\ &\widehat{\mathrm{x}}_{k+1|k+1} = \widehat{\mathrm{x}}_{k+1|k} + \mathrm{W}\left[\mathrm{z}_{K+1} - \widehat{\mathrm{z}}_{k+1|k}\right] \end{aligned} \tag{29}$$

Here, the pose of the target is used as a measurement rather than as image features, as commonly used in the literature; see, for example, [28] and [20]. An approach similar to that presented here is taken in [24]. This approach simplifies the structure of the filter which facilitates a computationally more efficient implementation. In particular, the dimension of the matrix **H** does not depend on the number of matched features in each frame but it remains constant during the tracking sequence.

5.3.5.2 Initial Pose Estimation

The initialization step uses the ideas proposed in [25]. During training, each image is projected as a point to the eigenspace and the corresponding pose of the object is stored with each point. For each object, we have used 96 training images (eight rotations for each angle on four different depths). One of the reasons for choosing this low number of training images is the workspace of the PUMA560 robot used. Namely, the workspace of the robot is quite limited and for our applications this discretization was satisfactory. To enhance the robustness with respect to variations in intensity, all images are normalized. At this stage, the size of the training samples is 100×100 pixel color images. The training procedure takes about 3 min on a Pentium III 550 running Linux.

Given an input image, it is first projected to the eigenspace. The corresponding parameters are found as the closest point on the pose manifold. Now, the wire-frame model of the object can be easily overlaid on the image. Since a low number of images is used in the training process, pose parameters will not accurately correspond to the input image. Therefore, a local refinement method is used for the final fitting (see Figure 11). The details are given in the next section.

During the training step, it is assumed that the object is approximately centered in the image. During task execution, the object can occupy an arbitrary part of the image. Since the recognition step delivers the image position of the object, it is easy to estimate the offset of the object from the image center and to compensate for it. In this way, the pose of the object relative to the camera frame can also be arbitrary.

An example of the pose initialization is presented in Figure 12. Here, the pose of the object in the training image (far left) was $X = -69.3$, $Y = 97.0$, $Z = 838.9$, $\varphi = 21.0$, $\psi = 8.3$, and $\gamma = -3.3$. After the fitting step, the pose was $X = 55.9$, $Y = 97.3$, $Z = 899.0$, $\varphi = 6.3$, $\psi = 14.0$, and $\gamma = 1.7$ (far right), showing the ability of the system to cope with significant differences in pose parameters.

5.3.5.3 Detection and Matching

When the estimate of the object's pose is available, the visibility of each edge feature is determined and internal camera parameters are used to project the model of the object

Figure 11
On the left is the initial pose estimated using the principal component analysis approach. On the right is the pose obtained by the local refinement method. (It is straightforward to estimate the desired velocity screw in the end-effector coordinate frame.)

Figure 12
Training image used to estimate the initial pose (far left) followed by the intermediate images of the fitting step.

onto the image plane. For each visible edge, a number of image points are generated along the edge. So-called tracking nodes are assigned at regular intervals in image coordinates along the edge direction. The discretization is performed using the Bresenham algorithm [29]. After this, a search is performed for the maximum discontinuity (nearby edge) in the intensity gradient along the normal direction to the edge. The edge normal is approximated with four directions: $\{-45, 0, 45, 90\}$ degrees.

In each point p_i along a visible edge, the perpendicular distance $d_i^{\perp}$ to the nearby edge is determined using a one-dimensional search. The search region is denoted by $\{S_i^j, j \in [-s, s]\}$. The search starts at the projected model point p_i and the traversal continues simultaneously in opposite search directions until the first local maximum is found. The size of the search region s is adaptive and inversely depends on the distance of the objects from the camera.

After the normal displacements are available, the method proposed in [18] is used. Lie group and Lie algebra formalism are used as the basis for representing the motion of a rigid body and pose estimation. The method is also related to the work presented in [30] and [31]. Implementation details can be found in [13].

5.3.5.4 Servoing Based on Projected Models

Once the pose of the object is available, any of the servoing approaches can be employed. In this section, we show how a model-based tracking system can be used for both image and position-based visual servoing.

5.3.5.4.1 Position-based servoing

Let us assume the following scenario. The task is to align the end-effector with respect to an object and maintain the constant pose when/if the object starts to move. It is assumed here that a model-based tracking algorithm is available and one stand-alone camera (not attached to the robot) is used during the execution of the tasks (see Figure 13).

Here, ${}^{o}X_G^*$ represents the desired pose between the object and the end-effector while ${}^{o}X_G$ represents the current (or initial) pose between them. To perform the task using the position-based servoing approach, the transformation between the camera and the robot

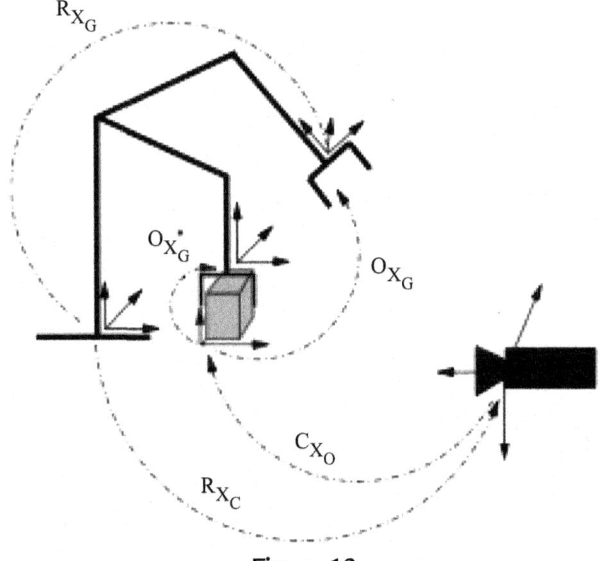

Figure 13
Relevant coordinate frames and their relationships for the "align-and-track" task.

coordinate frames, ${}^{C}X_{R}$, has to be known. The pose of the end-effector with respect to the robot base system, ${}^{R}X_{G}$, is known from the robot kinematics. The model-based visual tracking system estimates the pose of the object relative to the camera coordinate system, ${}^{C}X_{O}$.

Let us assume that the manipulator is controlled in the end-effector frame. According to Figure 13, if ${}^{o}X_{G} = {}^{o}X_{G}^{*}$ then ${}^{R}X_{G} = {}^{R}X_{G}^{*}$. The error function to be minimized may then be defined as the difference between the current and the desired end-effector pose:

$$\begin{aligned} \Delta^{R}t_{G} &= {}^{R}t_{G} - {}^{R}t_{G}^{*} \\ \Delta^{R}\theta_{G} &= {}^{R}\theta_{G} - {}^{R}\theta_{G}^{*} \end{aligned} \tag{30}$$

Here, ${}^{R}t_{G}$ and ${}^{R}\theta_{G}$ are known from the forward kinematics equations and ${}^{R}t_{G}^{*}$ and ${}^{R}\theta_{G}^{*}$ have to be estimated. The homogeneous transformation between the robot and desired end-effector frame is given by

$${}^{R}X_{G}^{*} = {}^{R}X_{C}\,{}^{C}X_{O}\,{}^{O}X_{G}^{*}. \tag{31}$$

The pose between the camera and the robot is estimated offline and the pose of the object relative to the camera frame is estimated using the model-based tracking system presented in Section 5.3.3. Expanding the transformations in Eqn (31) we obtain.

$${}^{R}t_{G}^{*} = {}^{R}R_{C}\,{}^{C}\widehat{R}_{O}\,{}^{O}t_{G}^{*} + {}^{R}R_{C}\,{}^{C}\widehat{t}_{O} + {}^{R}t_{C} \tag{32}$$

where ${}^{C}\widehat{R}_{O}$ and ${}^{C}\widehat{t}_{O}$ represent predicted values obtained from the tracking algorithm. A similar expression can be obtained for the change in rotation by using the addition of angular velocities (see Figure 13; [27]):

$${}^{R}\Omega_{G}^{*} = {}^{R}\Omega_{C} + {}^{R}R_{C}\,{}^{C}\widehat{\Omega}_{O} + {}^{R}R_{C}\,{}^{C}\widehat{R}_{O}\,{}^{O}\Omega_{G}^{*}. \tag{33}$$

Assuming that ${}^{R}R_{C}$ and ${}^{C}\widehat{R}_{O}$ are slowly-varying functions of time, integration of ${}^{R}\Omega_{G}^{*}$ gives [32]:

$${}^{R}\theta_{G}^{*} \approx {}^{R}\theta_{C} + {}^{R}R_{C}\,{}^{C}\widehat{\theta}_{O} + {}^{R}R_{C}\,{}^{C}\widehat{R}_{O}\,{}^{O}\theta_{G}^{*}. \tag{34}$$

Substituting Eqns (32) and (34) into Eqn (30) yields

$$\begin{aligned} \Delta^{R}t_{G} &= {}^{R}t_{G} - {}^{R}t_{C} - {}^{R}R_{C}\,{}^{C}\widehat{t}_{O} - {}^{R}R_{C}\,{}^{C}\widehat{R}_{O}\,{}^{O}t_{G}^{*} \\ \Delta^{R}\theta_{G} &\approx {}^{R}\theta_{G} - {}^{R}\theta_{C} - {}^{R}R_{C}\,{}^{C}\widehat{R}_{O}\,{}^{O}\theta_{G}^{*} \end{aligned} \tag{35}$$

which represents the error to be minimized:

$$e = \begin{bmatrix} \Delta^{R}t_{G} \\ \Delta^{R}\theta_{G} \end{bmatrix} \tag{36}$$

Figure 14
A sequence of a six-degree-of-freedom visual control. From an arbitrary starting position (upper left), the end-effector is controlled to a pre-defined reference position with respect to the target object (upper right). When the object starts moving, the visual system tracks the pose of the object. The robot is then controlled in a position-based framework to remain a constant pose between the gripper and the object frame.

After the error function is defined, a simple proportional control law is used to drive the error to zero. The velocity screw of the robot is defined as:

$$\dot{q} \approx Ke \tag{37}$$

Using the estimate of the object's pose and defining the error function in terms of pose, all six degrees of freedom of the robot are controlled as shown in Figure 14.

5.3.5.4.2 Image-based servoing

In many cases the change in the pose of the object is significant and certain features are likely to become occluded. An example of such motion is presented in Figure 15. A model-based tracking system allows us to use image positions of features (for example, end-points of an object's edges) even if they are not currently visible due to the pose of the object. An image-based approach can therefore successfully be used throughout the whole servoing sequence. In addition, since a larger number of features are available as well as the depth of the object, the feedback from one camera is sufficient for performing the tasks.

In our example, the error function, $\mathbf{e}(\mathbf{f})$, is defined which is to be regulated to zero. The vector of the final (desired) image point feature positions is denoted by f^* and the vector of their current positions by f^c. The task consists of moving the robot such that the

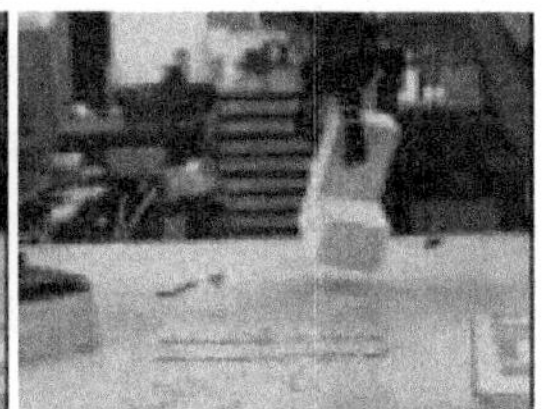

Figure 15
Start pose, destination pose and two intermediate poses in an image-based visual servo approach.

Euclidian norm of the error vector ($f^* - f^c$) decreases. Hence, we may constrain the image velocity of each point to exponentially reach its goal position with time. The desired behavior is then

$$\dot{f} = Ke(f) = K(f^* - f^c) \tag{38}$$

where **K** is a diagonal gain matrix that controls the convergence rate of the visual servoing. From [1]; a robot velocity screw may be computed using

$${}^{G}\dot{q} = KJ^{\dagger}(q)e(f) \tag{39}$$

where $\mathbf{J}^{\dagger}$ is a (pseudo-)inverse of the image Jacobian.

5.3.6 Example Tasks

In this section we show additional examples of how the presented visual systems were used for robotic tasks.

5.3.6.1 Mobile Robot Grasping

Here, we consider the problem of real manipulation in a realistic environment—a living room. Similarly to the previous example, we assume that a number of pre-defined grasps is given and a suitable grasp is generated depending on the current pose of the object. The experiment shows an XR4000 platform with a hand-mounted camera. The task is to approach the dinner table where there are several objects. The robot is instructed to pick up a package of rice having an arbitrary placement. Here, distributed control architecture [33] is used for integration of the different methods into a fully operational system. To perform the task, object recognition, voting-based two-dimensional tracking and model-based three-dimensional tracking are used. The details of the system implementation are reported in [2]. The results of a mission with the integrated system are outlined below.

The sequence in Figure 16 shows the robot starting at the far end of the living room, moving toward a point where a good view of the dinner table can be obtained. After the robot is instructed to pick up the rice package, it locates it in the scene using the system presented in Section 5.3.3.1. After that, the robot moves closer to the table keeping the rice package centered in the image using the two-dimensional tracking system presented in Section 5.3.4. Finally, the gripper is aligned with the object and grasping is performed. The details about the alignment can be found in [34].

5.3.6.2 Model-Based Tracking for Slippage Detection

The ability of a robotic system to generate both a *feasible* and a *stable* grasp adds to its robustness. By a *feasible* grasp, a kinematically feasible grasp is considered; by a *stable*

Figure 16
An experiment where the robot moves up to the table, recognizes the rice box, approaches it, picks it up and hands it over to the user.

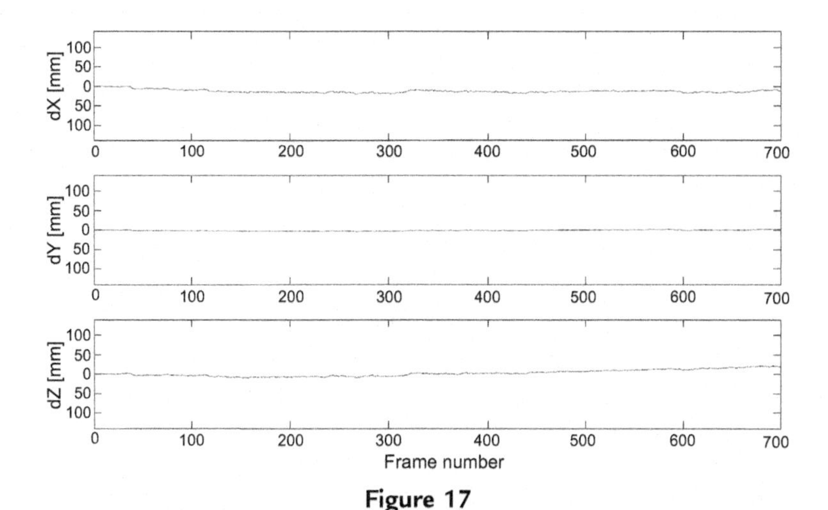

Figure 17
Change in translation between the gripper and object frames during a stable grasp. The change in *dX*, *dY* and *dZ* is very small and mostly less than 10 mm.

grasp, a grasp for which the object will not twist or slip relative to the end-effector. Here, the latter issue is considered. Aside from picking up the object, a task for the robot may also be to place the object at some desired position in the workspace. If that is the case, after the object is grasped, a *task monitoring* step may be initiated. The basic idea is that, even if the planed grasp was considered stable, when the manipulator starts to move the object may start to slide. If the grasp is stable, the relative transformation between the manipulator (gripper) and the object frames should be constant. Since tactile sensing is not available to us at this stage, our vision system is used to track the object held by the robot and estimate its pose during the placement task. The estimated pose of the object is then used to estimate the change between the object and the hand coordinate frames. For a stable grasp, this change should be ideally zero or very small.

In Figures 18 and 19, the variation in the hand/object transformation is presented, showing two cases obtained during a stable grasp and a grasp where the object slid from the hand. Comparing the figures, we see a significant difference in the transformation plots. The most significant change is observed for d*X* and d*Z*, which is for d*X* approximately 100 mm and for d*Z* almost 150 mm. These changes are caused by the object being removed from the gripper in Figure 19 (see Figure 17).

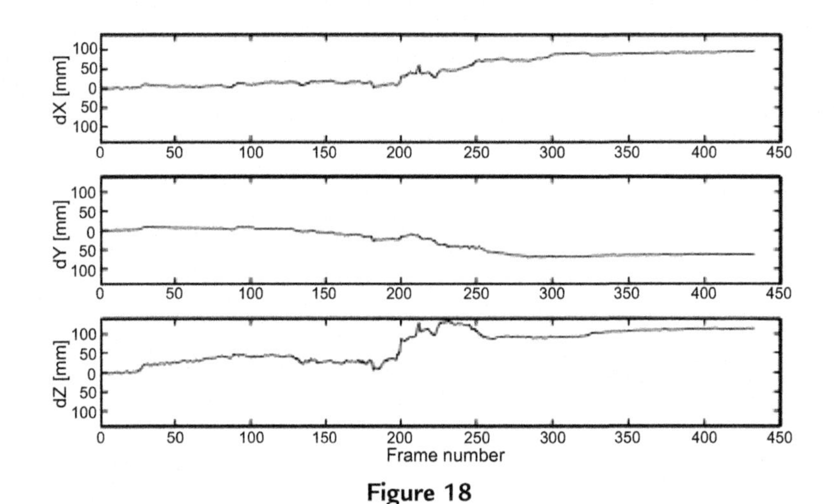

Figure 18

Change in translation between the gripper and object frames when the object was removed from the gripper. Compared to the plot for a stable grasp (Figure 18), the change is approximately 100 mm for the d*X* component and 150 mm for the d*Z* component.

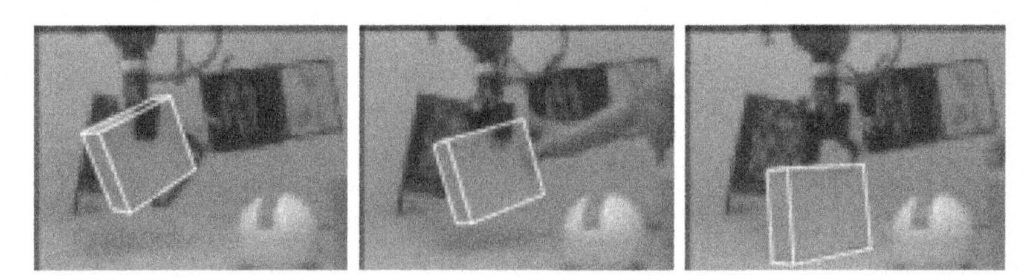

Figure 19

Removing the object from the gripper; the object is tracked during the whole sequence. The pose of the object is used to estimate the change in object/gripper transformation. The results are presented in Figure 19.

5.3.7 Conclusion

Due to the real-world variability, it is not enough to consider only control level robustness. It is equally important to consider how image information that serves as input to the control process can be used so as to achieve robust and efficient control. In visual servoing, providing a robust visual feedback is crucial to the performance of the overall system. It is well known that no single visual feature is robust to variations in geometry, illumination, camera motion and coarse calibration. In addition, no single visual servoing technique can easily be tailored to cope with large-scale variations. For the construction of realistic systems, there is consequently a need to formulate methods for integration of a range of different servoing techniques. The techniques differ in terms of the underlying control space (position/image/hybrid) and type of visual features used to provide the necessary measurements (2D/3D). We argue that each of these techniques must carefully consider how models, multiple cues and fusion can be utilized to provide the necessary robustness.

In this chapter, we have taken a first step toward the design of such a system by presenting a range of different vision-based techniques for object detection and tracking. Particular emphasis has been put on the use of computationally tractable methods in terms of real-time operation in realistic settings. The efficiency has in particular been achieved through careful consideration of task level constraints.

Through adaptation of a task oriented framework where the utility of different cues are considered together with methods for integration, it is possible to achieve desired robustness. Task constraints here allow dynamic selection of cues (in terms of weights), integration and associated control methodology so as to allow visual servoing over a wide range of work conditions. Such an approach is, for example, needed in the case of mobile manipulation of objects in service robotics applications. The use of the developed methodology has been demonstrated with a number of examples to illustrate the system's performance.

An open problem in the presented methodology is the optimal balance between different visual servoing approaches as well as the amount and type of information shared between them. An integral part of the future research will consider these issues.

Acknowledgment

This research has been sponsored by the Swedish Foundation for Strategic Research through the Center for Autonomous Systems. The funding is gratefully acknowledged.

References

[1] S. Hutchinson, G.D. Hager, P.I. Corke, A tutorial on visual servo control, IEEE Trans. Rob. Autom. 12 (5) (1996) 651–670.
[2] L. Petersson, P. Jensfelt, D. Tell, M. Strandberg, D. Kragic, H. Christensen, Systems integration for realworld manipulation tasks, in: IEEE International Conference on Robotics and Automation, ICRA 2002, vol. 3, 2002, pp. 2500–2505.
[3] D. Kragic, H.I. Christensen, Survey on Visual Servoing for Manipulation, Technical report, ISRN KTH/NA/P–02/01–SE, Computational Vision and Active Perception Laboratory, Royal Institute of Technology, Stockholm, Sweden, January 2002. http://www.nada.kth.se/cvap/cvaplop/lop-cvap.html.
[4] A. Bicchi, V. Kumar, Robotic grasping and contact: a review, in: Proceedings of the IEEE International Conference on Robotics and Automation, ICRA'00, 2000, pp. 348–353.
[5] K.B. Shimoga, Robot grasp synthesis algorithms: a survey, Int. J. Rob. Res. 15 (3) (1996) 230–266.
[6] S. Edelman (Ed.), Representation and Recognition in Vision, MIT Press, Cambridge, MA, 1999.
[7] D. Roobaert, M. Zillich, J.-O. Eklundh, A pure learning approach to background-invariant object recognition using pedagocical support vector learning, in: Proceedings of IEEE Computer Vision and Pattern Recognition, CVPR'01, December, Kauai, Hawaii, vol. 2, 2001, pp. 351–357.
[8] Y. Bar-Shalom, Y. Li, Estimation and Tracking: Principles, Techniques and Software, Artech House, Boston, MA, 1993.
[9] D.M. Blough, G.F. Sullivan, Voting using predispositions, IEEE Trans. Reliab. 43 (4) (1994) 604–616.
[10] J.K. Rosenblatt, C. Thorpe, Combining multiple goals in a behavior-based architecture, in: IEEE International Conference on Intelligent Robots and Systems, IROS'95, vol. 1, 1995, pp. 136–141.
[11] H.I. Christensen, D. Kragic, F. Sandberg, in: G. Hager, H.I. Christensen, H. Bunke, R. Klein (Eds.), Vision for Interaction, Sensor Based Intelligent Robots, Lecture Notes in Computer Science, vol. 2238, Springer Verlag, Berlin, 2001, pp. 51–73.
[12] J. Shi, C. Tomasi, Good features to track, in: Proceedings of the IEEE Computer Vision and Pattern Recognition, CVPR'94, 1994, pp. 593–600.
[13] D. Kragic, Visual Servoing for Manipulation: Robustness and Integration Issues (Ph.D. Thesis), Computational Vision and Active Perception Laboratory (CVAP), Royal Institute of Technology, Stockholm, Sweden, 2001.
[14] K. Toyama, G. Hager, Incremental focus of attention for robust visual tracking, in: Proceedings of the Computer Society Conference on Computer Vision and Pattern Recognition, CVPR'96, 1996, pp. 189–195.
[15] G.D. Hager, K. Toyama, The XVision system: a general-purpose substrate for portable real-time vision applications, Comput. Vision Image Understanding 69 (1) (1996) 23–37.
[16] L. Bretzner, Multi-scale Feature Tracking and Motion Estimation (Ph.D. Thesis), CVAP, NADA, Royal Institute of Technology, KTH, 1999.
[17] Y. Hu, R. Eagleson, M.A. Goodale, Human visual servoing for reaching and grasping: the role of 3D geometric features, in: Proceedings of the IEEE International Conference on Robotics and Automation, ICRA'99, vol. 3, 1999, pp. 3209–3216.
[18] T.W. Drummond, R. Cipolla, Real-time tracking of multiple articulated structures in multiple views, in: Proceedings of the 6th European Conference on Computer Vision, ECCV'00, vol. 2, 2000, pp. 20–36.
[19] N. Giordana, P. Bouthemy, F. Chaumette, F. Spindler, Two-dimensional model-based tracking of complex shapes for visual servoing tasks, in: M. Vincze, G. Hager (Eds.), Robust Vision for Vision-based Control of Motion, IEEE Press, Piscataway, NJ, 2000, pp. 67–77.
[20] V. Gengenbach, H.-H. Nagel, M. Tonko, K. Schäfer, Automatic dismantling integrating optical flow into a machine-vision controlled robot system, in: Proceedings of the IEEE International Conference on Robotics and Automation, ICRA'96, vol. 2, 1996, pp. 1320–1325.
[21] D. Lowe, Perceptual Organization and Visual Recognition. Robotics: Vision, Manipulation and Sensors, Kluwer Academic, Dordrecht, 1985.

[22] D. Koller, K. Daniilidis, H.H. Nagel, Modelbased object tracking in monocular image sequences of road traffic scenes, Int. J. Comput. Vision 10 (3) (1993) 257–281.

[23] M. Vincze, M. Ayromlou, W. Kubinger, An integrating framework for robust real-time 3D object tracking, in: Proceedings of the 1st International Conference on Computer Vision Systems, ICVS'99, 1999, pp. 135–150.

[24] P. Wunsch, G. Hirzinger, Real-time visual tracking of 3D objects with dynamic handling of occlusion, in: Proceedings of the IEEE International Conference on Robotics and Automation, ICRA'97, vol. 2, 1997, pp. 2868–2873.

[25] S.K. Nayar, S.A. Nene, H. Murase, Subspace methods for robot vision, IEEE Trans. Rob. Autom. 12 (5) (1996) 750–758.

[26] P. Wunsch, S. Winkler, G. Hirzinger, Real-Time pose estimation of 3D objects from camera images using neural networks, in: Proceedings of the IEEE International Conference on Robotics and Automation, ICRA'97, April, Albuquerque, NM, vol. 3, 1997, pp. 3232–3237.

[27] J.J. Craig, Introduction to Robotics: Mechanics and Control, Addison Wesley, Reading, MA, 1989.

[28] E. Dickmanns, V. Graefe, Dynamic monocular machine vision, Mach. Vision Appl. 1 (1988) 223–240.

[29] J.D. Foley, A. van Dam, S.K. Feiner, J.F. Hughes (Eds.), Computer Graphics—Principles and Practice, Addison-Wesley, Reading, MA, 1990.

[30] C. Harris, Tracking with rigid models, in: A. Blake, A. Yuille (Eds.), Active Vision, MIT Press, Cambridge, MA, 1992, pp. 59–73.

[31] R.L. Thompson, I.D. Reid, L.A. Munoz, D.W. Murray, Providing synthetic views for teleoperation using visual pose tracking in multiple cameras, IEEE Syst. Man, Cybern. Part A 31 (1) (2001) 43–54.

[32] W. Wilson, C.C. W. Hulls, G.S. Bell, Relative end-effector control using cartesian position based visual servoing, IEEE Trans. Rob. Autom. 12 (5) (1996) 684–696.

[33] L. Petersson, D. Austin, H.I. Christensen, DCA: a distributed control architecture for robotics, in: Proceedings of the IEEE International Conference on Intelligent Robots and Systems IROS'2001, October, Maui, Hawaii, vol. 4, 2001, pp. 2361–2368.

[34] D. Tell, Wide Baseline Matching with Applications to Visual Servoing (Ph.D. Thesis), Computational Vision and Active Perception Laboratory (CVAP), Royal Institute of Technology, Stockholm, Sweden, 2002.

CHAPTER 5.4

Implementation of Cognitive Controls for Robots

Kazuhiko Kawamura, Stephen M. Gordon, Palis Ratanaswasd, Christopher S. Garber, Erdem Erdemir
Center for Intelligent Systems, Vanderbilt University, Nashville, TN, USA

Chapter Outline

Household Service Robotics. http://dx.doi.org/10.1016/B978-0-12-800881-2.00019-0

5.4.1 Introduction

As the need to control complex systems increases, it is important to look beyond engineering and computer science for new ways to control robots. For example, humans have the capacity to receive and process enormous amount of sensory information from the environment, exhibiting integrated sensorimotor associations as early as two years old [1]. A good example of such sensorimotor intelligence by adults is the well-known Stroop test [2]. Most goal-oriented robots currently perform only those or similar tasks that they were programmed for, and very little emerging behaviors are exhibited. What is needed is an alternative paradigm for behavior learning and task execution. Specifically, we see cognitive flexibility and adaptability in the brain as desirable design goals for the next generation of intelligent robots.

At the HAM Workshop in 2006, a concept of human cognitive control [3] and a multi-agent-based, hybrid cognitive architecture for robots [4] were presented. In this chapter, we will present the progress made during the last year on the cognitive architecture and control, working memory (WM) training, and a self-motivated, internal state-based action selection mechanism.

5.4.2 Cognitive Control for Robots

Engineers have long used control systems utilizing feedback loops to control mechanical systems. Figure 1 illustrates a class of adaptive/learning control systems [5]. Limitations of model-based control led to a generation of *intelligent control techniques* such as fuzzy control, neuro computing, and reconfigurable control.

The human brain is known to process a variety of stimuli in parallel, ignoring noncritical stimuli to execute the task in hand, and to learn new tasks with minimum assistance.

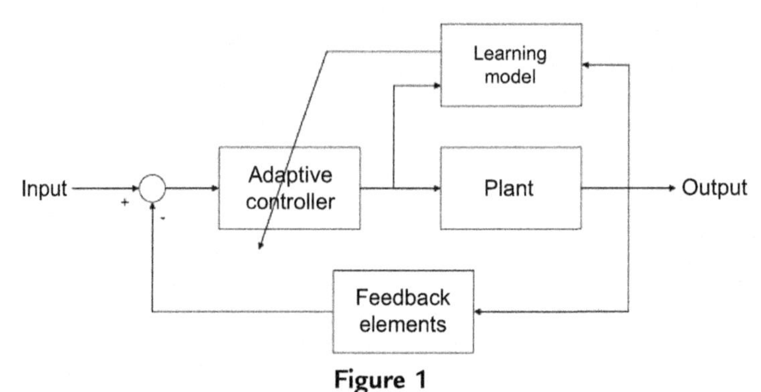

Figure 1
An adaptive control system [5].

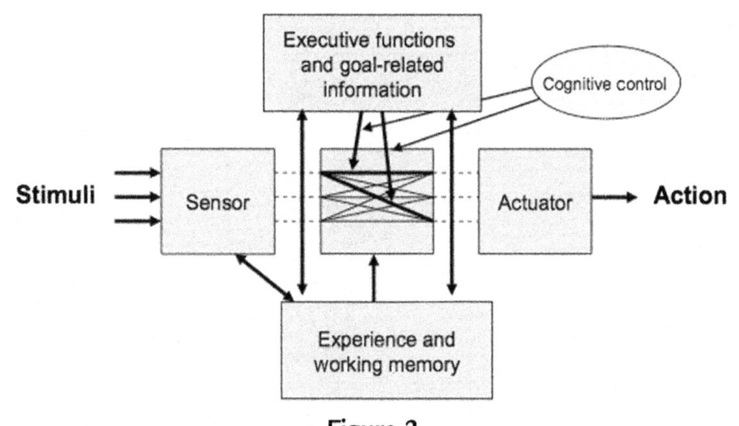

Figure 2
Model of cognitive control, modified from Miller et al. [6].

This process, known as *executive* or *cognitive control*, is unique to humans and a handful of animals [6]. Figure 2 illustrates a conceptual model of cognitive control that we are using to realize robust behavior generation and learning for our humanoid robot.

As the complexity of a task grows, so do the software complexities necessary to process sensory information and to purposefully control actions. Development and maintenance of complex or large-scale software systems can benefit from domain-specific guidelines that promote code reuse and integration through software agents. Information processing in our humanoid robot ISAC (Intelligent Soft Arm Control) is integrated into a multi-agent-based software architecture based on the Intelligent Machine Architecture (IMA) [7]. IMA is designed to provide guidelines for modular design and allows for the development of subsystems from perception modeling to behavior control through the collections of IMA agents and associated memories, as shown in Figure 3.

For any learning system, memory plays an important role. As Gazzaniga et al. state, "Learning has an outcome, and we refer to that as *memory*. To put it another way, learning happens when a memory is created or is strengthened by repetition" [1, p. 302]. ISAC's memory structure is divided into three classes: short-term memory (STM), long-term memory (LTM), and the working memory system (WMS). STM holds sensory information of the current environment in which ISAC is situated. LTM holds learned behaviors, semantic knowledge, and past experience. WMS holds task-specific information called "chunks" and streamlines the information flow to the cognitive processes during the task execution. STM is implemented using a sparse sensory data structure called the Sensory EgoSphere (SES). It was inspired by the egosphere concept defined by Albus [8] and serves as a spatiotemporal STM [9]. LTM stores information such as *skills learned* and *experience gained* for future recall.

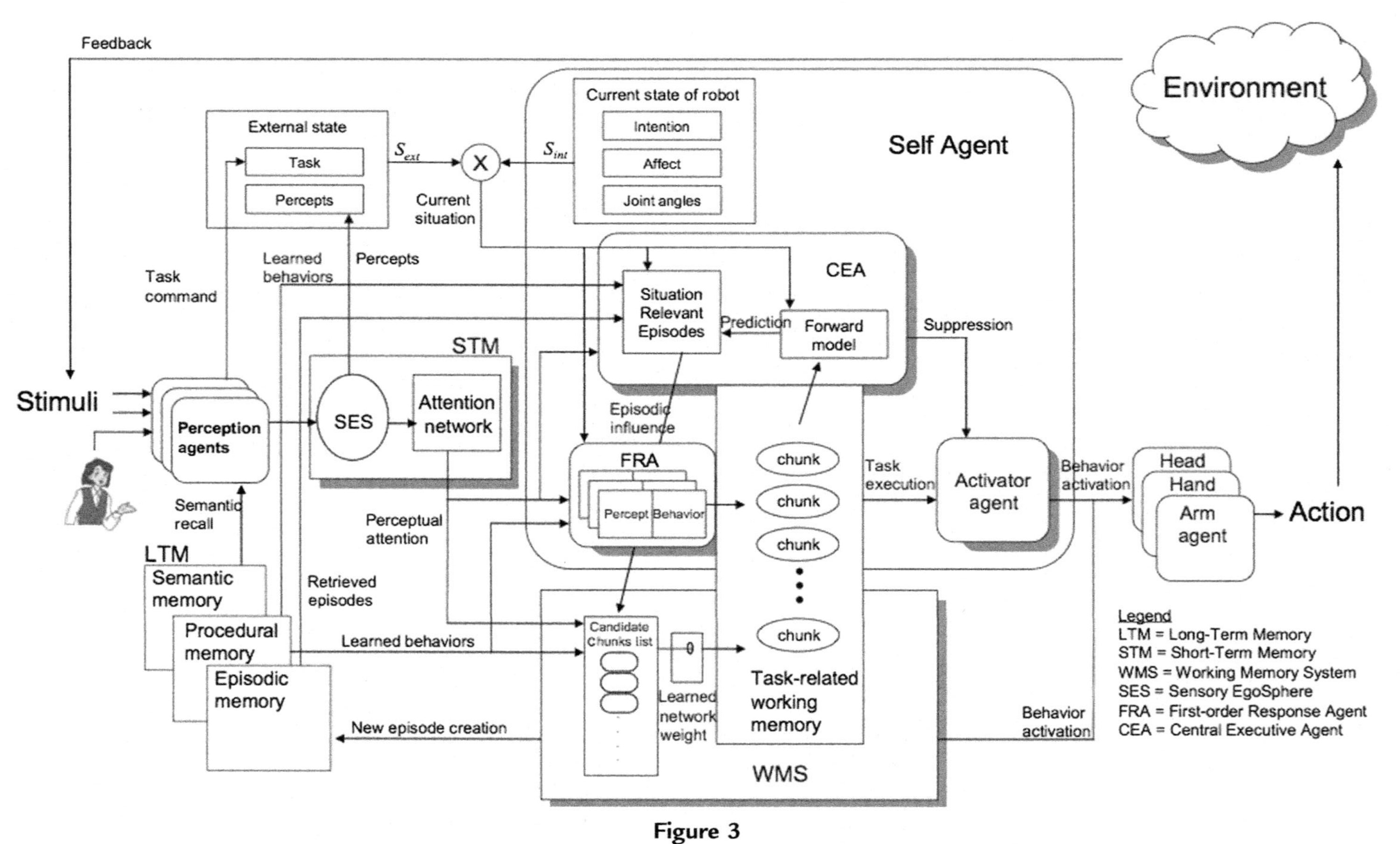

Figure 3
Multi-agent-based cognitive robot architecture.

5.4.3 The Working Memory System

5.4.3.1 Central Executive and Working Memory System

Cognitive functions ISAC can perform, i.e., cognitive control, are modeled after Baddeley's human working memory model [10]. In his model, the "central executive" controls two "working memory" systems: the phonological loop and the visuospatial sketch pad. Cognitive control functions are currently implemented using STM, LTM, the Attention Network, and WMS. As discussed, STM handles sensor-based percepts. These percepts can be assigned *focus of attention* (FOA) or *gating* by the Attention Network [11]. This happens as a result of associated knowledge (such as emotional salience) with the sensed percepts. FOA-based percepts are then passed to the WMS as candidate task-related chunks.

"Biological working memory represents a limited-capacity store for retaining information over the short term and for performing mental operations on the contents of this store" [12]. This type of memory system is said to be closely tied to task learning and execution [12]. Inspired by this, we have implemented the WM structure into ISAC to provide the embodiment necessary for exploring the critical issues of task execution and learning. Our hypothesis is that this integration will lead to a more complex, but realistic robotic learning system involving perceptual systems, actuators, reasoning, attention, emotion, and STM and LTM structures.

5.4.3.2 Working Memory Training Experiments for Percept–Behavior Association Learning

Within the ISAC architecture, learning how to respond to novel tasks is done through an untrained WM system. When a novel stimulus is present, this system explores different responses and, over time, learns what information from STM and LTM should be focused on to best execute the novel task. As the system learns, the trained WM is stored in episodic memory and can be retrieved in the future for quick, reliable, task execution. Experiments have been performed to utilize this task-learning portion of the ISAC cognitive architecture [13]. These experiments relate to tasks for which ISAC had no previous experience with the situation at the time of training. Figure 4 shows the task learning loop involving the WM within the cognitive architecture. Figure 5 shows sample configurations for the behaviors used in this section and the following sections.

The training conducted utilizing the WMS represented initial trial-and-error responses to novel tasks. For each experiment, the WMS initialized an untrained instance of WM. This was required because current computational limits only allowed a trained WM to be used for similar types of tasks. For instance, the tasks *reach to the bean bag* and *track the*

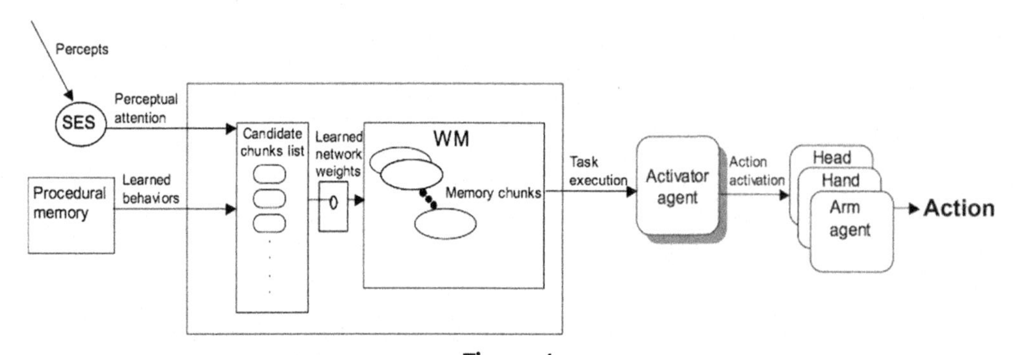

Figure 4
Task-learning loop in Intelligent Soft Arm Control cognitive architecture.

LEGO toy are similar because each task requires one percept and one behavior. The task of interacting with a person, however, is not similar to these two and would require a separately trained WM. Novel tasks were those for which a trained WM could not be found. The interpolation and execution of behaviors in this experiment, and those described later, were performed using a modification of the Verbs and Adverbs algorithm [14], discussed further in Appendix 2.

Thus far, WM has been successfully trained to perform the types of tasks discussed above, tasks involving one percept and one behavior. The example used in the remainder of this section will be, *reach to the bean bag*, as shown in Figure 6(a, b). In our trials, two bean bags were present, and the WMS was required to choose one. Preliminary results for training this WM have been presented in [13].

During the initial training of this WM, a high exploration percentage (15%) was used that helped avoid local maxima. In other words, 15% of the time, the system continued to explore random actions even after a solution had been found. In addition, a reward rule was provided that rewarded the WM for chunk selection based on the success of the current trial. Once the WM began to converge on the appropriate responses, the

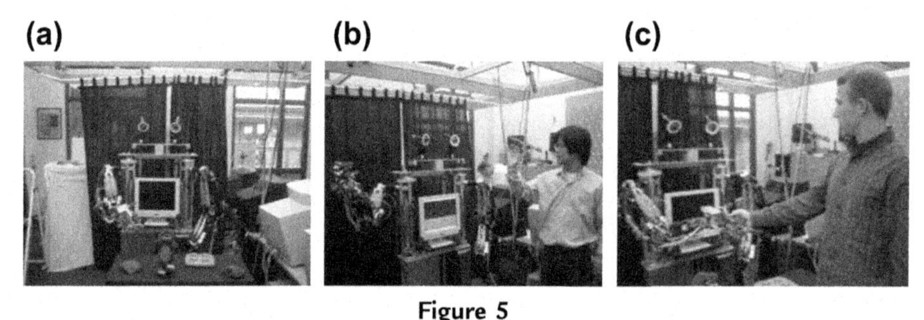

Figure 5
Sample configurations for behaviors used (a) reach, (b) wave, (c) handshake.

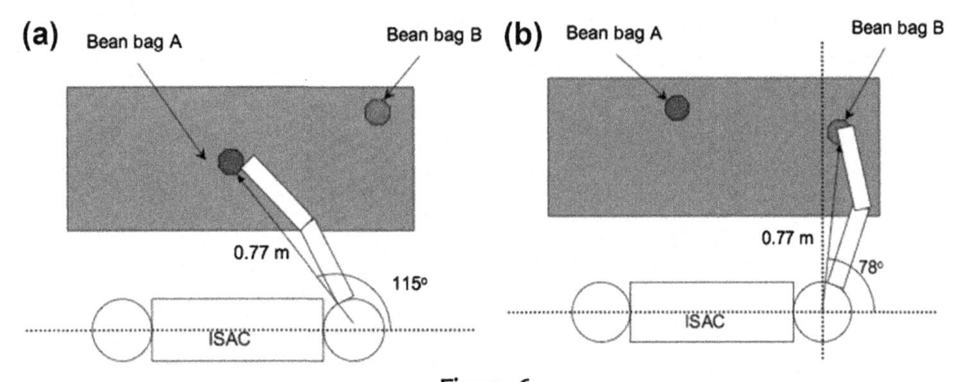

Figure 6
Sample configurations for reaching (a, b).

other cognitive processes could then begin recording and storing episodes. However, currently a human trainer performed the action of deciding when to record episodes. Among the items stored in these episodes was the current WM. Therefore, when a similar situation is encountered in the future, ISAC could not only recall matching similar episodes but also retrieve the now-trained WM used during those episodes.

5.4.3.3 System Performance

The performance of the WM during training was evaluated using that WM's specified reward rules. During the learning process, reward was given based on the following three criteria:

1. Did the behavior chunk chosen successfully accomplish the task?
2. Did the percept chunk chosen successfully accomplish the task?
3. What was the difference between similar performances (e.g., reaching to the nearest bean bag rather than the farthest one)?

Reward criterion 3 was implemented to allow differentiation between similar choices. An example of this is the *reach to the bean bag* task represented in Figure 6. Note that two bean bags are present, and a reach to either one would accomplish the task. However, it may be desirable to have ISAC understand that when the *reach to the bean bag* command is given, the intention of the instructor is actually to have ISAC reach to the nearest one, or perhaps always the blue or red one. Prior to task execution, the WMS had no understanding of this intention, but our experiments have shown that within approximately 20 trials (Figures 7 and 8), WMS learns this relationship. Furthermore, reward criterion 3 can be changed without notice. When the WMS fails to receive a reward when a reward was expected, it began exploring alternative choices. In other words, it detected that the instructor's intention had changed and attempted to learn the new intention.

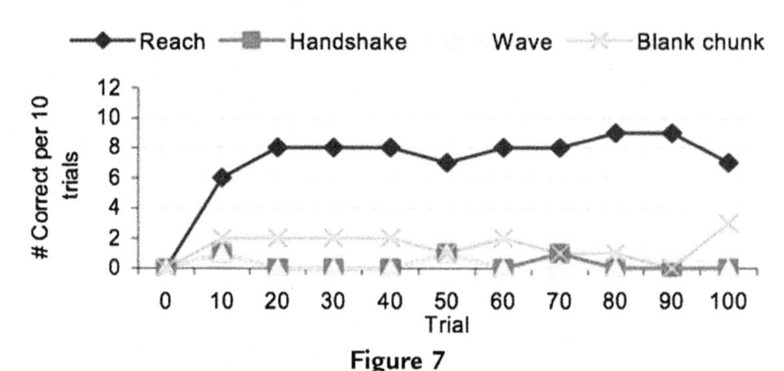

Figure 7

Learning to associate the *reach* command with the correct behavior.

When the reward criterions were met, discrete positive reward was given to the system. No preference (i.e., reward of 0) was given if the system did not choose correctly. Implementing the exploration percentage encouraged exploration even after learning had been accomplished. This measure helped avoid local maximum.

Initial trials were performed in simulation to speed up the testing phase of this percept–behavior learning. The simulation removed the time-bottleneck of generating and performing behaviors. If the WMS desired to act on an object within the workspace, it was assumed that ISAC would be able to perform the desired action and reward was given accordingly. Appendix 3 shows the contents of STM and LTM systems and some sample contents of WM during training.

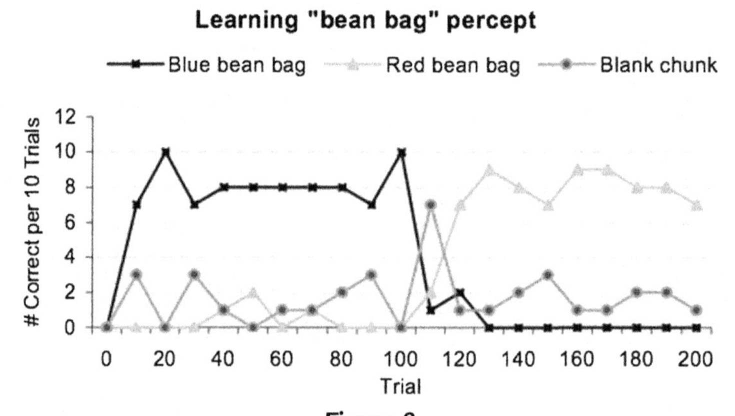

Figure 8

Learning to choose the correct *bean bag*.

In these trials, WMS was allowed to choose two "chunks" from the STM and LTM systems to accomplish the task. However, the WM was not restricted to choosing exactly one percept and one behavior. If the WM chooses to focus on two percepts, two behaviors, or chooses not to load enough chunks then a behavior or percept was necessarily chosen at random. When there was not a behavior (or percept) chunk present, a random number generator was used to fill in the missing chunk. This ensured that an action was always performed. The reasoning for this was to encourage the WMS to make choices. Without this safeguard, the WMS would begin avoiding the decision by not loading any chunks during the trials where the WMS was consistently making incorrect choices. This was a behavior inherent in the learning networks used to create the WM system. Randomly filling in the blank chunks allowed the system to continue exploration during these trials.

To graphically demonstrate the ability of WMS better, training trials were also conducted that only required WM to learn one chunk (percept or behavior) at a time. Figure 7 shows the learning curve for the behavior *reach* for the command *reach to the bean bag*. Prior to these trials, ISAC was taught a number of behaviors including three right-arm behaviors, *reach*, *handshake*, and *wave*. Within 20 trials, the WMS learned to associate the command *reach* with the appropriate behavior. Figure 8 shows the same curve for learning the *bean bag* percept. Again, within 20 trials the WMS had learned not only to associate a bean bag with the command, but also that the *blue bean bag* was the intended bean bag. After 100 trials, the WMS quit receiving a reward for the *blue bean bag*, the intention had changed to the *red bean bag* and within 20 more trials this intention had been learned.

5.4.4 The Role of CEA and FRA for Task Switching

5.4.4.1 The First-Order Response Agent (FRA) and the Central Executive Agent (CEA)

Figure 9 depicts the key IMA agents within the Self Agent. The Central Executive Agent (CEA), which is responsible for cognitive control during task execution, invokes behaviors necessary for performing the given task. CEA operates in accord to *intention*, which the Intention Agent interprets from a task command. Decision making in CEA is mediated by *affect* which is managed by the Affect Agent. The Activator Agent invokes head and arm agents to execute actions. The First-Order Response Agent (FRA) is responsible for generating both routine and reactive responses. The term *first-order responses* refers to responses of the system that are not generated from the cognitive process. This term was also used by Shanahan [15] in regard to responses generated reactively by the physical system, which are in contrast to responses generated by the higher-order loop that represent "imagination" in this work.

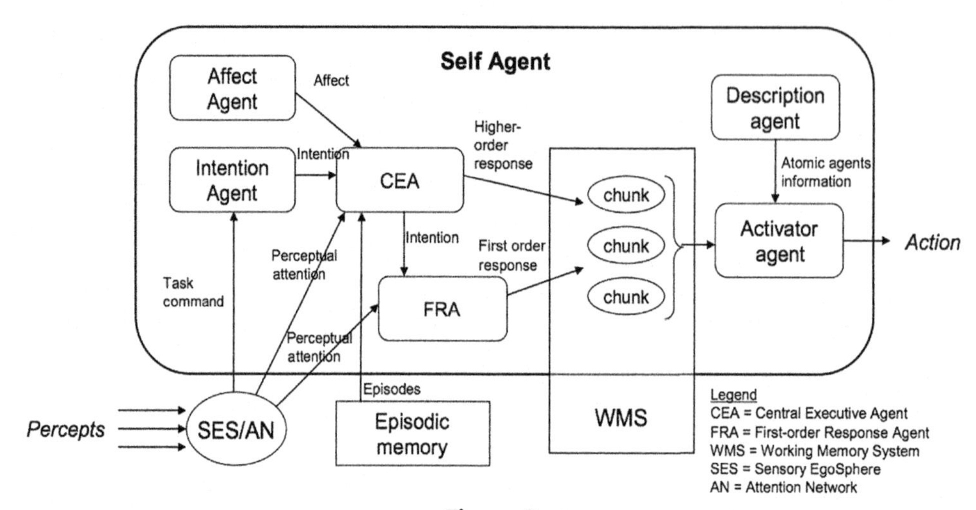

Figure 9
Structure of the self agent.

First, reactive responses are handled by FRA by invoking corresponding behaviors when certain percepts receive ISAC's attention. This concept is inspired by the schema theory where the system responds to certain stimuli by performing certain actions [16]. The associations between percept–behavior in FRA are provided as initial knowledge. FRA is implemented as a multithreaded process in which each stimulus–response pair is given its own separate running thread. As salient percepts on SES are put in FOA by the Attention Network [11], each thread compares the system's current most salient percept with that particular thread's percept from the percept–behavior pair. FRA posts both the matched percept and the behavior onto the WM as chunks when a match is found. The Activator Agent then takes the chunks from WM and distributes to Atomic Agents in the system for behavior execution.

FRA has one thread that is responsible for routine responses. This thread invokes corresponding behaviors when certain situations are recognized according to the percepts in FOA and the current task. The recognized situation causes FRA to retrieve the learned skill associated with the situation from LTM. A learned skill contains the behavior needed to perform a particular task. Note that the current task could be assigned externally by a human or internally generated by self-motivation (see Section 5 for self-motivated decision making). FRA posts the behavior found in the retrieved learned skill and the percept in FOA into WM as chunks, where the Activator Agent uses the chunks similarly to the case of reactive responses. However, the routine response thread will be subsumed when any one of reactive response threads are active. This phenomenon is similar to subsumption of behaviors in Brook's Subsumption Architecture [17].

5.4.4.2 FRA and Task-Switching Experiment

A two-part experiment was conducted to validate how FRA can handle the routine and reactive responses. Figure 10 shows the IMA agents and memory components utilized in the experiment.

5.4.4.2.1 Routine response experiment

The first part of the experiment was conducted to validate the capability to execute a task using a routine response and the ability to maintain the task context after a reactive response is invoked.

Experimental steps.

1. ISAC actively monitors the environment.
2. Barney doll is placed within the field of view causing ISAC to recognize it, i.e., the situation, and to decide to play with the doll according to its innate knowledge.
3. When someone claps their hands, ISAC detects the location of the sound using the sound localization algorithm described in Appendix 4.
4. ISAC stops executing the task and saccades toward the source of the sound.
5. Because the task context is still active in the WM, ISAC goes back to the task after the reactive response is completed.

5.4.4.2.2 Task-switching experiment

The second part of the experiment was conducted to validate the functionality of FRA to switch tasks when a new situation is recognized when an event occurs.

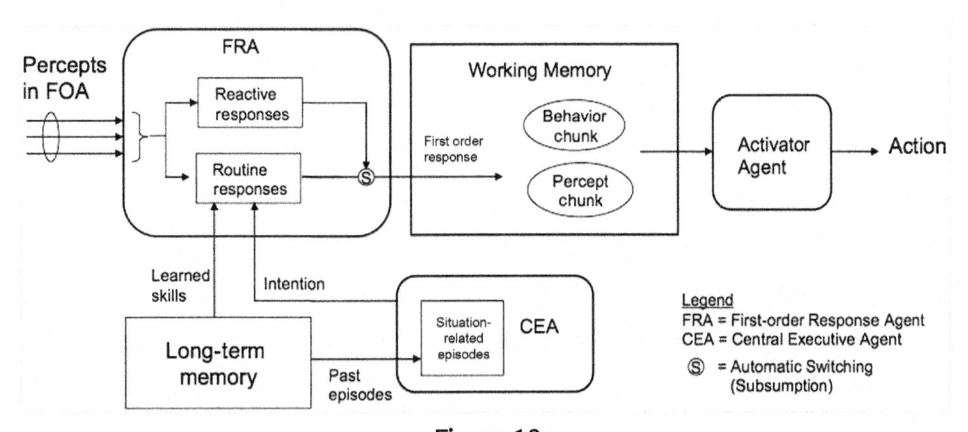

Figure 10
First-order response in Intelligent Soft Arm Control cognitive architecture.

Experimental steps.

1. ISAC continues the task from the above experiment.
2. Someone enters the room and approaches ISAC.
3. When a motion is detected at the door, using the motion detection method described in Appendix 4, ISAC stops executing the current task, fixates on the detected motion, and tracks the motion with the cameras.
4. When the motion enters the workspace, ISAC recalls a similar learned experience, thus executes the *handshake* behavior instead of going back to the previous task.

Figure 11 shows the lab view during the experiment.

This experiment shows a setup of ISAC cognitive architecture to perform simple cognitive control during task execution. In this setup, we focus on FRA, which was used to execute a task using first-order responses. To help in providing simple cognitive control, CEA was used to generate a task internally and make a decision for task switching. In this section, we would like to evaluate the FRA-based operation of the system using the following criteria:

1. The ability to switch back and forth between reactive and routine responses.
2. The ability to use routine responses to execute tasks.
3. The ability to switch tasks based on situational change.

In this experiment, the system has the ability to switch between reactive and routine responses seamlessly without losing its attention from the task. Reactive responses caused the system to immediately suspend the current task and attend to the particular percept that triggers the response. Reactive responses provided by FRA serve as a nontask-oriented mechanism which helps the system to become aware of other events that happen in the environment that may require attention; therefore, the system should respond within a short period of time after an event happens. A short delay is expected, however, because of the complexity of the detection algorithms and the propagation delay time in communication between agents. In the experiment, a set of clapping sounds was present at

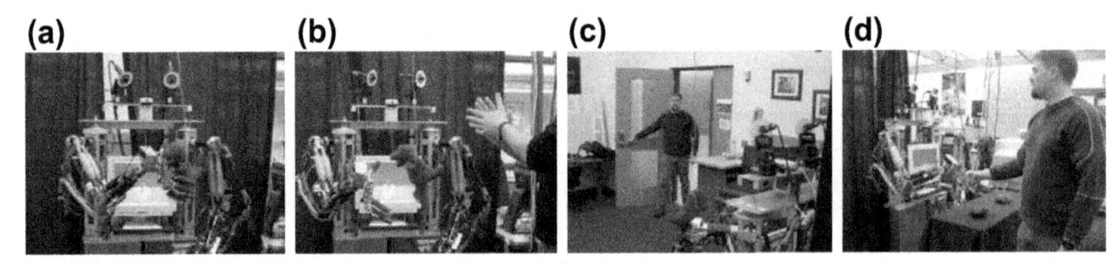

Figure 11

Lab views during experiments—Intelligent Soft Arm Control (a) played with Barney, (b) responded to clapping sound, (c) detected motion, (d) shook hands with the person.

Table 1: Response and resuming times to loud noise reaction

Trial	Angles (degrees)	Response Time (ms)	Resuming Time (ms)
1	64.07	102	153
2	−38.65	906	143
3	18	105	135
4	−49	594	141
5	−2.65	716	139

various angles where 0° was directly in front of ISAC. Table 1 summarizes the amount of time that the system took to respond after the clapping sounds were heard and the amount of time the system takes to resume the previous action after the reaction responses were completed.

Both parts of the experiment in this section show that FRA can be used to execute simple tasks successfully using knowledge about the task in LTM. Two tasks used in this experiment were *to play with Barney doll* and *to handshake the person.* Note that in this experiment, the tasks were not given by a person but instead were internally generated by CEA. The task information was passed between CEA and FRA using a shared memory slot. This method allows both agents to communicate very fast. FRA executed a task as soon as CEA posted the task to the memory slot.

The task-switching experiment demonstrates the capability of the system to switch tasks based on situational change. The decision to switch tasks comes from CEA based on the situation. In this experiment, the system did not go back to the previous task because the robot detected that a person was moving toward it and entering the workspace. Due to a strong association between the event and a task in the learned experience, CEA decided to switch to a handshake with the person instead of going back to the previous task.

The performance evaluation in this section is performed based on partial results from the experiment. The final performance evaluation will be completed before the workshop.

5.4.5 Self-Motivated, Internal State-Based Action Selection Mechanism

Cognitive robots may face complex situations where they cannot rely on the state of the environment alone to make decisions. For example, the internal state of *affect* is shown to play an important role in the human decision making [15,18]. In our architecture, affect is considered a part of the internal state of the system. It is maintained by the Affect Agent, which keeps track of the current affective level of ISAC. Similar to the work of Shanahan [15], affect interacts with CEA by running in parallel, influencing FOA, and offering mediating task execution by influencing the probabilistic decision-making model in the CEA [19]. However, unlike Shanahan's work, affect in our architecture does not offer executive veto power. This veto power is kept by CEA.

The total state of the system is represented by two sets of state variables, *external* and *internal*. External state variables are represented by percepts and are placed in the FOA. Internal state variables include ISAC hardware parameters such as joint angle positions as well as variables such as intention and affect (Figure 3). These variables, S_{ext} and S_{int}, combine to form the overall situation.

$$Situation\ (S_{total}) = S_{ext} \times S_{int} \tag{1}$$

Only a portion S_{total} is required for the work discussed in this section. The internal state variable used is S_{affect}. The important external variables are those represented by $S_{percepts}$ and S_{task}. As will be discussed below, S_{affect} is going to be determined quantitatively from these external variables.

In the ISAC cognitive architecture, task switching is based on the appearance of *events,* time-critical salient changes in the external state. The response for the current event depends not only on the current situation and the internal state, but also on the past experiences ISAC has encountered. The past experience of an event is stored as an "episode" within the episodic memory. We call the ability for ISAC to make decisions using the current state variables and past experiences (episodes) "situation-based action selection" (Figure 12).

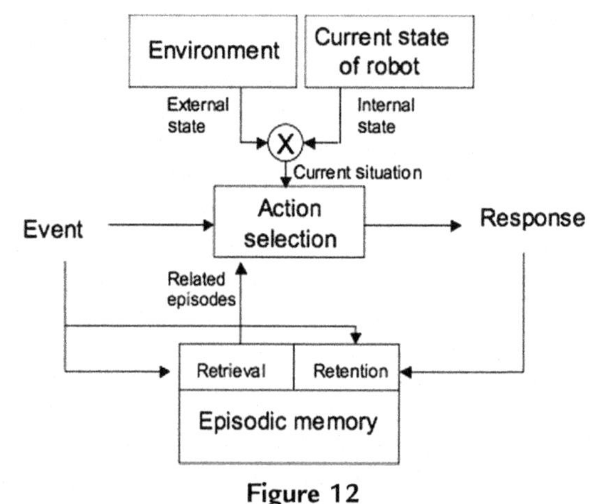

Figure 12
Situation-based action selection mechanism.

5.4.5.1 Affect Agent and Excitement Meter

Robots operating in the real world often encounter situations where more than one course of action could be considered appropriate. When ISAC encounters a situation for which two separate episodes are retrieved from the episodic memory, where each episode involves a different response to the situation, it is necessary to choose one in order to continue task execution. In any cognitive robot, this choice should not be made based on hard-coded *if-then* rules. Rather it should be *mediated* by past experiences and internal motivations. This mediation enables ISAC to make its own choices to deal with competing situations. Toward the goal of internally motivated task mediation, we have begun developing a means of allowing ISAC to make self-motivated decisions based on its own preference, or the affect [18]. In our architecture, the Affect Agent determines affectual associations with the current situation and provides suggestions to the CEA that impact decision making. The suggestions made to the CEA indicate which choices would lead to higher or lower affectual states. The suggestions cause an increase (or decrease) in the probability that a particular course of action is chosen. In other words, the Affect Agent tells the CEA to increase (or decrease) the probability that an action is chosen. Because the CEA can ignore the input from the Affect Agent, this input is regarded as a suggestion.

The CEA system implements a probabilistic model when making decisions with conflicting goals [19]. Past experience from episodic memory is used to fill in these probabilities. When episodes are retrieved, a list of possible actions is created, and each action is assigned a priority, p_j. The probability that action, A_j is chosen, is calculated as the priority of that action divided by the sum of the priorities of all actions. Further details of this probabilistic model are discussed in Appendix 1. The Affect Agent modifies these probabilities by using the affect associated with a particular set of stimuli to proportionately change the priority associated with that set of stimuli or action. By updating this priority, the probability that action is chosen is increased (or decreased). For example, if two sets of stimuli are present, the probability that the stimulus with the highest associated affect is chosen is increased. Likewise, the probability of choosing the second stimulus (the least affectual stimulus) is decreased.

In Figure 13, the current situation is input into the Affect Agent which calculates a new value for excitement and feeds this value back into the internal state. The current situation is passed on to the CEA, which uses that information to retrieve similar episodes from episodic memory and create an action list. Probabilities are assigned to the actions in the action list, and these probabilities are updated by the affect from the Affect Agent.

Prior work involving the Affect Agent [19] used predetermined, fixed affect vectors. It is important to note that there are many potential affective state variables for ISAC, but the

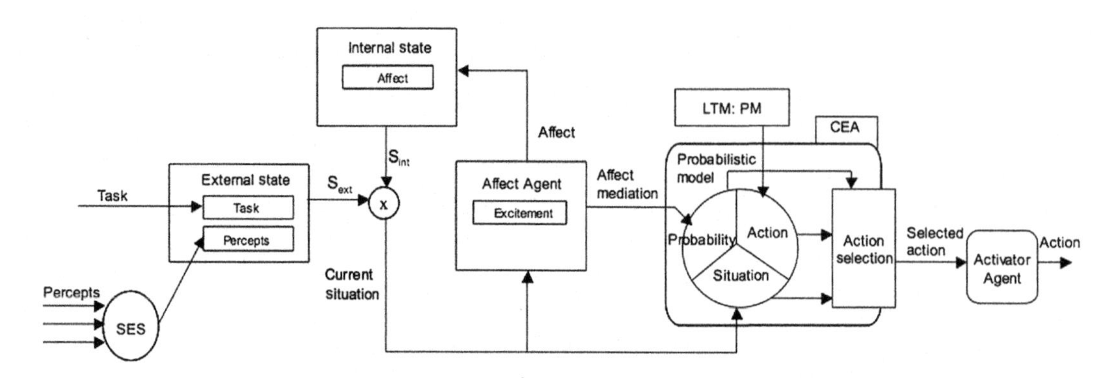

Figure 13
The role of the Affect Agent in the Intelligent Soft Arm Control Architecture.

current model will focus on only a single affect variable *excitement*. Our model of the Affect Agent (Figure 13) uses the following function for determining the affective state proposed by Anderson [20] but has also been suggested by Picard [18].

$$Excitement = Ae^{-Bt} \tag{2}$$

where

$$A = f(S_{ext}) \tag{3}$$

$$B = g(S_{ext}) \tag{4}$$

The values of A and B are parameters that are determined by the external state, S_{ext}. For example, situations associated with the action *reach to the bean bag* may retrieve a low value for A but a high value for B, indicating that these situations are not very exciting and that they decay rapidly. The variables (A, B) are functions of ISAC's current state, but they also represent one level of learning within this model. For instance, in a given situation the values of (A, B) can be modified to relate that the particular situation should no longer be deemed as exciting or, in fact, is to be considered more exciting the next time it is encountered. This is done by increasing or decreasing the stored values of (A, B) for particular situations.

Equation (2) assumes that ISAC's excitement level is continuous. As discussed above, task switching is based on the appearance of *events*. Therefore, it is more appropriate to use an *event-based hybrid system* [21] to calculate affect. For example, the action *reaching to the bean bag* is a continuous action. However, the *appearance* (*or disappearance*) of the bean bag percept is a discrete event. In order for the Affect Agent to deal with this, a second system (whose input is the external state and whose output is a discrete affect variable—D) is used as a switching mechanism from one situation to another, as shown in Figure 14. This switching mechanism reinitializes the local, event-based time used by the Affect Agent.

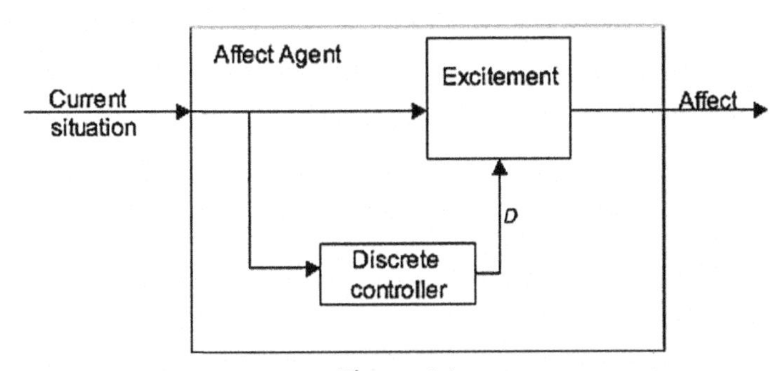

Figure 14
Hybrid structure of the Affect Agent.

Using our computational model of affect, the ISAC architecture is designed to mediate task switching to favor more exciting situations. The current implementation within the Affect Agent for the variable *excitement* is termed the *Excitement Meter.* This meter calculates the excitement response level to sets of stimuli and displays the history of the excitement value graphically on the "chest" monitor mounted on ISAC (Figure 15). The current value of the excitement is shown on the black axis. The previous values trail off to the right. Figure 15(a) shows the chest monitor as it normally appears. Figure 15(b) shows a close-up of the chest monitor when the Excitement Meter is displayed. Two jumps in excitement can be seen in the history section of the meter.

5.4.5.2 Excitement Meter Experiment

An experiment involving the *Excitement Meter* has been designed to validate the functionality of the Affect Agent and to demonstrate the ability of this system when using affect in cognitive decision making. In the designed experiment, competing sets of stimuli are presented to ISAC. The first set of stimuli is a command (from a human) to perform a task. The second set of stimuli is a number of toys that ISAC enjoys playing with. Each

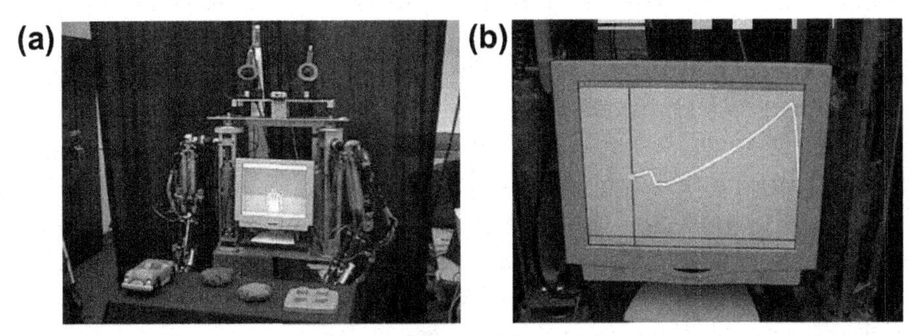

Figure 15
ISAC chest monitor (a, b).

set of stimuli will have an associated level of excitement, and this excitement level (along with past experiences and the current situation) will help CEA to choose what to focus on in order to either (a) perform the task on hand or (b) play with the toys.

Experimental Steps.

1. A pair of toys that ISAC can recognize are placed on a table in front of ISAC.
2. ISAC recalls a past episode which involved playing with the toys and subsequently begins to play with the toys.
3. A person enters the room and asks ISAC to perform a task. The task is encoded using a keyword search and is posted onto SES.
4. CEA recognizes this new stimulus (i.e., task command) and the Excitement Meter calculates the excitement associated with this situation.
5. The Excitement Meter passes the current excitement associations to CEA, where the decision is made to switch the task or not.
6. Based on the decision by the CEA, ISAC selects the appropriate action.

5.4.5.3 System Performance

Unlike many physical systems, the performance of a cognitive system must be evaluated on how it decides which action is *appropriate* to take, and not on the merit of *right* or *wrong* choices. The decision made by CEA involves a certain degree of uncertainty that it is, in fact, the best choice at that time. The probabilistic model that is used to make these decisions is influenced by the excitement associated with these choices, which is in turn derived by the Affect Agent. The candidate criteria used to evaluate system performance are as follows:

1. The degree that the response exhibited by the computational model was mediated by the level of the Excitement Meter.
2. The degree of influence on task execution by the probabilistic decision making derived from this computational model.

Prior to conducting the experiment described above, it was necessary to define initial weightings (A, B) for the various sets of stimuli that can be detected. The Excitement Meter was designed so that it was initially more excited by "fun" tasks such as *playing with toys*, or *listening to music* over-performing work (i.e., *executing commands*). The value of A for *playing with toys* was set to twice the value of A for *executing commands*. However, B for *executing commands* was set to half the value of B for *playing with toys*. Using this design, when the experiment was initially run, ISAC chose to play with the toys rather than performing the command. Additionally, the excitement discrepancy initially encoded between the two tasks caused the probability that *playing with toys* was chosen to increase to 100%, and subsequently *executing commands* was decreased to 0%. Figure 16 shows the level of excitement associated with each set of stimuli during the experiment.

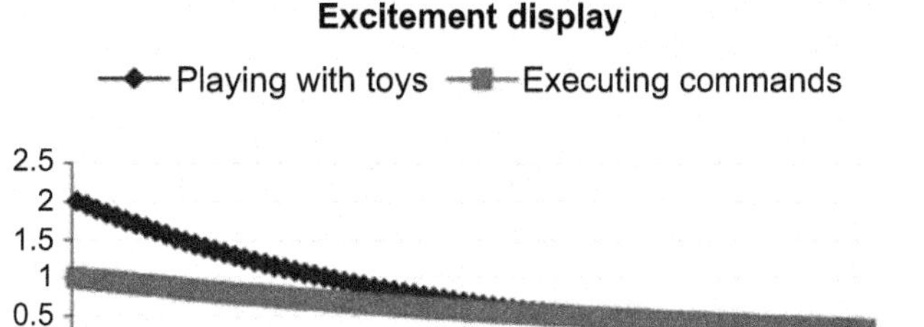

Figure 16
Excitement levels for competing tasks.

However, when this experiment was allowed to run for longer than ~3 min, ISAC's excitement level dropped sufficiently enough that it then became desirable to pay attention to the person and execute the command. Much of this behavior should be attributed to the fact that the current excitement associations are still heavily dependent on the initial weightings. Multiple trials over multiple days under varieties of circumstances have not yet been performed; therefore, the initial weights of (i, B) have not been modified by the system to reflect ISAC's personal experiences.

Performance criterion 1 was evaluated based on the effectiveness of Eqn (2). For this experiment, the computational model worked as intended in representing affect. The fact that certain excitement associations initialize higher or lower and decay faster or slower is the reason why Eqn (2) was chosen. Since the parameters (A, B) can be retained and updated like standard network weights, this equation adds dynamic flexibility to this system by allowing it to learn excitement associations over time. This can be done by rewarding particular choices and punishing others. Therefore, over time, the stimulus *executing commands* may begin to initialize higher than *playing with toys* based on ISAC's own experience.

Performance criterion 2 was evaluated based on the decisions made by the CEA. Using the weights (A, B) in Eqn (2), the CEA could choose to ignore or accept the suggestions made by the Affect Agent. The values (A, B) represented the strength of the input from the Affect Agent. When the CEA ignores the suggestions, it does not allow the probabilistic model to be updated by the Affect Agent. In order to evaluate the performance criterion, in this experiment, the CEA was forced to accept the suggestions made by the Affect Agent. Therefore, when both sets of stimuli were present the suggestions passed to the CEA by the Affect Agent caused the probability that *playing with toys* would be chosen to increase to 100% and, conversely, *executing the command* to decrease to 0%. However, due to this drastic change in the action probabilities, the initial decision making within the

CEA required no decision. However, later in the experiment a decision was necessary: whether to continue to *play with the toys* or switch to the more exciting *execute the command* task. Figure 16 shows that when this decision was necessary, the associated excitement levels with each task were approximately equal. Likewise, the CEA chose to switch tasks, at this point in the experiment, 50% of the time.

It is important to note that, for completeness, a third choice to *do nothing* should have been made available to ISAC. Future experiments will include this option. Also important in future work is the incorporation of other affect variables and their role in influencing each other. For example, *fear* of a negative reward for not performing a task could negatively influence excitement. *Joy* for being successfully able to *play with the toys* could positively influence excitement, possibly overriding *fear*. In addition, these new variables should also influence the probability that an action is selected. Future work must incorporate updating the parameters (A, B) over longer periods of time, possibly several days. These experiences, over time, are the keys to creating a dynamic affectual system that is individual to ISAC and, to a certain extent, unpredictable by humans.

5.4.6 Future Plans

In order to execute a task more robustly, CEA must monitor and evaluate the expected outcome of task execution according to the behavior–percept combination and interrupt task execution, if necessary, in real-time. We are currently looking into one real-time Windows operating system offered by a German company, KUKA (http://www.kuka-controls.com). In addition, ISAC requires past experience as an episode to be used in cognitive control. However, the current representation of episode is too simplistic. We plan to investigate a more robust episodic memory representation and retrieval. Another part of our future work includes expanding the capability of the system to handle more complex tasks that involve multiple percepts and behaviors.

Future work with the self-motivated, internal state-based action selection requires several key components. First, more internal state variables need to be added to the Affect Agent. Such variables as *fear*, *happiness*, *pain*, etc. would better enable ISAC to make decisions in more complex situations. How to incorporate the variables into the probabilistic decision-making model CEA currently uses and how to incorporate the effect these variables have on each other are two important issues to be looked at. Second, when to update the values (A, B) in the excitement Eqn (2) to reflect a change in ISAC's preference is an important issue with the Excitement Meter. These variables represent a key feature of the cognitive system, i.e., the ability to modify its own preferences based on experience. Third, the question "When should or should not CEA accept input from the Affect Agent?" (i.e., the change in the "probabilistic model" needs to be examined closer). Currently, the decision is based on the strength of the excitement association. A more robust solution integrating past experience, knowledge of the current situation, and other affect variables is required. Lastly, the hybrid system nature of the Affect Agent needs further investigation. Currently, events were sets of stimuli defined a priori. The real world, however, does not conform to a predefined structure, and it is important to understand how to organize low-level events (such as the appearance of a percept) into higher-level events that trigger complex cognitive control mechanisms.

5.4.7 Conclusions

During the past decade, we have seen major advances in the integration of sensor technologies, artificial intelligence, and machine learning into a variety of system designs and operations. A next challenge will be the integration of human-like cognitive controls into system designs and operations.

This chapter described our efforts to develop the next generation of robots with robust sensorimotor intelligence using a multi-agent-based cognitive control. Experiments conducted so far validated the effectiveness of our design.

Appendix 1. Spatial Attention and Action Selection

Humans pay attention by emphasizing the locations of percepts with high saliency. This process is known as spatial attention [22]. In our architecture, spatial attention is realized through the assigned FOA by the Attention Network [11]. The percepts in attention then are brought into the WM for further processing. During an event, past episodes are retrieved from Episodic Memory using cues such as percepts and task information. Given the event and current state, the system will need to make a decision using the action performed during these past episodes. The actions will be extracted from episodes, sorted, and given priority based on cues such as affective values and rewards of the episode from which they are extracted.

The action selection process retrieves episodes using cues such as the affect value. Retrieved episodes then are assigned probabilities as follows: Let p_j be the priority for the j-th action, and $p_T = \sum_{j=1}^{N} p_j$ where N is the number of the retrieved actions, then action A_j will be assigned the probability.

$$P[A_j] = \frac{p_j}{p_T}; j = 1, 2, \ldots, N, \text{where} \sum_{j=1}^{N} P[A_j] = 1. \tag{5}$$

The action selection process then will be performed probabilistically as follows: The unit interval [0,1], representing the summation of $P[A_j]$, is partitioned into N regions, and the

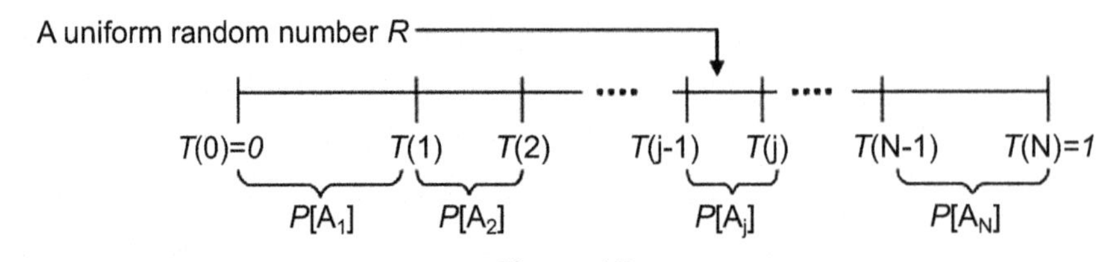

Figure 17
Probabilistic action selection process.

j-th region has a width of $P[A_j]$. A uniform random number R is generated, $0 \leq R \leq 1$. Let $T(0) = 0$. For each region j, compute the boundary of $P[A_j]$ as $T(j-1)$ to $T(j)$, where

$$T(j) = P[A_k] \tag{6}$$

If $T(j-1) \leq R \leq T(j)$, select the j-th action. Figure 17 illustrates the action selection process currently used.

Appendix 2. Verbs and Adverbs for Behavior Execution

The Verbs and Adverbs algorithm is a motion interpolation technique originally developed for computer graphics by Rose et al. [14]. In this technique, motion exemplars are used to construct verbs that can be interpolated across different spaces of the motion represented by the adverbs. An important aspect in storing and reusing a motion for a verb is the identification of the key times [23,14] of the motion. The key times represent significant structural breaks in the particular motion. For the Verbs and Adverbs technique to function properly, individual motions for the same verb must have the same number of key times, and each key time must have the same significance across each motion. Figure 18 shows key times for three example motions. The example motions are recordings of the same motion, three different times. This information is used to create the verb, *handshake*. The key times in this example are derived by analyzing the motions using a technique called Kinematic Centroid [24]. The x-axis represents the normalized point index for each motion. The y-axis represents the Euclidian distance of the kinematic centroid of the arm from the base of the arm.

Each verb can have any number of adverbs, each of which relate to a particular space of the motion. For example, the verb *reach* could have two adverbs: the first related to the direction of the reach and the second related to the distance from ISAC's origin that the particular motion is to extend. Extending this example, adverbs could be added to include features from any other conceivable space of the motion, such as the strength of the motion or the speed of the motion. Stored in the LTM are the verb exemplars and the adverb parameters for each verb. New motions such as reaching or handshaking are interpolated by ISAC at run time using the new (desired) adverb values. One important

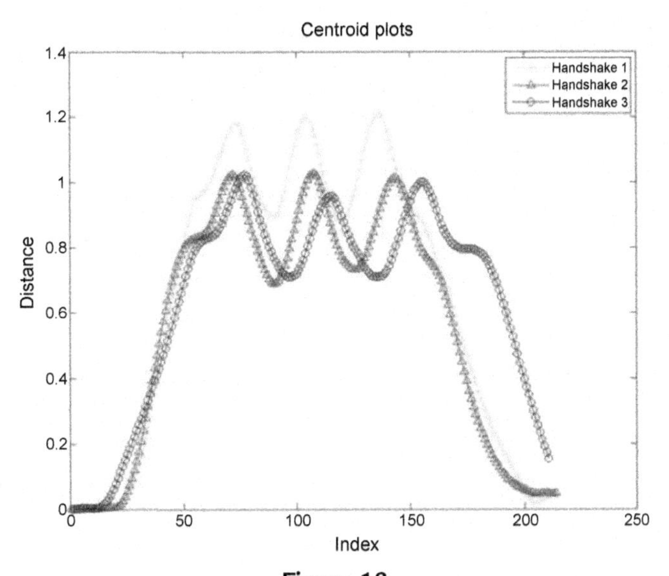

Figure 18
Example motions and key times [23].

point, in our system, new motions are never extrapolated. This is due to the fact that extrapolated motions can potentially lead to undesirable (or unachievable) arm configurations. Currently, ISAC is using the Verbs and Adverbs algorithm for three behaviors: reach, handshake, and wave.

Appendix 3. Memory Contents during Work Memory Training

Table 2 shows the contents of STM and LTM during the experiment discussed in Section 5.4.3.3. During the experiment, two bean bags were present in front of ISAC. Additionally, three behaviors had been trained and placed in LTM. This information was encoded into WM "chunks," void data structures. The WM then chose from these chunks and the contents of WM-guided task execution. In other words, if the chunks reach and *blue bean bag* were present then ISAC reached to the blue bean bag. Table 3 shows example contents

Table 2: Memory contents during simulation training

SES	LTM
1 Bean bag: location = (Figure 6(b)), type = blue	1 Reach
2 Bean bag: location = (Figure 6(a)), type = red	2 Handshake
	3 Wave

SES: Sensory EgoSphere; LTM: long-Term Memory.

Table 3: Working memory contents during simulation training

	WM Contents			
Trial#:	**1**	**2**	**3**	**4**
Chunk 1	Blue bean bag	Red bean bag	Wave	Handshake
Chunk 2	Reach	Blue bean bag	Blue bean bag	Red bean bag
Random:	NA	Wave	NA	NA
Reward:	20	0	0	0

of WM during four of the training trials. When one percept and one behavior chunk were not present, the missing chunk(s) were filled in at random.

Appendix 4. Perception Encoding Used in FRA Experiment

4.1 Sound Localization

Within the ISAC cognitive system, the location of a loud noise, such as hand clapping, is used as a means of focusing attention and a cause for invoking the reactive response in FRA. The location of the sound source is detected using sound localization. The basic configuration of the sound localization system used includes a pair of microphones, located a finite distance apart as illustrated below. The sound waves arrive at the microphones at different times. Using this time difference, the probability of the clapping sound found at any given location in the environment is calculated, and the result is sent to the SES (Figure 19).

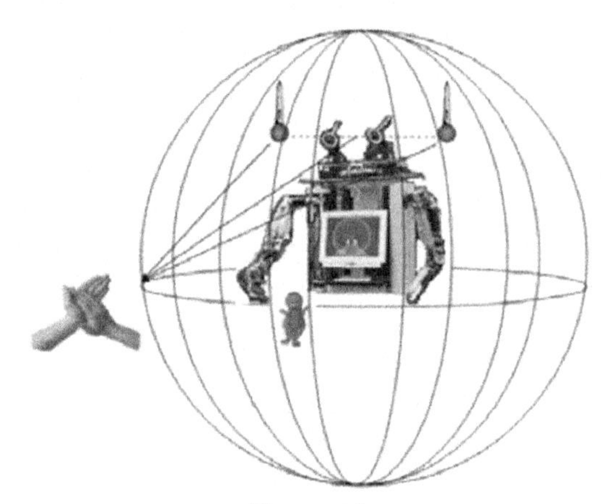

Figure 19
Sound localization.

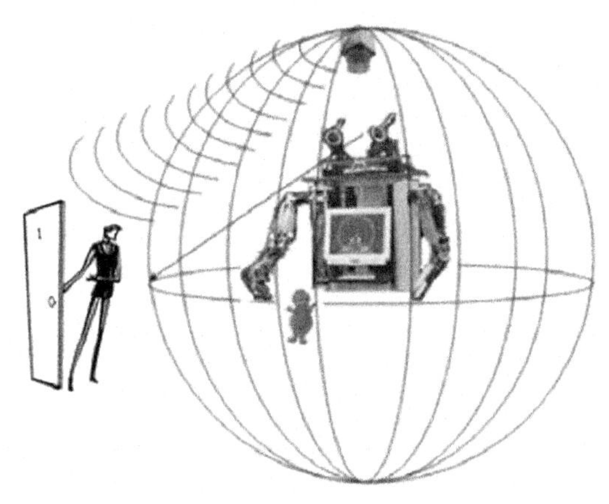

Figure 20
Motion detection.

4.2 Motion Detection

Motion detection is performed by the motion detection percept agent that utilizes a laser range finder mounted above the cameras on the head. The laser range finder is positioned so planar scans of the area in front of the robot can be obtained from various angles. Successive scans from the same angle are subtracted to obtain a temporal difference map. The positions of nonzero values along the temporal difference map, the angle of the planar scan, and the range data are used to indicate the position of motion in front of ISAC, which the information is posted to the SES (Figure 20).

Acknowledgments

Special thanks to Flo Wahidi, and the graduate students at the Center for Intelligent Systems. This work is supported in part under an NSF grant EIA0325641, "ITR: A Biologically Inspired Adaptive Working Memory System for Efficient Robot Control and Learning."

References

[1] M.S. Gazzaniga, Cognitive Neuroscience. The Biology of the Mind, second ed., Norton & Co., New York, 2002, p. 615.
[2] J.R. Stroop, Studies of interference in serial verbal reactions, J. Exp. Psychol. 18 (1935) 643–661.
[3] M.M. Botvinick, T.S. Braver, D.M. Barch, C.S. Carter, J.D. Cohen, Conflict monitoring and cognitive control, Psychol. Rev. 108 (3) (2001) 624–652.
[4] K. Kawamura, R.A. Peters II, R. Bodenheimer, N. Sarkar, J. Park, A. Spratley, K.A. Hambuchen, Multiagent-based cognitive robot architecture and its realization, Int. J. Hum. Robot. 1 (1) (2004) 65–93.
[5] J.M. Mendel, K.S. Fu (Eds.), Adaptive, Learning and Pattern Recognition Systems: Theory and Applications, Academic Press, New York, 1970.
[6] E.K. Miller, C.A. Erickson, R. Desimone, Neural mechanisms of visual working memory in prefrontal cortex of the macaque, J. Neurosci. 16 (16) (1996) 5154–5167.

[7] T. Pack, D.M. Wilkes, K. Kawamura, A software architecture for integrated service robot development, in: 1997 IEEE International Conference on Systems, Man, and Cybernetics, vol. 4, October 1997, pp. 3774–3779.
[8] J.S. Albus, Outline for a theory of intelligence, Proc. IEEE Trans. Syst., Man, Cybern. Syst. 21 (3) (1991) 473–509.
[9] R.A. Peters II, K.A. Hambuchen, K. Kawamura, D.M. Wilkes, The sensory egosphere as a short-term memory for humanoids, in: Proc. IEEE-RAS Int. Conf. on Humanoid Robots, Waseda University, Tokyo, November 22–24, 2001, pp. 451–459.
[10] A.D. Baddeley, Working Memory, Clarendon Press, Oxford, 1986.
[11] K.A. Hambuchen, Multi-modal Attention and Binding Using a Sensory Egosphere (Ph.D. Dissertation), Vanderbilt, Nashville, TN, May 2004.
[12] J.L. Phillips, D. Noelle, A biologically inspired working memory framework for robots, in: Proc. 27th Annual Conf. of the Cognitive Science Society, 2005, pp. 1750–1755.
[13] S.M. Gordon, J. Hall, System integration with working memory management for robotic behavior learning, in: Proc. 5th Int. Conf. on Development and Learning, Bloomington, IN, June 2006.
[14] C. Rose, M.F. Cohen, B. Bodenheimer, Verbs and adverbs: multidimensional motion interpolation, IEEE Comput. Graph. Appl. 18 (5) (September–October, 1998) 32–40.
[15] M.P. Shanahan, A cognitive architecture that combines internal simulation with a global workspace, Conscious. Cogn. 15 (2006) 433–449.
[16] M.A. Arbib, Perceptual structures and distributed motor control, in: V.B. Brooks (Ed.), Handbook of Physiology-The Nervous System II: Motor Control, American Physiological Society, Bethesda, MD, 1981, pp. 1449–1480.
[17] R.A. Brooks, A robust layered control system for a mobile robot, IEEE J. Robot. Autom. RA-2 (1) (1986) 14–23.
[18] R.W. Picard, Affective Computing, MIT Press, Cambridge, MA, 1997.
[19] P. Ratanaswasd, C. Garber, A. Lauf, Situation-based stimuli response in a humanoid robot, in: Proc. from 5th Int. Conf. on Development and Learning, Bloomington, IN, June 2006.
[20] J.R. Anderson, D. Bothell, M. Byrne, S. Douglas, C. Lebiere, Y. Qin, An integrated theory of the mind (ACT-R), Psychol. Rev. 111 (4) (2004) 1036–1060.
[21] M. Branicky, Analyzing continuous switching systems: theory and examples, in: Proc. American Control Conference, Baltimore, MD, June 1994.
[22] A. Cohen, R. Shoup, Response selection processes for conjunctive targets, J. Exp. Psychol. Hum. Percept. Perform. 26 (2002) 391–411.
[23] A.W. Spratley II, Verbs and Adverbs as the Basis for Motion Generation in Humanoid Robots (M.S. Thesis), Vanderbilt University, Nashville, TN, August 2006.
[24] O.C. Jenkins, M.J. Mataric, Automated derivation of behavior vocabularies for autonomous humanoid motion, in: 2nd Int. Joint Conf. on Autonomous Agents and Multiagent Systems, Melbourne, Australia, July 14–18, 2003.

PART 6

Human–Robot Interaction

CHAPTER 6.1

The State of the Art in Human–Robot Interaction for Household Services

Dan Xu[1], Huihuan Qian[1,2], Yangsheng Xu[1,2]

[1]The Chinese University of Hong Kong, Hong Kong SAR, China; [2]Shenzhen Institutes of Advanced Technology, Chinese Academy of Sciences, Shenzhen, China

Chapter Outline

Human–robot interaction (HRI) studies on how humans collaborate and interact with robots in an effective and natural way [1,2] have received considerable attention in the robot research community, especially with regard to the design and development of household service robots, owing to their highly interactive characteristics in domestic service activities. Many researchers have proposed a variety of principles, algorithms, and systems to achieve an effective and natural HRI in the domestic environment. In this chapter, we first give a brief illustration of the state of the art for HRI, divided into several categories, and then elaborate on some selected cases with regard to the design and implementation of HRI systems.

Goodrich et al. conducted an HRI survey [3] with two general categories: the remote interaction, in which the humans and robots are separated spatially or even temporally, and the proximate interaction, in which the humans and robots are colocalized. In contrast to their classification, which was formulated from an interaction perspective, we present our classification for the state of the art HRI according to the types of sensors being used.

6.1.1 Tactile HRI Systems

Humans interact with robots through physical contact in both experimental and operational environments [4,5]. There are many potential motivations for the detection of human contact with the robot, such as the consideration of safe robot operations around humans, and robot behavior needs to be guided by humans. The tactile human–robot interaction

Household Service Robotics. http://dx.doi.org/10.1016/B978-0-12-800881-2.00020-7

(tHRI) systems are primarily developed to intervene in the execution of robot behavior, contributing to, and assisting with the development of robot behavior through human contacts. Samples of such systems include the following.

Kosuge et al. [6,7] developed a female dance partner robot, referred to as MS Dancer, shown in Figure 1(a), which can cooperate and dance with a male human by physically interacting with him. MS Dancer has an omnidirectional mobile base, a body force sensor, and a control architecture, "CAST." The robot platform is also used for the analysis and research of physical HRI.

To ensure that robots can interact with humans and environments through whole-body contact, Yoshikai et al. [8] developed a humanoid robot with soft sensor flesh, shown in Figure 1(b). It uses its hardware abilities to collect and respond to rich, tactile sensor information in designing interaction behavior.

Huggable [9] is a companion robot designed and developed by the MIT Media Lab, as shown in Figure 1(c). Its full-body "sensitive skin," silent voice-coil actuators, and embedded processor allow it to realize active relational and affective tactile interactions with humans.

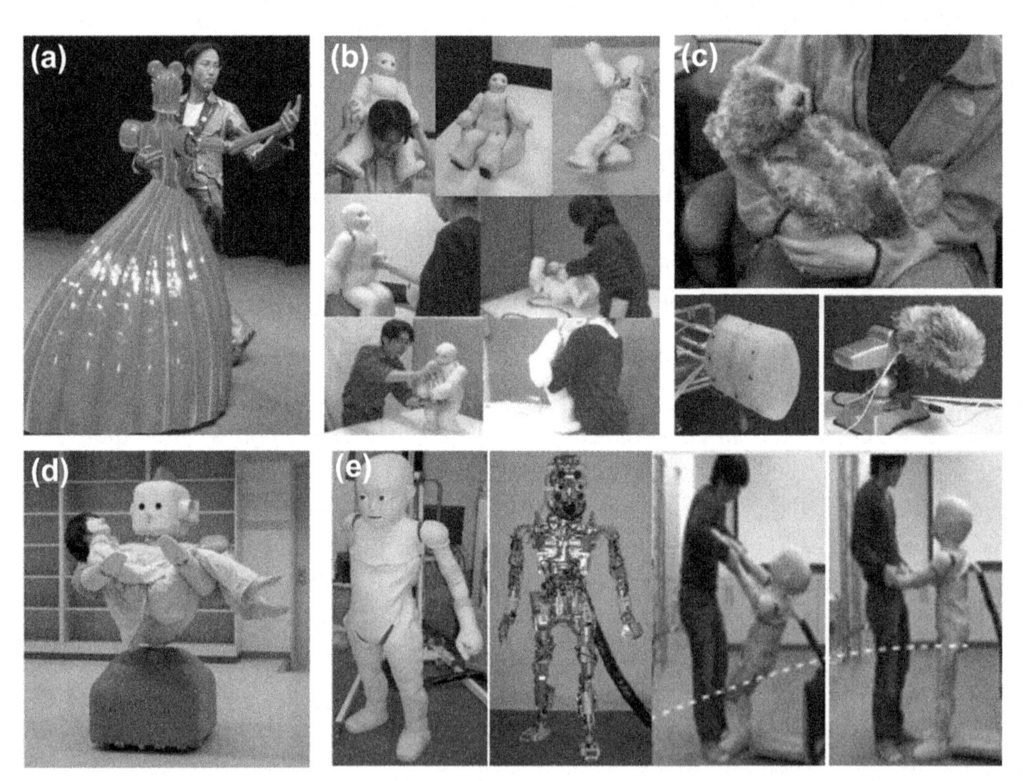

Figure 1

Examples of tactile human–robot interaction. (a) A female dance partner robot named MS Dancer. (b) A humanoid robot with soft sensor flesh. (c) A companion robot with full-body "sensitive skin" from MIT media Lab. (d) Robot "RI-MAN" with tactile sensor system. (e) Assisted standing-up and walking tasks with a humanoid robot.

To assist in the nursing of elderly people, Mukai et al. [10] developed a tactile sensor system within a human-interactive robot named "RI-MAN," shown in Figure 1(d). With this system, RI-MAN can lift a dummy human weighing 16 kg in its arms.

Ikemoto et al. [11] proposed an effective machine-learning algorithm for physical interaction application scenarios. Two interaction tasks—an assisted standing-up task and an assisted walking task—are performed through close-contact interaction with a humanoid robot (see Figure 1(e)) to demonstrate the effectiveness of the proposed algorithm.

6.1.1.1 Vision-Based HRI Systems

Natural interactions between humans and robots are the ultimate goal, and as a crucial aspect of such interactions, vision-based human–robot interaction (vHRI) has been widely studied by many computer vision and robotics researchers since the late 1990s. Many effective interaction systems have been developed and we selectively introduce them in three popular categories: gesture recognition-based HRI, facial expression recognition-based HRI, and activity understanding-based HRI.

Gesture recognition-based HRI. The visual representation and recognition of gestures such as hand, body, sign language, and command gestures can be used to accomplish a natural HRI [12,13]. Riek et al. experimentally studied the problems that exist in the interactions between humans and robots using cooperative gestures with a gesture interaction system deployed within a humanoid robot [14], as shown in Figure 2(a). Lee et al. developed an automatic gesture recognition system for intelligent HRI in which whole-body gestures can be seen and recognized to enable the robot to perform a series of complex tasks [15]. The arm gesture recognition interface developed by Waldherr et al. [16] can guide a mobile robot in carrying out an interactive cleanup task in changing-light

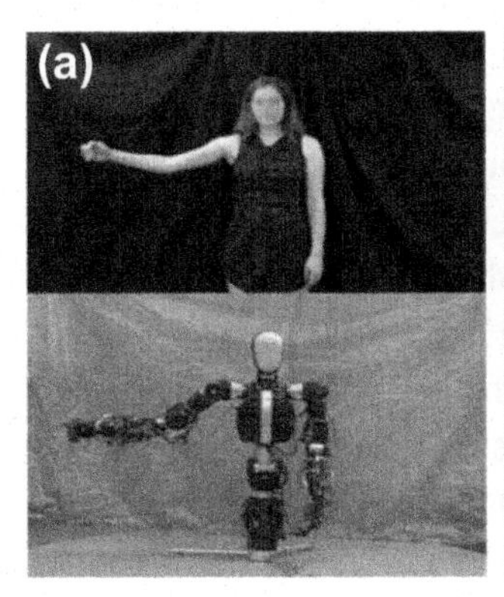

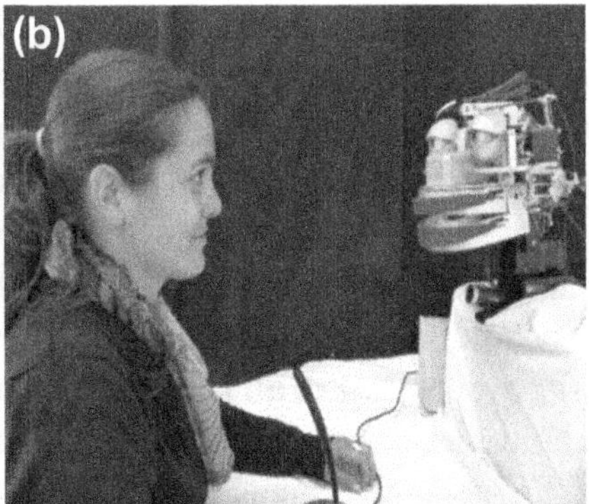

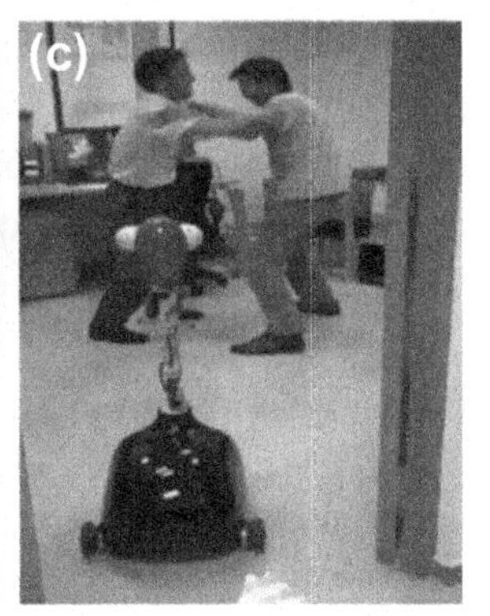

Figure 2
Examples of vision-based human–robot interaction. (a) Shows a cooperative gesture interaction system within a humanoid robot. (b) Shows a human/robot dialogue system based on human facial expression perception. (c) Shows a household service robot for detecting abnormal activities.

environments. A proposed pointing gesture recognition system [17] combines head orientation information to significantly improve the performance of the pointing gesture recognition, which can be used in many aspects of HRI such as indicating objects and locations in manipulation tasks and making the robot change the direction in which it is moving. At RoboCup 2009, Correa et al. demonstrated a real-time hand gesture recognition system for HRI [18]. The system allows interaction with a service robot using both static and dynamic hand gestures in dynamic environments.

Facial expression recognition-based HRI. Facial expression recognition-based HRI is one of the main research highlights in the field of HRI, especially affective HRI. Because a facial expression can naturally convey useful information related to human emotions, such as happiness, anger, or sadness, and to a lack of emotion, the emotional states of humans can be detected by the robot for interactive activities via the use of effective facial expression recognition interfaces. A famous facial action coding system built by Ekman et al. [19] describes and recognizes the facial expression of a human through a series of basic coded emotion units including joy, anger, surprise, disgust, fear, and sadness. Based on this system, researchers have proposed many improved algorithms and systems for the measurement and analysis of facial expressions [20–23]. To study the effect of facial expression on HRI, Gonsior et al. [24] constructed an HRI system, as shown in Figure 2(b), that allows a robot to carry out a dialogue with a human through his/her perceived facial expressions. Lee et al. proposed a method to detect the face and facial features under complex environmental conditions such as variations in pose, changes in illumination, and cluttered backgrounds for real-life HRI applications [25]. To accelerate the research and development of facial expression-based HRI, the MIT media lab has collected naturalistic and spontaneous facial expressions over the Internet to provide a public facial expression data set as a benchmark [26].

Activity understanding-based HRI. Activity understanding-based HRI demands that the robot be able to understand the meanings of human activities in interactive environments and launch interactions with humans accordingly, which is a higher level of interaction in the field of HRI. Kelley et al. proposed an approach that allowed the robot to detect and track a human based on computer vision techniques and then observe and analyze the human's intentions by modeling the robot's experience and interactions with the world [27]. Another activity detection system developed by Sung et al. [28] uses a Microsoft Kinect as the input sensor and is able to detect and recognize 12 different activities in different locations in the household, including kitchens, living rooms, and offices. To assist in the nursing of elderly people, Wu et al. [29] built a vision-based surveillance system deployed on a household service robot that can detect abnormal activities of "running" and "falling down" for moving humans in the domestic environment, as shown in Figure 2(c). To overcome the limitations of using a single camera, Fiore et al. proposed a multicamera human-activity monitoring system [30], which can track pedestrians across a location of interest and perceive a set of human activities under real-life conditions.

6.1.1.2 Speech-Based HRI Systems

With the rapid progress of automatic speech-recognition techniques [31–34], speech-based human–robot interaction (sHRI) has attracted increasing attention from the robotics research community. The researchers have developed many speech-based HRI systems that cover a wide range of application scenarios, and we briefly introduce several of these in this subsection.

To provide more robust models of language understanding for natural HRI, Cantrell et al. described an integral natural language understanding architecture for HRI [35], and the capabilities of the system were demonstrated through two experiments on the spoken instruction understanding of robots, including semantic ambiguities and incremental understanding with back-channel feedback.

Breazeal and Aryananda proposed an approach to recognize four distinct prosodic patterns to represent communication intent, including praise, prohibition, attention, and comfort, to preverbal infants, and this approach was integrated into the "emotion" system of a robot to enable humans to directly manipulate the robot's affective states [36].

For affective robot–child interaction, expressive speech synthesis and recognition are considered enabling techniques. A new speech synthesizer was developed by Yimazyildiz et al. [37] to allow the robot to synthesize expressive nonsense speech and then work toward effective affective interactions with children.

Kriz et al. developed speech-based HRI systems based on robot-directed speech to study the conceptualizations of robots [38,39]. Shown in Figure 3, is an experimental scenario in which the human asks the robot to fetch certain objects through robot-directed speech.

Figure 3
Speech-based interactions with AIBO.

In RoboCup 2008, Doostdar et al. proposed a speaker-independent speech-recognition system [40], using off-the-shelf technology and simple additional approaches, which can obtain high recognition accuracy under experimental conditions of loud noise and meets the needs of the mobile-service robot working in human environments.

6.1.2 Summary of Case Studies

The remainder of this chapter elaborates on five case studies of the design, implementation, and application of HRIs.

- Video-based HRI system for inexperienced users
  To ensure that robot designs are suitable for interactions with inexperienced users, Walters et al. proposed a new video-based HRI (VHRI) methodology [41], as illustrated in Figure 4. Unlike the previous VHRI methodology, which was used only in constrained HRI environments, the proposed methodology extends HRI scenarios with a series of events via a multinational approach. The study based on the proposed VHRI methodology provides a valuable means to obtain early user feedback and can also give useful guidance for future robot designs to improve the interaction experience.
- Socially assistive HRI system for older adults
  To help improve quality of life, Fasola et al. developed a socially assistive HRI system (as shown in Figure 5) to engage older adults in physical exercise [42]. The system design incorporates research insights with the intrinsic motivation from areas of psychology and maintains engagement through personalized social interaction. Two user studies were conducted to test the motivation theory in practice with the system.

Figure 4
Video-based human—robot interaction system.

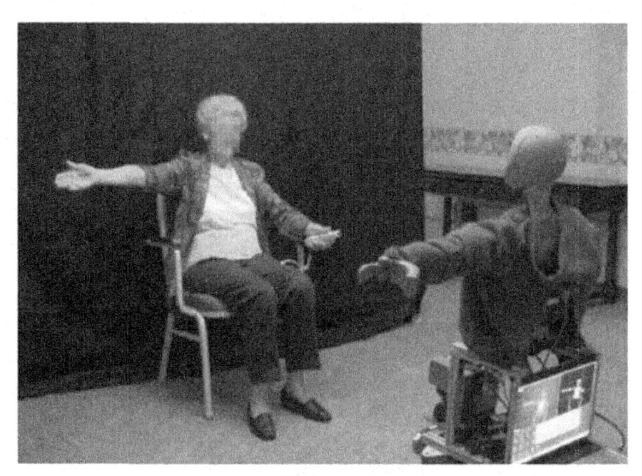

Figure 5
Socially assistive robot.

- Human–robot symbiotic system
  Kawamura et al. developed a human–robot symbiotic system that combines a multiagent-based robot control architecture with short- and long-term memory structures for the robot brain [43]. The multiagent control architecture comprises a human agent that serves as an internal active representation of people in the robot's environment and a self-agent that serves to monitor sensor signals, communicate between agents, and make decisions. The system was deployed through an intelligent robot named ISAC, shown in Figure 6. Two applications of human-guided object learning through shared attention- and situation-based acknowledgment were carried out to illustrate how the robot interacts with the human.

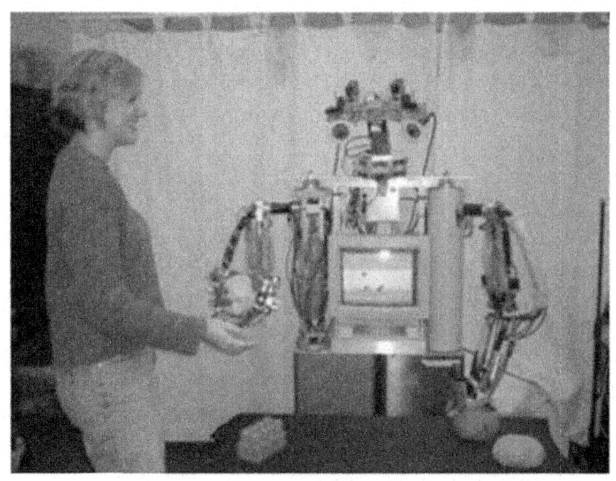

Figure 6
Vanderbilt's humanoid robot, ISAC.

References

[1] A. Steinfeld, T. Fong, D. Kaber, Common metrics for human–robot interaction, in: 1st ACM SIGCHI/SIGART Conference on Human–Robot Interaction, 2006, pp. 33–40.
[2] T. Fong, C. Thorpe, C. Baur, Collaboration, dialogue, human robot interaction, Robot. Res. (2003) 255–266.
[3] M.A. Goodrich, A.C. Schultz, Human–robot interaction: a survey, Found. Trend. Hum. Comput. Interact. 1 (3) (2007) 203–275.
[4] Brenna D. Argall, A.G. Billard, A survey of tactile human_robot interactions, Robot. Auton. Syst. 58 (10) (2010) 1159–1176.
[5] S. Haddadin, A. Albu Schäffer, G. Hirzinger, Safe physical human-robot interaction: measurements, analysis and new insights, in: International Symposium on Robotics Research (ISRR2007), 2007, pp. 439–450.
[6] K. Kosuge, T. Hayashi, Y. Hirata, R. Tobiyama, Dance partner robot-MS DanceR, IEEE Intell. Robot. Syst. (IROS) 4 (2003) 3459–3464.
[7] T. Takeda, K. Kosuge, Y. Hirata, HMM-based dance step estimation for dance partner robot-MS DanceR, in: IEEE/RSJ International Conference on Intelligent Robots and Systems, 2005, pp. 3245–3250.
[8] T. Yoshikai, M. Hayashi, Y. Ishizaka, T. Sagisaka, M. Inaba, Behavior integration for whole-body close interactions by a humanoid with soft sensor flesh, in: 7th IEEE-RAS International Conference on Humanoid Robots, 2007, pp. 109–114.
[9] W.D. Stiehl, J. Lieberman, C. Breazeal, L. Basel, L. Lalla, M. Wolf, The design of the huggable: a therapeutic robotic companion for relational, affective touch, in: AAAI Fall Symposium on Caring Machines: AI in Eldercare, 2006.
[10] T. Mukai, M. Onishi, T. Odashima, S. Hirano, Z. Luo, Development of the tactile sensor system of a human-interactive robot RI-MAN, IEEE Trans. Robot. 24 (2008) 505–512.
[11] S. Ikemoto, H. Ben Amor, T. Minato, B. Jung, H. Ishiguro, Physical human–robot interaction, IEEE Robot. Autom. Mag. 19 (2012) 24–35.
[12] H.-D. Yang, A.-Y. Park, S.-W. Lee, Gesture spotting and recognition for human–robot interaction, IEEE Trans. Robot. 23 (2) (2007) 256–270.
[13] C. Breazeal, Social interactions in HRI: the robot view, IEEE Trans. Syst. Man Cybern. C 34 (2) (2004) 181–186.
[14] L.D. Riek, T.-C. Rabinowitchy, P. Bremnerz, A.G. Pipez, M. Fraserz, P. Robinson, Cooperative gestures: effective signaling for humanoid robots, in: 5th ACM/IEEE International Conference on Human–Robot Interaction (HRI), 2010, pp. 61–68.
[15] S.-W. Lee, Automatic gesture recognition for intelligent human–robot interaction, in: 7th IEEE International Conference on Automatic Face and Gesture Recognition, 2006, pp. 645–650.
[16] S. Waldherr, A gesture based interface for human–robot interaction, Auton. Robot. 9 (2) (2000) 151–173.
[17] K. Nickel, R. Stiefelhagen, Visual recognition of pointing gestures for human–robot interaction, Image Vision Comput. 25 (12) (2007) 1875–1884.
[18] M. Correa, J. Ruiz-del-Solar, R. Verschae, J. Lee-Ferng, N. Castillo, Real-time hand gesture recognition for human robot interaction. RoboCup 2009: Robot Soccer World Cup XIII, 2010, pp. 46–57.
[19] P. Ekman, W.V. Friesen, J.C. Hager, Facial Action Coding System: The Manual on CD ROM, A Human Face, Salt Lake City, 2002.
[20] Y-li Tian, T. Kanade, J.F. Cohn, Recognizing action units for facial expression analysis, IEEE Trans. Pattern Anal. Mach. Intell. 23 (2) (2001) 97–115.
[21] M.S. Bartlett, G. Littlewort, M. Frank, C. Lainscsek, I. Fasel, J. Movellan, Fully automatic facial action recognition in spontaneous behavior, in: 7th International Conference on Automatic Face and Gesture Recognition, 2006, pp. 223–230.

[22] G. Donato, M.S. Bartlett, J.C. Hager, P. Ekman, T.J. Sejnowski, Classifying facial actions, Adv. Neural Inf. Process. Syst. (1996) 823–829.

[23] I.A. Essa, Coding, analysis, interpretation, and recognition of facial expressions, IEEE Trans. Pattern Anal. Mach. Intell. 19 (7) (1997) 757–763.

[24] B. Gonsior, S. Sosnowski, C. Mayer, J. Blume, B. Radig, D. Wollherr, K. Kuhnlenz, Improving aspects of empathy and subjective performance for HRI through mirroring facial expressions, in: IEEE RO-MAN, 2011, pp. 350–356.

[25] T. Lee, S.-K. Park, M. Park, An effective method for detecting facial features and face in human–robot interaction, Inform. Sci. 176 (21) (2006) 3166–3189.

[26] D. Mcduff, R. El Kaliouby, T. Senechal, M. Amr, J.F. Cohn, R. Picard, Affectiva-MIT facial expression dataset (AM-FED): naturalistic and spontaneous facial expressions collected "In-the-Wild". IEEE Conference on Computer Vision and Pattern Recognition Workshops (CVPRW), 2013, pp. 881–888.

[27] R. Kelley, A. Tavakkoli, C. King, Understanding human intentions via Hidden Markov Models in autonomous mobile robots, in: 3rd ACM/IEEE International Conference on Human-robot Interaction (HRI), 2008, pp. 367–374.

[28] J. Sung, C. Ponce, B. Selman, A. Saxena, Human activity detection from RGBD images. Plan, Activity, and Intent Recognition, 2011.

[29] X. Wu, H. Gong, P. Chen, Z. Zhong, Y. Xu, Surveillance robot utilizing video and audio information, J. Intell. Robot. Syst. 55 (4–5) (2009) 403–421.

[30] L. Fiore, D. Fehr, R. Bodor, A. Drenner, G. Somasundaram, N. Papanikolopoulos, Multi-camera human activity monitoring, J. Intell. Robot. Syst. 52 (1) (2008) 5–43.

[31] D. Povey, Discriminative Training for Large Vocabulary Speech Recognition, vol. 79, Cambridge University, 2004.

[32] H. Jiang, Confidence measure for speech recognition: a survey, Speech Commun. 45 (4) (2005) 455–470.

[33] M.A. Anusuya, S.K. Katti, Speech recognition by machine: a review, Int. J. Comput. Sci. Inform. Security 64 (4) (1976) 501–531.

[34] D. Ververidis, C. Kotropoulos, Emotional speech recognition: resources, features, and methods, Speech Commun. 48 (9) (2006) 1162–1181.

[35] R. Cantrell, M. Scheutz, P. Schermerhorn, X. Wu, Robust spoken instruction understanding for HRI, in: 5th ACM/IEEE International Conference on Human-robot Interaction (HRI), 2010, pp. 275–282.

[36] C. Breazeal, L. Aryananda, Recognition of affective communicative intent in robot-directed speech, Auton. Robot. 12 (1) (2002) 83–104.

[37] S. Yimazyildiz, W. Mattheyses, Y. Patsis, W. Verhelst, Expressive speech recognition and synthesis as enabling technologies for affective robot-child communication, in: Advances in Multimedia Information Processing-PCM, Springer Berlin Heidelberg, 2006, pp. 1–8.

[38] S. Kriz, G. Anderson, J.G. Trafton, M. Bugajska, Robot-directed speech as a means of exploring conceptualizations of robots, in: 4th ACM/IEEE International Conference on Human-robot Interaction (HRI), 2009, pp. 271–272.

[39] S. Kriz, G. Anderson, J.G. Trafton, Robot-directed speech: using language to assess first-time users' conceptualizations of a robot, in: 5th ACM/IEEE International Conference on Human-robot Interaction (HRI), 2010, pp. 267–274.

[40] M. Doostdar, S. Schiffer, G. Lakemeyer, A robust speech recognition system for service-robotics applications. RoboCup 2008: Robot Soccer World Cup XII, 2009, pp. 1–12.

[41] M.L. Walters, et al., Evaluating the robot personality and verbal behavior of domestic robots using video-based studies, Adv. Robot. 25 (18) (2011) 2233–2254.

[42] J. Fasola, M.J. Mataric, Using socially assistive human–robot interaction to motivate physical exercise for older adults, Proc. IEEE 100 (8) (2012) 2512–2526.

[43] K. Kawamura, T.E. Rogers, K.A. Hambuchen, D. Erol, Towards a human–robot symbiotic system, Robot. Comput. Integ. Manuf. 19 (6) (2003) 555–565.

CHAPTER 6.2

Evaluating the Robot Personality and Verbal Behavior of Domestic Robots Using Video-Based Studies[1]

Michael L. Walters[1], Manja Lohse[2], Marc Hanheide[3], Britta Wrede[4], Dag Sverre Syrdal[1], Kheng Lee Koay[1], Anders Green[5], Helge Hüttenrauch[6], Kerstin Dautenhahn[1], Gerhard Sagerer[2], Kerstin Severinson-Eklundh[5]

[1]Centre for Computer Science and Informatics Research, University of Hertfordshire, Hatfield, Hertfordshire, UK; [2]Research Institute for Cognition and Robotics (CoR-Lab), Hybrid Society Group, Bielefeld University, Bielefeld, Germany; [3]The Applied Informatics Group (Angewandte Informatik), Technical Faculty of Bielefeld University, Bielefeld, Germany; [4]Head of the Applied Informatics Group, Bielefeld University, Bielefeld, Germany; [5]KTH, Royal Institute of Technology, School for Computer Science and Communication, Stockholm, Sweden; [6]Department of Communication, Media and IT, Södertörns University, Huddinge, Sweden

Chapter Outline

[1] Evaluating the robot personality and verbal behavior of domestic robots using video-based studies, Michael L. Walters, Manja Lohse, Marc Hanheide et al., Advanced Robotics, 2011, reprinted by permission of the publisher (Taylor & Francis Ltd, http://www.tandf.co.uk/journals).

Household Service Robotics. http://dx.doi.org/10.1016/B978-0-12-800881-2.00021-9

6.2.1 Introduction

Robots currently find use as toys (e.g., Aibo [1], FurFriends [2], Lego Mindstorms [3]), for cleaning (e.g., Roomba [4]) and health-care (e.g., Paro [5]), and inexperienced users are expected to interact and control these machines. The robots should carry out tasks effectively and users should also want to interact with them on a long-term basis. When developing robots for use by non-technically-orientated users, design decisions should be evaluated by potential users. Therefore, user studies should be part of the development cycle, providing useful suggestions for improvements in both the technical and interaction capabilities of the robots.

Human–robot interaction (HRI) user studies typically employ live human–robot experiments in which humans and robots interact in various experimentally controlled scenarios [6,7]. Live HRI trials are generally complicated and expensive to run, and normally test a relatively small sample of possible users. An evaluation approach originally proposed in [8,9] is a video-based HRI (VHRI) methodology for user studies as a supporting methodology to live HRI user trials (cf. discussions in Dautenhahn [10] concerning different methodological approaches toward HRI user studies). In VHRI studies, interactive robot behaviors are recorded on videotape, which is then shown to many viewers who are then asked to rate the behaviors they watch. The methodology enables researchers to conduct studies with a large sample of participants in a relatively short time. We chose this approach as a pilot methodology to evaluate the prototype domestic robot BIelefeld RObot companioN (BIRON) [11–13]. This chapter presents the results of a multi-national study with 234 participants from Germany, the UK and Sweden (this chapter is an expanded version, including new experimental data, of content that has previously been presented at the *17th IEEE International Symposium on Robot and Human Interactive Communication (RoMan 08)* [14]).

The practical aim of HRI as a discipline is to design robots that are effective, efficient and usable for all interaction roles the user might require. Previously, theories and methods from psychology and human–computer interaction (HCI) have often been applied to HRI research [15,16] in order to evaluate systems with potential users in realistic conditions, including scenarios, environments and tasks [17,18]. Several approaches in this direction have been taken for robotics [19,20], although depending on the robot task and the context it is used in, it is often difficult to conduct live user studies. The effort to run full-scale live HRI trials can be large and the number of participants is typically relatively small. Live HRI trials, often with prototype robots and human operator controlled systems (cf. [21] regarding Wizard of Oz methods in HRI), may show different behaviors with different users, which may cause concerns regarding comparability of the data. The chapter is organized as follows: Section 6.2.2 introduces the VHRI methodology, Section 6.2.3 details the experiments, Section 6.2.4 presents the results of the user studies, Section 6.2.5 discusses the results and Section 6.2.6 provides the conclusions of the chapter.

6.2.2 VHRI Methodology

In the design of products and systems, video has long been used as a valid medium for visualizing, prototyping and user-testing a wide range of products [22]. Video based user trials provide a complementary methodology to conduct studies with many participants and improved comparability. We have also previously developed and verified the use of VHRI as a methodology for performing HRI user trials and the methodology chosen for this study was adapted from what we have employed in our previous study [8,9]. Note, in the previous studies the HRI situations studies were quite constrained (e.g., limited to approach directions in a simple fetch and carry task [8,9] or focusing on robot appearance and cues in an attention-seeking task [23,24]), but others have also adopted VHRI-based methodologies for HRI studies, including robot gender representations [25], user perceptions of robot heads [26] and human-looking robots [27]. The main advantages of using video-based methods for HRI are: (1) to reach larger numbers of participants as they are quicker to administer, (2) easily incorporate participant's ideas and views into later video trials simply by recording extra or replacement scenes into the video-based scenarios, (3) carry out trials exposing groups of participants to an HRI scenario simultaneously, (4) prototype proposed live trial scenarios to avoid wasted effort and test initial assumptions, and (5) allow greater control for standardized of methodologies (i.e., exactly the same robot behaviors, exact trial instructions, etc.). Additional compelling reasons also supported the VHRI methodology in the present study. First, our focus was solely on the evaluation of the interaction and video-based studies are particularly suitable to research user experiences rather than technical evaluations. Second, the HRI trial could be easily conducted with many participants at a time (e.g., in a university course) in three different countries: University of Bielefeld (UB), Germany; University of Hertfordshire (UH), UK; and Kungliga Tekniska Högskolan (KTH), Sweden. The robot effectively was brought to the participants, which would be difficult to organize in live trials in the different countries. Furthermore, all participants judged exactly the same robot behavior and the language is easily dubbed. Thus, comparability between groups is very high.

6.2.2.1 Methodology Development: Live and VHRI Comparison Studies

In our previous studies where the VHRI methodology was originally developed and verified, the results obtained from participants who viewed a video recording of another person participating in interactions with a robot were found to be comparable to those obtained from participants in the same live interactions. For more complete details on these studies see [8,9], but a summary of aspects relevant to the development of the VHRI methodology is provided here:

Initially, a pilot study was performed as a limited exploratory investigation to assess the potential for the comparability of people's perceptions from live and VHRI trials. Fifteen participants took part in live HRI trials and VHRI trials in which the scenario for both trials was identical, involving a robot fetching an object and carrying it to them using different approach directions. Findings from the pilot trials indicated moderate to high levels of agreement for participants' preferences and opinions for both the live and VHRI trials.

In order to verify these pilot trial findings, and to extend the investigation, a larger scale series of comparative live and VHRI trials were performed, with a large sample set of 42 participants and a wider range of HRI situations involving a robot approaching human participants. In this previous main study, additional controlled conditions included the human participants sitting in an open space, sitting at a table, standing in an open space and standing against a wall. The participant experienced the robot approaching from various directions for each of these contexts in HRI trials that were both live and video-based. There was a high degree of agreement between the results obtained from both the live and video-based trials using the same scenarios. The main findings from both types of trial methodologies were: humans strongly did not like a direct frontal approach by a robot, especially while sitting (even at a table) or while standing with their back to a wall. An approach from the front left or front right was preferred. When standing in an open space a frontal approach was acceptable and although a rear approach was not usually the most preferred, it was generally acceptable to participants if physically more convenient. Significant comparable results were also obtained for both sets of trials with regard to robot approach speed and distance.

Overall, the findings from these two sets of experiments supported the use of the VHRI methodology for developing and trying out new innovative studies that are in the pilot phase of testing. Naturally, it is appreciated that there are numerous limitations of using video footage for HRI studies and it should be appreciated that they are not a replacement for live HRI studies. It is expected that the more contingent the interactions between a robot and participant in a given trial, the less suitable VHRI trials would be, due to the increased importance of aspects of embodiment, dynamics and contingency of interaction. However, for the particular research questions that we considered in this current study, the contingency of robot and human movements played a less crucial role and, therefore, justified our choice of a VHRI trial methodology.

6.2.3 Experiments

6.2.3.1 Artifact

The robot used for the trials is called BIRON (Figure 1) and is based on a Pioneer PeopleBot platform. A Sony EVI D-31 pan–tilt color camera was mounted on top of the robot at a height of 142 cm to acquire images of the upper-body part of interacting

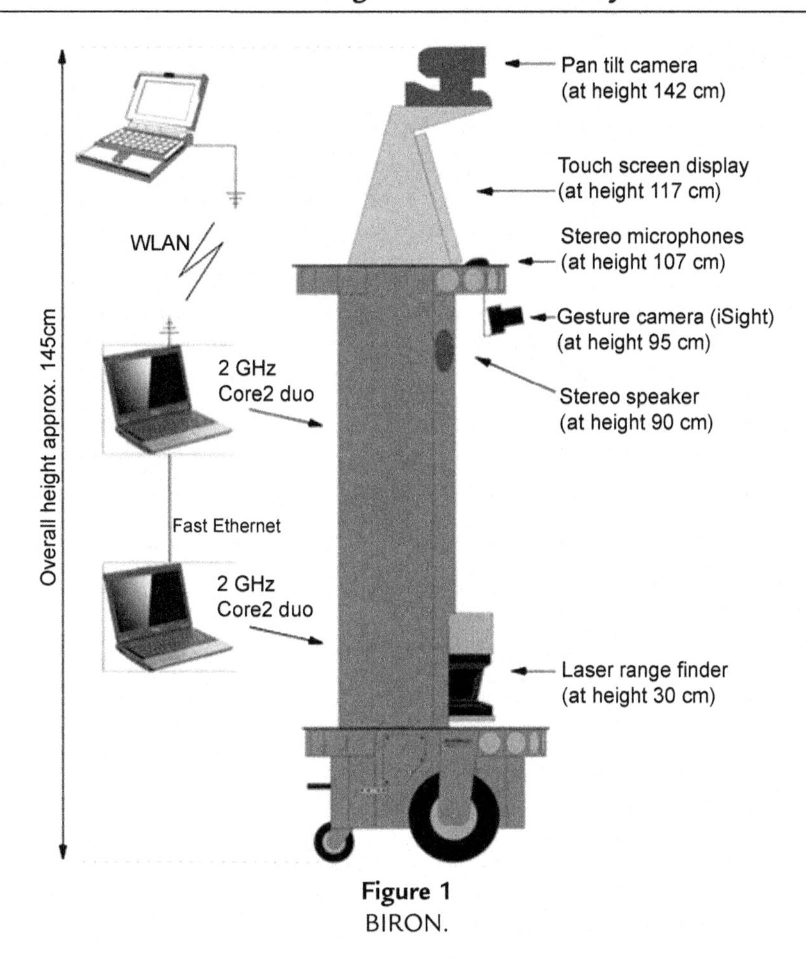

Figure 1
BIRON.

humans and referenced objects. An additional camera captured hand movements in order to recognize deictic references. A pair of AKG far-field microphones were located just below the touch screen display at a height of 107 cm to enable BIRON to localize speakers. A SICK laser range finder mounted at the front at a height of 30 cm measured distances to the user's legs and for navigation.

6.2.3.2 Scenario

The development of BIRON was framed by a home-tour scenario, which envisioned how household robots might adjust to a new user's home. The new environment is explored together, with the inexperienced customer teaching important objects and places to the robot. Therefore, a home-tour robot must exhibit capabilities for natural interaction, including understanding of spoken utterances, co-verbal deictic reference [28], verbal output, referential feedback, and person attention and following [29]. Previous experiments have indicated that robot personality has a major influence on HRIs [30–33]. Once a robot enters a person's home, its personality is important because if a user dislikes the robot, they will reject it. We, therefore, aim at developing a range of behaviors that allow the robot system to adapt to the users' preferences. In the related work cited above, users' perceptions of robot personality were normally influenced by different robot appearances. However, as Duffy [34] found for HRI and Walker et al. [35] for virtual agents, speech might also influence human–machine interaction and perceptions even more than appearance. According to Eysenck [36], extrovert personalities are described as sociable, friendly, talkative and outgoing. Introverts are quite introspective and prefer to be with small groups of people. We tried to model these behaviors in the verbal behavior of the robot and in the way it follows a person when entering a room. We, therefore, developed two different interactive behaviors (labeled here "Extrovert" and "Introvert") based on the two different verbal interaction styles and movement patterns of the robot.

6.2.3.3 Research Questions

The main research questions addressed in the present study are:

- Do participants recognize differences between the two robot behaviors (Extrovert and Introvert)?
- Which of the behaviors do participants prefer?
- Is the robot displaying Extrovert behaviors rated as being more friendly, intelligent and/or polite than the one displaying Introvert behaviors?

6.2.3.4 Implementation of Robot Behaviors

- *Extrovert robot verbal behavior.* The robot initiated conversations and volunteered information as to its capabilities. The robot initiated the first greeting with the user, asked if it could do anything and also provided more chatty replies when interacting with the user (e.g., "I see the TV, it is very nice", etc.)
- *Introvert robot verbal behavior.* The robot did not initiate conversations and only volunteered information when asked explicitly. The robot passively answered the users initial greeting, and only provided brief informational answers to the users questions and when interacting with the user (e.g., "I see the TV now"). At a certain point in the video scenario, the user showed the robot the kitchen. The robot demonstrated two ways in which it and the user might negotiate entering the room.

The robots' room entry movements and spatial interaction behaviors were also associated with E and I verbal behaviors. The extrovert robot (E) showed more autonomy, while the introverted robot (I) required detailed instructions from the user to negotiate entry to the kitchen. We do not focus here on how the robot behaviors are rated by people with

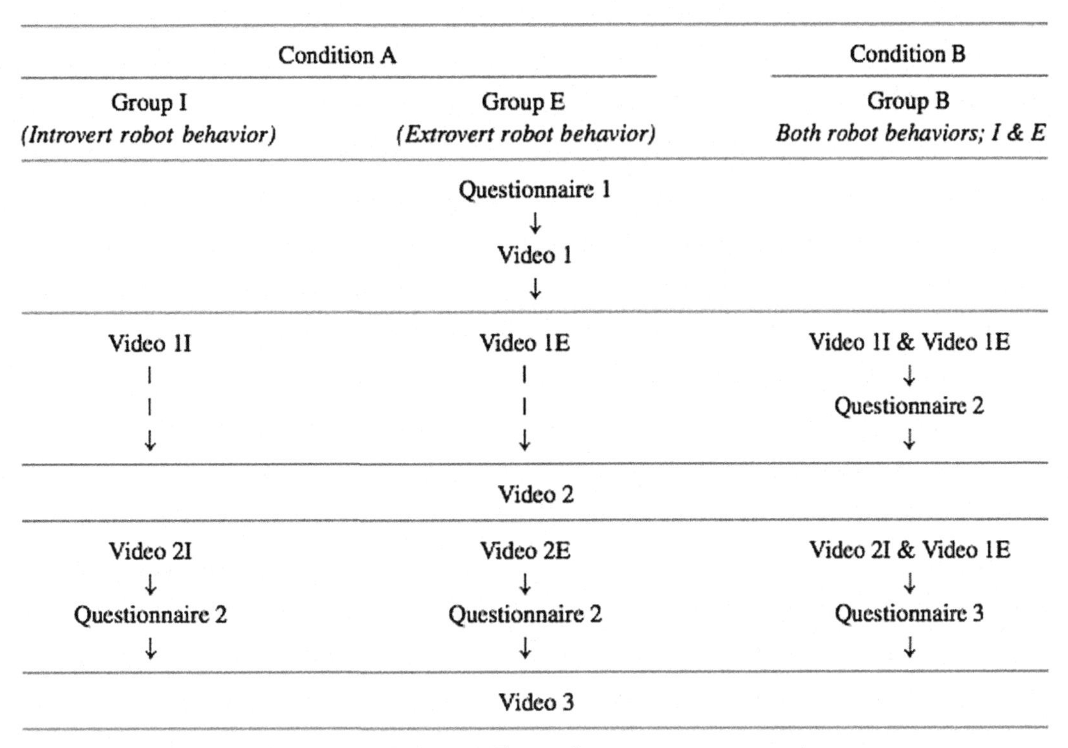

Figure 2
Experimental procedure.

different personalities because we first wanted to verify whether the robot behaviors were perceived as being distinct from each other. To test this, participants were divided into three groups, the first two groups (Condition A) each watched one robot behavior only (E or I), and the last group watched both (Condition B). The three groups were necessary to test in-group as well as intergroup differences. All participants of a group (e.g., a course cohort) watched the videos together. Figure 2 displays the experimental procedure for each group. The whole experiment took about 25–30 min for the short conditions (A) and 35–40 min with the longer one (B). The development of the videos and questionnaires is described in Section 6.2.3.5.

6.2.3.5 Videos

The VHRI trial videos were shot in a real apartment (Figure 3) with the robot system set up and functional, and an actor playing the part of the user. The robot movements were controlled by a human operator purely for convenience when shooting the videos to avoid necessary time-consuming resets and re-initialization of the robot if it was running autonomously. However, all the robot behaviors exhibited in the video were identical to

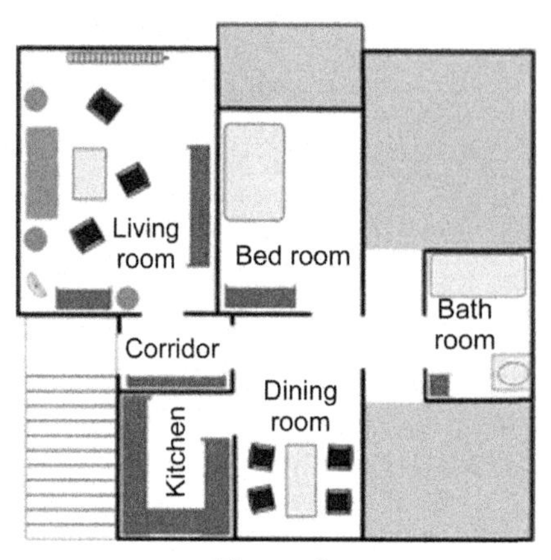

Figure 3
Robot apartment.

those previously demonstrated by the robot being fully autonomous. The videos were made following guidelines gained from running previous VHRI trials [8,9,24] that specified the mix of first- and third-person views to be shown, the forms of editing and effects allowed, and provide a standard formula for VHRI videos to follow: an initial wide-angle view of the HRI area was shown to establish the initial spatial relationship of robot and actor(s) and provide an overview of the scenario and the HRI. Then a series of first- and third-person views, showing the action primarily from the user's point of view in order to enhance the viewer's perception that they are in the middle of the action (Figure 4). There were no first-person views from the robot in order to reinforce the viewer's empathy with the human, not the robot. All action was shown happening in "real-time" with no quick cuts made from edited sequences to artificially enhance the interest of the video. Where a cut-away was made to signify a passing of time, a fade-out then a

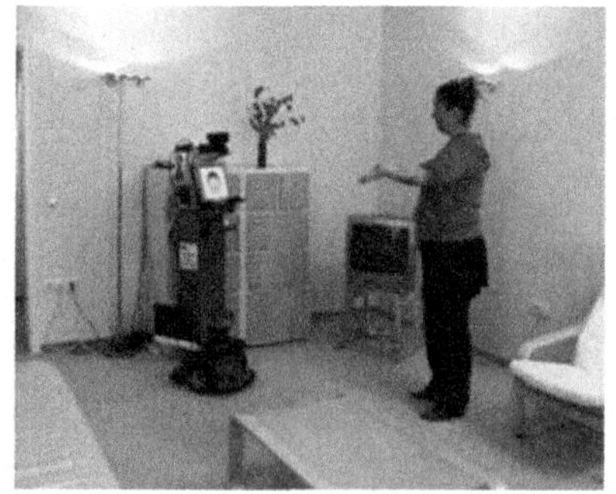

Figure 4
First-person and third-person views of the scene.

fade-in transition was employed. A subtitle to explain what happened during the period cut by the fade transition was also acceptable.

The video was composed of three main parts. Video 1 showed the introduction to the scenario with the robot being delivered and assembled by a mechanic, and incidentally provided additional information about the scenario of a domestic robot that can easily be purchased, set up and employed by inexperienced users. While this video was the same for all conditions, two different robot behaviors were recorded for most of the home tour (Videos 1I and 1E and 2I and 2E, see Figure 2).

Videos 1I and 1E presented a user (played by a professional actress) greeting the robot and showing objects in the living room, with the robot displaying different verbal behaviors. Video 2 was identical for all conditions and presented the robot on its way from the living room to the dining room. Videos 2I and 2E showed the user guiding the robot into the kitchen, again displaying two different E and I behaviors. The final Video 3 was identical for all groups and showed BIRON autonomously driving back to the living room.

The two robot behaviors consisted of different verbal and movement interaction patterns. Robot behavior I, labeled "Introverted", was designed to be less proactive. The robot in this condition waited until it was addressed by the user before talking. Apart from that, the robot talked little and used brief sentences, which significantly shortened the interaction (Video 1I). When the user guided the system through a door into the kitchen it needed to be steered directly by voice commands (Video 2I).

Robot behavior E was labeled "Extroverted". When the actress entered the living room, the robot addressed her instead of waiting for her to start the conversation. Moreover, the extrovert BIRON was more talkative. The robot uttered longer sentences, which were also more elaborate (Video 1E). In this condition, the robot entered the kitchen autonomously. It simply followed the user instead of waiting for instructions (Video 2E). The following example illustrates the difference between the extrovert and introvert verbal behavior:

Introvert (I):	User: Hello. Robot: Hello.
Extrovert (E):	Robot: Hello. My name is BIRON. What's your name? User: I'm Tina. Robot: Nice to meet you Tina.

6.2.3.6 Measures

Before they watched the videos, all groups completed a first questionnaire, which gained basic demographic data including age, course of study and gender, and participants rated their experience with computers and robots. They also indicated which robots they knew out of a list of 10. The participants then watched the videos.

People rating only one robot behavior (Condition A) watched all the videos of their condition at once, apart from the autonomous return of the robot to the living room (Video 3). Before watching this final video, they answered the second questionnaire. Participants rating both behaviors (Condition B) watched the interaction in the living room, answered the second questionnaire, watched the guiding to the kitchen and the interaction in the kitchen, answered the third questionnaire, and then finally watched the robot return to the living room. The questionnaires for this group contained the same items as those for the other two conditions, but participants answered a set of questions for each robot behavior. Sequence effects in Condition B (both behaviors) cannot be excluded since the videos were only shown to one group in each country. Thus, counterbalancing was not possible. Before the actual user studies were run, a pre-test was conducted to identify any problems in the design of the study, as described in Section 6.2.3.7.

6.2.3.7 Pre-Test

The pre-test was conducted in German with 54 students in three different courses. Students were divided into three groups where all three conditions were tested. The pre-test brought some insights that helped to improve the videos and the questionnaires. An advantage identified by Woods et al. [9] is that single video scenes can easily be changed or replaced. After the pre-test, this was advantageous regarding shortening the overall length of the video.

Participants in the pre-test watched the robot travel back to the living room before they filled in the final questionnaire. The robot traveled back autonomously to a room for which it had previously learned the location. This was an intelligent behavior and it turned out to overshadow the differences between the two robot behaviors. Therefore, a change was made in that participants filled in the second questionnaire before being shown the final Video 3. The first version of the questionnaire contained several open questions ("Which robots do you know?"; "Name adjectives to describe the robot") and after the pre-test these questions were replaced by Likert [37] scales to save time and to provide easily comparable answers. In the pre-test, each participant listed robots they knew, and the eight most frequently named robots were Aibo, Kismet, Mars Explorer, Asimo, Soccer Robot, Lego Mindstorms, Roomba and R2D2. We included these robots plus BIRON and a "service robot for the home" in the main trial questionnaire to explore whether people were familiar with the domain studied in these trials. We also analyzed the adjectives people used in the pre-test to describe the robot behavior. Categories containing words with synonymous meaning were created, and afterward we selected a word that best described each group and also a paired word with an opposite meaning. In the new questionnaire participants had to rate 14 adjective pairs on a five-point rating scale. This further increased comparability between participants and decreased the time to answer the questionnaire. The scale consisted of adjectives that were chosen as appropriate to divide between the two behaviors tested (active, passive; interested, indifferent; talkative, quiet) and others that might result from the perception of different robot personalities (intelligent, stupid; predictable, unpredictable; consistent, inconsistent, fast, slow; polite, impolite; friendly, unfriendly; obedient, disobedient; interesting (diversified), boring; attentive, inattentive). Other terms investigated the general usefulness of the robot (useful, useless; practical, impractical).

6.2.4 Results

The results include data acquired in a study with 200 participants in Germany ($N = 109$) and the UK ($N = 91$). For logistical reasons, a study with smaller participant numbers ($N = 34$) and slightly different procedures (mainly subtitles instead of sound dubbing) was run for the study in Sweden, so the data from this study is considered separately. All participants were assigned to one of the controlled experimental conditions: Condition A, viewed only one video, either of robot I (Introverted, $N = 62$) or E (Extroverted, $N = 72$); Condition B, viewed both I and E robot videos sequentially ($N = 66$).

The UB/UH participants' mean age was 23.95 years (31 for the Swedish study), 108 were male, 92 female. All the German participants were students, whereas for the UK sample, 10 people belonged to the academic staff. Overall, 46.5% had a background in computer science (Germany: 30%, UK: 66%). The rest came from other disciplines (linguistics, German studies, media science, psychology, business and health communication). All had some experience working with computers (mean = 3.97 on a scale of 1 (no experience at all) to 5 (a lot of experience)), but most had little experience of interacting with robots (mean = 1.65 on a scale of 1 (no experience at all) to 5 (a lot of experience)). The majority indicated they knew some robots (mean = 3.94 out of 10; minimum = 0, maximum = 10, SD = 2.8), the best-known being: R2D2 (66.5%), Aibo (62.5%), Mars Explorer (49.5%), Soccer Robots (46%) and Asimo (45%). Only 14.5% knew BIRON.

We analyzed the questions "How much do you like the robot?" and "How satisfied are you with the robot's behavior?" to find out whether participants actually noticed a difference between the robot behaviors and if one was preferred. Table 1 presents the combined UH/UB participants' ratings (Swedish ratings are indicated in brackets). No significant intercultural differences were found, which supports the assumption that videos with dubbed (or subtitled) language can be shown in various countries. It should be kept in mind that the sample was quite homogeneous as all countries were Western European. Table 1 illustrates that participants showed a significant preference for the extrovert robot's behavior (E). Both questions ("How much do you like the robot?" and "How satisfied are you with the robot's behavior?") were answered in favor of behavior E.

Table 1: Likability and satisfaction with robot behavior (mean on a scale of 1 (very low) to 5 (very high) for Conditions A and B (questionnaires two and 3))

	Condition A: Only Robot I Video or Robot E Video Shown to Participants		Condition B: Both Robot (I and E) Videos Shown to Participants			
			Questionnaire 1		Questionnaire 3	
Robot	I	E	I	E	I	E
	$N = 62$	$N = 72$	$N = 66$			
	($N = 13$)	($N = 13$)				
Likability	2.46 (2.62)	3.27 (2.38)	2.20	3.18	2.33	2.29
Satisfaction	2.45 (3.23)	2.88 (3.15)	2.23	3.12	2.42	2.30

Note: Swedish data are shown in brackets.

This finding is supported by 95.2% of the participants in Condition B indicating that they noticed a difference between robot behaviors in the second video for robot behaviors I and E. However, this does not hold true for the third rating of group B that judged both robots after the kitchen entry scene. Even though 63.5% noticed a difference between robot behaviors I and E, the ratings of the likability and satisfaction with the robot behavior did not differ. In order to explore the underlying reasoning and understanding of the robot behaviors that led to these satisfaction and likability ratings, one-way analysis of variance (ANOVA) was performed on the adjectives used to describe the robot for the UH/UB groups that rated only one robot behavior (Table 2, Condition A: likability ratings for I and E robot types: $F = 21.278$; d.f. = 1.130; Cohen's $f = 0.41$; $p < 0.001$; satisfaction with robot behaviors I and E: $F = 5.917$; d.f. = 1.132; Cohen's $f = 0.22$; $p = 0.016$). Paired t-tests were performed for the group that rated both the extrovert and introvert behaviors (Condition B: questionnaires 2 and 3, see Tables 3 and 4; likability ratings: mean difference = −0.969 (SD = 0.951), $dz = -1.02$, $T = -8.231$, d.f. = 64, $p < 0.001$; satisfaction ratings: mean difference = −0.877 (SD = 0.875), $dz = -1.00$, $T = -8.079$; d.f. = 64; $p < 0.001$). The differences between the ratings of behaviors I and E showed an interesting pattern. For Condition A, behavior E was rated as being more significantly active, talkative and interested, suggesting that the modeling of behavior E as more extrovert was successful. A similar pattern was found for the first scene in Condition B, supporting this finding. On the other hand, the differences between robot behaviors I and E in Condition A were not significant for Attentiveness, Consistency, Predictability, Obedience, Practicality and Usefulness.

For the first scene for Condition B, a similar pattern emerged, with some differences. While there were no significant differences between behaviors I and E in terms of Consistency, Obedience and Practicality, behavior E was rated as more Useful, while behavior I was rated as more Predictable.

Table 2: One-way ANOVA of ratings of robot behaviors for UB and UH data for Condition A (mean on a scale of 1 (not at all) to 5 (very much), *F*-value (d.f. = 1132) and significance)

Attribute	Mean I	Mean E	*F*-value	Effect Size (Cohen's *f*)	Significance
Active	2.30	2.89	12.247	0.31	0.001[b]
Talkative	2.00	2.93	26.145	0.47	<0.001[b]
Interested	2.87	3.26	5.36	0.2	0.022[a]
Attentive	3.54	3.59	0.105	0.03	0.747
Fast	1.61	2.00	6.813	0.23	0.010[a]
Consistent	3.22	3.31	0.361	0.05	0.549
Predictable	3.35	3.44	0.256	0.05	0.614
Polite	4.05	4.31	3.149	0.15	0.078
Friendly	3.56	4.03	9.218	0.27	0.003[b]
Obedient	4.16	4.31	0.894	0.08	0.346
Diversified	2.12	2.69	9.218	0.3	0.003[b]
Intelligent	2.98	3.34	4.433	0.18	0.037[a]
Practical	2.10	2.27	1.026	0.08	0.313
Useful	2.18	2.13	0.087	0.03	0.768

[a]<0.05.
[b]<0.005.

Table 3: *t*-Test for paired samples for ratings of robot behaviors in Condition B (mean on a scale of 1 (not at all) to 5 (very much), *t* (d.f. = 65)) and significances for ratings from Questionnaire 2 (verbal behaviors)

Attribute	Mean I	Mean E	*t*	Effect Size (dz)	Significance (Two-Tailed)
Active	2.23	3.85	−12.734	−1.530	<0.001[b]
Talkative	1.97	4.02	−15.086	−1.860	<0.001[b]
Interested	2.41	3.89	−10.841	−1.330	<0.001[b]
Attentive	2.68	3.50	−7.083	−0.880	<0.001[b]
Fast	2.02	2.58	−4.511	−0.550	<0.001[b]
Consistent	3.30	3.39	−0.760	−0.010	0.45
Predictable	3.42	3.06	2.168	0.270	0.034[a]
Polite	2.98	4.12	−7.855	−0.970	<0.001[b]
Friendly	2.88	4.03	−9.114	−1.120	<0.001[b]
Obedient	3.55	3.56	−0.136	−0.020	0.89
Diversified	1.88	3.02	−8.550	−1.220	<0.001[b]
Intelligent	2.65	3.55	−7.421	−0.910	<0.001[b]
Practical	2.21	2.35	−1.732	−0.210	0.09
Useful	2.21	2.45	−2.248	−0.280	0.026[a]

[a]<0.05.
[b]<0.005.

Table 4: *t*-Test for paired samples for rating of robot behaviors in Condition B (mean on a scale of 1 (not at all) to 5 (very much), *t* (d.f. = 65)) and significances for ratings from Questionnaire 3 (non-verbal behaviors)

Attribute	Mean I	Mean E	*t*	Effect Size (dz)	Significance (Two-Tailed)
Active	2.50	2.48	0.12	0.15	0.9
Talkative	2.18	2.33	1.07	0.13	0.29
Interested	2.70	2.61	0.73	0.09	0.47
Attentive	3.09	2.86	2.00	0.24	0.050[a]
Fast	1.89	1.92	−0.28	−0.03	0.78
Consistent	3.20	2.95	2.90	0.36	0.005[b]
Predictable	3.47	3.06	2.92	0.36	0.005[b]
Polite	3.21	3.02	1.66	0.20	0.1
Friendly	2.94	3.03	−0.90	−0.11	0.37
Obedient	3.80	3.59	1.87	0.23	0.07
Diversified	2.03	2.05	−0.16	0.02	0.88
Intelligent	2.58	2.53	0.38	0.05	0.7
Practical	2.17	2.06	1.07	0.13	0.29
Useful	2.12	2.00	1.43	0.18	0.16

[a]<0.05.
[b]<0.005.

The main difference in the adjective rating between Conditions A and B can be found in the kitchen entry scene. Here, behavior I was rated as more Attentive, Consistent and Predictable. There are no other significant differences between the behaviors. There were also no significant differences in the overall ratings of the kitchen entry scene. This could be due to individual differences in the participant sample, particularly in terms of how robot autonomy is perceived [38], as the two behaviors differed only in the degree of verbal control needed by the user to directly supervise the robot's movements.

Politeness was rated very highly in both conditions (E and I). Only the direct comparison of group B shows that the extrovert behavior was judged as being significantly more polite. Altogether, the results indicate that the verbal behavior of the robot is a powerful means to model robot personality traits. Even though a clear preference for robot behavior E was found, the behavior only had a small effect on the perceived usefulness of the system. In all the conditions, people did not rate the robot as being very useful or practical (see Tables 2 and 3). Reasons for this are (1) that BIRON did not perform any manipulative tasks in the video because the study focused on more general behavior and (2) the robot used has no kind of manipulator to actually provide manipulation services in the household (e.g., picking up glasses).

The findings from the Swedish study are generally supportive of those from the UK and German studies (see Tables 5 and 6). Essentially, the direction of preference is the same in

Table 5: One-way ANOVA of ratings of robot behaviors for the Swedish (KTH) data for conditions I and E (means on a scale of 1 (not at all) to 5 (very much), F-value (d.f. = 1.12), and significance)

Property	KTH Robot I (*n* = 13)	KTH Robot E (*n* = 13)	*F*-value	Significance
Active[a]	2.38	3.46	6.391	0.018[a]
Talkative[a]	3.00	4.00	6.5	0.018[a]
Interested[a]	3.15	4.08	5.878	0.023[a]
Attentive[a]	2.77	4.08	7.572	0.011[a]
Fast[a]	1.54	2.62	8.110	0.009[a]
Consistent	3.69	3.54	0.230	0.635
Predictable	3.85	3.15	3.197	0.086
Polite	4.15	4.54	3.333	0.080
Friendly[b]	3.38	4.46	17.552	0.0003[b]
Obedient	4.15	4.54	2.113	0.159
Interesting (diversified)	2.77	3.62	2.988	0.097
Intelligent	2.69	3.15	0.864	0.362
Practical[a]	2.08	3.15	5.227	0.031
Usable[b]	2.00	3.62	11.605	0.002[b]

[a]<0.05.
[b]<0.005.

Table 6: The introverted (I) and extroverted (E) mean robot attribute ratings from the Swedish (KTH) mean ratings compared to the combined UK (UH) and German (UB) mean attribute ratings

	KTH Robots (*n* = 13)			UH/UB Robots (*n* = 200)		
Property	I	E	Significant Differences	I	E	Significant Differences
Active	0.00	3.80	<0.05	2.30	2.89	<0.005
Talkative	0.00	4.00	<0.05	2.00	2.93	<0.005
Interested	0.00	3.90	<0.05	2.87	3.26	<0.05
Attentive	2.90	2.90	<0.05	3.54	3.59	
Fast	0.00	2.40	<0.05	1.61	2.00	<0.05
Consistent	3.80	3.60		3.22	3.31	
Predictable	3.60	3.40		3.35	3.44	
Polite	4.10	4.30		4.05	4.31	
Friendly	0.00	4.10	<0.005	3.56	4.03	<0.005
Obedient	4.30	4.00		4.16	4.31	
Interesting (diversified)	0.00	3.00		2.12	2.69	<0.005
Intelligent	0.00	3.00		2.98	3.34	<0.05
Practical	2.50	2.90		2.10	2.27	
Usable	2.60	2.90	<0.005	2.18	2.13	

Differences.

the Swedish sample with some exceptions (i.e., if the difference between behaviors I and E is negative in the UB/UH sample, it tends to be the same in the KTH sample). The exceptions between the two sample sets are with the attributes Useful, Attentive and Practical, Attentive and Diversified (Interesting). The differences within the UH/UB sample for Attentive are not significant. Note, due to confusion in translation, comparison of the KTH ratings for the attribute Diversified (Interesting) is not reliable. It should be noted that the KTH groups viewing the I and E robot types were somewhat different in age (26 and 36 years, respectively), but to see if this is related to the differences observed for ratings for Useful and Practical would require further investigation. However, for all other significant differences in the UH/UB sample, the KTH sample seems to display a similar trend.

6.2.5 Discussion

The video-based HRI study methodology has been further developed from [8,9] and has demonstrated a major advantage of reaching many participants ($N = 234$) in geographically distant places in a very short time. Across the three countries that participated in the studies, similar trends in the findings have been identified. This suggests that responses to the different robot behaviors were independent of any cultural differences between these samples. It should be noted, however, that all three samples were Northern European, so the generalizability of these findings to other regional cultures will require further research [39]. Relating these findings back to the original research questions:

- *Do participants recognize differences between the two robot behaviors (Extrovert and Introvert)?* Participants who viewed videos of BIRON did notice differences between the introvert (I) and extrovert (E) behaviors. Traits like intelligence, interest, friendliness and diversity were more strongly associated with extrovert behavior, which is also true for human–human interaction. It appears that these particular differences in trait attribution were also primarily due to dialogue design, as they did not appear for the kitchen entry scene.
- *Which of the behaviors do participants prefer?* Participants overall preferred the robot Extrovert (E) for both Condition A (which rated the whole sequence with just one behavior type) and Condition B (which showed two sets of sequences with both robot behaviors). For the kitchen entry scene there were no significant differences between the robot behaviors in terms of Likability and Satisfaction. The Introvert robot was however, rated as more Attentive, Predictable and Consistent by participants in Condition B. Interestingly, robot behavior I in this scene was rated as more competent, autonomous and requiring much less direct control from the user to navigate into the kitchen than behavior E. One explanation for this finding might be that the lack of any

significant differences between the behaviors on these ratings reflects individual differences, especially in regards to robot autonomy [38]. It could also be argued that due to the nature of the video medium, the differences in dialogue between the two behaviors were more salient to the participants than that of the robot's actual movement.

- *Is the robot displaying Extrovert behaviors rated as being more friendly, intelligent and/or polite than the one displaying Introvert behaviors?* When considering the results for Condition A (in which participants only viewed one robot behavior), there is a strong possibility that the extrovert robot's "chatty" dialogue style might have mitigated any annoyance at its less efficient navigation. This suggests that tolerance to technical shortcomings in general can be mitigated to some extent by the robot through initiating a "chattier" dialogue. However, while the findings related to the door-crossing/kitchen entry scene do highlight a possible shortcoming of the VHRI method, the results overall nevertheless gives helpful insights into users' preferences. These can usefully be used to guide the current robot system design and implementation, and successfully provide early feedback from potential users.

6.2.6 Conclusions

We have gained insight into modeling robot behaviors and how people can perceive differences in robot personality by using different verbal behaviors Whereas, most previous research using the VHRI methodology has primarily investigated robot appearance and how it influenced peoples' perceptions of robots [8,24,26,27], we have found that the verbal behavior of the robot can also be a powerful means to create aspects of perceived robot personality. Robot designers should take account that a particular robot's verbal behavior can greatly influence users' perceptions of the personality of the robot. Ideally, the personality should match both the task of the robot and the user's individual preferences, therefore, there is no one "perfect" personality for all robots. However, in the home tour context, the extrovert robot was preferred by most of the participants in all of the three countries that took part in the study. However, how these preferences relate to the user's own personalities is an open question that is left for future HRI research to address.

The VHRI methodology is not a replacement for live HRI trials but, by using well-established video-editing techniques, can provide potential users with a realistic experience of future proposed and prototype robotic technology developments. It can provide an early indication to developers regarding users' perceptions and ratings of the acceptability of both robotic behavioral and appearance attributes. Perhaps the most valuable outcome from these VHRI user trials is that indications of user preferences can be gained at a very early stage of robot technology development. In fact, developers of robot technology can gain early feedback from users even before the first prototype systems are fully working. This makes it much less likely that there will have to be expensive and fundamental changes at later stages of robot technical system development in the light of full-scale live user HRI trials.

Acknowledgments

The work described in this chapter was conducted within the EU Integrated Project COGNIRON ("The Cognitive Robot Companion"; www.cogniron.org) and was funded by the European Commission Division FP6-IST Future and Emerging Technologies under contract FP6-002020.

References

[1] Sony Europe Ltd, Aibo, 2009. http://support.sony-europe.com/aibo.
[2] Tiger Electronics Ltd, FurFriends, 2009. www.hasbro.com/shop/browse//FurReal-Friends.
[3] Lego Plc, Lego Robotics, 2009. http://shop.lego.com/ByCategory/Leaf.aspx?cn=389&d=292.
[4] iRobot Plc, Roomba, 2009. http://www.iroboteurope.co.uk.
[5] Paro Robots Inc, Paro, 2009. http://www.parorobots.com.
[6] A. Green, E.A. Topp, H. Huttenrauch, Measuring up as an intelligent robot — on the use of high-fidelity simulations for human–robot interaction research, in: Proc. Performance Metrics for Intelligent Systems Workshop, Gaithersburg, MD, 2006, pp. 247–252.
[7] M.L. Walters, S.N. Woods, K.L. Koay, K. Dautenhahn, Practical and methodological challenges in designing and conducting human–robot interaction studies, in: Proc. AISB'05 Symp. on Robot Companions Hard Problems and Open Challenges in Human–Robot Interaction, London, 2005, pp. 110–119.
[8] S.N. Woods, M.L. Walters, K.L. Koay, K. Dautenhahn, Comparing human robot interaction scenarios using live and video based methods: towards a novel methodological approach, in: Proc. 9th Int. Workshop on Advanced Motion Control, Istanbul, 2006, pp. 750–755.
[9] S.N. Woods, M.L. Walters, K.L. Koay, K. Dautenhahn, Methodological issues in HRI: a comparison of live and video-based methods in robot to human approach direction trials, in: Proc. 15th IEEE Int. Symp. on Robot and Human Interactive Communication, Hatfield, 2006, pp. 51–58.
[10] K. Dautenhahn, Methodology and themes of human–robot interaction: a growing research field, Int. J. Adv. Robotic Syst. 4 (2007) 103–108.
[11] J.F. Maas, T. Spexard, J. Fritsch, B. Wrede, G. Sagerer, BIRON, what's the topic? — A multimodal topic tracker for improved human–robot interaction, in: Proc. IEEE Int. Workshop on Robot and Human Interactive Communication, Hatfield, 2006, pp. 26–32.
[12] S. Li, B. Wrede, G. Sagerer, A dialog system for comparitive user studies on robot verbal behavior, in: Proc. 15th IEEE Int. Symp. on Robot and Human Interactive Communication, Hatfield, 2006, pp. 129–134.
[13] M. Lohse, K. Rohlfing, B. Wrede, G. Sagerer, Try something else. When users change their discursive behavior, in: Proc. IEEE Int. Conf. on Robotics and Automation, Pasadena, 2008, pp. 3481–3486.
[14] M. Lohse, M. Hanheide, B. Wrede, M.L. Walters, K.L. Koay, D.S. Syrdal, A. Green, H. Hüttenrauch, K. Dautenhahn, G. Sagerer, K. Severinson Eklundh, Evaluating extrovert and introvert behaviour of a domestic robot — a video study, in: Proc. 17th IEEE Int. Symp. on Robot and Human Interactive Communication, Munich, 2008, pp. 488–493.
[15] P.H. Kahn, H. Ishiguro, B. Friedman, T. Kanda, N.G. Freier, R.L. Severson, J. Miller, What is a human?— toward psychological benchmarks in the field of human–robot interaction, Interact. Studies 8 (2007) 363–390.

[16] B. Scasselatti, Using robots to study abnormal social development, in: Proc. 5th Int. Workshop on Epigenetic Robotics, Nara, Japan, 2005, pp. 11–14 (in Japanese).
[17] D.J. Mayhew, Principles and Guidelines in Software User Interface Design, Prentice Hall, Englewood Cliffs, NJ, 1991.
[18] B. Shneiderman, Designing the User Interface: Strategies for Effective Human–Computer Interaction, third ed., Addison Wesley, Reading, MA, 2003.
[19] H.A. Yanco, J.L. Drury, J. Scholtz, Beyond usability evaluation: analysis of human–robot interaction at a major robotics competition, Human–Comp. Interact. 19 (2004) 117–149.
[20] M. Lohse, M. Hanheide, A. Green, H. Hüttenrauch, B. Wrede, G. Sagerer, K. Severinson Eklundh, BIRON, This Is a Table! — A Corpus in Multimodal Human–Robot, Interaction, Technical Report, University of Bielefeld, Bielefeld, 2008.
[21] A. Green, H. Huttenrauch, K. Severinson Eklundh, Applying the wizard of Oz framework to co-operative service discovery and configuration, in: Proc. 13th IEEE Int. Workshop on Robot and Human Interactive Communication, Kurashiki, 2004, pp. 575–580.
[22] S. Ylirisku, J. Buur, Designing with Video, Springer, Berlin, 2007.
[23] M.L. Walters, K. Dautenhahn, R. Te Boekhorst, K.L. Koay, Exploring the design space of robot appearance and behaviour in an attention-seeking 'living room' scenario for a robot companion, in: Proc. IEEE-Artificial Life, Honolulu, HI, 2007, pp. 341–347.
[24] M.L. Walters, D.S. Syrdal, K. Dautenhahn, R. Te Boekhorst, K.L. Koay, Avoiding the uncanny valley — robot appearance, personality and consistency of behavior in an attention-seeking home scenario for a robot companion, J. Autonomous Robots 24 (2008) 159–178.
[25] J. Carpenter, J.M. Davis, N. Erwin-Stewart, T.R. Lee, J.D. Bransford, N. Vye, Gender representation and humanoid robots designed for domestic use, Int. J. Soc. Robotics 1 (2009) 261–265.
[26] K.F. MacDorman, Subjective ratings of robot video clips for human likeness, familiarity, and eeriness: an exploration of the uncanny valley, in: Proc. Con. on Cognitive Science, Workshop on Android Science, Vancouver, BC, 2006, pp. 26–29.
[27] C. Ho, K.F. MacDorman, Z.D. Pramono, A human emotion and the uncanny valley: a GLM, MDS, and Isomap analysis of robot video ratings, in: Proc. 3rd ACM/IEEE Int. Conf. on Human–Robot Interaction, Amsterdam, 2008, pp. 169–176.
[28] A. Haasch, S. Hohenner, S. Hüwel, M. Kleinehagenbrock, S. Lang, I. Toptsis, G.A. Fink, J. Fritsch, B. Wrede, G. Sagerer, BIRON — The bielefeld robot companion, in: Proc. Int. Workshop on Advances in Service Robotics, Stuttgart, 2004, pp. 27–32.
[29] J. Fritsch, M. Kleinehagenbrock, S. Lang, T. Plötz, G.A. Fink, G. Sagerer, Multi-modal anchoring for human–robot interaction, robotics and autonomous systems, Robotics Autonomous Syst. 43 (2003) 133–147.
[30] R. Gockley, J. Forlizzi, R. Simmons, Interactions with a moody robot, in: Proc. 1st ACM SIGCHI/SIGART Conf. on Human–Robot Interaction, New York, 2006, pp. 186–193.
[31] A. Tapus, M.J. Matarić, User personality matching with Hands-Off robot for post-stroke rehabilitation therapy, in: Proc. 10th Int. Symp. on Experimental Robotics, Rio de Janeiro, 2006, pp. 24–31.
[32] S.N. Woods, K. Dautenhahn, J. Schulz, The design space of robots: investigating children's views, in: Proc. 13th IEEE Int. Workshop on Robot and Human Interactive Communication, Kurashiki, 2004, pp. 47–52.
[33] S.N. Woods, K. Dautenhahn, C. Kaouri, R. Te Boekhorst, K.L. Koay, Is this robot like me? Links between human and robot personality traits, in: Proc. IEEE-RAS Int. Conf. on Humanoid Robots, Tsukuba, 2005, pp. 375–380.
[34] B.R. Duffy, Anthropomorphism and the social robot, Robotics Autonomous Syst. 42 (2003) 177–190.
[35] M.A. Walker, J.E. Cahn, S.J. Whittaker, Improvising linguistic style: social and affective bases for agent personality, in: Proc. 1st Int. Conf. on Autonomous Agents, Marina del Rey, CA, 1997, pp. 96–105.
[36] H.J. Eysenck, Biological dimensions of personality, in: L.A. Pervin (Ed.), Handbook of Personality: Theory and Research, Guilford, New York, 1990, pp. 244–276.

[37] R. Likert, A technique for the measurement of attitudes, Arch. Psychol. 22 (1932) 1–55.
[38] D.S. Syrdal, K. Dautenhahn, K.L. Koay, M.L. Walters, The negative attitudes towards robots scale and reactions to robot behaviour in a live human–robot interaction study, in: Proc. New Frontiers in HRI Symposium, Edinburgh, 2009, pp. 109–115.
[39] G. Cortellessa, M. Scopelliti, L. Tiberio, G. Koch Svedberg, A. Loutfi, F. Pecora, Cross-cultural evaluation of domestic assistive robots, in: Proc. AAAI Fall Symp. on AI in Eldercare: New Solutions to Old Problems, Arlington, VA, 2008. FS-08–02.

CHAPTER 6.3

Using Socially Assistive Human–Robot Interaction to Motivate Physical Exercise for Older Adults[1]

Juan Fasola, Maja J. Matarić
University of Southern California, Los Angeles, CA, USA

Chapter Outline

[1] © 2013 IEEE. Reprinted, with permission, from Juan Fasola and Maja J. Matarić, Using Socially Assistive Human–Robot Interaction to Motivate Physical Exercise for Older Adults, Proceedings of the IEEE, Vol. 100, No. 8, pp. 2512–2526, 2012, IEEE.

Household Service Robotics. http://dx.doi.org/10.1016/B978-0-12-800881-2.00022-0

6.3.1 Introduction

An aging population is increasing the demand for healthcare services worldwide. By the year 2050, the number of people over the age of 85 will have increased fivefold, according to recent estimates [1], and the shortfall of nurses is already becoming an issue [2–4]. Regular physical exercise has been shown to be effective at maintaining and improving the overall health of elderly individuals [5–8]. Physical fitness is associated with higher functioning in the executive control processes [9], correlated with less atrophy of frontal cortex regions [10], and with improved reaction times [11] compared with the sedentary. Social interaction, and specifically high perceived interpersonal social support, has also been shown to have a positive impact on general mental and physical wellbeing [12], in addition to reducing the likelihood of depression [13–16]. Among the many healthcare services that will need to be provided, physical exercise therapy, social interaction, and companionship can be addressed by socially assistive robotics technology.

A socially assistive robot (SAR) is a system that employs hands–off interaction strategies, including the use of speech, facial expressions, and communicative gestures, to provide assistance in accordance with the particular healthcare context [17]. Previous SAR work from our research laboratory includes systems that were developed and tested for stroke patients [18,19], Alzheimer's patients [20], children with autism spectrum disorder [21,22], as well as healthy adults [23] and healthy elderly adults [24].

This chapter focuses on the design methodology, implementation details, and user study evaluations of a SAR system that aims to motivate and engage elderly users in physical exercise as well as social interaction to help address the physical and cognitive healthcare needs of the growing elderly population. SAR systems equipped with such motivational, social, and therapeutic capabilities have the potential to facilitate elderly individuals to live independently in their own homes, to enhance their quality of life, and to improve their overall health.

The rest of this chapter is organized as follows. In the next section, we discuss the related work in the area of assistive robotics for the elderly. Section 6.3.3 presents our SAR system approach and design methodologies. Section 6.3.4 introduces our SAR humanoid robot platform along with the implementation details of our SAR exercise system. In Sections 6.3.5 and 6.3.6, we discuss two user studies conducted with our system to investigate and evaluate the effects of different motivational techniques, to test system effectiveness, and to obtain user feedback. We conclude the chapter with a summary of the key research contributions of this work.

6.3.2 Related Work

6.3.2.1 Robots for the Elderly

Literature that addresses the area of assistive robotics for the elderly is limited. Representative work includes robots that focus on providing assistance for functional needs, such as mobility aids and navigational guides. Dubowsky et al. developed a robotic cane/walker device designed to help individuals by functioning as a mobility aid that provides physical support when walking as well as guidance and health monitoring of a user's basic vital signs [25]. Montemerlo et al. designed and pilot tested a robot that escorts elderly individuals in an assisted living facility, reminds them of their scheduled appointments, and provides informational content such as weather forecasts [26].

Researchers have also investigated the use of robots to help address the social and emotional needs of the elderly, including reducing depression and increasing social interaction with peers. Wada et al. studied the psychological effects of a stuffed seal robot, Paro, used to engage seniors at a day service center. The study found that Paro, which was always accompanied by a human handler, was able to consistently improve the moods of elderly participants who had spent time petting it and engaging with it over the course of a six-week period [27]. Kidd et al. used Paro in another study that found it to be useful as a catalyst for social interaction. They observed that seniors who participated with the robot in a group were more likely to interact socially with each other when the robot was present and powered on, than when it was powered off or absent [28].

Perhaps the most related robotic system for the elderly to our SAR exercise system is the work of Matsusaka et al., who developed an exercise demonstrator robot, TAIZO, to aid human demonstrators teaching simple arm exercises to a training group [29]. However, this robot was not autonomous: it was controlled via key input or voice by the lead human demonstrator, and did not have any sensors for perceiving the users. Hence, the system did not provide any real-time feedback, active guidance, or personalized training.

6.3.2.2 Social Agent Coaches

Social agents that aim to assist individuals in health-related tasks such as physical exercise have been developed in both the human–computer interaction (HCI) and human–robot interaction (HRI) communities. Bickmore and Picard developed a computer-based virtual relational agent that served as a daily exercise advisor by engaging the user in conversation and providing educational information about walking for exercise, asking about the user's daily activity levels, tracking user progress over time while giving feedback, and engaging the user in relational dialog [30]. Kidd and Breazeal developed a tabletop robot to serve as a daily weight-loss advisor, which interacted through a touchscreen interface, tracked user progress and the user–robot relationship state over time, and was tested in a six-week field study with participants at home [31]. French et al. designed and explored the use of a virtual coach to assist manual wheelchair drivers by providing advice and guidance to help users avoid hazardous forms of locomotion [32].

These systems are similar to our SAR exercise system in the manner in which they provide feedback (from a social agent), and with the exception of French's work, in the activity being monitored (physical exercise). However, our system differs from all in that the agent, a robot in our case, not only provides active guidance, feedback, and task monitoring, but is also directly responsible for instructing and steering the task. Hence, our agent is both an administrator and active participant in the health-related activity, resulting in a unique characteristic for the system: the social interaction between the robot and the user is not only useful for maintaining user engagement and influencing intrinsic motivation, but is also an instrumental necessity in achieving the physical exercise task.

6.3.3 SAR Approach

In designing our system to help address the physical exercise needs of the elderly population, we followed the design methodology which asserts that the SAR agent must possess: (1) the ability to influence the user's intrinsic motivation to perform the task; and (2) the ability to personalize the social interaction to maintain user engagement in the task and build trust in the task-based human–robot relationship. The following elaborates on the importance of both of these qualities in the context of providing healthcare interventions, as well as details how each was incorporated into our SAR exercise system.

6.3.3.1 Intrinsic Motivation

Motivation is a fundamental tool in establishing adherence to a therapy regimen or task scenario and in promoting behavior change. There are two forms of motivation: intrinsic motivation, which comes from within a person, and extrinsic motivation, which comes from sources external to a person. Extrinsic motivation, though effective for short-term task compliance, has been shown to be less effective than intrinsic motivation for long-term task compliance and behavior change [33].

Intrinsic motivation, however, can be, and often is, affected by external factors. In a task scenario, the instructor (a SAR, in our case) can impact the user's intrinsic motivation through verbal feedback. Praise, for example, is considered a form of positive feedback and has the potential to increase the user's intrinsic motivation for performing the task, whereas criticism, a form of negative feedback, tends to negatively impact the user's intrinsic motivation [34,35]. The effect of positive feedback, however, is closely tied to the user's own perceived competence at the task. Once the user believes he is competent at the task, additional praise no longer affects his intrinsic motivation. Our SAR exercise system provides positive feedback to the user in the form of praise upon correct completion of the given exercises, and never gives negative feedback so as to avoid diminishing intrinsic motivation to engage in the exercise task.

Indirect competition, wherein the user is challenged to compete against an ideal outcome, has also been shown to increase user enjoyment on an otherwise noncompetitive task [36]. For example, when the user is shown her high score on the task, her intrinsic motivation for the task tends to increase, as she strives to better her previous performance. Thus, in a task scenario, it is important that the task instructor continually report to the user his/her performance scores during the task, for motivational purposes. Our robot exercise instructor implements this strategy by reporting the user's personal high scores during two of the three exercise games played.

Verbal feedback provided to the user by the instructor certainly plays an important role in task-based motivation, but the task itself and how it is presented to the user perhaps plays an even more significant role. Csikszentmihalyi's research suggests that "when one engages in an optimally challenging activity with respect to one's capacities, there is a maximal probability for task-involved enjoyment or flow" [37]. He also states that intrinsically motivated activities are those characterized by enjoyment. Simply put, people are "intrinsically motivated under conditions of optimal challenge" [38]. If a task is below the optimal challenge level, it is too easy for the user and results in boredom. Alternatively, if the task is above the optimal challenge level, it is too hard and causes the user to get anxious or frustrated. Therefore, an instructor that oversees user performance in a task scenario must be able to continually adjust the task to meet the appropriate needs of the user in order to increase or maintain intrinsic motivation to perform the task. We have incorporated these guidelines for achieving the optimal challenge level for the user into our SAR exercise system. For example, the exercise games are changed at regular intervals to prevent the user from getting bored or frustrated with any one of them. In addition, the Memory game, discussed in the next section, challenges the user with progressively more difficult exercise sequences based on the user's performance level.

Another task characteristic with the potential to influence user enjoyment is the incorporation of direct user input. Studies have shown that tasks that support user autonomy and self-determination lead to increased intrinsic motivation, self-esteem, creativity, and other related variables among the participants [39], all of which are important for achieving task adherence and long-term behavior change. Self-determination, represented in the task in the form of choice of activity [40], choice of difficulty level

[39], and choice of rewards [41], has been shown to either increase or be less detrimental to intrinsic motivation than similar task conditions that do not involve choice. In the context of our SAR exercise system, user choice is a very interesting research question and one that we investigated with a user study to test the role of choice in the exercise scenario. The study design and results are presented in Section 6.3.6.

6.3.3.2 Social Interaction and Personalization

Many social intricacies contribute to the foundation of a meaningful relationship, both in HCI (as detailed by Bickmore and Picard [30]) and in HRI, these include empathy, humor, references to mutual knowledge, continuity behaviors, politeness, and trust, among others. We place great importance on these relationship building tools; therefore, we integrated each, in one form or another, into the social interaction component of our robot exercise instructor.

Our primary focus was on eliminating the perceived repetitiveness of the robot's verbal instructions/comments. We believe that if the robot is perceived by the user as repetitive and hence predictable, this can lead to a decrease in the perception of the robot's intelligence by the user, and ultimately to a loss of trust in the robot's helpfulness in motivating exercise. We therefore placed special attention on adding variety to the robot's utterances. Toward this end, the robot always drew from a list of phrases that emphasized the same point when speaking to the user, choosing randomly at run time. For example, there were more than 10 different ways in which the robot could praise the user (e.g., "Awesome!," "Nice job!," "Fantastic!"). Furthermore, if the robot did need to repeat itself exactly, for example when providing the same feedback comment during one of the exercise games, it added filler words to the given phrase, such as the user's name or the word "try" or both (e.g., "Try to raise your left arm," "John, raise your left arm").

Adding the user's name to the interaction dialog was an important part of our system design, not only to add variability, but also for its relationship building effect [30]. The robot always used the user's name at the first greeting, and also when bidding farewell at the end of a session. Having the robot refer to the user by name is an important part of personalizing the interaction, along with providing direct feedback specific to the individual user's performance level and performance history during the games, and referencing mutual knowledge. Our SAR exercise system introduced continuity by having the robot refer to previous sessions with the user upon introduction, reference planned future sessions at the end of interaction, and refer to past exercise performance, such as when reporting previous high scores.

6.3.4 Robot Exercise System

In this section, we present the design and implementation details of the SAR exercise system, including the motivation behind the types of exercise routines, the humanoid robot platform, the different exercise games, and the robot's visual user arm motion recognition procedure.

6.3.4.1 System Overview

The exercise scenario consists of a SAR whose purpose is to instruct, evaluate, and encourage users to perform simple exercises. The scenario is one-on-one; the robot focuses its attention on the user in order to provide timely and accurate feedback, and to maximize the effectiveness of the exercise session for the user. In the setup, the user is seated in a chair in front of the robot; the user and the robot face each other. A black curtain is used as a backdrop to facilitate the visual perception of the user's arm movements. The exercise setup is shown in Figure 1.

During the exercise sessions, the robot asks the user to perform simple seated arm gesture exercises. The range of the robot's arm motion in the exercises is restricted to the sides of the body in order to maximize the accuracy of the robot's visual detection of the user's arms. This type of seated exercise, called "chair exercise" or "chair aerobics," is commonly practiced in senior living facilities and provides grounding for our exercise system. Chair exercises are highly regarded for their accessibility to those with low mobility [5–8], for their safety as they reduce the possibility of injury due to falling from improper balance [5,8], and for their health benefits such as improved flexibility [5,7], muscle strength [5,7,8], ability to perform everyday tasks [5,7,8], and even memory recall [6].

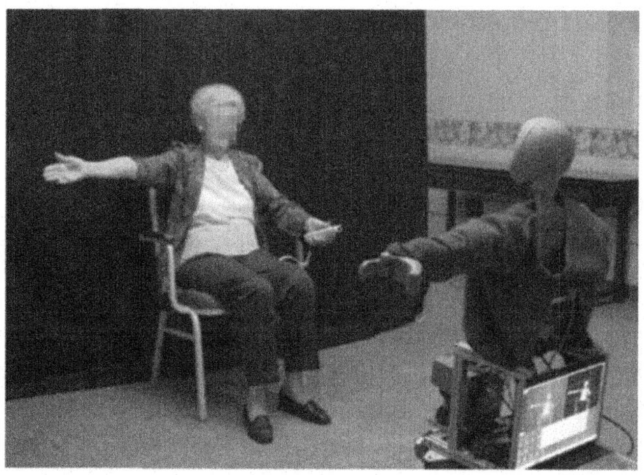

Figure 1
Exercise setup with user and robot facing each other.

The user is able to communicate with the robot through a wireless button control interface, the popular Wiimote remote controller, which communicates via Bluetooth with the button labels modified to suit our system. There are two buttons available to the user to respond to prompts from the robot, labeled "yes" and "no," and one button for the user to request a rest break at any time during the interaction.

It is important to note that the robot conducts the exercise sessions, evaluates user performance, and gives the user real-time feedback completely autonomously, without human operator intervention at any time during the exercise sessions.

6.3.4.2 Robot Platform

To address the role of the robot's physical embodiment, we used Bandit, a biomimetic anthropomorphic robot platform that consists of a humanoid torso (developed with BlueSky Robotics) mounted on a MobileRobots Pioneer 2DX mobile base. The torso contains 19 controllable degrees of freedom (DOF): six DOF arms (x2), one DOF gripping hand (x2), two DOF pan/tilt necks, one DOF expressive eyebrow, and two DOF expressive mouths. The robot is shown in Figure 2.

A standard USB camera is located at the waist of the robot, and used to capture the user's arm movements during the exercise interaction, allowing the robot to provide appropriate performance feedback to the user.

The robot's speech is generated by the commercially available NeoSpeech text-to-speech engine [42] and a speaker on the robot outputs the synthesized voice to the user. The robot's lip movements are synchronized with the robot's speech so that the lips open at the start and close at the end of spoken utterances.

6.3.4.3 Exercise Games

Three exercise games are available in our system: the Workout game, the Imitation game, and the Memory game. During an exercise session, the user is given the opportunity to play all three games, and often can play each more than once within the session duration. The following is a description of each game in detail.

1. *Workout Game*: In this game, the robot fills the role of a traditional exercise instructor by demonstrating the arm exercises with its own arms, and asking the user to imitate. The robot gives the user feedback in real time, providing corrections when appropriate (e.g., "Raise your left arm and lower your right arm" or "Bend your left forearm inward a little"), and praise in response to each successful imitation (e.g., "Great job!" or "Now you've got the hang of it"). In monitoring user performance, the robot compares the user's current arm angles as detected by the vision module to those of the specified goal

Figure 2
Robot platform used in the experiments.

arm angles to determine performance accuracy. The comparison procedure is robust to user fatigue and variations in the range of motion; it relies more on the user's current hand positions and forearm angles than on the absolute differences between the user's arm angles and the target angles.

2. *Imitation Game*: In this game, the roles of the user and the robot from the Workout game are reversed; the user becomes the exercise instructor showing the robot what to do. The robot encourages the user to create his/her own arm gesture exercises, and imitates user movements in real time.

As the roles of the interaction are reversed, with the robot relinquishing control of the exercise routine to the user, the robot no longer provides instructive feedback on the exercises. However, the robot does continue to speak and engage the user by means of encouragement and general commentary. For example, if the robot detects that the user is not moving, it encourages the user to create new gestures by saying, for instance, "Mary, try and come up with your own gestures and I'll imitate you." In addition, the robot makes general comments about the game or the user, such as "You're a good instructor, Mary" or "This is my favorite game, thanks for the workout."

3. *Memory Game*: In this game, the user is challenged to learn a sequence of different arm gestures. The goal of the game is for the user to try and memorize ever-longer sequences, and thus compete against his/her own high score. The sequence is determined at the start of the game and does not change for the duration of the game. The arm gesture poses used for each position in the sequence are chosen at random at run time, and there is no inherent limit to the sequence length, thereby making the game challenging for users at any skill level.

The robot starts out by showing the first two positions of the sequence and asks the user to perform them while it provides feedback. Once the user has successfully repeated the first two gestures with the help of the robot, the user is asked to repeat the sequence again from the beginning, this time without demonstration or verbal feedback from the robot. Once the gestures are completed without help, the robot shows the next two gestures in the sequence, and the user is again asked to perform the entire sequence from the beginning (now four gestures in length). As the user continues to successfully memorize all shown gestures, the robot continues to show the user two more (six, then eight gestures, and so on), and the game progresses in difficulty.

The robot helps the user to keep track of the sequence by counting along with each correct gesture, and reminding the user of the poses when it detects errors (e.g., "Oh, that's too bad! Here is gesture five again"). The robot also reports to the user his/her current high score (i.e., the number of gestures remembered correctly) in an attempt to motivate improvement upon past performance.

6.3.4.4 Vision Module

In order to monitor user performance and provide accurate feedback during the exercise routines, the robot must be able to recognize the user's arm gestures. To accomplish this, we developed a vision module that recognizes the user's arm gestures/poses in real time, with minimal requirements for the surrounding environment and none for the user.

Several different approaches have been developed to accomplish tracking of human motion, both in 2-D and 3-D, including skeletonization methods [43,44], gesture recognition using probabilistic methods [45], and color-based tracking [46], among others. We opted to create an arm pose recognition system that takes advantage of our simplified exercise setup in order to achieve real-time results without imposing any markers on the user.

To simplify visual recognition of the user, a black curtain was used to provide a static and contrasting background for fast segmentation of the user's head and hands, the most important features of the arm pose recognition task, independent of the user's skin tone.

The arm pose recognition algorithm works by first segmenting the original grayscale camera frame into a black and white image by applying a single threshold over the image; white pixels are assumed to form part of the user's body. The algorithm determines the final arm angles after localizing the user's hand and elbow locations using a heuristic procedure which takes as input the extrema points of the segmentation image. Example detection results are displayed in Figure 3. Additional details regarding the visual recognition procedure can be found in [24].

The development of the SAR exercise system and visual recognition procedure predated the availability of the Microsoft Kinect [47]. Future implementations of the system will utilize Kinect-type 3-D vision technology and do away with the curtain and the planar limits of the motions. Nevertheless, the 2-D nature of the exercises was not noted as an issue by any of the participants in our user studies.

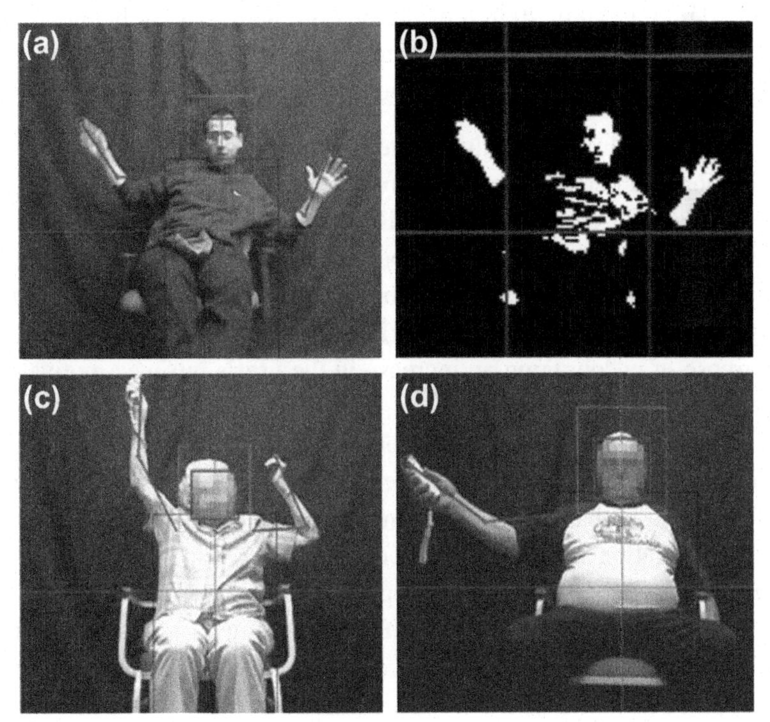

Figure 3
(a), (c), and (d) Example face and arm angle detection results superimposed over original grayscale camera frames. (b) Segmented image of camera frame shown in (a).

6.3.5 Motivation Study I: Praise and Relational Discourse Effects

We designed and conducted an intrinsic motivation study to investigate the role of praise and relational discourse (politeness, humor, empathy, etc.) in the robot exercise system. Toward that end, the study compared the effectiveness and participant evaluations of two different coaching styles used by our system to motivate elderly users to engage in physical exercise. This section discusses the study methods employed, the subjective and objective measures that were evaluated, and the outcomes of the study and system evaluation with elderly participants.

6.3.5.1 Study Design

The study consisted of two conditions, relational and non-relational, to explore the effects of praise and communicative relationship-building techniques on a user's intrinsic motivation to engage in the exercise task with the SAR coach. The study design was within our subject content; participants saw both conditions, one after the other, and the order of appearance of the conditions was counterbalanced among the participants. Each condition lasted 10 min, totaling 20 min of interaction, with surveys being administered after both sessions to capture participant perceptions of each study condition independently. The following describes the two conditions in greater detail.

1. *Relational Condition*: In this condition, the SAR exercise coach employs all of the social interaction and personalization approaches described in Section 6.3.3. Specifically, the robot always gives the user praise upon correct completion of a given exercise gesture (an example of positive feedback) and provides reassurance in the case of failure (an example of empathy). The robot also displays continuity behaviors (e.g., by referencing past experiences with the user), humor, and refers to the user by name, all with the purpose of encouraging an increase in the user's intrinsic motivation to engage in the exercise session.
2. *Non-relational Condition*: In this condition, the SAR exercise coach guides the exercise session by providing instructional feedback as needed (e.g., user score, demonstration of gestures, verbal feedback during gesture attempts, etc.), but does not employ explicit relationship building discourse of any kind. Specifically, the robot does not provide positive feedback (e.g., praise) in the case of successful user completion of an exercise gesture, nor does it demonstrate empathy (e.g., reassurance) in the case of user failure. The SAR coach also does not display continuity behaviors, humor, or refer to the user by name. This condition represents the baseline condition of our SAR exercise system, wherein the robot coach does not employ any explicit motivational techniques to encourage an increase in the user's intrinsic motivation to engage in the task.

6.3.5.2 Participant Statistics

We recruited elderly individuals to participate in the study through a partnership with be.group, an organization of senior living communities in Southern California, using flyers and word-of-mouth. Thirteen participants responded and successfully completed both conditions of the study. The sample population consisted of 12 female participants (92%) and one male participant (8%). Participants' ages ranged from 77 to 92, and the average age was 83 (S.D. = 5.28). Half of the participants (n = 7) engaged in the relational condition in the first session, whereas the other half (n = 6) engaged first in the non-relational condition.

6.3.5.3 Measures

Survey data were collected at the end of the first and second sessions in order to analyze participant evaluations of the robot and of the interaction with the exercise system in both conditions. The same evaluation surveys were used for each session to allow for objective comparisons between the two conditions.

In addition to these evaluation measures, at the end of the last exercise session, we administered one final survey asking the participants to directly compare the two study conditions (labeled "first" and "second") according to 10 evaluation categories. This survey allowed us to obtain a general sense of the participants' preferences regarding the different SAR approaches and hence gauge their respective motivational capabilities.

Objective measures were also collected to evaluate user performance and compliance in the exercise task.

The following describes the specific evaluation measures captured in the post session surveys, and the objective measures captured during the exercise sessions.

1. *Evaluation of Interaction*: Two dependent measures were used to evaluate the interaction with the robot exercise system. The first measure was the enjoyableness of the interaction, collected from participant assessments of the interaction according to six adjectives: enjoyable; interesting; fun; satisfying; entertaining; boring; and exciting (Cronbach's $\alpha = 0.93$). Participants were asked to rate how well each adjective described the interaction on a ten-point scale, anchored by "describes very poorly" (1) and "describes very well" (10). Ratings for the adjective "boring" were inverted to keep consistency with the other adjectives that reflect higher scores as being more positive. The enjoyableness of the interaction was measured to gain insight into the user's motivation level to engage in the task, because, as Csikszentmihalyi states, intrinsically motivating activities are characterized by enjoyment [37]. The second measure was the perceived value or usefulness of the interaction. Participants were asked to evaluate how

well each of the following four adjectives described the interaction: useful; beneficial; valuable; and helpful (Cronbach's $\alpha = 0.95$). The same ten-point scale anchored by "describes very poorly" (1) and "describes very well" (10) was used in the evaluation. The perceived usefulness of the system was measured to estimate user acceptance and trust of the system in helping to achieve the desired health goals, which is necessary for the system to be successful in the long term.

2. Evaluation of Robot: The companionship of the robot was measured based on participant responses to nine 10-point semantic differential scales concerning the following robot descriptions: bad/good; not loving/loving; not friendly/ friendly; not cuddly/ cuddly; cold/warm; unpleasant/pleasant; cruel/kind; bitter/sweet; and distant/close (Cronbach's $\alpha = 0.86$). These questions were derived from the Companion Animal Bonding Scale of Poresky et al. [48]. The companionship of the robot was measured to assess potential user acceptance of the robot as an in—home companion, thereby demonstrating the capability of the system toward uses in independent living/aging -in-place facilities.

To assess the perceptions of the capabilities of the system in motivating exercise, we measured participant evaluations of the robot as an exercise coach. Participant evaluations of the robot as an exercise coach were gathered from a combination of the participants' reported level of agreement toward two coaching-related statements, and responses to three additional questions. The two statements and three questions were, respectively: I think Bandit is a good exercise coach; I think Bandit is a good motivator of exercise; How likely would you be to recommend Bandit as an exercise partner to your friends? How much would you like to exercise with Bandit in the future? How much have you been motivated to exercise while interacting with Bandit? (Cronbach's $\alpha = 0.88$). The two statements were rated on a ten-point scale anchored by "very strongly disagree" (1) and "very strongly agree" (10), and the three question items were each measured according to a ten-point scale anchored by "not at all" (1) and "very much" (10).

To quantify the effectiveness of the robot's social capabilities, we measured the social presence of the robot. Social presence is defined as the feeling that mediates how people respond to social agents; it strongly influences the relative success of a social interaction [49]. In essence, the greater the social presence of the robot, the more likely the interaction is to be successful. The social presence of the robot was measured by a ten-point scale anchored by "not at all" (1) and "very much" (10) using questionnaire items established from Jung and Lee [50] (e.g., While you were exercising with Bandit, How much did you feel as if you were interacting with an intelligent being?) (Cronbach's $\alpha = 0.82$).

3. *Direct Comparison of Conditions*: The 10 evaluation categories assessed by the direct-comparison survey, which asked participants to choose between the first or second exercise sessions, were as follows: enjoy more; more useful; better at motivating

exercise; prefer to exercise with; more frustrating; more boring; more interesting; more intelligent; more entertaining; choice from now on. Analysis of the direct-comparison data serves primarily to support and confirm the results obtained from the within-subjects analysis of the dependent measures across study conditions.

4. *User Performance Measures*: To help assess the effectiveness of the SAR exercise system in motivating exercise among the participants, we collected nine different objective measures during the exercise sessions regarding user performance and compliance in the exercise task. Most of the objective measures were captured during the Workout game, wherein the robot guides the interaction similar to a traditional exercise coach. These measures include the average time to gesture completion (from the moment the robot demonstrates the gesture, to successful user completion of the gesture), number of seconds per exercise completed, number of failed exercises, number of movement prompts by the robot to the user due to lack of arm movement, and feedback percentage. The feedback percentage measure refers to the fraction of gestures, out of the total given, where the robot needed to provide verbal feedback to the user regarding arm positions in order to help guide the user to correct gesture completion.

We also recorded the maximum score over all sessions, average maximum score among users, and average time per gesture attempt in the Memory game. For the Imitation game, the only measure captured was again the number of movement prompts by the robot due to lack of user arm movement.

6.3.5.4 Hypotheses

Based on the related research on the positive effects of praise and relational discourse on intrinsic motivation, discussed in Section 6.3.3, seven hypotheses were established for this study.

- *Hypothesis 1*: Participants will evaluate the enjoyableness of their interaction with the relational robot more positively than their interaction with the non-relational robot.
- *Hypothesis 2*: Participants will evaluate the usefulness of their interaction with the relational robot more positively than their interaction with the non-relational robot.
- *Hypothesis 3*: Participants will evaluate the companionship of the relational robot more positively than that of the non-relational robot.
- *Hypothesis 4*: Participants will evaluate the relational robot more positively as an exercise coach than the non-relational robot.
- *Hypothesis 5*: There will be no significant difference between participant evaluations of the social presence of the relational robot and non-relational robot. The reasoning behind this hypothesis is that people's sense of social presence is largely determined by the embodiment type and perceived intelligence of the social agent, which is assumed to be more or less equal in the two robot conditions.

- *Hypothesis 6*: Participants will report a clear preference for the relational robot over the non-relational robot when asked to directly compare both exercise sessions.
- *Hypothesis 7*: There will be no significant difference in participant exercise performance when interacting with either the relational or non-relational robot. This hypothesis is based on the assumption that, due to the short-term nature of the study and novelty of the system, performance measures will be approximately equal between robot conditions.

6.3.5.5 Results

1. *Evaluation of Interaction Results*: Participants who engaged with the relational robot in their first session, rated the non-relational condition on average 22% lower than the relational condition in terms of enjoyment ($M_R = 7.5$ vs. $M_{NR} = 5.9$), and 23% lower in terms of usefulness ($M_R = 7.5$ vs. $M_{NR} = 5.8$). Similarly, the participants who instead engaged with the non-relational robot in their first session also expressed a greater preference for interacting with the relational robot by rating the relational condition on average 10% higher than the non-relational condition in terms of enjoyment ($M_{NR} = 7.6$ vs. $M_R = 8.4$), and 7% higher in terms of usefulness ($M_{NR} = 7.5$ vs. $M_R = 8.0$).

Altogether, 85% of the participants (11 of 13) rated the relational condition higher than the non-relational condition in terms of enjoyment, and 77% of the participants (10 of 13) rated the relational condition higher in terms of usefulness than the non-relational condition.

To test for significant differences among the participant evaluations of the study conditions, we performed a Wilcoxon signed-rank test on the data to analyze matched pairs from the sample population's evaluations of both study conditions according to the dependent measures. Supporting Hypothesis 1, the results show that the participants evaluated the interaction with the relational robot as significantly more enjoyable/entertaining than the interaction with the non-relational robot ($W [12] = 4$, $p < 0.005$), and as somewhat more valuable/useful than the interaction with the non-relational robot, although not to a significant degree ($W [12] = 15.5$, $p < 0.10$), hence Hypothesis 2 was not supported by the data. For illustration purposes, Figure 4(a) shows the average participant ratings of the enjoyableness and usefulness of the interaction for both study conditions.

2. *Evaluation of Robot Results*: Participants who engaged in the relational condition in their first session rated the non-relational robot on average 11% lower than the relational robot in terms of companionship ($M_R = 7.4$ versus $M_{NR} = 6.5$), 11% lower as an exercise coach ($M_R = 7.7$ vs. $M_{NR} = 6.9$), and 1% lower in terms of social presence

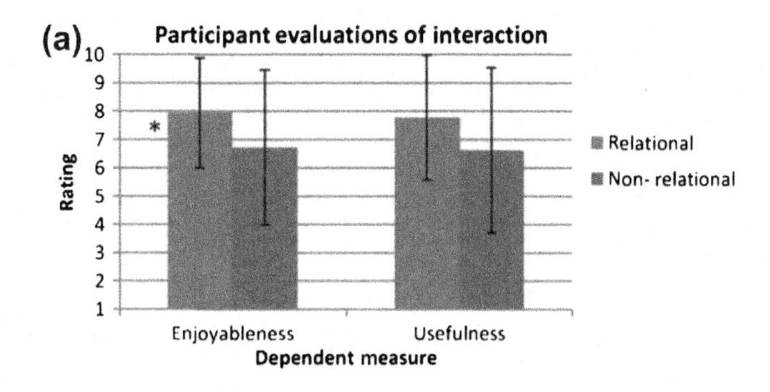

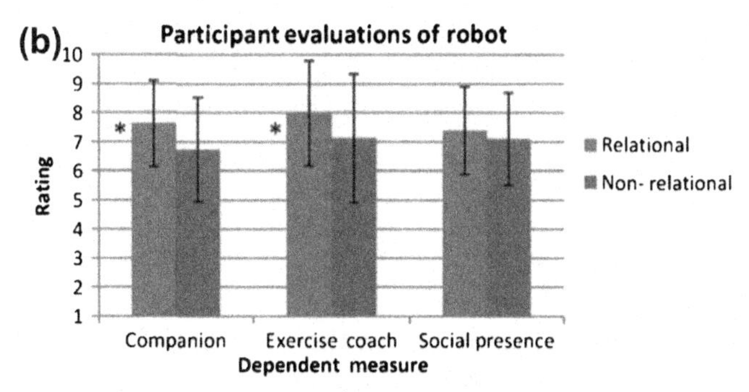

Figure 4
(a) Plot of participant evaluations of the interaction, in terms of enjoyableness and usefulness, for both study conditions; (b) plot of participant evaluations of the robot (as a companion, exercise coach, and level of social presence) for both study conditions. Note: significant differences are marked by asterisks (*).

> ($M_R = 7.2$ vs. $M_{NR} = 7.1$). Greater positive scores for the relational robot were also reported by the participants who instead engaged first in the non-relational condition, having rated the relational robot on average 14% higher than the non-relational robot in terms of companionship ($M_{NR} = 6.9$ vs. $M_R = 7.9$), 10% higher as an exercise coach ($M_{NR} = 7.4$ vs. $M_R = 8.2$), and 8% higher in terms of social presence ($M_{NR} = 6.9$ vs. $M_R = 7.5$).

Altogether, 77% of the participants (10 of 13) rated the relational robot higher than the non-relational robot in terms of companionship, 77% of the participants (10 of 13) rated the relational robot more positively as an exercise coach, and the comparative ratings of social presence between the robot conditions were approximately equal, as 54% of participants (7 of 13) reported higher social presence for the relational robot.

We again analyzed the data to test for significant differences among participant evaluations across the two robot conditions by performing a Wilcoxon signed-rank test. The results show that the participants rated the relational robot as a significantly better companion than the non-relational robot ($W[13] = 14$, $p < 0.05$), supporting Hypothesis 3, and as a significantly better exercise coach than the non-relational robot ($W[11] = 7$, $p < 0.02$),in support of Hypothesis 4. As expected, there was no significant difference in the participant evaluations of social presence between both robot conditions ($W[12] = 28.5$, $p > 0.2$), confirming Hypothesis 5, with both robots receiving equally high ratings. The average participant ratings of both robot conditions for all three dependent measures are shown in Figure 4(b).

3. *Direct Comparison Results*: At the end of the final exercise session, participants were asked to directly compare both robot conditions with respect to 10 different evaluation categories; results are provided in Table 1. It is important to note that the study conditions were labeled as "first session" and "second session" on the survey. These labels would correspond to either the relational condition or non-relational condition, depending on the order of the conditions in which each participant engaged, and were chosen to avoid any potential bias in the survey items.

The results support Hypothesis 6 by demonstrating that, regardless of the order of condition presentation, the participants expressed a strong preference for the relational robot over the non-relational robot. Specifically, the relational robot received 82% of the positive trait votes versus 16% for the non-relational robot, with the remaining 2% shared equally between them. Other notable results include the high number of participants who rated the relational robot as more enjoyable (10 votes, 77%), better at motivating exercise (11 votes, 85%), more useful (11 votes, 85%), and the robot they would choose to exercise with in the future (11 votes, 85%). In contrast, the non-relational robot received a high number of votes for being more frustrating (10 votes, 77%) and more boring (10 votes, 77%) than the relational robot.

4. *User Exercise Performance Statistics*: The collected statistics regarding participant performance in the exercise task were very encouraging as they demonstrated a

Table 1: Participant responses to direct comparison survey items

	Relational	Non-Relational	Batch Equal
Enjoy	10 (77%)	3 (23%)	0 (0%)
More intelligent	11 (85%)	2 (15%)	0 (0%)
More useful	11 (85%)	2 (15%)	0 (0%)
Prefer to exercise with	11 (85%)	2 (15%)	0 (0%)
Better at motivating	11 (85%)	2 (15%)	0 (0%)
More frustrating	3 (23%)	10 (77%)	0 (0%)
More boring	2 (15%)	10 (77%)	1 (8%)
More interesting	10 (77%)	2 (15%)	1 (8%)
More entertaining	10 (77%)	2 (15%)	1 (8%)
Choice from now on	11 (85%)	2 (15%)	0 (0%)

consistently high level of user exercise performance and compliance with the exercise task. As expected, and in support of Hypothesis 7, there were no significant differences found in participant performance between the two study conditions, with both conditions reporting equally high performance among the participants. For example, the average gesture completion time for participants in the relational condition was 2.45 s (S.D. = 0.65), compared to 2.46 s (S.D. = 0.78) for participants in the non-relational condition ($W\,[13] = 37$, $p > 0{:}2$). Given the lack of significant difference in user performance between the two conditions, the statistics presented in this section refer to the participant performance across all exercise sessions of the study.

User compliance and performance in the Workout game were high. The average gesture completion time was 2.46 s (S.D. = 0.70), and the overall exercise performance averaged 5.21 s per exercise (S.D. = 1.0), which also includes time taken for verbal praise, feedback, and score reporting from the robot. The low percentage of necessary corrective feedback, averaging 7.4%, zero failures, and zero movement prompts during the entire study, are all very encouraging results, as they suggest that the participants were consistently motivated to do well on the exercises throughout the interaction.

A summary of all statistics regarding user performance, including those from the Memory and Imitation games, can be found in Table 2.

6.3.5.6 Discussion

The results of the study show a strong user preference for the relational robot over the non-relational robot, demonstrating the positive effects of praise and relational discourse in a healthcare task-oriented HRI scenario, and supporting all of our hypotheses with the exception of Hypothesis 2, which missed reaching significance by a small margin. Participants rated the relational robot significantly higher than the non-relational robot in

Table 2: Participant exercise performance statistics

Objective Measure	Avg. (std.)
Time to gesture completion	2.46 (0.70)
Seconds per exercise	5.21 (1.00)
Feedback percentage	7.4% (4.8%)
Number of failed gestures	0
Number of movement prompts$_w$	0
Maximum score	6
Average maximum score	3.08 (1.12)
Time per gesture attempt (seconds)	8.57 (4.11)
Number of movement prompts	0.26 (0.53)

terms of enjoyableness, companionship, and as an exercise coach. Comments made by participants after the study further illustrate the positive response to the relational robot, including "It's nice to hear your name, it's personal. I felt more positive reinforcement," and from another participant "The robot encourages you, compliments you; that goes a long way." These results provide significant insight into how people respond to SARs, and confirm the positive influence that praise and relational discourse have on intrinsic motivation. These are of particular importance for the healthcare domain, where effectiveness in social interaction, relationship building, and gaining user acceptance and trust are all necessary in ultimately achieving the desired health outcomes of the therapeutic interventions.

The effectiveness of the SAR exercise system was also demonstrated by the outcomes of the study. Not only did the participants rate the interaction with our robot coach as highly enjoyable/entertaining, suggesting they were intrinsically motivated to engage in the exercise task, but they also consistently engaged in physical exercise throughout the interaction, as demonstrated by the gathered user performance statistics. These results are very encouraging, as they clearly show that the system was successful in motivating elderly users to engage in physical exercise, thereby confirming its effectiveness and achieving the primary goal of the system.

6.3.6 Motivation Study II: User Choice and Self-Determination

As discussed in Section 6.3.3, allowing the user to gain a sense of self-determination within a task, for example from choice of activity, has been shown to increase or be less detrimental to intrinsic motivation when compared to similar task conditions that do not involve choice [39,40]. To investigate the role of choice and user autonomy in influencing user intrinsic motivation in the robot exercise system, as well as to further test and validate the effectiveness of our system, we conducted a second user study with elderly participants.

6.3.6.1 Study Design

The study consisted of two conditions, choice and no choice, designed to test user preferences regarding choice of activity. The conditions differed only in the manner in which the three exercise games (Workout, Imitation, Memory) were chosen during the exercise sessions. As in the first study, the design was within our subject content; each participant engaged in both conditions one after the other, with the order of appearance counterbalanced among the participants. Each condition lasted 10 min, totaling 20 min of interaction. The following are descriptions of each condition in greater detail.

1. *Choice Condition*: In this condition, the user is given the choice of which game to play at specific points in the interaction. The robot prompts the user to press the "Yes" button

upon hearing the desired game, and then calls out the names of each of the three game choices. After the user has made a choice, the chosen game is played for a duration ranging from 1 to 2 min in length. Then, the robot asks the user if he would like to play a different game. Depending on the user's response, the robot either continues playing the same game for another 1–2 min, or prompts the user again to choose the game to play next.

2. *No Choice Condition*: In this condition, the robot chooses which of the three games to play at the specified game change intervals (every 1–2 min). The robot always changes games, to try to minimize any user frustration, as in this condition the robot is unaware of the user's game preferences. For simplicity, in this condition, the robot always chooses to first play the Workout game, followed by the Imitation and then Memory games, then cycles through them again in the same order.

6.3.6.2 Participant Statistics

We recruited elderly individuals to participate in the study again through our partnership with the be.group senior living organization. Eleven individuals participated in the first trial of the study, which was subsequently expanded to include 13 additional participants. Therefore, a total of 24 participants were recruited and successfully completed both conditions of the study. Half of the participants engaged in the choice condition in their first session, whereas the other half engaged first in the no choice condition. The sample population consisted of 19 female participants (79%) and five male participants (21%). Participants' ages ranged from 68 to 89, and the average age was 77 (S.D. = 5.76).

6.3.6.3 Measures

As in the first study, survey data were collected at the end of the first and second sessions in order to analyze participant evaluations of the interaction with the exercise system in both conditions. The same evaluation surveys were used for each session to allow for objective comparison between the two conditions.

We administered an additional questionnaire at the end of the last session, asking the participants about their preferences regarding choice in the exercise system, in addition to various other opinion items for further evaluation of the exercise system.

The following describes the specific evaluation measures captured in the post session surveys.

1. *Evaluation of Interaction*: The two dependent measures used to evaluate the interaction with the robot exercise system were the same as in the previous study, namely the enjoyableness of the interaction, and the perceived value or usefulness of the interaction. The ratings scales and survey items for each measure also remained the same.

2. *User Preferences Regarding Choice*: Three questionnaire items were used to assess participant preferences and opinions regarding choice in the exercise system (direct user input in choosing the exercise games). The first item asked participants to state their session preference, labeled as "first" or "second," which referred to either the choice or no choice conditions, depending on each participant's session ordering. The ordinal labels were again chosen, as in the previous study, to avoid any bias in the survey item. The second item asked the participants about user choice, specifically whether they preferred to choose the exercise games to be played, or whether they preferred to let the robot choose instead. This question is similar to the first item but in more direct terms. Last, the third item asked the participants about added enjoyment due to user choice, specifically asking whether having the ability to choose which game to play added to their enjoyment of the interaction.
3. *Evaluation of SAR System*: The last seven questionnaire items were used to obtain additional feedback on the user perceptions of and feelings toward the SAR exercise system. The first four of these items asked participants to rate, respectively: their perception of the robot's intelligence, their perception of the robot's helpfulness, the level of importance they put on their participation in the exercise sessions with the robot, and their mood in general during the exercise sessions. The rating scales were five-point Likert scales, anchored by "not at all" (1) and "very" (5) (e.g., "not at all intelligent" and "very intelligent"). The question regarding user mood during the sessions contained a modified scale, where the mood options ranged from "irritated/frustrated" (1) to "happy/joyful" (5), with the medium range being "normal" (3). Participants were also asked to report their favorite game, least favorite game, and to state their choice of the robot description that best fit among four available options: companion, exercise instructor, game conductor, none of these.

6.3.6.4 Hypotheses

Based on the related research on the positive effects of user choice and autonomy on intrinsic motivation, discussed in Section 6.3.3, five hypotheses were established for this study.

- *Hypothesis 1*: Participants will evaluate the enjoyableness of their interaction in the choice condition more positively than their interaction in the no choice condition.
- *Hypothesis 2*: Participants will evaluate the usefulness of their interaction in the choice condition more positively than their interaction in the no choice condition.
- *Hypothesis 3*: Participants will report a clear preference for the choice condition over the no choice condition when asked to directly compare both exercise sessions.
- *Hypothesis 4*: Participants will report a clear preference for choosing the exercise games themselves, as opposed to having the robot choose which games to play during the interaction.

- *Hypothesis 5*: Participants will report feeling an increase in the enjoyment of the exercise task when given the opportunity to choose which games to play during the interaction.

6.3.6.5 Results

1. *Evaluation of Interaction Results*: The evaluation of interaction survey items was introduced after the first trial of the study; therefore the results presented here for the two interaction measures were analyzed from the data gathered solely from the 13 participants of the expanded study. Nevertheless, all other survey results presented were gathered from all 24 participants of the study.

Participants who engaged in the choice condition in the first session rated the no choice condition on average 7% higher than the choice condition in terms of enjoyment ($M_C = 7.5$ vs. $M_{NC} = 8.0$), and 2% lower in terms of usefulness ($M_C = 8.7$ vs. $M_{NC} = 8.5$). In slight contrast, the participants who instead engaged in the no choice condition in their first session rated the choice condition on average 4% higher than the no choice condition in terms of enjoyment ($M_{NC} = 8.5$ vs. $M_C = 8.8$), and 5% higher in terms of usefulness ($M_{NC} = 9.0$ vs. $M_C = 9.5$).

Altogether, there was no clear participant preference for one condition over the other, as 62% of the participants (8 of 13) rated the no choice condition higher than the choice condition in terms of enjoyment, and 62% of the participants (8 of 13) rated the choice condition higher in terms of usefulness than the no choice condition.

We performed a Wilcoxon signed-rank test on the data and found no significant differences between participant evaluations of the two study conditions, neither with respect to the enjoyableness ($W[13] = 28.5$, $p > 0.2$), nor the usefulness of the interaction ($W[13] = 30.5$, $p > 0.2$). Thus, Hypotheses 1 and 2 were not supported by the data. Nevertheless, participant ratings for the enjoyableness ($M = 8.18$, $S.D. = 1.67$) and usefulness ($M = 8.95$, $S.D. = 1.63$) of the interaction across both conditions were very positive, with scores even higher than those seen in the previous study. These high evaluations of the SAR exercise system further illustrate the effectiveness of the system in instructing and motivating elderly users to exercise.

2. *User Preferences Regarding Choice Results*: The survey results regarding session preference indicated that 42% of the participants (10 of 24) preferred the no choice condition, 33% of the participants (8 of 24) preferred the choice condition, and 25% of the participants (6 of 24) expressed no preference for one condition over the other. Figure 5(a) plots the participants' stated preferences of study conditions. The varied participant condition preferences indicate no clear preference for one over the other, and thus Hypotheses 3 was not supported. Concerning user choice in the exercise

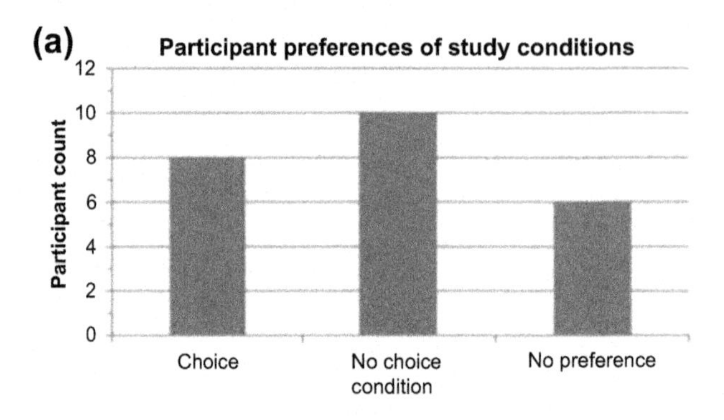

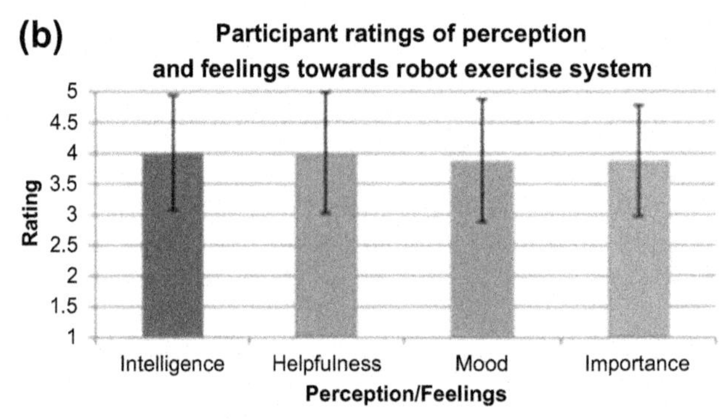

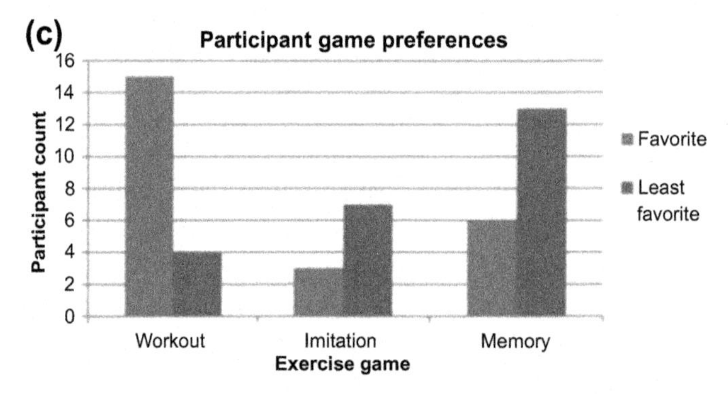

Figure 5

Graphs of: (a) the participants' preferences of study condition; (b) the participants' ratings in response to survey questions on their perception of the robot's intelligence, helpfulness, their mood during sessions, and how important the sessions were to them; and (c) the participants' preferences of exercise game.

system, 62% of participants (15 of 24) reported preferring to let the robot choose the games to play, with the remaining 38% of participants (9 of 24) preferring to choose the games themselves. The slight preference among participants for having the robot choose countered the reasoning of Hypothesis 4, which was not supported.

It is interesting to note that even though most participants preferred letting the robot decide which games to play, almost all of the participants, 92% (22 of 24), reported increased enjoyment of the task when given the opportunity to choose the exercise game to play. This result supports Hypothesis 5 and is consistent with the literature on the effects of user choice on intrinsic motivation [39,40].

3. *Evaluation of SAR System Results*: The results of the survey questions regarding participant perceptions and feelings toward the SAR exercise system are very encouraging; the participants rated the robot highly in terms of intelligence (M= 4.0, S.D. = 0.93) and helpfulness (M = 4.0, S.D. = 0.97), attributed a moderately high level of importance to the exercise sessions (M = 3.87, S.D. = 0.89), and reported their mood throughout the sessions to be normal-to-moderately pleased (M = 3.87, S.D. = 0.99). These results are important because positive user perceptions of the agent's intelligence and helpfulness are a key part of establishing trust in the human–robot relationship. This, along with a positive user mood and user-attributed importance to the therapeutic task, are in turn important for establishing and maintaining user intrinsic motivation. These are all key components for achieving long-term success in any SAR setting. An illustration of the results is shown in Figure 5(b).

Regarding the exercise games, the participants largely favored (62%, 15 of 24) the Workout game over the others, wherein the robot serves as a traditional exercise coach, with the Memory game being chosen most often as the participants' least favorite game (54%, 13 of 24). Figure 5(c) summarizes the participants' game preferences.

The description most chosen by the participants as the best fit for the robot was that of an exercise instructor (67%, 16 of 24), not surprisingly, as opposed to that of a game conductor (25%, 6 of 24) or companion (8%, 2 of 24). While all of the descriptions represent characteristics of the robot in one form or another, the primary selection of an exercise coach by the participants illustrates the perception of the robot as an agent that they can trust and that is capable of helping, rather than simply entertaining.

6.3.6.6 Discussion

The results of the study showed no clear preference for one condition over the other, as the user enjoyment level of the interaction was reported to be equally high for both conditions, with or without user choice of activity. The high participant evaluations regarding the enjoyableness and usefulness of the interaction, the intelligence and helpfulness of the robot, and positive user mood and attributed importance to the exercise sessions, further validate the SAR system's effectiveness in motivating elderly users to engage in physical exercise.

The relatively mixed condition preferences among participants, or rather the lack of a clear preference for the choice condition, seem somewhat counterintuitive given the positive effect that choice and user autonomy have been shown to have on task-based enjoyment [39,40]. One possible explanation for the mixed preferences may be that, since the robot's role in the interaction was that of an exercise instructor, some participants might have felt it was the robot's duty to determine the exercise regimen, and hence were comfortable relinquishing the choice of exercise games. Another possible explanation may be that the enjoyment derived from choosing the games did not outweigh the enjoyment derived from relaxation due to the reduced responsibility of not having to choose the games. Both explanations seem plausible, as some of the participants reported preferring the robot to have the "responsibility" of steering the task. A third explanation may be that, given the short-term nature of the study, some participants may have needed more experience with the robot system before they felt confident enough to make task-based decisions themselves.

It is interesting to note that, even though the condition preferences were varied and nearly half of participants preferred letting the robot decide which games to play, all participants at one point or another during the study took advantage of having greater control in the choice condition. Specifically, when given the option by the robot to change games, all participants at some point either chose to continue playing the same game they were playing, or chose to avoid playing a game they did not want to play. Neither of these cases could occur in the no choice condition, as the robot was unaware of the user's current game preferences.

This observation speaks to the value of user preference within the task scenario, suggesting that a hybrid approach that includes both user and robot decision making, personalized and tuned automatically for each user, might ultimately be the best solution for achieving a fluid and enjoyable task interaction for all users. For example, for users who prefer greater robot responsibility and input in the SAR-based task, the robot can recommend the "best" choices given the current task conditions and situation, giving the user an informed choice. Alternatively, for users who prefer greater control only once they've gained enough experience, the robot can initially make all task-based decisions until the user is ready and confident in making choices. For users who have a clear preference regarding who should make task-based decisions during interaction, the chosen strategy can be implemented continually throughout the sessions. Clearly, no single fixed user-choice strategy is appropriate for all users; users have varied preferences regarding choice, and those preferences may even change over time. Therefore, it is important that the strategy employed in SAR systems regarding user choice and autonomy be continually adapted to the specific user engaged in the interaction, thus personalizing the therapeutic intervention.

6.3.7 Conclusion

In this chapter, we have presented the design methodology, implementation, and evaluation of a SAR that is capable of interacting with elderly users and engaging them in physical exercise in a seated aerobic exercise scenario. Methods for influencing an individual's intrinsic motivation to perform a task, including verbal praise, relational discourse, and user choice, were implemented and evaluated in two separate user studies with elderly participants.

The results of the first motivation study showed a strong participant preference for the relational robot over the non-relational robot in terms of enjoyableness of the interaction, companionship, and as an exercise coach, in addition to demonstrating similar evaluations of both robots in terms of usefulness of interaction and social presence. These results illustrate the positive effects of motivational relationship building techniques, namely praise and relational discourse, on participant perceptions of the social agent and interaction in a health-related task scenario, and ultimately on user intrinsic motivation to engage in the task. The results of the second motivation study showed varying participant preferences regarding user choice within the exercise system, suggesting the need for customizable interactions automatically tailored to accommodate the personal preferences of the individual users.

The SAR exercise system was very well received, as demonstrated by both user studies, with high participant evaluations regarding the enjoyableness and usefulness of the interaction, companionship, social presence, intelligence, and helpfulness of the robot coach, and the positive mood and attributed importance of the exercise sessions. The system was also found to be effective in motivating consistent physical exercise throughout the interaction, according to various objective measures, including average gesture completion time, seconds per exercise, and feedback percentage.

The overall acceptance of the SAR exercise system by elderly users, as evidenced by the outcomes of two user studies evaluating the motivation capabilities and effectiveness of the system, is very encouraging and illustrates the potential of the system to help the elderly population to engage in physical exercise to achieve beneficial health outcomes, to facilitate independent living, and ultimately to improve quality of life.

Acknowledgments

The authors would like to thank the be.group (formerly Southern California Presbyterian Homes), its staff, and participating residents, with special thanks to A. Mehta, K. Reed, A. Escobedo, and M. Anderson for their help in the preparation and execution of the robot exercise system studies.

This work was supported by a grant from the Robert Wood Johnson Foundation's Pioneer Portfolio through its national program, "Health Games Research: Advancing Effectiveness of Interactive Games for Health," and the National Science Foundation under Grants IIS-0713697, CNS-0709296, and IIS-1117279. The authors are with the University of Southern California, Los Angeles, CA 90089, USA (e-mail: fasola@usc.edu; mataric@usc.edu).

References

[1] U.S. Census Bureau, U.S. Census Bureau Report (Issue Brief No. CB09-97), 2009 Available from: http://www.census.gov/population/international/.
[2] American Association of Colleges of Nursing, Nursing Shortage Fact Sheet, 2010 Available from: http://www.aacn.nche.edu/media/FactSheets/NursingShortage.htm.
[3] American Health Care Association, Summary of 2007 AHCA Survey Nursing Staff Vacancy and Turnover in Nursing Facilities, July 2008.
[4] P. Buerhaus, Current and future state of the US nursing workforce, J. Am. Med. Assoc. 300 (20) (2008) 2422–2424.
[5] E.E. Baum, D. Jarjoura, A.E. Polen, D. Faur, G. Rutecki, Effectiveness of a group exercise program in a long-term care facility: a randomized pilot trial, J. Am. Med. Dir. Assoc. 4 (2003) 74–80.
[6] D. Dawe, R. Moore-Orr, Low-intensity, range-of-motion exercise: invaluable nursing care for elderly patients, J. Adv. Nurs. 21 (1995) 675–681.
[7] M.D. McMurdo, L.M. Rennie, A controlled trial of exercise by residents of old people's homes, Age Ageing 22 (1993) 11–15.
[8] V.S. Thomas, P.A. Hageman, Can neuromuscular strength and function in people with dementia be rehabilitated using resistance-exercise training? Results from a preliminary intervention study, J. Gerontol. A Biol. Sci. Med. Sci. 58 (8) (August 2003) 746–751.
[9] S. Colcombe, A. Kramer, Fitness effects on the cognitive function of older adults, Psychol. Sci. 14 (2003) 125–130.
[10] S.J. Colcombe, A.F. Kramer, K.I. Erickson, P. Scalf, E. McAuley, N.J. Cohen, Cardiovascular fitness, cortical plasticity, and aging, in: Proc. Nat. Acad. Sci. USA, vol. 101, 2004, pp. 3316–3321.
[11] W. Spirduso, P. Clifford, Replication of age and physical activity effects on reaction and movement time, J. Gerontol. 33 (1978) 26–30.
[12] Z.B. Moak, A. Agrawal, The association between perceived interpersonal social support and physical and mental health: results from the national epidemiological survey on alcohol and related conditions, J. Public Health 32 (2010) 191–201.
[13] L.K. George, D.G. Blazer, D.C. Hughes, N. Fowler, Social support and the outcome of major depression, Br. J. Psychiatr. 154 (1989) 478–485.
[14] E. Paykel, Life events, social support and depression, Acta Psychiat. Scand. 89 (1994) 50–58.
[15] E. Stice, J. Ragan, P. Randall, Prospective relations between social support and depression: differential direction of effects for parent and peer support? J. Abnorm. Psychol. 113 (2004) 155–159.
[16] S.A. Stansfeld, G.S. Rael, J. Head, M. Shipley, M. Marmot, Social support and psychiatric sickness absence: a prospective study of British civil servants, Psychol. Med. 27 (1997) 35–48.
[17] D.J. Feil-Seifer, M.J. Matarić, Defining socially assistive robotics, in: Proc. Int. Conf. Rehabil. Robot, Chicago, IL, June 2005, pp. 465–468.
[18] M.J. Matarić, J. Eriksson, D.J. Feil-Seifer, C.J. Winstein, Socially assistive robotics for post-stroke rehabilitation, J. Neuroeng. Rehabil. 4 (5) (February 2007).
[19] A. Tapus, C. Tapus, M.J. Matarić, User-robot personality matching and robot behavior adaptation for post-stroke rehabilitation therapy, Intell. Serv. Robot 1 (2) (April 2008) 169–183.
[20] A. Tapus, C. Tapus, M. Matarić, The use of socially assistive robots in the design of intelligent cognitive therapies for people with dementia, in: Proc. IEEE Int. Conf. Rehabil. Robot, Kyoto, Japan, 2009, pp. 924–929.
[21] D.J. Feil-Seifer, M.J. Matarić, Towards the integration of socially assistive robots into the lives of children with ASD, in: Human-Robot Interaction Workshop on Societal Impact: How Socially Accepted Robots Can Be Integrated in Our Society, San Diego, CA, March 2009.
[22] D.J. Feil-Seifer, M.J. Matarić, Using proxemics to evaluate human-robot interaction, Osaka, Japan, in: Proc. Int. Conf. Human-robot Interaction, March 2010, pp. 143–144 (Poster Paper).

[23] J. Fasola, M.J. Matarić, Robot motivator:Increasing user enjoyment and performance on a physical/cognitive task, in: Proc. IEEE Int. Conf. Develop. Learn, Ann Arbor, MI, August 2010, pp. 274–279.
[24] J. Fasola, M.J. Matarić, Robot exercise instructor: a socially assistive robot system to monitor and encourage physical exercise for the elderly, in: Proc. 19th IEEE Int. Symp. Robot Human Interactive Commun., September 2010, pp. 416–421. Viareggio, Italy.
[25] S. Dubowsky, F. Genot, S. Godding, H. Kozono, A. Skwersky, H. Yu, L. Shen Yu, PAMM-A robotic aid to the elderly for mobility assistance and monitoring, in: Proc. IEEE Int. Conf. Robot. Autom, vol. 1, April 2000, pp. 570–576.
[26] M. Montemerlo, J. Pineau, N. Roy, S. Thrun, V. Verma, Experiences with a mobile robotic guide for the elderly, in: Proc. AAAI Nat. Conf. Artif. Intell., Edmonton, AB, Canada, August 2002, pp. 587–592.
[27] K. Wada, T. Shibata, T. Saito, K. Tanie, Analysis of factors that bring mental effects to elderly people in robot assisted activity, in: Proc. Int. Conf. Intell. Robots Syst., vol. 2, October 2002, pp. 1152–1157.
[28] C. Kidd, W. Taggart, S. Turkle, A sociable robot to encourage social interaction among the elderly, in: Proc. Int. Conf. Robot. Autom., Orlando, FL, May 2006, pp. 3972–3976.
[29] Y. Matsusaka, H. Fujii, T. Okano, I. Hara, Health exercise demonstration robot TAIZO and effects of using voice command in robot-human collaborative demonstration, in: Proc. 18th IEEE Int. Symp. Robot Human Interactive Commun., September–October 2009, pp. 472–477.
[30] T.W. Bickmore, R.W. Picard, Establishing and maintaining long-term human-computer relationships, ACM Trans. Comput. Hum. Interact. 12 (2) (June 2005) 293–327.
[31] C.D. Kidd, C. Breazeal, Robots at home: understanding long-term human-robot interaction, in: Proc. IEEE/RSJ Int. Conf. Intell. Robots Syst., September 2008, pp. 3230–3235.
[32] B. French, D. Tyamagundlu, D. Siewiorek, A. Smailagic, D. Ding, Towards a virtual coach for manual wheelchair users, in: Proc. Int. IEEE Symp. Wearable Comput., September 2008, pp. 77–80.
[33] R.A. Dienstbier, G.K. Leak, Effects of Monetary Reward on Maintenance of Weight Loss: An Extension of the Overjustification Effect, American Psychological Association Convention, Washington, DC, 1976.
[34] R.J. Vallerand, G. Reid, On the causal effects of perceived competence on intrinsic motivation: a test of cognitive evaluation theory, J. Sport Psychol. 6 (1984) 94–102.
[35] R.J. Vallerand, Effect of differential amounts of positive verbal feedback on the intrinsic motivation of male hockey players, J. Sport Psychol. 5 (1983) 100–107.
[36] R.S. Weinberg, J. Ragan, Effects of competition, success/failure, and sex onintrinsic motivation, Res. Quart. 50 (1979) 503–510.
[37] M. Csikszentmihalyi, Beyond Boredom and Anxiety, Jossey-Bass, San Francisco, CA, 1975.
[38] E. Deci, R. Ryan, Intrinsic Motivation and Self-determination in Human Behavior, Plenum, New York, 1985, pp. 29, 318, 322.
[39] C.D. Fisher, The effects of personal control, competence, and extrinsic reward systems on intrinsic motivation, Organ. Behav. Hum. Perform. 21 (1978) 273–288.
[40] M. Zuckerman, J. Porac, D. Lathin, R. Smith, E.L. Deci, On the importance of self-determination for intrinsically motivated behavior, Pers. Soc. Psychol. Bull. 4 (1978) 443–446.
[41] R.B. Margolis and C.R. Mynatt. "The effects of self and externally administered reward on high base rate behavior," Bowling Green State University, unpublished manuscript, 1979.
[42] NeoSpeech Text-to-speech, 2009. Available from: http://www.neospeech.com.
[43] O.C. Jenkins, C. Chu, M.J. Matarić, Nonlinear Spherical Shells for Approximate Principal Curves Skeletonization, 2004. Univ. Southern California Ctr. Robot. Embedded Syst., Tech. Rep. CRES-04–004.
[44] H. Fujiyoshiand, A. Lipton, Real-time human motion analysis by image skeletonization, in: Proc. Workshop Appl. Comput. Vis., Princeton, NJ, October 1998, pp. 15–21.
[45] S. Waldherr, S. Thrun, R. Romero, D. Margaritis, Template-based recognition of pose and motion gestures on a mobile robot, in: Proc. Nat. Conf. Artif. Intell, Madison, WI, 1998, pp. 977–982.
[46] C.R. Wren, A. Azarbayejani, T. Darrell, A.P. Pentland, Pfinder: realtime tracking of the human body, IEEE Trans. Pattern Anal. Mach. Intell. 19 (7) (July 1997) 780–785.
[47] Microsoft Kinect, 2010. Available from: http://www.xbox.com/kinect.

[48] R.H. Poresky, C. Hendrix, J.E. Mosier, M. Samuelson, Companion animal bonding scale: internal reliability and construct validity, Psychol. Rep. 60 (1987) 743–746.
[49] K.M. Lee, Presence, Explicated, Commun. Theory 14 (1) (2004) 27–50.
[50] Y. Jung, K.M. Lee, Effects of physical embodiment on social presence of social robots, in: Proc. Presence, 2004, pp. 80–87.
[51] J. Wainer, D.J. Feil-Seifer, D.A. Shell, M.J. Matarić, Embodiment and Human-Robot Interaction: a task-based perspective, in: Proc. IEEE Int. Workshop Robot Human Interactive Commun, Jeju Island, Korea, August 2007, pp. 872–877.

CHAPTER 6.4

Toward a Human–Robot Symbiotic System[1]

Kazuhiko Kawamura, Tamara E. Rogers, Kimberly A. Hambuchen, Duygun Erol
Center for Intelligent Systems, Vanderbilt University, Nashville, TN, USA

Chapter Outline

6.4.1 Introduction

The robotics field has evolved from industrial robots in the 1960s to nontraditional branches such as medical robots and search and rescue robots in the 2000s. One area that is gaining popularity among robotic researchers is anthropomorphic robots or humanoid robots [1,2]. Increasing popularity reflects a recent announcement in a new journal called the International Journal of Humanoid Robotics (IJHR). The inaugural issue of IJHR is

---

[1] Reprinted from Robotics and Computer-Integrated Manufacturing, Kazuhiko Kawamura, Tamara E. Rogers, Kimberly A. Hambuchen, and Duygun Erol, Toward a human–robot symbiotic system, Vol. 19, pp. 555–565, Copyright (2013), with permission from Elsevier.

Household Service Robotics. http://dx.doi.org/10.1016/B978-0-12-800881-2.00023-2

expected in 2004 and the Center for Intelligent Systems was asked to contribute an article to this inaugural issue [3]. At the Cognitive Robotics Laboratory of Vanderbilt University, we have been developing a humanoid robot called the Intelligent Soft-Arm Control (ISAC) (Figure 1) since 1995. Originally ISAC was designed to assist the physically disabled [4], but gradually became a general-purpose humanoid robot to work with a human as a partner or an assistant at home or in a factory [5]. We have developed a multi-agent architecture for parallel, distributed robot control [6] based on a unique design philosophy [7] as described in Section 2 of the paper, along with a robust human-robot interface [8]. Unlike many humanoid research groups in the world who put more emphasis on human-like motion control and efficient walking pattern generation, our group places emphasis on the cognitive aspects of the humanoid. The research described herein is to report recent progress on developing two agents, the Human Agent and the Self Agent, plus memory structures that enable ISAC to learn new skills. The Human Agent is the humanoid's internal representation of the human. It includes information about the location, activity, and state of the human, as determined through observations and conversations. The Self Agent is the humanoid's internal representation of itself. It provides the system with a sense of self-awareness concerning the performance of hardware, behaviors and tasks. Our approach to robot memory structures is through short- and long-term memories called the Sensory EgoSphere (SES) and the Procedural Memory (PM), respectively. The SES is a data structure that encapsulates short-term memory for the humanoid in a time-varying, spatially indexed database interfacing the environment with a geodesic hemisphere [9]. It allows ISAC to maintain a spatially indexed map of relative sensory data in its environment. PM is a data structure that encapsulates both primitive and meta behaviors and forms a basis to learn new behaviors and tasks.

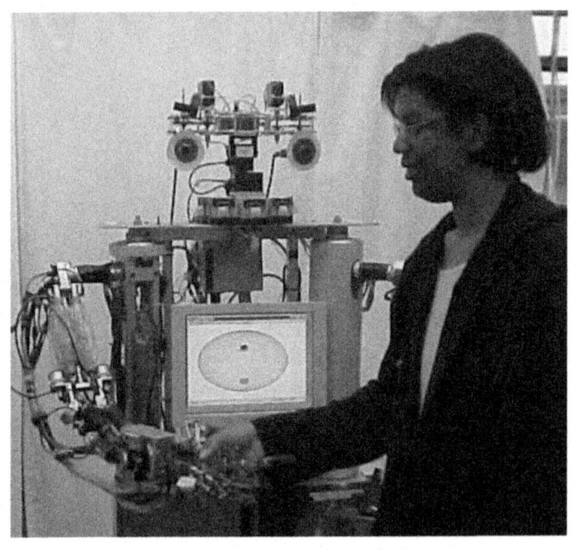

Figure 1
Vanderbilt's humanoid robot, ISAC.

6.4.2 Framework for Human–Humanoid Interaction

Our philosophy for software design for an intelligent machine such as a humanoid is to integrate both the human and the humanoid in a unified multi-agent based framework [10]. As such, we have grouped aspects of Human–Humanoid Interaction (HHI) into the following three categories: *Physical*, *Sensory*, and *Cognitive*. *Physical* aspects are those that pertain to the structure or body of the human and the humanoid. These aspects cover the essential physical features and manipulation capabilities needed for robust interaction. *Sensory* aspects refer to the channels through which the human and the humanoid gain information about each other and the world. *Cognitive* aspects describe those concerned with the internal workings of the system. For the humans these include the mind and affective state; for the humanoid these include the reasoning and abilities to communicate its intentions. It is difficult to consistently determine cognitive aspects of humans. Therefore our application is limited to cases where both the human and the humanoid intend to achieve a common goal. We are also interested in giving the humanoid its own emotional or affective module to make HHI more socially pleasant [11].

ISAC has been equipped with sensors such as cameras, microphones, and infrared sensors for capturing communication modes (see Appendix: ISAC System Information). We use Microsoft Speech engines for detecting human speech, and have implemented a sound-localization system [12]. An infrared motion detector provides ISAC with a means of sensing human presence. A face detector returns the location of a face and the level of confidence for the detection. A simple finger–point detector locates a person's fingertip.

Likewise, we are developing techniques for ISAC to give feedback to people. ISAC can exploit the sensory modalities that people are accustomed to using, such as seeing and hearing. In addition to speaking to people, ISAC can physically manipulate its body and arms. For example, we are developing display behaviors, such as gesturing, that ISAC can use to communicate its intention. We are also developing the use of a visual display of the SES that can be projected on ISAC's monitor located in the middle of its' body (see: Figure 2). The interface is based on a multi-agent based architecture [13,14] called the Intelligent Machine Architecture (IMA) developed at Vanderbilt [6]. Figure 3 illustrates the overall IMA agent structure with the short and long-term memory structures. In the sequel, two cognitive agents that form the nucleus of human-ISAC interactions will be described.

6.4.2.1 The Human Agent

The Human Agent [15] is a virtual agent, realized as a collection of IMA agents, which serves as an internal *active representation* of people in the robot's environment. As an *active representation* [16], it is able to detect, represent and monitor people. It also

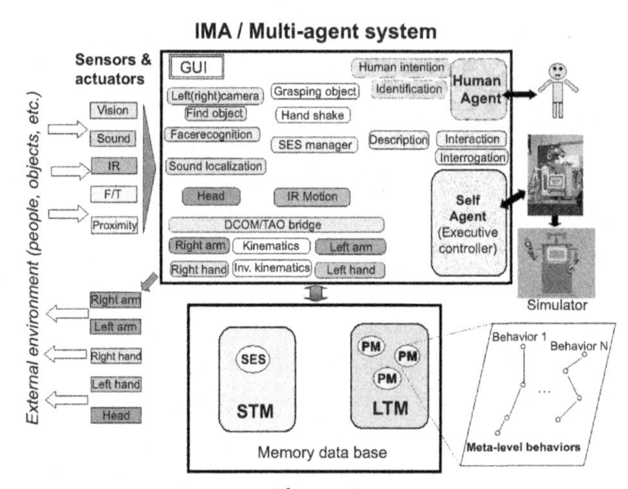

Figure 2
Intelligent machine architecture agents and memory structures.

facilitates human-robot interaction by determining appropriate interaction behaviors. As shown in Figure 4, five IMA agents perform the core functions of the Human Agent. These agents are grouped as two compound IMA agents, the Monitoring Agent and the Interaction Agent, which address the two main roles of the Human Agent. The Human Agent receives input concerning the human from various atomic agents that detect the aforementioned *physical* aspects, such as a person's face, speech, location, etc. The Human Agent then communicates with various supporting agents which are each responsible for functions such as detecting features of people, interfacing with memory data structures, or reporting human status information to the Self Agent.

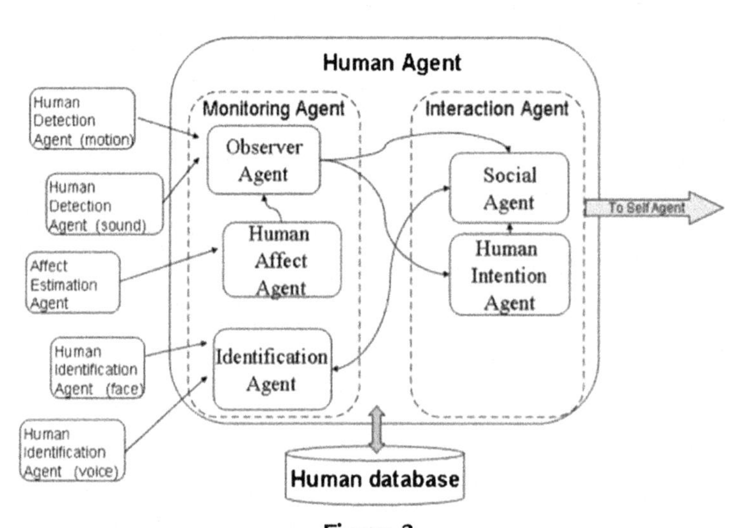

Figure 3
The Human Agent and supporting atomic agents.

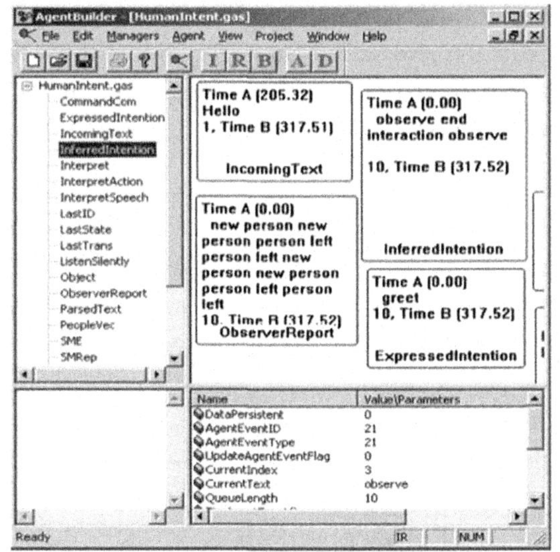

Processes Inferred Intention and Expressed Intention

* Expressed intention is mapped from speech keywords
* Inferred Intention is mapped from observations of people
  - Uses report information from Observer Agent

Histories are cataloged in IntentHistory table in human database

Figure 4

The Human intention agent, shown in an AgentBuilder window.

6.4.2.1.1 Monitoring Agent

This agent monitors human features, including physical features and the emotional or affective state. To reflect that people typically interact through various modalities, our approach integrates several technologies such as speech recognition and speaker identification, face detection and recognition, sound localization, and motion detection, each of which contribute to an awareness of humans and their actions in the environment. The Monitoring Agent includes three atomic IMA agents, the Observer, Identification, and Human Affect Agents. Certain atomic agents, categorized as Human Detection Agents (HDAs) in Figure 4, each detect a feature of the human, such as the location of a face, a voice, etc., and report this to the Observer Agent. Performance data on various HDAs are shown in Table 1. Two of the HDAs are the Face Detection Agent and the Sound Localization Agent. Face detection is template-based and returns location and confidence

Table 1: Human detection agents

Technology	Accuracy
Sound localization	89% labeling source 6 out of region
IR motion	Reports 72% of time person is there
Human face (detection)	Finds present faces 60%
Human face (recognition)	Correct identity 77%
Speaker ID	Collect further data
Speech recognition	Using Microsoft Speech SDK 5.1
Word valence	(depends on speech recognition – no real measure of accuracy)

of detection. The Sound Localization Agent determines the direction of the source based on the relationship between the energy of the two stereo channels. In turn, the Observer Agent integrates this information to detect and track the human during the interaction and is currently able to integrate data to represent that there are 0, 1, or 2 people in the environment.

The Identification Agent identifies the human in the environment and is used to personalize interactions. It receives information from Human Identification Agents, including the Speaker Identification and Face Recognition agents. Each of these agents employs forced-choice classifiers to match inputs to corresponding data in the library of known people. For speaker identification, we have implemented a simple speaker identification routine using a forced-choice classifier that operates on a library of known speakers stored in the human database. The face recognition algorithm utilizes Embedded Hidden-Markov Models and the Intel OpenSource Computer Vision Library [17]. Location and identity information is posted on the SES, a short-term sensory-event memory of ISAC.

The Human Affect Agent, in a similar design paradigm to the above Observer and Identification Agents, will receive inputs from various agents, called Affect Estimation Agents, which each represent a feature of human affect. Currently, this agent receives input from a Word Valence Agent, which monitors the human's speech for words known to contain positive or negative expression. The goal of this agent is to provide ISAC with knowledge of the person's affective state to ultimately influence the interaction method.

6.4.2.1.2 Interaction Agent

While the Monitoring Agent performs passive roles of observing people, the Interaction Agent may be considered as handling the more pro-active functions of the interaction. Two roles are (1) handling communication via an Interaction Agent and (2) modeling the interaction with the human via the Social Agent.

The Human Intention Agent, shown in Figure 5 in an AgentBuilder window, handles verbal and non-verbal communication between the humanoid and the human. The AgentBuilder program is a development tool for the IMA. It provides the interface for the design, testing, and execution of IMA agents. It processes two types of intention from people, namely *expressed* or *inferred.* The Human Agent maps speech directed toward the robot into an expressed intention, based on mapping of keywords. If the person greets or requests a task of the robot, this is considered to be an *expressed* intention. Other intentions, however, represent what the robot can infer based on the actions of the human and are labeled *inferred* intentions. For example when a person leaves the room ISAC assumes that the person no longer intends to interact and, therefore, can reset its expectations. An initial suite of human intentions include the intent to communicate (which is used to eliminate speech or sounds with no communicative intent), intent to

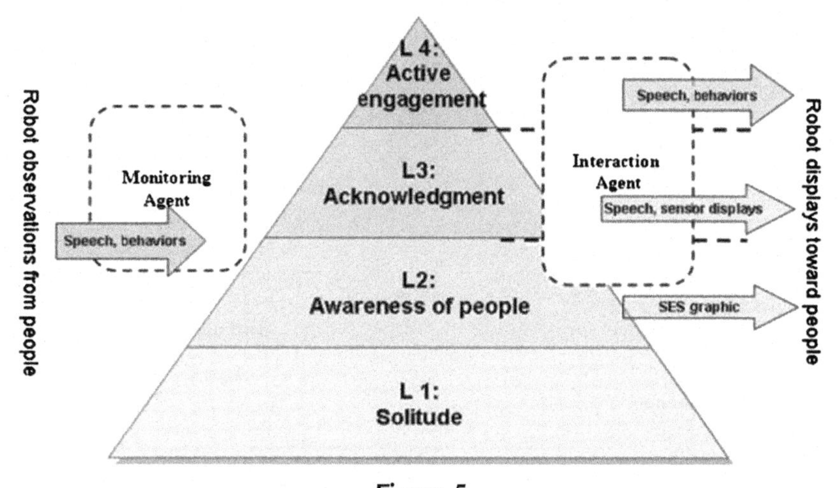

Figure 5
Levels of interaction within the Human Agent.

interact with ISAC, intent for ISAC to perform a task, intent to end interaction. Based on observations of interactions between ISAC and various people, we have included a special case intent to observe.

The Social Agent contains a rule set for social interaction which enables the robot to interact with people as a function of the overall context and, therefore, more naturally. The method of action selection is based on a set of social rules developed for ISAC to interact with people. The rule base is a production system and operates on features such as the level of interaction, current human intention, etc. and provides suggestions of appropriate behaviors to the Self Agent for consideration. The Social Agent constantly monitors the state of certain variables that represent external information, such as the number of people, current person, and new intentions. The Self Agent then interprets this suggestion in the context of its own current state, e.g., current intention, status, tasks, etc.

Figure 6 shows the model of the levels of interaction engagement, represented by a numerical value, that we have developed as the basis for modeling social interaction. These levels progress the robot from a state of no interaction to an ultimate goal of completing a task with (or for) a person. Level 1, Solitude, corresponds to when ISAC does not detect anyone in the environment. In this situation, ISAC may choose actions to actively attract people with whom to interact. Level 2, Awareness of People, corresponds to a stage, often short time, when ISAC is aware of people around it, and has not interacted with them. Level 3, Acknowledgment, is the phase when the robot actively acknowledges the presence of a person. This is performed if the person is approaching ISAC for the first time or if the person is interrupting an ongoing interaction. Level 4, Active Engagement, represents that stage of active interaction.

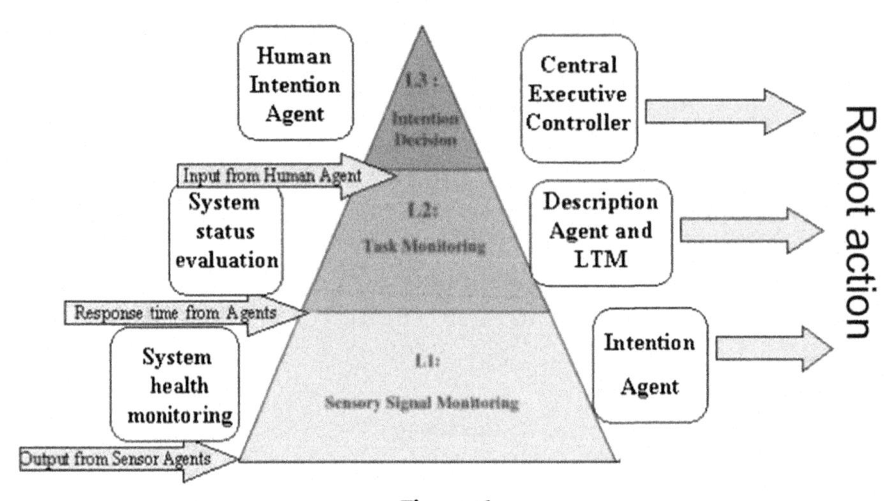

Figure 6
Levels of monitoring and action in the Self Agent.

6.4.2.1.3 Human database

The Human Database is a repository of information about people with whom the robot has previously interacted. It provides two services. The first is to store the personal data that the robot uses to recognize and to identify people, images, and voices. Secondly, the Human Database stores data on previous interactions — time, type and individual preferences of interactions or activities. To represent individuals, the database keeps an identification-indexed record of people the robot has met (see Table 2). As the robot learns about a person (presently performed offline), the database contains an indicator that the person has been added to the face and/or voice recognition libraries. These libraries will store the actual processed exemplars for recognition.

The second database table stores task-related information about the interactions that the robot has had with people called representative data (see Table 3). This data is stored by time and is a log of what interactions ISAC has had. This table is the source for determining an individual's most recent or most frequent tasks. It may also be used to determine the robot's most recent or frequent tasks across all people or on a given day. This data will be used to determine personal trends of the current individual and personalize the data.

Table 2: Example of known people

Index	Name	First meeting	Last meting	Face learned	Voice learned
1	Tamara	2001-12-31	2002-06-18 11:33:34	Yes	Yes
2	Kim	2001-12-31	2002-04-26 19:47:23	Yes	Yes
3	Xinyu	2001-12-31	2002-06-20 10:15:32	Yes	Yes

Table 3: Example of intention histories database table

Index	Name	Intent	Time
1	Tamara	ColorGame	2002-06-18 11:33:34
2	Xinyu	Recognize	2002-06-20 09:35:54
3	Xinyu	HandShake	2002-06-20 09:37:23
4	Xinyu	ColorGame	2002-06-20 09:39:43

6.4.2.2 The Self Agent

The Self Agent is a cognitive agent responsible for monitoring sensor signals, agent communications, and high-level (i.e., cognitive) decision-making aspects of ISAC (Figure 7). Sensory signal monitoring currently involves component-level fault monitoring and diagnosis using a Kalman Filter.

The Self Agent then integrates failure information from sensors and maintains information about the task-level status of the humanoid. Cognitive aspect covered by the Self Agent includes recognition of the human's intention it receives from the Human Agent and to select appropriate actions by activating meta behaviors within the Long Term Memory (LTM). Figure 2 shows the structure of the Self Agent and supporting atomic agents.

The Intention Agent will determine the intended action of the humanoid based on the intention of the human, the state of the humanoid, and whether this action would conflict with the humanoid's current activities. If there is no conflict, the action can be performed, otherwise, the Intention Agent must resolve the conflict. Conflict resolution is handled based on the relative priority of the conflicting actions. Because ISAC currently can only

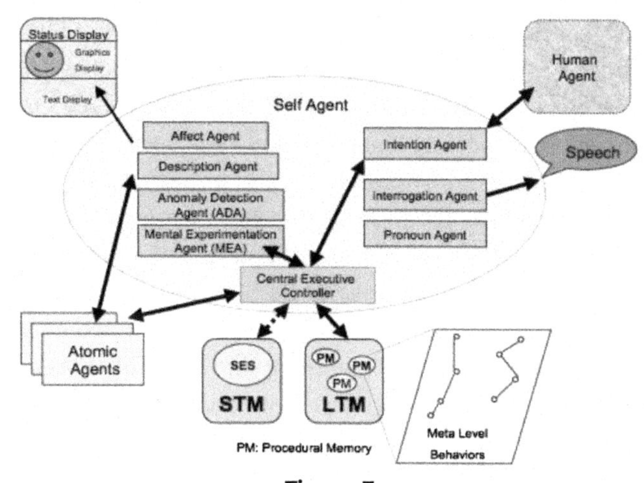

Figure 7
The Self Agent and supporting Atomic Agents.

perform a limited number of actions, we use a table ranking possible actions by priority. When the current action's priority is higher than the new intended actions, then the current task will be paused. ISAC will explain to the person why the new action cannot be performed and will resume current activity. If the new action's priority is higher, then the current action will be halted and the new action will begin.

The Central Executive Controller (CEC) coordinates the various PM structures stored in the LTM that encapsulate ISAC's behaviors. These behaviors are used in a plan constructed by the CEC to perform a task. The goal of the generated plan is determined according to the input from the Intention Agent, then the constructed plan is put into action. The execution of the plan is monitored by the Self Agent and in the case of an abnormal execution, the Self Agent is responsible for solving the impasse by generating a novel plan.

6.4.3 Robot Memory Data Structures

To monitor the vast amounts of data and procedures a robot acquires when interacting with people, a robot should maintain both short-term and long-term memory capacities. These memory structures should store and manipulate sensory, declarative and procedural information. Two data structures have been developed for ISAC to address these issues: the SES provides ISAC with a short-term memory store for sensory events while Procedural Memories provide ISAC with a long-term memory for performing behaviors.

6.4.3.1 STM: Sensory EgoSphere

The SES is a data structure that serves as a short-term memory for a robot [9]. The SES was inspired by the egosphere as defined by Albus [18,19]. The original egosphere was conceived as a topological sphere surrounding an entity onto which external or internal events are projected. Albus proposed multiple egospheres for different components of a system, i.e., head egosphere, camera egosphere, body egosphere. The objective of the SES is to store sensory information, both exteroceptive and proprioceptive, that the robot detects. The SES is structured as a geodesic sphere that is indexed by azimuth and elevation angles and is centered at a robot's origin. With ISAC, the head's pan tilt origin serves as the center of the SES. The geodesic dome provides an interface for projection of data while a database provides storage for a description of projected data. Sensory processing agents in the system send their outputs to the SES for short-term storage. The SES then projects these outputs onto its geodesic interface. The projection vector is a unit vector from the center of the sphere in the direction of the robot's pan and tilt angles. The information and data that was sent by the sensory processor is then entered into the database with a timestamp reference.

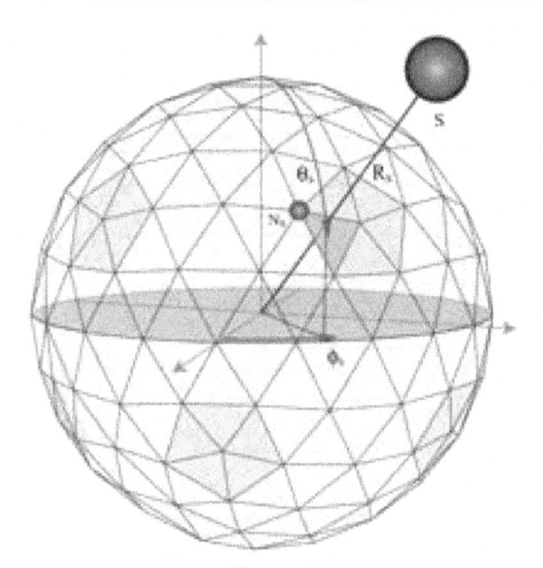

Figure 8
Projection of objects onto the SES.

Figure 8 shows the projection of an object onto the SES. In this view, ISAC's head is located in the center of the sphere. The projection vector occurs at a given azimuth, elevation from the center of the sphere which correlates to a specific pan, tilt location of the head. Once this object is projected onto the sphere, a distance measure is performed to find the vertex on the sphere closest to the projection spot. This vertex, or node, becomes the registration node for the specific object. The SES enters the object, any information reported about the object and a timestamp into the database at the location of its registration node. This process occurs each instance that a sensory processing agent sends its output to the SES.

The SES can also retrieve information stored in its database. An agent can request retrieval of data from the SES in one of three methods: by data name, data type and location of the data. For the first two methods, the SES simply queries the database using the name or type of the data. To retrieve data using a location, the requesting agent must supply the SES with a specific pan, tilt angle pair and a search neighborhood. The search neighborhood specifies how many nodes from the center node to retrieve data. The center node is defined as the closest node to the given pan, tilt angle pair. The SES calculates which nodes are included in the neighborhood and then queries for data from these nodes. Figure 9 shows ISAC, the SES and the registered location of events ISAC has detected.

Since the SES is a short-term memory component, old data is continuously removed from the sphere. The timestamps of each registered data are checked against pre-defined time limits. If any of the timestamps exceed the limits, the data is removed from its registration node.

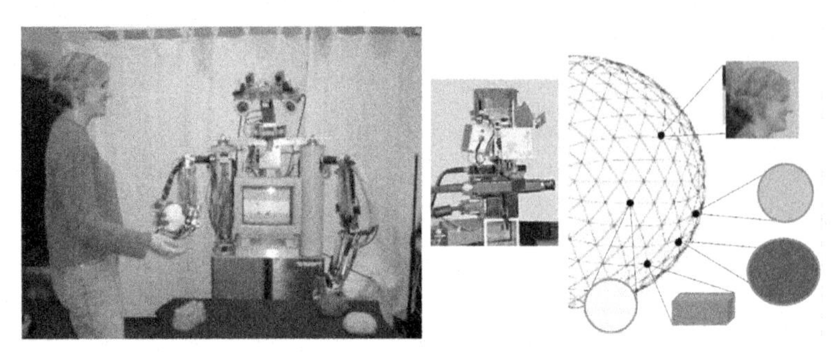

Figure 9
Projection of objects and events onto the SES.

6.4.3.1.1 SES graphical interface

A graphical display of the SES is an interface for assisting the human to understand what ISAC is sensing in the environment. The prototype display (Figure 10) shows ISAC's view of the SES. Each time an object is posted to the SES or old records are purged from the database, the SES sends out all contents in ISAC's short-term memory to the display agent. The display shows icons of the types of data posted and labels the icons with the data names, thus the human sees what ISAC senses. This view provides the person with insight into malfunctions in the system [22], such as when a sensory event occurred, but was not detected by ISAC.

6.4.3.2 Long-Term Memory: Procedural Memory

LTM is a data structure which contains the primitive and *meta*-level behaviors that will be combined later to perform the desired tasks. We call an LTM unit member a PM. The PM encapsulates primitive and *meta*-level behaviors (Figure 11). These behaviors are derived

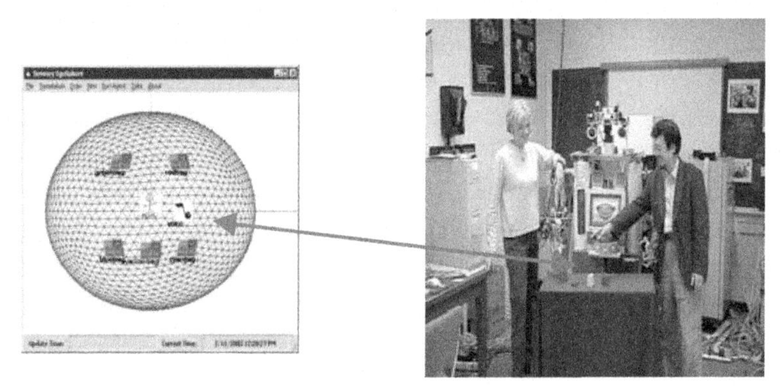

Figure 10
Graphical display of the Sensory EgoSphere.

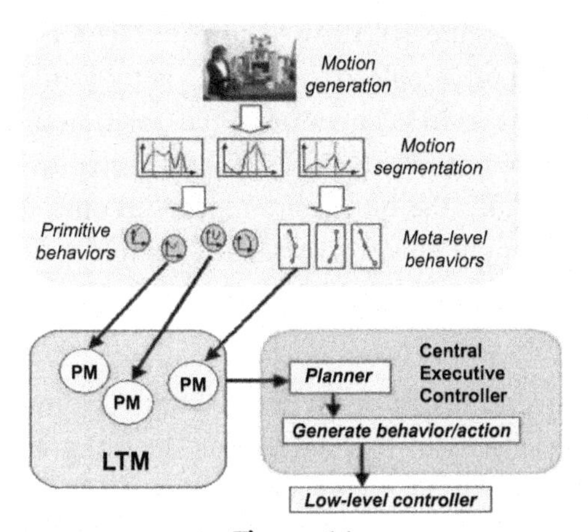

Figure 11
Behavior representation in LTM.

using the spatio-temporal Isomap method proposed by Jenkins and Mataric [20]. A short description of how it was used to generate LTM units is shown.

Motion data are collected from the tele-operation of ISAC. The motion streams collected are then segmented into a set of motion segments as shown in Figure 11. The central idea in the derivation of behaviors from motion segments is to discover spatio-temporal structure in a motion stream. This structure can be estimated by extending a nonlinear dimension reduction method called Isomap [21] to handle motion data with spatial and temporal dependencies. Isomap method was originally developed to handle a large number of spatially distributed data. It was extended by Jenkins and Mataric to generate spatial and temporal motions generated by human motions. Spatio-temporal Isomap dimension reduction, clustering and interpolation methods are applied to the motion segments to produce primitive behaviors (Figure 11). *Meta*-level behaviors are formed by further application of the spatio-temporal Isomap method and linking component primitives with transition probabilities.

We are currently developing an agent called the CEC which generates a task-level behavior or action through a combination of a planning mechanism, goals, intentions and beliefs.

6.4.4 Current Applications

6.4.4.1 Demonstrations

6.4.4.1.1 Demonstration in human-guided object learning through shared attention

This demonstration through finger-pointing utilizes the HHI framework of Figure 3 to allow people to direct ISAC's attention to objects in its workspace. In the current demonstration, represented in Figure 10, ISAC is directed by a human to look at several objects (assorted color blocks) on a table. When ISAC is told to look at the green block, ISAC looks for the pointed finger. ISAC then takes the position of the green block and registers the location and name onto the SES. ISAC is directed to look at a red block and repeats the previous actions. After the blocks are registered onto the SES, ISAC returns to its initial position (looking straight ahead). Then ISAC is told to look at one of the previously taught objects. ISAC retrieves the object named from the SES and saccades to the location given in the SES.

The application involves several levels of the framework for directing attention to known and unknown objects. Two aspects of this interaction, learning and recalling object locations, are shown schematically in Figure 10. To direct ISAC's attention to an unknown object (Figure 12(a)), the application begins with speech from the human directing the robot's attention to an object. The Human Agent activates the Human Finger Agent and parses the name of the object. The Human Finger Agent finds a pointed finger to fixate on the object. At this point, the Human Agent sends the object name and location to the SES Manager, which registers it on the SES. To direct ISAC's attention to a known object that is registered on the SES, the human tells ISAC to find the desired object. The Human Agent translates this speech command into text and sends the object name to the SES Manager to initiate a search of the SES using the name. When the object is retrieved, the SES Manager returns the location of the object to the Human Agent. In the case of a person requesting ISAC to recall the location of an object, the flow, depicted in Figure 12(b), is as follows. The person's request is parsed by the Human Agent into text describing the object of interest. This information is forwarded to the SES to retrieve the object's location. When this information is returned, the world coordinates of the object are sent to the Right Arm Agent and the robot performs a pointing gesture to point to the requested object.

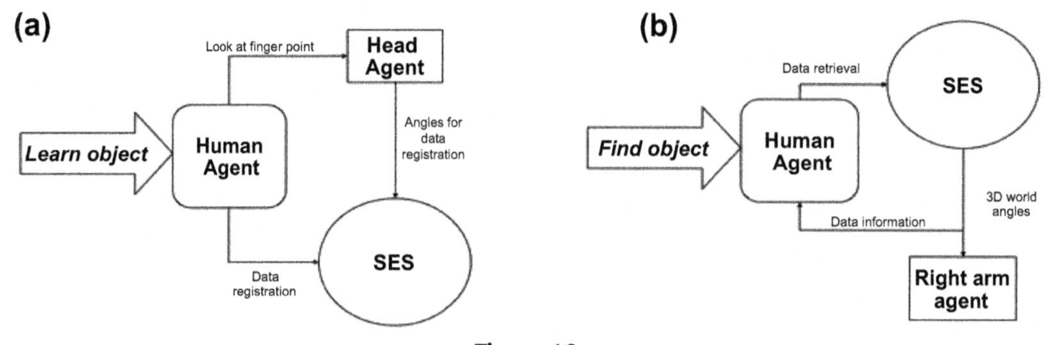

Figure 12
Schematic for (a) learn object and (b) find object.

6.4.4.1.2 Demonstration in situation-based acknowledgment

In this demo, ISAC processes the intentions of the human, resolves them with its own intentions and abilities, and communicates to the person if there is a problem with the request. The scenario begins as a person approaches ISAC and gains its attention. The Human Agent determines that the person has an intention to interact with ISAC. If ISAC is unoccupied at the time, ISAC begins its interaction behaviors by turning toward the person and initiating a greeting and identification sequence. Once interaction is established, ISAC begins a social dialogue. After greeting, ISAC may respond to a person's task request (an intention for ISAC to do something) if it is within ISAC's abilities. If a second person approaches ISAC and attempts to gain its attention, the Human Agent will notify the Self Agent that there is a new person with a pending intention. The Self Agent then must resolve the current human intention with its own current intention. If the second human intention is not of sufficient priority to override ISAC's current task, then ISAC will then pause its current interaction, turn to the interrupter, and apologize for being busy (Figure 13(b)). ISAC can then return its attention to the first person and resume its previous interaction (Figure 13(a)). There may also be a situation where the request of the interrupting person actually has higher priority than the task ISAC is currently performing. In this case the Self Agent determines to switch to the new task after giving an explanation to the current person.

6.4.4.1.3 Future enhancement

To demonstrate the role of LTM in conjunction with STM, we are planning to apply the methodology shown in Figure 11 to the Hand Shaking Demo [15]. A set of *meta*-level handshaking behaviors will allow ISAC to adapt its behavior when a new person comes and greets ISAC from unexpected directions [22].

6.4.5 Conclusion

Realization of general-purpose humanlike robots with adult-level intelligence continues to be the goal of many robotic researchers. We have already seen major advances in the last five years in humanoid robotics research and expect this progress to accelerate. In particular, we may see a major breakthrough in terms of new behavior and task learning within the next several years. We expect ISAC to adapt its behaviors and learn new tasks based on its self-reflection mechanism and external interactions with humans and the environment in limited domains. A growing number of robotics researchers believe that intelligence is an emerging property of an autonomous agent such as humans and robots and that behavior learning requires the coupling of mind, body and the environment. It is our goal to make human-robot interaction less hard-coded and more adaptive and socially acceptable through cognitive robots.

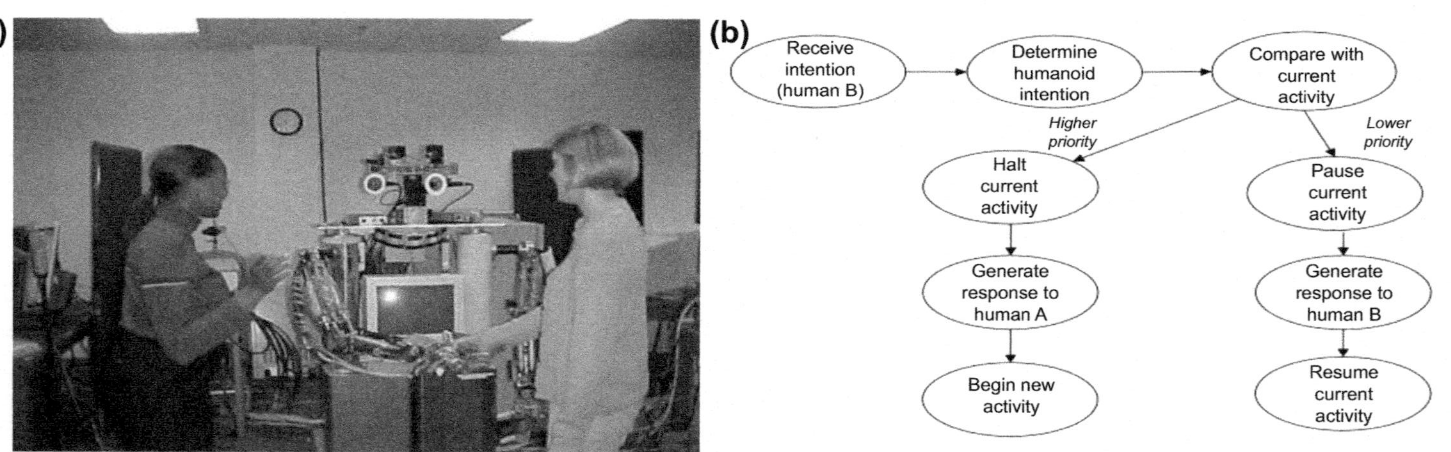

Figure 13
(a) ISAC responding to an interruption; (b) role of Intention Processing during an interruption.

Acknowledgments

This work has been partially funded through a DARPA grant (Grant #DASG60-01-1-0001) and a NASA GSRP (Grant #NGT9-52-JSC). The authors would like to thank colleagues and graduate students in the Center for Intelligent Systems at Vanderbilt University for their contributions.

References

[1] Humanoid Robotics, IEEE Intell. Syst. App. 15 (4) (July–August 2000) (special issue).
[2] T. Fukuda, et al., How far away is "artificial man"? IEEE Robot. Autom. Mag. (March 2001) 66–73.
[3] X. Ming, private communication (April 15, 2003).
[4] K. Kawamura, S. Bagchi, M. Iskarous, M. Bishay, Intelligent robotic systems in service of the disabled, IEEE Trans. Rehabil. Eng. 3 (1) (1995) 14–21.
[5] K. Kawamura, D.M. Wilkes, T. Pack, M. Bishay, J. Barile, Humanoids: future robots for home and factory, in: Proc. 1st Int. Symp. on Humanoid Robots, 1996, pp. 53–62.
[6] R.T. Pack, D.M. Wilkes, K. Kawamura, A software architecture for integrated service robot development, IEEE Trans. Syst., Man, Cybern. (SMC) (1997) 3774–3779.
[7] K. Kawamura, R.T. Pack, M. Bishay, M. Iskarous, Design philosophy for service robots, Robot. Auton. Syst. 18 (1996) 109–116.
[8] K. Kawamura, A. Alford, K. Hambuchen, M. Wilkes, Towards a unified framework for human-humanoid interaction, in: Proc. 1st IEEE-RAS Int. Conf. on Humanoid Robots (Humanoids' 00), 2000.
[9] R.A. Peters II, K.E. Hambuchen, K. Kawamura, D.M. Wilkes, The Sensory EgoSphere as a short-term memory for humanoids, in: Proc. 2nd IEEE-RAS Int. Conf. on Humanoid Robots (Humanoids 2001), 2001, pp. 451–459.
[10] K. Kawamura, R.A. Peters II, M.W. Wilkes, W.A. Alford, T.E. Rogers, ISAC: foundations in human-humanoid interaction, IEEE Intell. Syst. App. 15 (4) (July–August 2000) 38–45.
[11] C. Breazeal, Designing Sociable Robots, MIT Press, 2002.
[12] A.S. Sekman, M. Wilkes, K. Kawamura, An application of passive human-robot interaction: human tracking-based on attention distraction, IEEE Trans. Syst. Man Cybern. A 32 (2) (2002) 248–259.
[13] M. Minsky, The Society of Mind, Simon and Schuster, New York, 1986.
[14] J. Ferber, Multi-agent Systems: An Introduction to Distributed Artificial Intelligence, Addison-Wesley, Harlow England, 1999.
[15] K. Kawamura, T.E. Rogers, X. Ao, Development of a human agent for a multi-agent based human-robot interaction, AAMAS (2002) 1379–1386.
[16] R. Bajcsy, Active perception, Proc. IEEE 76 (8) (August 1988) 996–1005.
[17] A. Nefian, M. Hayes, Face recognition using an embedded HMM, in: IEEE Conf. on Audio and Video-Based Biometric Person Authentication, 1999, pp. 19–24.
[18] J.S. Albus, Outline for a theory of intelligence, IEEE Trans. Syst. Man Cybern. 21 (3) (1991) 473–509.
[19] J.S. Albus, A.M. Meystel, Engineering of Mind: An Introduction to the Science of Intelligent Systems, John Wiley & Sons, 2001.
[20] O.C. Jenkins, M.J. Mataric, Automated derivation of behavior vocabularies for autonomous humanoid motion, Accepted Paper, in: 2nd Int. Joint Conf. on Autonomous Agents and Multiagent Systems, 2003.
[21] J.B. Tenenbaum, V. de Silva, J.C. Langford, A global geometric framework for nonlinear dimensionality reduction, Science 290 (5500) (2000) 2319–2323.
[22] D. Erol, J. Park, E. Turkay, K. Kawamura, O.C. Jenkins, M.J. Mataric, Motion generation for humanoid robots with automatically derived behaviors, in: Proc. IEEE International Conference on Systems, Man, and Cybernetics (SMC), Washington, DC, October 2003, pp. 1816–1821.

Index

Note: Page numbers followed by "f" and "t" indicate figures and tables, respectively.

D

S

T

U

V

W